T0269228

Applied Mathematical Sciences

Volume 81

Applied Mathematical Sciences

1. *John:* Partial Differential Equations, 4th ed.
2. *Sirovich:* Techniques of Asymptotic Analysis.
3. *Hale:* Theory of Functional Differential Equations, 2nd ed.
4. *Percus:* Combinatorial Methods.
5. *von Mises/Friedrichs:* Fluid Dynamics.
6. *Freiberger/Grenander:* A Short Course in Computational Probability and Statistics.
7. *Pipkin:* Lectures on Viscoelasticity Theory.
9. *Friedrichs:* Spectral Theory of Operators in Hilbert Space.
11. *Wolovich:* Linear Multivariable Systems.
12. *Berkovitz:* Optimal Control Theory.
13. *Bluman/Cole:* Similarity Methods for Differential Equations.
14. *Yoshizawa:* Stability Theory and the Existence of Periodic Solution and Almost Periodic Solutions.
15. *Braun:* Differential Equations and Their Applications, 3rd ed.
16. *Lefschetz:* Applications of Algebraic Topology.
17. *Collatz/Wetterling:* Optimization Problems.
18. *Grenander:* Pattern Synthesis: Lectures in Pattern Theory, Vol I.
20. *Driver:* Ordinary and Delay Differential Equations.
21. *Courant/Friedrichs:* Supersonic Flow and Shock Waves.
22. *Rouche/Habets/Laloy:* Stability Theory by Liapunov's Direct Method.
23. *Lamperti:* Stochastic Processes: A Survey of the Mathematical Theory.
24. *Grenander:* Pattern Analysis: Lectures in Pattern Theory, Vol. II.
25. *Davies:* Integral Transforms and Their Applications, 2nd ed.
26. *Kushner/Clark:* Stochastic Approximation Methods for Constrained and Unconstrained Systems
27. *de Boor:* A Practical Guide to Splines.
28. *Keilson:* Markov Chain Models—Rarity and Exponentiality.
29. *de Veubeke:* A Course in Elasticity.
30. *Sniatycki:* Geometric Quantization and Quantum Mechanics.
31. *Reid:* Sturmian Theory for Ordinary Differential Equations.
32. *Meis/Markowitz:* Numerical Solution of Partial Differential Equations.
33. *Grenander:* Regular Structures: Lectures in Pattern Theory, Vol. III.
34. *Kevorkian/Cole:* Perturbation methods in Applied Mathematics.
35. *Carr:* Applications of Centre Manifold Theory.
36. *Bengtsson/Ghil/Källén:* Dynamic Meteorology: Data Assimilation Methods.
37. *Saperstone:* Semidynamical Systems in Infinite Dimensional Spaces.
38. *Lichtenberg/Lieberman:* Regular and Stochastic Motion.
39. *Piccini/Stampacchia/Vidossich:* Ordinary Differential Equations in R^n.
40. *Naylor/Sell:* Linear Operator Theory in Engineering and Science.
41. *Sparrow:* The Lorenz Equations: Bifurcations, Chaos, and Strange Attractors.
42. *Guckenheimer/Holmes:* Nonlinear Oscillations, Dynamical Systems and Bifurcations of Vector Fields.
43. *Ockendon/Tayler:* Inviscid Fluid Flows.
44. *Pazy:* Semigroups of Linear Operators and Applications to Partial Differential Equations.
45. *Glashoff/Gustafson:* Linear Optimization and Approximation: An Introduction to the Theoretical Analysis and Numerical Treatment of Semi-Infinite Programs.
46. *Wilcox:* Scattering Theory for Diffraction Gratings.
47. *Hale et al.:* An Introduction to Infinite Dimensional Dynamical Systems—Geometric Theory.
48. *Murray:* Asymptotic Analysis.
49. *Ladyzhenskaya:* The Boundary-Value Problems of Mathematical Physics.
50. *Wilcox:* Sound Propagation in Stratified Fluids.
51. *Golubitsky/Schaeffer:* Bifurcation and Groups in Bifurcation Theory, Vol. I.
52. *Chipot:* Variational Inequalities and Flow in Porous Media.
53. *Majda:* Compressible Fluid Flow and Systems of Conservation Laws in Several Space Variables.
54. *Wasow:* Linear Turning Point Theory.

(continued following index)

George W. Bluman   Sukeyuki Kumei

Symmetries and Differential Equations

With 21 Illustrations

Springer-Verlag
New York Berlin Heidelberg
London Paris Tokyo Hong Kong

George W. Bluman
Department of Mathematics
University of British Columbia
Vancouver, British Columbia V6T 1Y4
Canada

Sukeyuki Kumei
Faculty of Textile Science
and Technology
Shinshu University
Ueda, Nagano 386
Japan

Mathematics Subject Classification (1980): 22E225, 22E65, 22E70, 34-01, 34A05, 35-01, 35C05, 35F20, 35G20, 35K05, 35L05, 35Q20, 58F35, 58F37, 58G35, 58G37, 70H35

Library of Congress Cataloging-in-Publication Data
Bluman, George W.
Symmetries and differential equations / George W. Bluman, Sukeyuki Kumei.
p. cm. — (Applied mathematical sciences ; v. 81)
Includes bibliographical references and indexes.
ISBN 0-387-96996-9
1. Differential equations—Numerical solutions. 2. Differential equations, Partial—Numerical solutions. 3. Lie groups. I. Kumei, Sukeyuki. II. Title. III. Series: Applied mathematical sciences (Springer-Verlag New York Inc.) ; v. 81.
QA1.A647 vol. 81
[QA372]
510 s—dc20
[515'.35] 89-6386

Printed on acid-free paper.

Camera-ready copy prepared using LaTEX.
Printed and bound by R.R. Donnelley & Sons, Harrisonburg, Virginia.
Printed in the United States of America.

9 8 7 6 5 4 3 2 1

ISBN 0-387-96996-9 Springer-Verlag New York Berlin Heidelberg
ISBN 3-540-96996-9 Springer-Verlag Berlin Heidelberg New York

Preface

In recent years there have been considerable developments in symmetry methods (group methods) for differential equations as evidenced by the number of research papers devoted to the subject. This is no doubt due to the inherent applicability of the methods to nonlinear differential equations. Symmetry methods for differential equations, originally developed by Sophus Lie, are highly algorithmic. They systematically unify and extend existing ad hoc techniques to construct explicit solutions for differential equations, most importantly for nonlinear differential equations. Often ingenious techniques for solving particular differential equations arise transparently from the group point of view, and thus it is somewhat surprising that symmetry methods are not more widely used.

A major portion of this book discusses work that has appeared since the publication of the book *Similarity Methods for Differential Equations*, Springer-Verlag, 1974, by G.W. Bluman and J.D. Cole. The present book includes a comprehensive treatment of Lie groups of transformations and thorough discussions of basic symmetry methods for solving ordinary and partial differential equations. No knowledge of group theory is assumed. Emphasis is placed on explicit computational algorithms to discover symmetries admitted by differential equations and to construct solutions resulting from symmetries.

This book should be particularly suitable for physicists, applied mathematicians, and engineers. Almost all of the examples are taken from physical and engineering problems including those concerned with heat conduction, wave propagation, and fluid flows. A preliminary version was used as lecture notes for a two-semester course taught by the first author at the University of British Columbia in 1987–88 to graduate and senior undergraduate students in applied mathematics and physics.

Chapters 1 through 4 encompass basic material. More specialized topics are covered in Chapters 5 through 7.

Chapter 1 introduces the basic ideas of group transformations and their connections with differential equations through a thorough treatment of dimensional analysis and generalizations of the well-known Buckingham Pi-theorem. This chapter should give the reader an intuitive grasp of the subject matter of the book in an elementary setting.

Chapter 2 develops the basic concepts of Lie groups of transformations and Lie algebras necessary in subsequent chapters. A Lie group of transformations is characterized in terms of its infinitesimal generators which form a Lie algebra.

Chapter 3 is concerned with ordinary differential equations. It is shown how group transformations are used to construct solutions and how to find group transformations leaving ordinary differential equations invariant. We present a reduction algorithm that reduces an nth order differential equation, admitting a solvable r-parameter Lie group of transformations, to an $(n-r)$th order differential equation plus r quadratures. We derive an algorithm to construct special solutions (invariant solutions) that are invariant under admitted Lie groups of transformations. For a first order differential equation such invariant solutions include separatrices and singular envelope solutions.

Chapter 4 is concerned with partial differential equations. It is shown how one finds group transformations leaving them invariant, how corresponding invariant solutions are constructed, and how group methods are applied to boundary value problems.

Chapter 5 discusses the connection between conservation laws and the invariance of Euler–Lagrange equations, arising from variational problems, under Lie groups of transformations. Various formulations of Noether's theorem are presented to construct such conservation laws. This leads to generalizing the concept of Lie groups of point transformations of earlier chapters to Lie–Bäcklund transformations that account for higher order conservation laws associated with partial differential equations that have solutions exhibiting soliton behavior. We present algorithms to construct recursion operators generating infinite sequences of Lie–Bäcklund symmetries.

In Chapter 6 it is shown how group transformations can be used to determine whether or not a given differential equation can be mapped invertibly to a target differential equation. Algorithms are given to construct such mappings when they exist. In particular, we give necessary and sufficient conditions for mapping a given nonlinear system of partial differential equations to a linear system of partial differential equations and for mapping a given linear partial differential equation with variable coefficients to a linear partial differential equation with constant coefficients.

In Chapter 7 the concept of Lie groups of transformations is generalized further to include nonlocal symmetries of differential equations. We present a systematic method for finding a special class of nonlocal symmetries that are realized as local symmetries of related auxiliary systems (potential symmetries). The introduction of potential symmetries significantly extends the applicability of group methods to both ordinary and partial differential equations. Together with the mapping algorithms developed in Chapter 6, the use of potential symmetries allows one to find systematically non-invertible mappings that transform nonlinear partial differential equations to linear partial differential equations.

Chapters 2 through 7 can be read independently of Chapter 1. The material in Chapter 2 is essential for all subsequent chapters but a reader only interested in scalar ordinary differential equations may omit Sections 2.3.3 to 2.3.5. Chapter 4 can be read independently of Chapter 3. A reader interested in conservation laws (Chapter 5) needs to know how to find Lie groups of transformations admitted by differential equations (Sections 3.2.3, 3.3.4, 4.2.3, 4.3.3). Chapter 6 can be read independently of Chapters 3 and 5.

Every topic is illustrated by examples. Almost all sections have many exercises. It is essential to do these exercises in order to obtain a working knowledge of the material. Each chapter ends with a Discussion section that puts its contents in perspective by summarizing major results, by referring to related works, and by introducing related material in subsequent chapters.

Within each section and subsection of a given chapter, definitions, theorems, lemmas, and corollaries are numbered separately as well as consecutively. For example, Definition 2.2.3-1 refers to the first definition and Theorem 2.2.3-1 to the first theorem in Section 2.2.3; Definition 1.4-1 refers to the first definition in Section 1.4. Exercises appear at the conclusion of a section; Exercise 1.3-4 refers to the fourth problem of Exercises 1.3.

We thank Greg Reid for helpful suggestions that improved Chapter 7, Alex Ma for his assistance, and Doug Jamison, Mei-Ling Fong, Sheila Hancock, Joanne Congo, Marilyn Lacate, Joan de Niverville, and Rita Sieber for their patience and care in typing various drafts of the manuscripts.

Vancouver, Canada

George W. Bluman
Sukeyuki Kumei

Contents

Introduction

In the latter part of the 19th century Sophus Lie introduced the notion of continuous groups, now known as Lie groups, in order to unify and extend various specialized solution methods for ordinary differential equations. Lie was inspired by lectures of Sylow given at Christiania (present-day Oslo) on Galois theory and Abel's related works. [In 1881 Sylow and Lie collaborated in a careful editing of Abel's complete works.] Lie showed that the order of an ordinary differential equation can be reduced by one, constructively, if it is invariant under a one-parameter Lie group of point transformations.

Lie's work systematically relates a miscellany of topics in ordinary differential equations including: integrating factor, separable equation, homogeneous equation, reduction of order and the methods of undetermined coefficients and variation of parameters for linear equations, solution of the Euler equation, and the use of the Laplace transform. Lie (1881) also indicated that for linear partial differential equations, invariance under a Lie group leads directly to superpositions of solutions in terms of transforms.

A *symmetry group of a system of differential equations* is a group of transformations which maps any solution to another solution of the system. In Lie's framework such a group depends on continuous parameters and consists of either point transformations (*point symmetries*) acting on the system's space of independent and dependent variables, or more generally, contact transformations (*contact symmetries*) acting on the space including all first derivatives of the dependent variables. Elementary examples of Lie groups include translations, rotations, and scalings. An autonomous system of first order ordinary differential equations, i.e. a stationary flow, essentially defines a one-parameter Lie group of point transformations. Unlike discrete groups, for example reflections, Lie showed that for a given differential equation the admitted continuous group of point transformations, acting on the space of its independent and dependent variables, can be determined by an explicit computational algorithm (*Lie's algorithm*).

In this book the applications of continuous groups to differential equations make no use of the global aspects of Lie groups. These applications use connected local Lie groups of transformations. Lie's fundamental theorems show that such groups are completely characterized by their *infinitesimal generators.* In turn these form a *Lie algebra* determined by structure constants.

Lie groups, and hence their infinitesimal generators, can be naturally *extended* or "*prolonged*" to act on the space of independent variables, dependent variables and derivatives of the dependent variables up to any finite order. As a consequence, the seemingly intractable nonlinear conditions of group invariance of a given system of differential equations reduce to linear homogeneous equations determining the infinitesimal generators of the group. Since these *determining equations* form an overdetermined system of linear homogeneous partial differential equations, one can usually determine the infinitesimal generators in closed form. For a given system of differential equations, the setting up of the determining equations is entirely routine. Symbolic manipulation programs exist to set up the determining equations and in some cases explicitly solve them [cf. Schwarz (1985, 1988), Kersten (1987)].

If a system of partial differential equations is invariant under a Lie group of point transformations, one can find, constructively, special solutions, called *similarity solutions* or *invariant solutions,* which are invariant under some subgroup of the full group admitted by the system. These solutions result from solving a reduced system of differential equations with fewer independent variables. This application of Lie groups was discovered by Lie but first came to prominence in the late 1950s through the work of the Soviet group at Novosibirsk, led by Ovsiannikov (1962, 1982). Invariant solutions can also be constructed for specific boundary value problems. Here one seeks a subgroup of the full group of a given partial differential equation which leaves boundary curves and conditions imposed on them invariant [cf. Bluman and Cole (1974)]. Such solutions include *self-similar* (*automodel*) *solutions* which can be obtained through *dimensional analysis* or more generally from invariance under groups of scalings. Connections between invariant solutions and separation of variables have been studied extensively by Miller (1977) and co-workers.

In a celebrated paper, Noether (1918) showed how the symmetries of an action integral (*variational symmetries*) lead constructively to conservation laws for the corresponding Euler–Lagrange equations. For example, conservation of energy follows from invariance under translation in time; conservation of linear and angular momenta, respectively, from invariances under translations and rotations in space. Such variational symmetries leave the Euler–Lagrange equations invariant. They can be determined through Lie's algorithm.

The applicability of symmetry methods to differential equations is further extended by considering invariance under *Lie–Bäcklund transformations* (*Lie–Bäcklund symmetries*). Here the infinitesimal generators depend on derivatives of the dependent variables up to any finite order. The possibility of the existence of such symmetries was recognized by Noether (1918). Lie–Bäcklund transformations are discussed in some detail in Olver (1986) and Ibragimov (1985). Lie–Bäcklund transformations cannot be represented in closed form by integrating a finite system of ordinary differential equations as is the case for Lie groups of point transformations. However their infinitesimal generators can be computed for a given differential equation by a simple extension of Lie's algorithm. The invariance of a partial differential equation under a Lie–Bäcklund symmetry usually leads to invariance under an infinite number of such symmetries connected by recursion operators [Olver (1977)]. The theory and computation of recursion operators are discussed comprehensively in Olver (1986). Lie–Bäcklund symmetries can be shown to account for the conserved Runge–Lenz vector for the Kepler problem and the infinity of conservation laws for the Korteweg–de Vries equation and other nonlinear partial differential equations exhibiting soliton behavior.

Another application of symmetry methods to differential equations is to discover related differential equations of simpler form. By comparing the Lie groups admitted by a given differential equation and another differential equation (target equation) one can find, constructively, necessary conditions for a mapping of the given equation to the target equation. If the target equation is characterized completely in terms of a Lie symmetry group then one can algorithmically determine if an invertible mapping exists between the equations. In particular, one can constructively answer such questions as: Can a given nonlinear system of partial differential equations be mapped invertibly to a linear system? Can a given linear partial differential equation with variable coefficients be mapped into one with constant coefficients?

One can extend the classes of symmetries admitted by differential equations beyond *local symmetries* (which include point, contact, and Lie–Bäcklund symmetries) to *nonlocal symmetries* by considering a system related to a given differential equation. Here one starts by finding a conservation law for the given differential equation. This leads to a related system through the introduction of auxiliary dependent variables (*potentials*). A Lie group of point transformations admitted by the system is a symmetry group of the given differential equation since it maps any solution of the given equation to another solution. For partial differential equations such symmetries are often nonlocal symmetries (*potential symmetries*). This extension of local symmetries to potential symmetries considerably widens the applicability of symmetry methods to the construction of solutions of both ordinary and partial differential equations.

1

Dimensional Analysis, Modelling, and Invariance

1.1 Introduction

In this chapter we introduce the ideas of invariance concretely through a thorough treatment of dimensional analysis. We show how dimensional analysis is connected to modelling and the construction of solutions obtained through invariance for boundary value problems for partial differential equations.

Often for a quantity of interest one knows at most the independent quantities it depends upon, say n in total, and the dimensions of all $n+1$ quantities. The application of dimensional analysis usually reduces the number of essential independent quantities. This is the starting point of modelling where the objective is to reduce significantly the number of experimental measurements. In the following sections we will show that dimensional analysis can lead to a reduction in the number of independent variables appearing in a boundary value problem for a partial differential equation. Most importantly we show that for partial differential equations the reduction of variables through dimensional analysis is a special case of reduction from invariance under groups of scaling (stretching) transformations.

1.2 Dimensional Analysis—Buckingham Pi-Theorem

The basic theorem of dimensional analysis is the so-called *Buckingham Pi-theorem,* attributed to the American engineering scientist Buckingham (1914, 1915a, b). General references on the subject include those of Birkhoff (1950), Bridgman (1931), Barenblatt (1979), Sedov (1959), and Bluman (1983a). A historical perspective is given by Görtler (1975). For a detailed mathematical perspective see Curtis, Logan, and Parker (1982).

The following assumptions and conclusions of dimensional analysis constitute the Buckingham Pi-theorem.

1.2.1 Assumptions Behind Dimensional Analysis

Essentially no real problem violates the following assumptions:

(i) A quantity u is to be determined in terms of n *measurable quantities* (variables and parameters), $(W_1, W_2, \ldots, W_n)$:

$$u = f(W_1, W_2, \ldots, W_n), \tag{1.1}$$

where f is an unknown function of $(W_1, W_2, \ldots, W_n)$.

(ii) The quantities $(u, W_1, W_2, \ldots, W_n)$ involve m *fundamental dimensions* labelled by $L_1, L_2, \ldots, L_m$. For example in a mechanical problem these are usually the mechanical fundamental dimensions L_1 = length, L_2 = mass, and L_3 = time.

(iii) Let Z represent any of $(u, W_1, W_2, \ldots, W_n)$. Then the *dimension of* Z, denoted by $[Z]$, is a product of powers of the fundamental dimensions, in particular

$$[Z] = L_1^{\alpha_1} L_2^{\alpha_2} \cdots L_m^{\alpha_m} \tag{1.2}$$

for some real numbers, usually rational, $(\alpha_1, \alpha_2, \ldots, \alpha_m)$ which are the dimension exponents of Z. The *dimension vector of* Z is the column vector

$$\boldsymbol{\alpha} = \begin{bmatrix} \alpha_1 \\ \alpha_2 \\ \vdots \\ \alpha_m \end{bmatrix}. \tag{1.3}$$

A quantity Z is said to be *dimensionless* if and only if $[Z] = 1$, i.e. all dimension exponents are zero. For example, in terms of the mechanical fundamental dimensions, the dimension vector of energy E is

$$\alpha(E) = \begin{bmatrix} 2 \\ 1 \\ -2 \end{bmatrix}.$$

Let

$$\mathbf{b}_i = \begin{bmatrix} b_{1i} \\ b_{2i} \\ \vdots \\ b_{mi} \end{bmatrix} \tag{1.4}$$

be the dimension vector of W_i, $i = 1, 2, \ldots, n$, and let

$$\mathrm{B} = \begin{bmatrix} b_{11} & b_{12} & \cdots & b_{1n} \\ b_{21} & b_{22} & \cdots & b_{2n} \\ \vdots & \vdots & & \vdots \\ b_{m1} & b_{m2} & \cdots & b_{mn} \end{bmatrix} \tag{1.5}$$

be the $m \times n$ *dimension matrix* of the given problem.

(iv) For any set of fundamental dimensions one can choose a *system of units* for measuring the value of any quantity Z. A change from one system of units to another involves a positive *scaling* of each fundamental dimension which in turn induces a scaling of each quantity Z. For example for the mechanical fundamental dimensions the common systems of units are MKS, c.g.s. or British foot-pounds. In changing from c.g.s. to MKS units, L_1 is scaled by 10^{-2}, L_2 is scaled by 10^{-3}, L_3 is unchanged, and hence the value of energy E is scaled by 10^{-7}. Under a change of system of units the value of a dimensionless quantity is unchanged, i.e. its value is *invariant* under an arbitrary scaling of any fundamental dimension. Hence it is meaningful to deem dimensionless quantities as large or small. The last assumption of dimensional analysis is that formula (1.1) acts as a dimensionless equation in the sense that (1.1) is invariant under an arbitrary scaling of any fundamental dimension, i.e. (1.1) is independent of the choice of system of units.

1.2.2 Conclusions from Dimensional Analysis

The assumptions of the Buckingham Pi-theorem stated in Section 1.2.1 lead to:

(i) Formula (1.1) can be expressed in terms of dimensionless quantities.

(ii) The number of dimensionless quantities is $k+1 = n+1-r(\mathrm{B})$ where $r(\mathrm{B})$ is the rank of matrix B. Precisely k of these dimensionless quantities depend on the measurable quantities $(W_1, W_2, \ldots, W_n)$.

(iii) Let

$$\mathbf{x}^{(i)} = \begin{bmatrix} x_{1i} \\ x_{2i} \\ \vdots \\ x_{ni} \end{bmatrix}, \quad i = 1, 2, \ldots, k, \tag{1.6}$$

represent the $k = n - r(\mathrm{B})$ linearly independent solutions $\mathbf{x}$ of the system

$$\mathrm{B}\mathbf{x} = 0. \tag{1.7}$$

Let

$$\mathbf{a} = \begin{bmatrix} a_1 \\ a_2 \\ \vdots \\ a_m \end{bmatrix} \tag{1.8}$$

be the dimension vector of u and let

$$\mathbf{y} = \begin{bmatrix} y_1 \\ y_2 \\ \vdots \\ y_n \end{bmatrix} \tag{1.9}$$

represent a solution of the system

$$\mathbf{B}\mathbf{y} = -\mathbf{a}. \tag{1.10}$$

Then formula (1.1) simplifies to

$$\pi = g(\pi_1, \pi_2, \ldots, \pi_k) \tag{1.11}$$

where π, π_i, are dimensionless quantities,

$$\pi = u W_1^{y_1} W_2^{y_2} \cdots W_n^{y_n}, \tag{1.12a}$$

$$\pi_i = W_1^{x_{1i}} W_2^{x_{2i}} \cdots W_n^{x_{ni}}, \quad i = 1, 2, \ldots, k, \tag{1.12b}$$

and g is an unknown function of its arguments. In particular (1.1) becomes

$$u = W_1^{-y_1} W_2^{-y_2} \cdots W_n^{-y_n} g(\pi_1, \pi_2, \ldots, \pi_k). \tag{1.13}$$

[In terms of experimental modelling formula (1.13) is "cheaper" than (1.1) by $r(\mathbf{B})$ orders of magnitude.]

1.2.3 Proof of the Buckingham Pi-Theorem

First of all,

$$[u] = L_1^{a_1} L_2^{a_2} \cdots L_m^{a_m}, \tag{1.14a}$$

$$[W_i] = L_1^{b_{1i}} L_2^{b_{2i}} \cdots L_m^{b_{mi}}, \quad i = 1, 2, \ldots, n. \tag{1.14b}$$

Next we use assumption (iv) and consider the invariance of (1.1) under arbitrary scalings of the fundamental dimensions by taking each fundamental dimension in turn. We scale L_1 by letting

$$L_1^* = e^{\epsilon} L_1, \quad \epsilon \in \mathbb{R}. \tag{1.15}$$

Then accordingly

$$u^* = e^{\epsilon a_1} u, \tag{1.16a}$$

$$W_i^* = e^{\epsilon b_{1i}} W_i, \quad i = 1, 2, \ldots, n. \tag{1.16b}$$

Equations (1.16a,b) define a one-parameter (ϵ) Lie group of scaling transformations of the $n+1$ quantities $(u, W_1, W_2, \ldots, W_n)$ with $\epsilon = 0$ corresponding to the identity transformation. This group is induced by the one-parameter group of scalings (1.15) of the fundamental dimension L_1.

From assumption (iv), formula (1.1) holds if and only if

$$u^* = f(W_1^*, W_2^*, \ldots, W_n^*),$$

i.e.,

$$e^{\epsilon a_1} u = f(e^{\epsilon b_{11}} W_1, e^{\epsilon b_{12}} W_2, \ldots, e^{\epsilon b_{1n}} W_n), \quad \text{for all } \epsilon \in \mathbb{R}. \tag{1.17}$$

Case I. $b_{11} = b_{12} = \cdots = b_{1n} = a_1 = 0$. Here L_1 is not a fundamental dimension of the problem or, in other words, formula (1.1) is *dimensionless with respect to* L_1.

Case II. $b_{11} = b_{12} = \cdots = b_{1n} = 0$, $a_1 \neq 0$. It follows that $u \equiv 0$, a trivial situation.

Hence it follows that $b_{1i} \neq 0$ for some $i = 1, 2, \ldots, n$. Without loss of generality we assume $b_{11} \neq 0$. We define new measurable quantities

$$X_{i-1} = W_i W_1^{-b_{1i}/b_{11}}, \quad i = 2, 3, \ldots, n, \tag{1.18}$$

and let

$$X_n = W_1. \tag{1.19}$$

We choose as the new unknown

$$v = u W_1^{-a_1/b_{11}}. \tag{1.20}$$

In terms of the quantities (1.18)–(1.20), formula (1.1) is equivalent to

$$v = F(X_1, X_2, \ldots, X_n) \tag{1.21}$$

where F is an unknown function of $(X_1, X_2, \ldots, X_n)$, and the group of transformations (1.16a,b) becomes

$$v^* = v, \tag{1.22a}$$

$$X_i^* = X_i, \quad i = 1, 2, \ldots, n-1, \tag{1.22b}$$

$$X_n^* = e^{\epsilon b_{11}} X_n, \tag{1.22c}$$

so that $(v, X_1, X_2, \ldots, X_{n-1})$ are *invariants* of (1.16a,b). Moreover the quantities $(v, X_1, X_2, \ldots, X_n)$ satisfy assumption (iii), and formula (1.21) satisfies assumption (iv). Hence

$$v = F(X_1, X_2, \ldots, X_{n-1}, e^{\epsilon b_{11}} X_n), \tag{1.23}$$

for all $\epsilon \in \mathbb{R}$. Consequently F is independent of X_n. Moreover the measurable quantities $(X_1, X_2, \ldots, X_{n-1})$ are products of powers of $(W_1, W_2, \ldots, W_n)$ and v is a product of u and powers of $(W_1, W_2, \ldots, W_n)$. Formula (1.1) reduces to

$$v = G(X_1, X_2, \ldots, X_{n-1}), \tag{1.24}$$

where $(v, X_1, X_2, \ldots, X_{n-1})$ are dimensionless with respect to L_1 and G is an unknown function of its $n-1$ arguments.

Continuing in turn with the other $m-1$ fundamental dimensions, we reduce formula (1.1) to a dimensionless formula

$$\pi = g(\pi_1, \pi_2, \ldots, \pi_k), \tag{1.25}$$

where $[\pi] = [\pi_i] = 1$, g is an unknown function of $(\pi_1, \pi_2, \ldots, \pi_k)$,

$$\pi = u W_1^{y_1} W_2^{y_2} \cdots W_n^{y_n}, \tag{1.26a}$$

and

$$\pi_i = W_1^{x_{1i}} W_2^{x_{2i}} \cdots W_n^{x_{ni}}, \tag{1.26b}$$

for some real numbers $\{y_j, x_{ji}\}$, $i = 1, 2, \ldots, k$; $j = 1, 2, \ldots, n$.

Next we show that the number of measurable dimensionless quantities is $k = n - r(\mathrm{B})$. This follows immediately since

$$[W_1^{x_1} W_2^{x_2} \cdots W_n^{x_n}] = 1$$

if and only if

$$\mathbf{x} = \begin{bmatrix} x_1 \\ x_2 \\ \vdots \\ x_n \end{bmatrix}$$

satisfies (1.7). Equation (1.7) has $k = n - r(\mathrm{B})$ linearly independent solutions $\mathbf{x}^{(i)}$ given by (1.6). The real numbers

$$\mathbf{y} = \begin{bmatrix} y_1 \\ y_2 \\ \vdots \\ y_n \end{bmatrix}$$

follow from setting

$$[u W_1^{y_1} W_2^{y_2} \cdots W_n^{y_n}] = 1,$$

leading to $\mathbf{y}$ satisfying (1.10). □

Note that the proof of the Buckingham Pi-theorem makes no assumption about the continuity of the unknown function f, and hence of g, with respect to any of their arguments.

1.2.4 Examples

(1) *The Atomic Explosion of 1945*

Sir Geoffrey Taylor (1950) deduced the approximate energy released by the first atomic explosion in New Mexico from motion picture records of J.E.

Mack declassified in 1947. But the amount of energy released by the blast was still classified in 1947! [Taylor carried out the analysis for his deduction in 1941.] A dimensional analysis argument of Taylor's deduction follows:

An atomic explosion is approximated by the release of a large amount of energy E from a "point." A consequence is an expanding spherical fireball whose edge corresponds to a powerful shock wave. Let $u = R$ be the radius of the shock wave. We treat R as the unknown and assume that

$$R = f(W_1, W_2, W_3, W_4) \tag{1.27}$$

where

$W_1 = E$, the energy released by the explosion,
$W_2 = t$, the elapsed time after the explosion takes place,
$W_3 = \rho_0$, the initial or ambient air density,
and
$W_4 = P_0$, the initial or ambient air pressure.

For this problem we use the mechanical fundamental dimensions. The corresponding dimension matrix is

$$\mathrm{B} = \begin{bmatrix} 2 & 0 & -3 & -1 \\ 1 & 0 & 1 & 1 \\ -2 & 1 & 0 & -2 \end{bmatrix}. \tag{1.28}$$

$r(\mathrm{B}) = 3$ and hence $k = n - r(\mathrm{B}) = 4 - 3 = 1$. The general solution of $\mathbf{B}\mathbf{x} = 0$ is $x_1 = -\frac{2}{5}x_4$, $x_2 = \frac{6}{5}x_4$, $x_3 = -\frac{3}{5}x_4$ where x_4 is arbitrary. Setting $x_4 = 1$, we get the measurable dimensionless quantity

$$\pi_1 = P_0 \left[\frac{t^6}{E^2\rho_0^3}\right]^{1/5}. \tag{1.29}$$

The dimension vector of R is

$$\mathbf{a} = \begin{bmatrix} 1 \\ 0 \\ 0 \end{bmatrix}. \tag{1.30}$$

The general solution of $\mathbf{B}\mathbf{y} = -\mathbf{a}$ is

$$\mathbf{y} = \frac{1}{5}\begin{bmatrix} -1 \\ -2 \\ 1 \\ 0 \end{bmatrix} + \mathbf{x} \tag{1.31}$$

where $\mathbf{x}$ is the general solution of $\mathbf{B}\mathbf{x} = 0$. Setting $\mathbf{x} = 0$ in (1.31), we obtain the dimensionless unknown

$$\pi = R\left[\frac{Et^2}{\rho_0}\right]^{-1/5}. \tag{1.32}$$

Thus from dimensional analysis

$$R = \left[\frac{Et^2}{\rho_0}\right]^{1/5} g(\pi_1) \tag{1.33}$$

where g is an unknown function of π_1.

Now we assume that $g(\pi_1)$ is continuous at $\pi_1 = 0$ so that $g(\pi_1) \simeq g(0)$ if $\pi_1 << 1$. Moreover, we assume that $g(0) \neq 0$. This leads to Taylor's approximation formula

$$R = At^{2/5} \tag{1.34}$$

where

$$A = \left(\frac{E}{\rho_0}\right)^{1/5} g(0). \tag{1.35}$$

Plotting $\log R$ versus $\log t$ for a light explosives experiment, one can determine that $g(0) \simeq 1$. Using Mack's motion picture for the first atomic explosion, Taylor plotted $\frac{5}{2}\log_{10} R$ versus $\log_{10} t$ with R and t measured in c.g.s. units. [See Figure 1.2.4-1 where the motion picture data is indicated by +.] This verified the use of the approximation $g(\pi_1) \simeq g(0)$ and led to an accurate estimation of the classified energy E of the explosion!

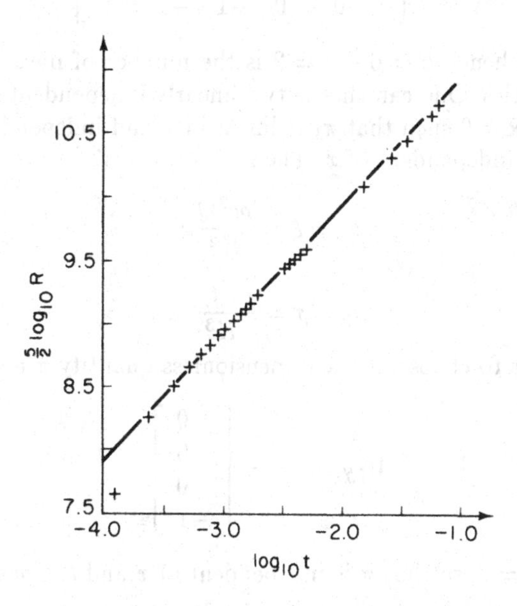

Figure 1.2.4-1

(2) *An Example in Heat Conduction Illustrating the Choice of Fundamental Dimensions*

Consider the standard problem of one-dimensional heat conduction in an "infinite" bar with constant thermal properties, initially heated by a point source of heat. Let u be the temperature at any point of the bar. We assume that

$$u = f(W_1, W_2, W_3, W_4, W_5, W_6) \tag{1.36}$$

where

$W_1 = x$, the distance along the bar from the point source of heat,
$W_2 = t$, the elapsed time after the initial heating,
$W_3 = \rho$, the mass density of the bar,
$W_4 = c$, the specific heat of the bar,
$W_5 = K$, the thermal conductivity of the bar,
$W_6 = Q$, the strength of the heat source measured in energy units per (length units)^2.

It is interesting to consider the effects of dimensional analysis in simplifying (1.36) with two different choices of fundamental dimensions.

Choice I (*Dynamical Units*). Here we let L_1 = length, L_2 = mass, L_3 = time, and L_4 = temperature. Correspondingly, the dimension matrix is

$$B_I = \begin{bmatrix} 1 & 0 & -3 & 2 & 1 & 0 \\ 0 & 0 & 1 & 0 & 1 & 1 \\ 0 & 1 & 0 & -2 & -3 & -2 \\ 0 & 0 & 0 & -1 & -1 & 0 \end{bmatrix}. \tag{1.37}$$

$r(B_I) = 4$ and hence $k = 6 - 4 = 2$ is the number of measurable dimensionless quantities. One can choose two linearly independent solutions $\mathbf{x}^{(1)}$ and $\mathbf{x}^{(2)}$ of $B_I\mathbf{x} = 0$ such that π_1 is linear in x and independent of t; π_2 is linear in t and independent of x. Then

$$\pi_1 = \xi = \frac{\rho c^2 Q}{K^2} x, \tag{1.38a}$$

$$\pi_2 = \tau = \frac{\rho c^3 Q^2}{K^3} t. \tag{1.38b}$$

It is convenient to choose as the dimensionless quantity π a solution of

$$B_I\mathbf{y} = -\mathbf{a} = \begin{bmatrix} 0 \\ 0 \\ 0 \\ -1 \end{bmatrix}$$

where $y_1 = y_2 = 0$, so that π is independent of x and t. Consequently

$$\pi = \frac{K^2}{Q^2 c} u. \tag{1.39}$$

Hence dimensional analysis with dynamical units reduces (1.36) to

$$u = \frac{Q^2 c}{K^2} F(\xi, \tau) \tag{1.40}$$

where F is an unknown function of ξ and τ.

Choice II (*Thermal Units*). Motivated by the implicit assumption that in the posed problem there is no conversion of heat energy to mechanical energy, we refine the dynamical units by introducing a thermal unit $L_5 =$ "calories." The corresponding dimension matrix is

$$\mathrm{B}_{II} = \begin{bmatrix} 1 & 0 & -3 & 0 & -1 & -2 \\ 0 & 0 & 1 & -1 & 0 & 0 \\ 0 & 1 & 0 & 0 & -1 & 0 \\ 0 & 0 & 0 & -1 & -1 & 0 \\ 0 & 0 & 0 & 1 & 1 & 1 \end{bmatrix}. \tag{1.41}$$

$r(\mathrm{B}_{II}) = 5$ and hence there is only one measurable dimensionless quantity. It is convenient to choose as dimensionless quantities

$$\pi_1 = \eta = \frac{x}{\sqrt{\kappa t}}, \quad \text{where } \kappa = \frac{K}{\rho c}, \tag{1.42a}$$

and

$$\pi = \frac{\sqrt{\rho c K t}}{Q} u. \tag{1.42b}$$

Thus dimensional analysis with thermal units reduces (1.36) to

$$u = \frac{Q}{\sqrt{\rho c K t}} G(\eta) \tag{1.43}$$

where G is an unknown function of η.

Note that equation (1.43) is a special case of equation (1.40) where

$$\eta = \frac{\xi}{\sqrt{\tau}} \quad \text{and} \quad F(\xi, \tau) = \frac{1}{\sqrt{\tau}} G\left(\frac{\xi}{\sqrt{\tau}}\right).$$

[In terms of thermal units each of the quantities, ξ, τ, $\frac{K^2 u}{Q^2 c}$ is not dimensionless.]

Obviously, if it is correct, equation (1.43) is a great simplification of equation (1.40). By conducting experiments or associating a properly-posed boundary value problem to determine u, one can show that thermal units are justified. In turn thermal units can then be used for other heat (diffusion) problems where the governing equations are not completely known.

Exercises 1.2

1. Use dimensional analysis to prove the Pythagoras theorem. [Hint: Drop a perpendicular to the hypotenuse of a right-angle triangle and consider the resulting similar triangles.]

2. How would you use dimensional analysis and experimental modelling to find the time of flight of a body dropped vertically from a height h?
3. Given that in c.g.s. units $\rho_0 = 1.3 \times 10^{-3}$, and $P_0 = 1.0 \times 10^6$, use Figure 1.2.4-1 to estimate the domain of π_1 and E.

1.3 Application of Dimensional Analysis to Partial Differential Equations

Consider the use of dimensional analysis where the quantities $(u, W_1, W_2, \ldots, W_n)$ arise in a boundary value problem for a partial differential equation which has a unique solution. Then the unknown u (the *dependent variable* of the partial differential equation) is the solution of the BVP and $(W_1, W_2, \ldots, W_n)$ denote all *independent variables* and *constants* appearing in the BVP. From the Buckingham Pi-theorem it follows that such a BVP can always be re-expressed in dimensionless form where π is a dimensionless dependent variable and $(\pi_1, \pi_2, \ldots, \pi_k)$ are dimensionless independent variables and dimensionless constants.

Say $(W_1, W_2, \ldots, W_\ell)$ are the ℓ independent variables and $(W_{\ell+1}, W_{\ell+2}, \ldots, W_n)$ are the $n - \ell$ constants appearing in the BVP. Let

$$\mathrm{B}_1 = \begin{bmatrix} b_{11} & b_{12} & \cdots & b_{1\ell} \\ b_{21} & b_{22} & \cdots & b_{2\ell} \\ \vdots & \vdots & & \vdots \\ b_{m1} & b_{m2} & \cdots & b_{m\ell} \end{bmatrix} \tag{1.44a}$$

be the dimension matrix of the independent variables and let

$$\mathrm{B}_2 = \begin{bmatrix} b_{1,\ell+1} & b_{1,\ell+2} & \cdots & b_{1n} \\ b_{2,\ell+1} & b_{2,\ell+2} & \cdots & b_{2n} \\ \vdots & \vdots & & \vdots \\ b_{m,\ell+1} & b_{m,\ell+2} & \cdots & b_{mn} \end{bmatrix} \tag{1.44b}$$

be the dimension matrix of the constants. The dimension matrix of the BVP is

$$\mathrm{B} = \left[\mathrm{B}_1 \ \vdots \ \mathrm{B}_2\right]. \tag{1.45}$$

A dimensionless π_i quantity is called a *dimensionless constant* if it does not depend on $(W_1, W_2, \ldots, W_\ell)$, i.e., in equation (1.26b), $x_{ji} = 0$, $j = 1, 2, \ldots, \ell$. A dimensionless π_i quantity is a *dimensionless variable* if $x_{ji} \neq 0$ for some $j = 1, 2, \ldots, \ell$. An important objective in applying dimensional analysis to a BVP is to reduce the number of independent variables. The rank of B_2, i.e. $r(\mathrm{B}_2)$, represents the reduction in the number of constants through dimensional analysis. Consequently the reduction in the number of independent variables is $\rho = r(\mathrm{B}) - r(\mathrm{B}_2)$. In particular the number of dimensionless measurable quantities is $k = n - r(\mathrm{B}) = [\ell - \rho] + [(n - \ell) - r(\mathrm{B}_2)]$ where $\ell - \rho$ of the π_i quantities are dimensionless independent variables and $(n - \ell) - r(\mathrm{B}_2)$ are dimensionless constants.

If $r(\mathrm{B}) = r(\mathrm{B}_2)$, then dimensional analysis reduces the given BVP to a dimensionless BVP with $(n - \ell) - r(\mathrm{B}_2)$ dimensionless constants. In this case the number of independent variables is not reduced. Nonetheless this is useful as a starting point for perturbation analysis.

If $\ell \geq 2$, $\ell - \rho = 1$, then the resulting solution of the BVP is called a *self-similar solution* or *automodel solution.*

1.3.1 EXAMPLES

(1) *Source Problem for Heat Conduction*

Consider the unknown temperature u of the heat conduction problem of Section 1.2.4 as the solution $u(x, t)$ of the BVP:

$$\rho c \frac{\partial u}{\partial t} - K \frac{\partial^2 u}{\partial x^2} = 0, \quad -\infty < x < \infty,\ t > 0, \tag{1.46a}$$

$$u(x, 0) = \frac{Q}{\rho c}\delta(x), \tag{1.46b}$$

$$\lim_{x \to \pm\infty} u(x, t) = 0. \tag{1.46c}$$

In equation (1.46b) $\delta(x)$ is the Dirac delta function.

The use of dimensional analysis with dynamical units reduces (1.46a–c) to

$$\frac{\partial F}{\partial \tau} - \frac{\partial^2 F}{\partial \xi^2} = 0, \quad -\infty < \xi < \infty,\ \tau > 0, \tag{1.47a}$$

$$F(\xi, 0) = \delta(\xi), \tag{1.47b}$$

$$\lim_{\xi \to \pm\infty} F(\xi, \tau) = 0, \tag{1.47c}$$

with u defined in terms of $F(\xi, \tau)$ by (1.40) and ξ, τ given by (1.38a,b). Consequently there is no essential progress in solving BVP (1.46a–c).

We now justify the use of dimensional analysis with thermal units to solve (1.46a–c) as follows:

First note that from equations (1.47a,c) we have

$$\frac{\partial}{\partial \tau}\int_{-\infty}^{\infty} F(\xi, \tau)d\xi = \int_{-\infty}^{\infty} \frac{\partial^2 F}{\partial \xi^2}(\xi, \tau)d\xi = 0.$$

Then from this equation and (1.47b) we get the conservation law

$$\int_{-\infty}^{\infty} F(\xi, \tau)d\xi = 1, \quad \text{valid for all } \tau > 0.$$

Consequently the substitution $F(\xi,\tau) = \frac{1}{\sqrt{\tau}}G(\frac{\xi}{\sqrt{\tau}})$, which results from using dimensional analysis with thermal units, reduces (1.47a–c) and hence (1.46a–c) to a BVP for an ordinary differential equation with independent variable $\eta = \frac{\xi}{\sqrt{\tau}}$ and dependent variable $G(\eta)$:

$$2\frac{d^2G}{d\eta^2} + \eta\frac{dG}{d\eta} + G = 0, \quad -\infty < \eta < \infty, \tag{1.48a}$$

$$\int_{-\infty}^{\infty} G(\eta)d\eta = 1, \tag{1.48b}$$

$$G(\pm\infty) = 0. \tag{1.48c}$$

This reduction of (1.46a–c) to a BVP for an ordinary differential equation is obtained much more naturally and easily in Section 1.4 from invariance of (1.46a–c) under a one-parameter group of scalings of its variables.

(2) *Prandtl–Blasius Problem for a Flat Plate*

Consider the Prandtl boundary layer equations for flow past a semi-infinite flat plate:

$$u\frac{\partial u}{\partial x} + v\frac{\partial u}{\partial y} = \nu\frac{\partial^2 u}{\partial y^2}, \tag{1.49a}$$

$$\frac{\partial u}{\partial x} + \frac{\partial v}{\partial y} = 0, \tag{1.49b}$$

$0 < x < \infty$, $0 < y < \infty$, with boundary conditions

$$u(x,0) = 0, \tag{1.49c}$$

$$v(x,0) = 0, \tag{1.49d}$$

$$u(x,\infty) = U, \tag{1.49e}$$

$$u(0,y) = U. \tag{1.49f}$$

In BVP (1.49a–f), x is the distance along the plate surface from its edge (tangential coordinate), y is the distance from the plat surface (normal coordinate), u is the x-component of velocity, v is the y-component of velocity, ν is the kinematic viscosity and U is the velocity of the incident flow [Figure 1.3.1-1].

Our aim is to calculate the shear at the plate (skin friction) $\frac{\partial u}{\partial y}(x,0)$ which leads to the determination of the viscous drag on the plate.

We look at the problem of determining $\frac{\partial u}{\partial y}(x,0)$ as defined through BVP (1.49a–f) from three analytical perspectives:

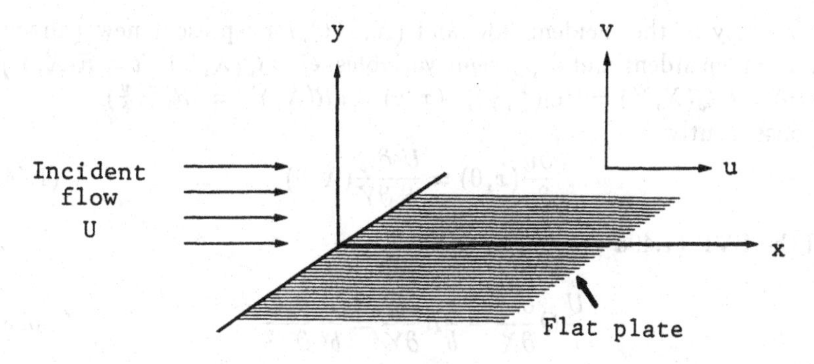

Figure 1.3.1-1

(i) *Dimensional Analysis.* From (1.49a–f) it follows that

$$\frac{\partial u}{\partial y}(x,0) = f(x, U, \nu) \tag{1.50}$$

with the unknown f to be determined as a function of measurable quantities x, U, ν. The fundamental dimensions are L = length and T = time. Then in terms of these fundamental dimensions:

$$\left[\frac{\partial u}{\partial y}(x,0)\right] = T^{-1}, \tag{1.51a}$$

$$[x] = L, \tag{1.51b}$$

$$[U] = LT^{-1}, \tag{1.51c}$$

$$[\nu] = L^2T^{-1}. \tag{1.51d}$$

Consequently $r(\mathrm{B}) = 2$. Dimensionless quantities are

$$\pi_1 = \frac{Ux}{\nu}, \tag{1.52a}$$

and

$$\pi = \frac{\nu}{U^2}\frac{\partial u}{\partial y}(x,0). \tag{1.52b}$$

Hence dimensional analysis leads to

$$\frac{\partial u}{\partial y}(x,0) = \frac{U^2}{\nu} g\left(\frac{Ux}{\nu}\right) \tag{1.53}$$

where g is an unknown function of $\frac{Ux}{\nu}$.

(ii) *Scalings of Quantities Followed by Dimensional Analysis.* Consider a linear transformation of the variables of (1.49a–f) given by $x = aX$, $y = bY$, $u = UQ$, $v = cR$ where (a, b, c) are undetermined positive constants, U is the velocity of the incident flow and (X, Y, Q, R) represent new (dimensional) independent and dependent variables: $Q = Q(X, Y)$, $R = R(X, Y)$; $u(x, y) = UQ(X, Y) = UQ(\frac{x}{a}, \frac{y}{b})$, $v(x, y) = cR(X, Y) = cR(\frac{x}{a}, \frac{y}{b})$.

Consequently

$$\frac{\partial u}{\partial y}(x, 0) = \frac{U}{b}\frac{\partial Q}{\partial Y}(X, 0) \tag{1.54}$$

and the BVP (1.49a–f) transforms to

$$\frac{U}{a}Q\frac{\partial Q}{\partial X} + \frac{c}{b}R\frac{\partial Q}{\partial Y} = \frac{\nu}{b^2}\frac{\partial^2 Q}{\partial Y^2}, \tag{1.55a}$$

$$\frac{U}{a}\frac{\partial Q}{\partial X} + \frac{c}{b}\frac{\partial R}{\partial Y} = 0, \tag{1.55b}$$

$0 < X < \infty$, $0 < Y < \infty$, with

$$Q(X, 0) = 0, \tag{1.55c}$$

$$R(X, 0) = 0, \tag{1.55d}$$

$$Q(X, \infty) = 1, \tag{1.55e}$$

$$Q(0, Y) = 1. \tag{1.55f}$$

From the form of (1.55a,b) it is convenient to choose (a, b, c) so that

$$\frac{U}{a} = \frac{c}{b} = \frac{\nu}{b^2}.$$

Hence we set $c = 1$, $b = \nu$, $a = U\nu$. As a result equations (1.55a,b) become cleared of constants:

$$Q\frac{\partial Q}{\partial X} + R\frac{\partial Q}{\partial Y} = \frac{\partial^2 Q}{\partial Y^2}, \tag{1.56a}$$

$$\frac{\partial Q}{\partial X} + \frac{\partial R}{\partial Y} = 0, \tag{1.56b}$$

$0 < X < \infty$, $0 < Y < \infty$. Moreover

$$\frac{\partial u}{\partial y}(x, 0) = \frac{U}{\nu}\frac{\partial Q}{\partial Y}(X, 0) = \frac{U}{\nu}\frac{\partial Q}{\partial Y}\left(\frac{x}{U\nu}, 0\right). \tag{1.57}$$

But now it follows that since $Q(X, Y)$ results from the solution of (1.56a,b), (1.55c–f), we have

$$\frac{\partial Q}{\partial Y}(X, 0) = h(X), \tag{1.58}$$

for some unknown function h. We apply dimensional analysis to (1.58):

$$\left[\frac{\partial Q}{\partial Y}\right] = LT^{-1}, \tag{1.59a}$$

$$[X] = L^{-2}T^2. \tag{1.59b}$$

Hence (1.58) reduces to

$$h(X) = \sigma X^{-1/2} \tag{1.60}$$

for some fixed dimensionless constant σ to be determined. Thus (1.53) simplifies further to $g(\frac{Ux}{\nu}) = \sigma(\frac{Ux}{\nu})^{-1/2}$ so that

$$\frac{\partial u}{\partial y}(x,0) = \sigma\left(\frac{U^3}{x\nu}\right)^{1/2}. \tag{1.61}$$

(iii) *Further Use of Dimensional Analysis on the Full BVP.* We now apply dimensional analysis to the BVP (1.56a,b), (1.55c–f), to reduce it to a BVP for an ordinary differential equation. It is convenient (but not necessary) to introduce a *potential (stream function)* $\psi(X,Y)$ from the form of (1.56b). Let $Q = \frac{\partial\psi}{\partial Y}$, $R = -\frac{\partial\psi}{\partial X}$. Then in terms of the single dependent variable ψ, BVP (1.56a,b), (1.55c–f), becomes:

$$\frac{\partial\psi}{\partial Y}\frac{\partial^2\psi}{\partial X\partial Y} - \frac{\partial\psi}{\partial X}\frac{\partial^2\psi}{\partial Y^2} = \frac{\partial^3\psi}{\partial Y^3}, \tag{1.62a}$$

$0 < X < \infty$, $0 < Y < \infty$, with

$$\frac{\partial\psi}{\partial Y}(X,0) = 0, \tag{1.62b}$$

$$\frac{\partial\psi}{\partial X}(X,0) = 0, \tag{1.62c}$$

$$\frac{\partial\psi}{\partial Y}(X,\infty) = 1, \tag{1.62d}$$

$$\frac{\partial\psi}{\partial Y}(0,Y) = 1. \tag{1.62e}$$

Moreover we get

$$\frac{\partial Q}{\partial Y}(X,0) = \frac{\partial^2\psi}{\partial Y^2}(X,0) = \sigma X^{-1/2}. \tag{1.63}$$

We apply dimensional analysis to simplify $\psi(X,Y)$. Since BVP (1.62a–e) has no constants, we have

$$\psi = F(X,Y), \tag{1.64}$$

for some unknown function F. We see that

$$[\psi] = [Y] = L^{-1}T, \tag{1.65a}$$

$$[X] = L^{-2}T^2. \tag{1.65b}$$

Consequently there is only one measurable dimensionless quantity. It is convenient to choose as dimensionless quantities

$$\pi_1 = \eta = \frac{Y}{\sqrt{X}}, \tag{1.66a}$$

and

$$\pi = \frac{\psi}{\sqrt{X}}. \tag{1.66b}$$

Hence

$$\psi(X,Y) = \sqrt{X}\,G(\eta) \tag{1.67}$$

where $G(\eta)$ solves a BVP for an ordinary differential equation which is obtained by substituting (1.67) into (1.62a–e). Moreover from (1.67) and (1.63) it follows that

$$\sigma = G''(0). \tag{1.68}$$

[A prime denotes differentiation with respect to η.] Note that

$$\frac{\partial\psi}{\partial Y} = G'(\eta),\ \frac{\partial\psi}{\partial X} = \frac{1}{2\sqrt{X}}[G - \eta G'],$$

$$\frac{\partial^2\psi}{\partial Y^2} = \frac{1}{\sqrt{X}}G'',\ \frac{\partial^3\psi}{\partial Y^3} = \frac{1}{X}G''',$$

$$\frac{\partial^2\psi}{\partial X\partial Y} = \frac{1}{2X}[-\eta G''];$$

$0 < X < \infty$, $0 < Y < \infty$ leads to $0 < \eta < \infty$; $Y = 0$ leads to $\eta = 0$; $Y \to \infty$ leads to $\eta \to \infty$; $X = 0$ leads to $\eta \to \infty$. Correspondingly BVP (1.62a–e) reduces to solving the third order ordinary differential equation known as the *Blasius equation* for $G(\eta)$,

$$2\frac{d^3G}{d\eta^3} + G\frac{d^2G}{d\eta^2} = 0, \quad 0 < \eta < \infty, \tag{1.69a}$$

with boundary conditions

$$G(0) = G'(0) = 0, \quad G'(\infty) = 1. \tag{1.69b}$$

The aim is to find $\sigma = G''(0)$.

A numerical procedure for solving BVP (1.69a,b) is the shooting method where one considers the auxiliary initial value problem

$$2\frac{d^3H}{dz^3} + H\frac{d^2H}{dz^2} = 0, \quad 0 < z < \infty, \tag{1.70a}$$

$$H(0) = H'(0) = 0, \quad H''(0) = A, \tag{1.70b}$$

for some initial guess A. One integrates out the IVP (1.70a,b) and determines that $H'(\infty) = B$ for some number, $B = B(A)$. One continues shooting with different values of A until B is close enough to 1.

It turns out that the invariance of (1.70a) and the initial conditions $H(0) = H'(0) = 0$ under a one-parameter family of scalings (one-parameter Lie group of scaling transformations) leads to determining σ with only one shooting:

The transformation

$$z = \frac{\eta}{\alpha}, \tag{1.71a}$$

$$H(z) = \alpha G(\eta), \tag{1.71b}$$

with $\alpha > 0$ an arbitrary constant, maps (1.70a,b) to (1.69a) with initial conditions

$$G(0) = G'(0) = 0, \quad G''(0) = \frac{A}{\alpha^3}. \tag{1.72}$$

Moreover $H'(\infty) = B$ implies that

$$G'(\infty) = \frac{B}{\alpha^2}. \tag{1.73}$$

Hence we pick α so that $\alpha^2 = B$, i.e. $\alpha = \sqrt{B}$. Then

$$\sigma = G''(0) = \frac{A}{B^{3/2}}. \tag{1.74}$$

One can show that

$$\sigma = 0.332\ldots\ . \tag{1.75}$$

Exercises 1.3

1. For the heat conduction problem (1.46a–c), show that $r(\mathrm{B}_2) = 4$ for both dynamical and thermal units.

2. Derive (1.47a–c).

3. Derive (1.48a–c).

4. The BVP (1.46a–c) in effect has only two constants: $\kappa = \frac{K}{\rho c}$ (diffusivity) and $\lambda = \frac{Q}{\rho c}$. Use dimensional analysis with dynamical units to reduce (1.46a–c) where now $W_1 = x$, $W_2 = t$, $W_3 = \kappa$, $W_4 = \lambda$.

5. Consider the *Rayleigh flow problem* [see Schlichting (1955)] where an infinite flat plate is immersed in an incompressible fluid at rest. The plate is instantaneously accelerated so that it moves parallel to itself with constant velocity U.

   Let u be the fluid velocity in the direction of U (x-direction). Let the y-direction be the direction normal to the plate. The situation is illustrated in Figure 1.3.1.

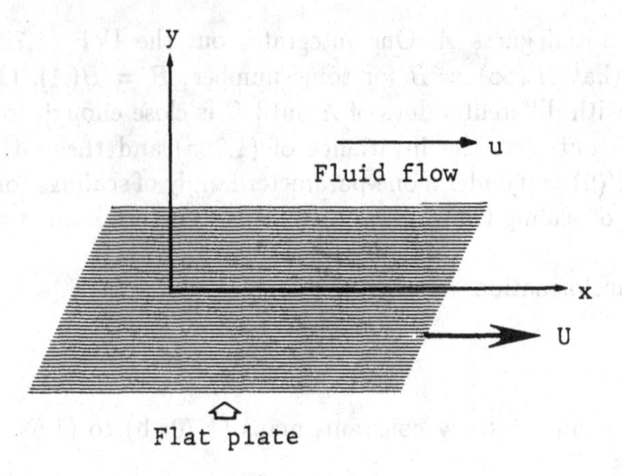

Figure 1.3.1

From symmetry considerations the Navier–Stokes equations governing this problem reduce to the viscous diffusion equation

$$\frac{\partial u}{\partial t} = \nu \frac{\partial^2 u}{\partial y^2}, \quad 0 < t < \infty,\ 0 < y < \infty, \tag{1.76a}$$

with boundary conditions

$$u(y, 0) = 0, \tag{1.76b}$$

$$u(0, t) = U, \tag{1.76c}$$

$$u(\infty, t) = 0. \tag{1.76d}$$

(a) Use dimensional analysis to simplify BVP (1.76a–d).

(b) Use scalings of quantities followed by dimensional analysis to further simplify (1.76a–d). Find the explicit self-similar solution $u(y, t)$ of (1.76a–d).

1.4 Generalization of Dimensional Analysis—Invariance of Partial Differential Equations Under Scalings of Variables

In both examples of Section 1.3.1 the use of dimensional analysis to reduce a BVP for a partial differential equation to a BVP for an ordinary differential equation is rather cumbersome and should make the reader feel uneasy. For the heat conduction problem the use of dimensional analysis depends on either making the right choice of fundamental dimensions (thermal units) or combining effectively the constants before using dynamical units [cf.

Exercise 1.3-4]. For the Prandtl–Blasius problem we used scaled variables before applying dimensional analysis.

A much easier way to accomplish such a reduction for a BVP is to consider the invariance property of the BVP under a one-parameter family of scalings (one-parameter Lie group of scaling transformations) when its variables are scaled but the constants of the BVP are not scaled. If the BVP is invariant under such a family of scaling transformations, then the number of independent variables is reduced constructively by one. We show that if, for some choice of fundamental dimensions, dimensional analysis leads to a reduction of the number of independent variables of a BVP, then such a reduction is always possible through invariance of the BVP under scalings applied strictly to its variables. [Recall that dimensional analysis involves scalings of *both* variables and constants.] Moreover there exist BVP's for which the number of independent variables is reduced from invariance under a one-parameter family of scalings of its variables but the number of independent variables is not reduced from the use of dimensional analysis for any *known* choice of fundamental dimensions. [One could argue that this is a way of determining new sets of fundamental dimensions!] Hence, for the purpose of reducing the number of independent variables of a BVP, invariance of a BVP under a one-parameter family of scalings of its variables is a generalization of dimensional analysis.

Zel'dovich (1956) [see also Barenblatt and Zel'dovich (1972) and Barenblatt (1979)] calls a *self-similar solution of the first kind* a solution of a BVP obtained by reduction through dimensional analysis and a *self-similar solution of the second kind* a solution to a BVP obtained by reduction through invariance under scalings of the variables when this reduction is not possible through dimensional analysis. The two examples of Section 1.3.1 show that these distinctions are somewhat blurred.

Before proving a general theorem relating dimensional analysis and invariance under scalings of variables, we consider the invariance property of the heat conduction problem (1.46a–c) under scalings of its variables.

Consider the family of scaling transformations

$$x^* = \alpha x, \tag{1.77a}$$

$$t^* = \beta t, \tag{1.77b}$$

$$u^* = \gamma u, \tag{1.77c}$$

where α, β, γ are arbitrary positive constants.

Definition 1.4-1. A transformation of the form (1.77a–c) *leaves BVP* (1.46a–c) *invariant* (*is admitted by BVP* (1.46a–c)) if and only if for any solution $u = \Theta(x,t)$ of (1.46a–c) it follows that

$$v(x^*,t^*) = u^* = \gamma u = \gamma\Theta(x,t) \tag{1.78}$$

solves the BVP

$$\rho c \frac{\partial v}{\partial t^*} - K \frac{\partial^2 v}{\partial x^{*2}} = 0, \quad -\infty < x^* < \infty, \; t^* > 0, \tag{1.79a}$$

$$v(x^*, 0) = \frac{Q}{\rho c}\delta(x^*), \tag{1.79b}$$

$$\lim_{x^* \to \pm\infty} v(x^*, t^*) = 0. \tag{1.79c}$$

[Implicitly it is assumed that the domain $-\infty < x^* < \infty$, $t^* > 0$ corresponds to the domain $-\infty < x < \infty$, $t > 0$; $t^* = 0$ corresponds to $t = 0$; $x^* \to \pm\infty$ corresponds to $x \to \pm\infty$, i.e. (1.77a–c) leaves the boundary of BVP (1.46a–c) invariant.]

Lemma 1.4-1. *If a scaling* (1.77a–c) *leaves BVP* (1.46a–c) *invariant, and* $u = \Theta(x,t)$ *solves* (1.46a–c), *then* $u = \gamma\Theta(\frac{x}{\alpha}, \frac{t}{\beta})$ *also solves* (1.46a–c).

Proof. See Exercise 1.4-1. □

In order that (1.77a–c) leaves BVP (1.46a–c) invariant, it is sufficient to leave each of these three equations separately invariant. Invariance of (1.46a), i.e. $u = \Theta(x,t)$ solves (1.46a) if and only if $v = \gamma\Theta(x,t)$ solves (1.79a), leads to $\beta = \alpha^2$ and invariance of (1.46b,c) leads to $\gamma = \frac{1}{\alpha}$. Hence the one-parameter ($\alpha > 0$) family of scaling transformations

$$x^* = \alpha x, \tag{1.80a}$$

$$t^* = \alpha^2 t, \tag{1.80b}$$

$$u^* = \frac{1}{\alpha} u, \tag{1.80c}$$

is admitted by (1.46a–c).

Clearly if $u = \Theta(x,t)$ solves (1.46a–c) then

$$v(x^*, t^*) = \Theta(x^*, t^*) = \Theta(\alpha x, \alpha^2 t) \tag{1.81}$$

solves (1.79a–c). Hence a transformation (1.80a–c) maps any solution $v = \Theta(x^*, t^*)$ of (1.79a–c) to a solution

$$v = \frac{1}{\alpha}\Theta(x,t) = \frac{1}{\alpha}\Theta\left(\frac{x^*}{\alpha}, \frac{t^*}{\alpha^2}\right)$$

of (1.79a–c) or, equivalently, maps any solution $u = \Theta(x,t)$ of (1.46a–c) to a solution $u = \frac{1}{\alpha}\Theta(\frac{x}{\alpha}, \frac{t}{\alpha^2})$ of (1.46a–c).

The solution of (1.46a–c) and hence (1.79a–c) is unique. As a result the solution $u = \Theta(x,t)$ of (1.46a–c) satisfies the functional equation (arising from the uniqueness of the solution to this BVP)

$$\Theta(x^*, t^*) = \frac{1}{\alpha}\Theta(x,t). \tag{1.82}$$

Such a solution of a partial differential equation, arising from invariance under a one-parameter Lie group of transformations, is called a *similarity solution* or *invariant solution.* The functional equation (1.82), satisfied by the invariant solution, is called the ***invariant surface condition.*** An invariant solution arising from invariance under a one-parameter Lie group of scalings such as (1.80a–c) is also called a *self-similar solution* or *automodel solution.*

From (1.80a,b), the invariant surface condition (1.82) satisfied by $\Theta(x,t)$ is

$$\Theta(\alpha x, \alpha^2 t) = \frac{1}{\alpha}\Theta(x,t). \tag{1.83}$$

In order to solve (1.83), let $z = \frac{x}{\sqrt{t}}$ and $\Theta(x,t) = \frac{1}{\sqrt{t}}\phi(z,t)$. Then in terms of z, t, $\phi(z,t)$, equation (1.83) becomes

$$\frac{1}{\sqrt{t}}\phi(z,t) = \frac{\alpha}{\sqrt{\alpha^2 t}}\phi(z,\alpha^2 t) = \frac{\phi(z,\alpha^2 t)}{\sqrt{t}}.$$

Hence $\phi(z,t)$ satisfies the functional equation

$$\phi(z,t) = \phi(z,\alpha^2 t) \quad \text{for any } \alpha > 0. \tag{1.84}$$

Thus $\phi(z,t)$ does not depend on t. This leads to the *invariant form* (*similarity form*)

$$u = \Theta(x,t) = \frac{1}{\sqrt{t}}F(z) \tag{1.85}$$

for the solution of BVP (1.46a–c); z is called the *similarity variable.* The substitution of (1.85) into (1.46a–c) leads to a BVP for an ordinary differential equation with unknown $F(z)$. The details are left to Exercise 1.4-2.

Now consider the following theorem connecting dimensional analysis and invariance under scalings of variables.

Theorem 1.4-1. *If the number of independent variables appearing in a BVP for a partial differential equation can be reduced by ρ through dimensional analysis, then the number of variables can be reduced by ρ through invariance of the BVP under a ρ-parameter family of scaling transformations of its variables.*

Proof. Consider the dimension matrices B, B_1 and B_2 defined by (1.44a,b), (1.45). Through dimensional analysis the number of independent variables of the given BVP is reduced by $\rho = r(B) - r(B_2)$.

An arbitrary scaling of any fundamental dimension is represented by the m-parameter family of scaling transformations

$$L_j^* = e^{\epsilon_j}L_j, \quad j = 1,2,\dots,m \tag{1.86}$$

where $(\epsilon_1, \epsilon_2, \dots, \epsilon_m)$ are arbitrary real numbers. Let the row vector

$$\epsilon = [\epsilon_1, \epsilon_2, \dots, \epsilon_m]. \tag{1.87}$$

The scaling (1.86) induces a scaling of the value of each measurable quantity W_i:

$$W_i^* = e^{\sum_{j=1}^m \epsilon_j b_{ji}} W_i = e^{(\epsilon \mathrm{B})_i} W_i, \quad i = 1, 2, \ldots, n, \tag{1.88}$$

where $(\epsilon \mathrm{B})_i$ is the ith component of the n-component row vector $\epsilon \mathrm{B}$; the value of u scales to

$$u^* = e^{\sum_{j=1}^m \epsilon_j a_j} u. \tag{1.89}$$

From assumption (iv) of the Buckingham Pi-theorem, the family of scaling transformations (1.88), (1.89), induced by the m-parameter family of scalings of the fundamental dimensions (1.86), leaves the given BVP invariant. Our aim is to find the number of essential parameters in the subfamily of transformations of the form (1.88), (1.89) for which the constants are all invariant, i.e. we aim to find the dimension of the vector space of all vectors $\epsilon = [\epsilon_1, \epsilon_2, \ldots, \epsilon_m]$ such that

$$W_i^* = W_i, \quad i = \ell + 1, \ell + 2, \ldots, n, \tag{1.90a}$$

and

$$W_j^* \neq W_j \quad \text{for some} \quad j = 1, 2, \ldots, \ell. \tag{1.90b}$$

Equation (1.90a) holds if and only if

$$\epsilon \mathrm{B}_2 = 0. \tag{1.91}$$

The number of essential parameters is the number of linearly independent solutions ϵ of (1.91) such that $\epsilon \mathrm{B}_1 \neq 0$.

It is helpful to introduce a few definitions and some notation:

Let A be a matrix linear transformation acting on vector space V such that if $\mathbf{v} \in V$ then $\mathbf{v}\mathrm{A}$ is the action of A on $\mathbf{v}$. The *null space* of A is the vector space $V_{(\mathrm{A})_N} = \{\epsilon \in V : \epsilon \mathrm{A} = 0\}$; the *range space* of A is the vector space $V_{(\mathrm{A})_R} = \{z : z = \epsilon \mathrm{A} \text{ for some } \epsilon \in V\}$; $\dim V$ is the dimension of the vector space V. It follows that

$$\dim V = \dim V_{(\mathrm{A})_R} + \dim V_{(\mathrm{A})_N}.$$

Consider the matrices B, B_1, and B_2 defined by (1.44a,b), (1.45). Let V be R^m, where m is the number of rows of each of these three matrices, so that $\dim V = m$. Then $\dim V_{(\mathrm{B})_N}$ is the number of linearly independent solutions ϵ of the set of equations $\epsilon \mathrm{B} = 0$, and $\dim V_{(\mathrm{B}_2)_N}$ is the number of linearly independent solutions ϵ of $\epsilon \mathrm{B}_2 = 0$. It follows that

$$\dim V_{(\mathrm{B}_2)_N} = m - r(\mathrm{B}_2); \quad \dim V_{(\mathrm{B})_N} = m - r(\mathrm{B}) = m - r(\mathrm{B}_2) - \rho.$$

Since $V_{(\mathrm{B}_2)_N(\mathrm{B}_1)_N} = V_{(\mathrm{B})_N}$, it follows that

$$\begin{aligned} \dim V_{(\mathrm{B}_2)_N} &= \dim V_{(\mathrm{B}_2)_N(\mathrm{B}_1)_N} + \dim V_{(\mathrm{B}_2)_N(\mathrm{B}_1)_R} \\ &= \dim V_{(\mathrm{B})_N} + \dim V_{(\mathrm{B}_2)_N(\mathrm{B}_1)_R}. \end{aligned}$$

Hence dim $V_{(B_2)_N(B_1)_R} = \rho$. But dim $V_{(B_2)_N(B_1)_R}$ is the number of linearly independent solutions ϵ of the system $\epsilon B_2 = 0$ such that $\epsilon B_1 \neq 0$. Hence the number of essential parameters is ρ, completing the proof of the theorem. □

Exercises 1.4

1. Prove Lemma 1.4-1.

2. Set up the BVP for $F(z)$ as defined by equation (1.85). Put this BVP in dimensionless form using

   (a) dynamical units;

   (b) thermal units. Explain.

3. Consider diffusion in a half-space with a concentration dependent diffusion coefficient which is directly proportional to the concentration of a substance $C(x,t)$. Initially and far from the front face $x = 0$ the concentration is assumed to be zero. The concentration is fixed on the front face. The aim is to find the concentration flux on the front face, $\frac{\partial C}{\partial x}(0,t)$. In special units $C(x,t)$ satisfies the BVP

$$\frac{\partial C}{\partial t} = \frac{\partial}{\partial x}\left(C\frac{\partial C}{\partial x}\right), \quad 0 < x < \infty,\ 0 < t < \infty, \tag{1.92a}$$

   where

$$C(x,0) = C(\infty,t) = 0; \quad C(0,t) = A. \tag{1.92b}$$

   (a) Exploit similarity to determine $\frac{\partial C}{\partial x}(0,t)$ as effectively as possible.

   (b) Use scaling invariance to reduce the BVP (1.92a,b) to a BVP for an ordinary differential equation.

   (c) Discuss a numerical procedure to determine $\frac{\partial C}{\partial x}(0,t)$ based on the scaling property of the reduced BVP derived in (b).

4. For boundary layer flow over a semi-infinite wedge at zero angle of attack, the governing partial differential equations are

$$u\frac{\partial u}{\partial x} + v\frac{\partial u}{\partial y} - U(x)\frac{dU}{dx} = \nu\frac{\partial^2 u}{\partial y^2},$$

$$\frac{\partial u}{\partial x} + \frac{\partial v}{\partial y} = 0, \quad 0 < x < \infty,\ 0 < y < \infty,$$

   with boundary conditions $u(x,0) = v(x,0) = 0$, $\lim_{y\to\infty} u(x,y) = U(x)$; $U(x) = Ax^{\ell}$ where A, ℓ are constants with $\ell = \frac{\beta}{2-\beta}$ corresponding to the opening angle $\pi\beta$ of the semi-infinite wedge. In this problem x is the distance from the leading edge on the wedge surface and y is the distance from the wedge surface [Figure 1.4.1].

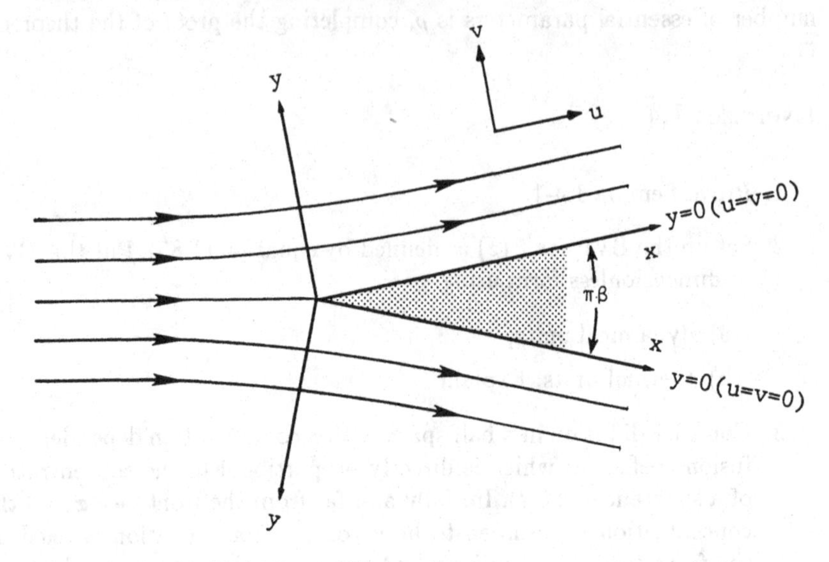

Figure 1.4.1

As for the Prandtl boundary layer equations (1.49a,b) introduce a stream function $\psi(x, y)$. Use scaling invariance to reduce the given problem to a BVP for an ordinary differential equation. Choose coordinates so that the Blasius equation arises if $\ell = 0$.

5. The following BVP for a nonlinear diffusion equation arises from a biphasic continuum model of soft tissue [Holmes (1984)]:

$$\frac{\partial^2 u}{\partial x^2} - K\left(\frac{\partial u}{\partial x}\right)\frac{\partial u}{\partial t} = 0, \quad 0 < x < \infty,\ 0 < t < \infty,$$

where K is a function of $\frac{\partial u}{\partial x}$, with boundary conditions $\frac{\partial u}{\partial x}(0, t) = -1$, $u(\infty, t) = u(x, 0) = 0$. Reduce this problem to a BVP for an ordinary differential equation.

6. Use invariance under scalings of the variables to solve the Rayleigh flow problem (1.76a–d).

7. Consider again the source problem for heat conduction in terms of the dimensionless form arising from dynamical units

$$\frac{\partial u}{\partial t} - \frac{\partial^2 u}{\partial x^2} = 0, \quad -\infty < x < \infty,\ t > 0,$$

$$u(x, 0) = \delta(x),$$

$$\lim_{x\to\pm\infty} u(x,t) = 0.$$

The use of scaling invariance with respect to the variables (1.80a–c) leads to the similarity form for the solution $u = \frac{1}{\sqrt{t}}G(\frac{x}{\sqrt{t}})$.

(a) Show that the problem is invariant under the one-parameter (β) family of transformations

$$x^* = x - \beta t,\ t^* = t,\ u^* = ue^{\frac{1}{2}\beta x - \frac{1}{4}\beta^2 t}, \tag{1.93}$$

for any constant β, $-\infty < \beta < \infty$.

(b) Check that t and $ue^{x^2/4t}$ are invariants of these transformations.

(c) Show that these transformations lead to the similarity form

$$u(x,t) = e^{-x^2/4t}H(t). \tag{1.94}$$

Hence show that invariance under scalings (1.80a–c) and the transformations (1.93) lead to the well-known fundamental solution

$$u = \frac{1}{\sqrt{4\pi t}}e^{-x^2/4t}.$$

1.5 Discussion

Dimensional analysis is necessary for ascertaining fundamental dimensions and consequent essential quantities which arise in a real problem in order to design proper model experiments. If a given problem can be described in terms of a boundary value problem (BVP) for a system of partial differential equations then dimensional analysis may lead to a reduction in the number of independent variables. Moreover if such a reduction exists, it can always be accomplished by considering the invariance properties of the BVP under scaling transformations applied only to its variables.

As will be seen in Chapter 4 the invariance properties of partial differential equations (or more particularly BVP's) under scalings of variables can be generalized to the study of the invariance properties of partial differential equations under arbitrary one-parameter Lie groups of point transformations of their variables. Moreover for a given differential equation such transformations are found algorithmically. [For example one can easily deduce transformations (1.93) and (1.94).] This follows from the properties of such transformations, most importantly their characterization by infinitesimal generators [see Chapter 2].

References on dimensional analysis specific to various fields include: de Jong (1967) [economics]; Sedov (1959), Birkhoff (1950) and Barenblatt (1979) [mechanics, elasticity, and hydrodynamics]; Venikov (1969) [electrical engineering]; Taylor (1974) [mechanical engineering]; Becker (1976) [chemical engineering]; Kurth (1972) [astrophysics]; Murota (1985) [systems analysis].

Examples of dimensional analysis and scaling invariance applied to BVP's appear in Sedov (1959), Birkhoff (1950), Barenblatt (1979), Dresner (1983), Hansen (1964), and Seshadri and Na (1985). Examples which use scalings to convert BVP's for ordinary differential equations to initial value problems appear in Klamkin (1962), Na (1967, 1979), Dresner (1983), and Seshadri and Na (1985).

2

Lie Groups of Transformations and Infinitesimal Transformations

2.1 Lie Groups of Transformations

In dimensional analysis the scaling transformations of the fundamental dimensions (1.86), the induced scaling transformations of the measurable quantities (1.88), the induced scaling transformations of all quantities (1.88), (1.89), and the induced scaling transformations preserving all constants (1.88), (1.91), are all examples of Lie groups of transformations. Scaling transformations are easily described in terms of their global properties as seen in Chapter 1. From the point of view of finding solutions to differential equations a general theory of Lie groups of transformations is unnecessary if transformations are restricted to scalings, translations, or rotations. However it turns out that much wider classes of transformations leave differential equations invariant including transformations composed of scalings, translations, and rotations. For the use and discovery of such transformations the notion of a Lie group of transformations is crucial—in particular the characterization of such transformations in terms of infinitesimal generators (which form a Lie algebra). This chapter introduces the basic ideas of Lie groups of transformations necessary in later chapters for the study of invariance properties of differential equations.

2.1.1 Groups

Definition 2.1.1-1. A *group* G is a set of elements with a law of composition ϕ between elements satisfying the following axioms:

(i) *Closure property:* For any element a and b of G, $\phi(a,b)$ is an element of G.

(ii) *Associative property:* For any elements a, b, and c of G,

$$\phi(a,\phi(b,c)) = \phi(\phi(a,b),c).$$

(iii) *Identity element:* There exists a unique identity element e of G such that for any element a of G,

$$\phi(a,e) = \phi(e,a) = a.$$

(iv) *Inverse element:* For any element a of G there exists a unique inverse element a^{-1} in G such that

$$\phi(a, a^{-1}) = \phi(a^{-1}, a) = e.$$

Definition 2.1.1-2. A group G is *Abelian* if $\phi(a, b) = \phi(b, a)$ holds for all elements a and b in G.

Definition 2.1.1-3. A *subgroup* of G is a group formed by a subset of elements of G with the same law of composition ϕ.

2.1.2 Examples of Groups

(1) G is the set of all integers with $\phi(a, b) = a + b$. Here $e = 0$ and $a^{-1} = -a$.

(2) G is the set of all positive reals with $\phi(a, b) = a \cdot b$. Here $e = 1$ and $a^{-1} = \frac{1}{a}$.

(3) G is the set of symmetries (set of transformations) which leave an equilateral triangle ABC invariant with both sides painted in the same color [Figure 2.1.2-1].

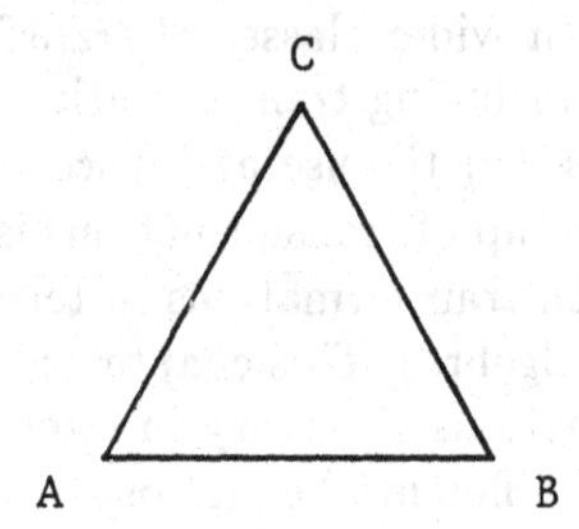

Figure 2.1.2-1

Here the group can be represented by all permutations of the vertices A, B, and C. The *identity element* $e = (1, 2, 3)$ corresponds to vertex 1 located at A, vertex 2 at B, and vertex 3 at C [Figure 2.1.2-2(a)]. The *rotation element* $R = (3, 1, 2)$ corresponds to a counterclockwise rotation of $\frac{2\pi}{3}$ radians of the configuration of Figure 2.1.2-1 about an axis coming out of the page through the center of the triangle [Figure 2.1.2-2(b)]; as a consequence vertex 3 is located at A, vertex 1 at B, and vertex 2 at C. The *flip element* $r = (3, 2, 1)$ represents rotation of the configuration of Figure 2.1.2-1 about the indicated perpendicular by π radians; as a consequence vertex 3 is located at A, vertex 2 at B, and vertex 1 at C [Figure 2.1.2-2(c)].

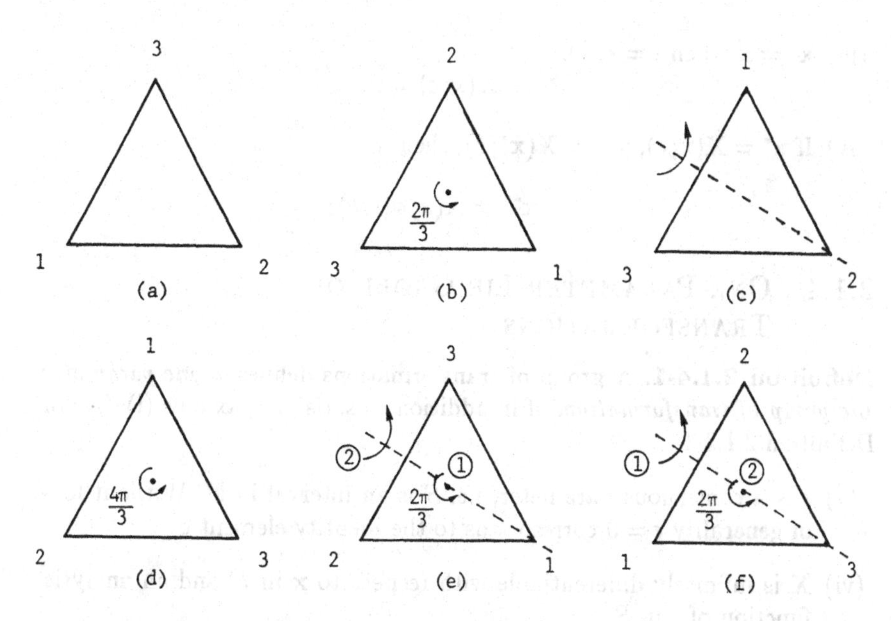

Figure 2.1.2-2. *Symmetry group of an equilateral triangle.* (a) identity, e; (b) rotation by $\frac{2\pi}{3}$, R; (c) flip, r; (d) rotation by $\frac{4\pi}{3}$, $\phi(R,R)$; (e) rotation by $\frac{2\pi}{3}$ followed by flip, $\phi(R,r)$; (f) flip followed by rotation by $\frac{2\pi}{3}$, $\phi(r,R)$.

The element $\phi(R,R) = R^2 = (2,3,1)$ represents a counterclockwise rotation of $\frac{4\pi}{3}$ radians of the configuration in Figure 2.1.2-1 [Figure 2.1.2-2(d)]. It is a composition of a counterclockwise rotation of $\frac{2\pi}{3}$ radians followed by a counterclockwise rotation of $\frac{2\pi}{3}$ radians. The composition element $\phi(R,r) = rR = (2,1,3)$ represents a counterclockwise rotation of $\frac{2\pi}{3}$ radians followed by a flip [Figure 2.1.2-2(e)]. The composition element $\phi(r,R) = Rr = (1,3,2)$ represents a flip followed by a counterclockwise rotation of $\frac{2\pi}{3}$ radians [Figure 2.1.2-2(f)]. It is left to Exercise 2.1-1 to prove that the symmetries of an equilateral triangle form a group with six elements. Note that this group is not *Abelian* since $\phi(r,R) \neq \phi(R,r)$.

2.1.3 Groups of Transformations

Definition 2.1.3-1. Let $\mathbf{x} = (x_1, x_2, \ldots, x_n)$ lie in region $D \subset \mathbb{R}^n$. The set of transformations

$$\mathbf{x}^* = \mathbf{X}(\mathbf{x};\epsilon), \tag{2.1}$$

defined for each $\mathbf{x}$ in D, depending on parameter ϵ lying in set $S \subset \mathbb{R}$, with $\phi(\epsilon,\delta)$ defining a law of composition of parameters ϵ and δ in S, forms a *group of transformations* on D if:

(i) For each parameter ϵ in S the transformations are one-to-one onto D, in particular $\mathbf{x}^*$ lies in D.

(ii) S with the law of composition ϕ forms a group G.

(iii) $\mathbf{x}^* = \mathbf{x}$ when $\epsilon = e$, i.e.

$$\mathbf{X}(\mathbf{x}; e) = \mathbf{x}.$$

(iv) If $\mathbf{x}^* = \mathbf{X}(\mathbf{x}; \epsilon)$, $\mathbf{x}^{**} = \mathbf{X}(\mathbf{x}^*; \delta)$, then

$$\mathbf{x}^{**} = \mathbf{X}(\mathbf{x}; \phi(\epsilon, \delta)).$$

2.1.4 One-Parameter Lie Group of Transformations

Definition 2.1.4-1. A group of transformations defines a *one-parameter Lie group of transformations* if in addition to satisfying axioms (i)–(iv) of Definition 2.1.3-1:

(v) ϵ is a continuous parameter, i.e. S is an interval in $\mathbb{R}$. Without loss of generality $\epsilon = 0$ corresponds to the identity element e.

(vi) $\mathbf{X}$ is infinitely differentiable with respect to $\mathbf{x}$ in D and an analytic function of ϵ in S.

(vii) $\phi(\epsilon, \delta)$ is an analytic function of ϵ and δ, $\epsilon \in S$, $\delta \in S$.

If one thinks of ϵ as a time variable and $\mathbf{x}$ as spatial variables then a one-parameter Lie group of transformations in effect defines a stationary flow. This will be shown in Section 2.2.1 but now can be partially seen as follows: Let

$$\mathbf{X}(\mathbf{x}; G) \tag{2.2}$$

define the evolution of $\mathbf{x}$ over all elements in G. This defines a curve γ_1 represented by Figure 2.1.4-1(a).

Now let $\mathbf{y} = \mathbf{X}(\mathbf{x}; \epsilon)$ represent a point on γ_1. Then $\mathbf{x}^{**} = \mathbf{X}(\mathbf{y}; \delta) = \mathbf{X}(\mathbf{x}; \phi(\epsilon, \delta))$ must lie on γ_1. Note that the self-intersecting curve γ_2 [Figure 2.1.4-1(b)] cannot represent the evolution defined by (2.2).

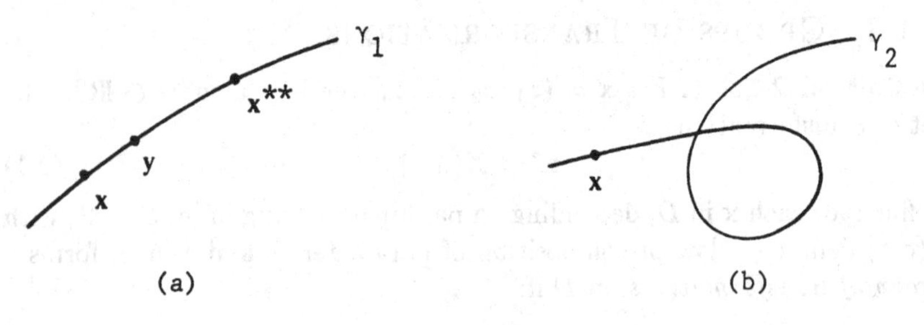

Figure 2.1.4-1

2.1.5 Examples of One-Parameter Lie Groups of Transformations

(1) *A Group of Translations in the Plane:*

$$x^* = x + \epsilon,$$

$$y^* = y, \quad \epsilon \in \mathbb{R}.$$

Here $\phi(\epsilon, \delta) = \epsilon + \delta$. This group corresponds to motions parallel to the x-axis. [In Figure 2.1.5-1 the curve γ represents the evolution $\mathbf{X}(\mathbf{x}_0; G)$.]

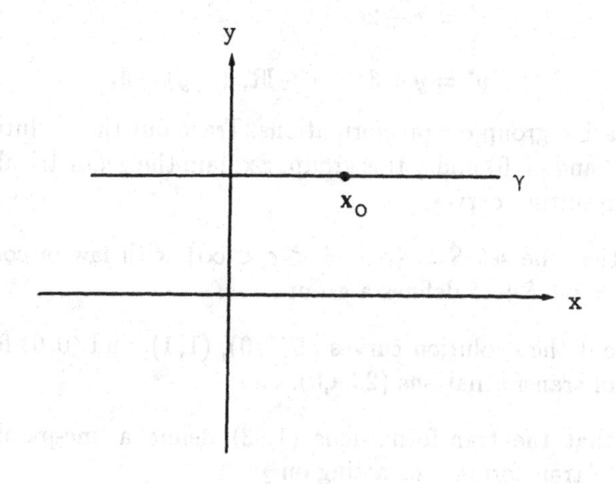

Figure 2.1.5-1

(2) *A Group of Scalings in the Plane:*

$$x^* = \alpha x,$$

$$y^* = \alpha^2 y, \quad 0 < \alpha < \infty.$$

Here $\phi(\alpha, \beta) = \alpha\beta$ and the identity element $e = 1$. This group of transformations can also be re-parametrized in terms of $\epsilon = \alpha - 1$:

$$x^* = (1 + \epsilon)x, \tag{2.3a}$$

$$y^* = (1 + \epsilon)^2 y, \quad -1 < \epsilon < \infty, \tag{2.3b}$$

so that the identity element is $e = 0$ with the law of composition of parameters given by

$$\phi(\epsilon, \delta) = \epsilon + \delta + \epsilon\delta. \tag{2.4}$$

Exercises 2.1

1. Show that the symmetries of an equilateral triangle, with both sides painted in the same color, form a group with six elements. What happens if both sides are painted in different colors?

2. Show that the curve γ_2 of Figure 2.1.4-1(b) cannot represent the transformations (2.2).

3. Show that the transformations

$$x^* = x + 2\epsilon, \tag{2.5a}$$

$$y^* = y + 3\epsilon, \quad \epsilon \in \mathbb{R}, \ (x,y) \in \mathbb{R}^2, \tag{2.5b}$$

define a Lie group of transformations. Trace out the evolution curves of $(0,0)$ and $(1,0)$ under this group. Explain the geometrical situation of the resulting curves.

4. Show that the set $S = \{\epsilon : -1 < \epsilon < \infty\}$ with law of composition $\phi(\epsilon, \delta) = \epsilon + \delta + \epsilon\delta$ defines a group.

5. Trace out the evolution curves of $(1,0)$, $(1,1)$, and $(0,0)$ for the Lie group of transformations (2.3a,b).

6. Show that the transformations (1.93) define a one-parameter Lie group of transformations acting on

   (a) (x,t)-space;

   (b) (x,t,u)-space.

7. Show whether or not each of the following families of transformations of the plane with parameter ϵ corresponds to a Lie group of transformations:

   (a) $x^* = x - \epsilon y$, $y^* = y + \epsilon x$;

   (b) $x^* = x + \epsilon^2$, $y^* = y$;

   (c) $x^* = x + \epsilon$, $y^* = \frac{xy}{x+\epsilon}$.

2.2 Infinitesimal Transformations

Consider a one-parameter (ϵ) Lie group of transformations

$$\mathbf{x}^* = \mathbf{X}(\mathbf{x}; \epsilon) \tag{2.6}$$

with identity $\epsilon = 0$ and law of composition ϕ. Expanding (2.6) about $\epsilon = 0$, we get (for some neighborhood of $\epsilon = 0$)

$$\begin{aligned}\mathbf{x}^* &= \mathbf{x} + \epsilon\left(\frac{\partial \mathbf{X}}{\partial \epsilon}(\mathbf{x};\epsilon)\bigg|_{\epsilon=0}\right) + \frac{\epsilon^2}{2}\left(\frac{\partial^2 \mathbf{X}}{\partial \epsilon^2}(\mathbf{x};\epsilon)\bigg|_{\epsilon=0}\right) + \ldots \\ &= \mathbf{x} + \epsilon\left(\frac{\partial \mathbf{X}}{\partial \epsilon}(\mathbf{x};\epsilon)\bigg|_{\epsilon=0}\right) + O(\epsilon^2). \qquad (2.7)\end{aligned}$$

Let

$$\boldsymbol{\xi}(\mathbf{x}) = \frac{\partial \mathbf{X}}{\partial \epsilon}(\mathbf{x};\epsilon)\bigg|_{\epsilon=0}. \qquad (2.8)$$

The transformation $\mathbf{x} + \epsilon\boldsymbol{\xi}(\mathbf{x})$ is called the *infinitesimal transformation* of the Lie group of transformations (2.6); the components of $\boldsymbol{\xi}(\mathbf{x})$ are called the *infinitesimals* of (2.6).

2.2.1 First Fundamental Theorem of Lie

The following lemma is useful:

Lemma 2.2.1-1.

$$\mathbf{X}(\mathbf{x};\epsilon + \Delta\epsilon) = \mathbf{X}(\mathbf{X}(\mathbf{x};\epsilon);\phi(\epsilon^{-1},\epsilon + \Delta\epsilon)). \qquad (2.9)$$

Proof.

$$\begin{aligned}\mathbf{X}(\mathbf{X}(\mathbf{x});\phi(\epsilon^{-1},\epsilon+\Delta\epsilon)) &= \mathbf{X}(\mathbf{x};\phi(\epsilon,\phi(\epsilon^{-1},\epsilon+\Delta\epsilon))) \\ &= \mathbf{X}(\mathbf{x};\phi(\phi(\epsilon,\epsilon^{-1}),\epsilon+\Delta\epsilon)) \\ &= \mathbf{X}(\mathbf{x};\phi(0,\epsilon+\Delta\epsilon)) \\ &= \mathbf{X}(\mathbf{x};\epsilon+\Delta\epsilon). \qquad \square\end{aligned}$$

Theorem 2.2.1-1 (First Fundamental Theorem of Lie). *There exists a parameterization $\tau(\epsilon)$ such that the Lie group of transformations (2.6) is equivalent to the solution of the initial value problem for the system of first order differential equations*

$$\frac{d\mathbf{x}^*}{d\tau} = \boldsymbol{\xi}(\mathbf{x}^*), \qquad (2.10a)$$

with

$$\mathbf{x}^* = \mathbf{x} \quad \textit{when} \quad \tau = 0. \qquad (2.10b)$$

In particular

$$\tau(\epsilon) = \int_0^\epsilon \Gamma(\epsilon')d\epsilon' \qquad (2.11)$$

where

$$\Gamma(\epsilon) = \frac{\partial\phi(a,b)}{\partial b}\bigg|_{(a,b)=(\epsilon^{-1},\epsilon)} \qquad (2.12)$$

and

$$\Gamma(0) = 1. \tag{2.13}$$

[ϵ^{-1} *denotes the inverse element to* ϵ.]

Proof. First we show that (2.6) leads to (2.10a,b), (2.11), (2.12). Expand the left-hand side of (2.9) in a power series in $\Delta\epsilon$ about $\Delta\epsilon = 0$ so that

$$\mathbf{X}(\mathbf{x};\epsilon+\Delta\epsilon) = \mathbf{x}^* + \frac{\partial \mathbf{X}(\mathbf{x};\epsilon)}{\partial\epsilon}\Delta\epsilon + O((\Delta\epsilon)^2) \tag{2.14}$$

where $\mathbf{x}^*$ is given by (2.6). Then expanding $\phi(\epsilon^{-1}, \epsilon+\Delta\epsilon)$ in a power series in $\Delta\epsilon$ about $\Delta\epsilon = 0$, we have

$$\begin{aligned}\phi(\epsilon^{-1}, \epsilon+\Delta\epsilon) &= \phi(\epsilon^{-1}, \epsilon) + \Gamma(\epsilon)\Delta\epsilon + O((\Delta\epsilon)^2)\\ &= \Gamma(\epsilon)\Delta\epsilon + O((\Delta\epsilon)^2)\end{aligned} \tag{2.15}$$

where $\Gamma(\epsilon)$ is defined by (2.12). Consequently, after expanding the right-hand side of (2.9) in a power series in $\Delta\epsilon$ about $\Delta\epsilon = 0$, we obtain

$$\begin{aligned}\mathbf{X}(\mathbf{x};\epsilon+\Delta\epsilon) &= \mathbf{X}(\mathbf{x}^*;\phi(\epsilon^{-1},\epsilon+\Delta\epsilon))\\ &= \mathbf{X}(\mathbf{x}^*;\Gamma(\epsilon)\Delta\epsilon + O((\Delta\epsilon)^2))\\ &= \mathbf{X}(\mathbf{x}^*;0) + \Delta\epsilon\Gamma(\epsilon)\left.\frac{\partial \mathbf{X}}{\partial\delta}(\mathbf{x}^*;\delta)\right|_{\delta=0} + O((\Delta\epsilon)^2)\\ &= \mathbf{x}^* + \Gamma(\epsilon)\xi(\mathbf{x}^*)\Delta\epsilon + O((\Delta\epsilon)^2).\end{aligned} \tag{2.16}$$

Equating (2.14) and (2.16) we see that $\mathbf{x}^* = \mathbf{X}(\mathbf{x};\epsilon)$ satisfies the initial value problem for the system of differential equations

$$\frac{d\mathbf{x}^*}{d\epsilon} = \Gamma(\epsilon)\xi(\mathbf{x}^*) \tag{2.17a}$$

with

$$\mathbf{x}^* = \mathbf{x} \quad \text{at} \quad \epsilon = 0. \tag{2.17b}$$

From (2.7) it follows that $\Gamma(0) = 1$. The parameterization $\tau(\epsilon) = \int_0^\epsilon \Gamma(\epsilon')d\epsilon'$ leads to (2.10a,b).

Since $\frac{\partial \xi}{\partial x_i}(\mathbf{x})$, $i = 1, 2, \ldots, n$, is continuous, it follows from the existence and uniqueness theorem for an initial value problem (IVP) for a system of first order differential equations, that the solution of (2.10a,b), and hence (2.17a,b), exists and is unique. This solution must be (2.6), completing the proof of Lie's First Fundamental Theorem. □

Lie's First Fundamental Theorem shows that the infinitesimal transformation contains the essential information determining a one-parameter Lie group of transformations. One can always re-parameterize a given group in terms of a parameter τ such that for parameter values τ_1 and τ_2 the law of composition becomes $\phi(\tau_1, \tau_2) = \tau_1 + \tau_2$. Lie's First Fundamental Theorem also shows that a one-parameter Lie group of transformations (2.6) defines a stationary flow given by (2.10a,b) and moreover any stationary flow (2.10a,b) defines a one-parameter Lie group of transformations.

2.2.2 EXAMPLES ILLUSTRATING LIE'S FIRST FUNDAMENTAL THEOREM

(1) *Group of Translations in the Plane*

For the groups of translations

$$x^* = x + \epsilon, \tag{2.18a}$$

$$y^* = y, \tag{2.18b}$$

the law of composition is $\phi(a,b) = a+b$, and $\epsilon^{-1} = -\epsilon$. Then $\frac{\partial\phi(a,b)}{\partial b} = 1$ and hence $\Gamma(\epsilon) \equiv 1$.

Let $\mathbf{x} = (x,y)$. Then the group (2.18a,b) is $\mathbf{X}(\mathbf{x};\epsilon) = (x+\epsilon, y)$. Thus $\frac{\partial \mathbf{X}}{\partial\epsilon}(\mathbf{x};\epsilon) = (1,0)$. Hence

$$\xi(\mathbf{x}) = \left.\frac{\partial \mathbf{X}}{\partial\epsilon}(\mathbf{x};\epsilon)\right|_{\epsilon=0} = (1,0).$$

Consequently (2.17a,b) become

$$\frac{dx^*}{d\epsilon} = 1, \quad \frac{dy^*}{d\epsilon} = 0, \tag{2.19a}$$

with

$$x^* = x, \ y^* = y \quad \text{at} \ \epsilon = 0. \tag{2.19b}$$

The solution of IVP (2.19a,b) is easily seen to be (2.18a,b).

(2) *Group of Scalings*

For the group of scalings

$$x^* = (1+\epsilon)x, \tag{2.20a}$$

$$y^* = (1+\epsilon)^2 y, \quad -1 < \epsilon < \infty, \tag{2.20b}$$

the law of composition is $\phi(a,b) = a+b+ab$, and $\epsilon^{-1} = -\frac{\epsilon}{1+\epsilon}$. Here $\frac{\partial\phi(a,b)}{\partial b} = 1+a$ and hence

$$\Gamma(\epsilon) = \left.\frac{\partial\phi(a,b)}{\partial b}\right|_{(a,b)=(\epsilon^{-1},\epsilon)} = 1+\epsilon^{-1} = \frac{1}{1+\epsilon}.$$

Let $\mathbf{x} = (x,y)$. Then the group (2.20a,b) is $\mathbf{X}(\mathbf{x};\epsilon) = ((1+\epsilon)x, (1+\epsilon)^2 y)$. Thus $\frac{\partial \mathbf{X}}{\partial\epsilon}(\mathbf{x};\epsilon) = (x, 2(1+\epsilon)y)$ and

$$\xi(\mathbf{x}) = \left.\frac{\partial \mathbf{X}}{\partial\epsilon}(\mathbf{x};\epsilon)\right|_{\epsilon=0} = (x, 2y).$$

As a result (2.17a,b) become

$$\frac{dx^*}{d\epsilon} = \frac{x^*}{1+\epsilon}, \quad \frac{dy^*}{d\epsilon} = \frac{2y^*}{1+\epsilon}, \tag{2.21a}$$

with

$$x^* = x, \; y^* = y \quad \text{at} \quad \epsilon = 0. \tag{2.21b}$$

The solution of IVP (2.21a,b) is of course (2.20a,b).

In terms of parameterization by

$$\tau = \int_0^\epsilon \Gamma(\epsilon')d\epsilon' = \int_0^\epsilon \frac{1}{1+\epsilon}d\epsilon' = \log(1+\epsilon),$$

the group (2.20a,b) becomes

$$x^* = e^\tau x, \tag{2.22a}$$

$$y^* = e^{2\tau} y, \quad -\infty < \tau < \infty, \tag{2.22b}$$

with law of composition $\phi(\tau_1, \tau_2) = \tau_1 + \tau_2$.

2.2.3 Infinitesimal Generators

In view of Lie's First Fundamental Theorem, from now on, without loss of generality, we assume that a one-parameter (ϵ) Lie group of transformations is parameterized such that its law of composition is $\phi(a,b) = a+b$ so that $\epsilon^{-1} = -\epsilon$ and $\Gamma(\epsilon) \equiv 1$. Thus in terms of its infinitesimals $\boldsymbol{\xi}(\mathbf{x})$ the one-parameter Lie group of transformations (2.6) becomes

$$\frac{d\mathbf{x}^*}{d\epsilon} = \boldsymbol{\xi}(\mathbf{x}^*), \tag{2.23a}$$

with

$$\mathbf{x}^* = \mathbf{x} \quad \text{at} \quad \epsilon = 0. \tag{2.23b}$$

Definition 2.2.3-1. The *infinitesimal generator* of the one-parameter Lie group of transformations (2.6) is the operator

$$\mathrm{X} = \mathrm{X}(\mathbf{x}) = \boldsymbol{\xi}(\mathbf{x}) \cdot \nabla = \sum_{i=1}^n \xi_i(\mathbf{x}) \frac{\partial}{\partial x_i} \tag{2.24}$$

where ∇ is the gradient operator,

$$\nabla = \left(\frac{\partial}{\partial x_1}, \frac{\partial}{\partial x_2}, \ldots, \frac{\partial}{\partial x_n} \right); \tag{2.25}$$

for any differentiable function $F(\mathbf{x}) = F(x_1, x_2, \ldots, x_n)$,

$$\mathrm{X}F(\mathbf{x}) = \boldsymbol{\xi}(\mathbf{x}) \cdot \nabla F(\mathbf{x}) = \sum_{i=1}^n \xi_i(\mathbf{x}) \frac{\partial F(\mathbf{x})}{\partial x_i}.$$

Note that $\mathrm{X}\mathbf{x} = \boldsymbol{\xi}(\mathbf{x})$.

It follows that a one-parameter Lie group of transformations, which from Lie's First Fundamental Theorem is "equivalent" to its infinitesimal transformation, is also "equivalent" to its infinitesimal generator. The following theorem shows that use of the infinitesimal generator (2.24) leads to an algorithm to find the explicit solution of IVP (2.23a,b).

Theorem 2.2.3-1. *The one-parameter Lie group of transformations* (2.6) *is equivalent to*

$$\begin{aligned}\mathbf{x}^* = e^{\epsilon \mathrm{X}}\mathbf{x} &= \mathbf{x} + \epsilon \mathrm{X}\mathbf{x} + \frac{\epsilon^2}{2}\mathrm{X}^2\mathbf{x} + \ldots \\ &= [1 + \epsilon \mathrm{X} + \frac{\epsilon^2}{2}\mathrm{X}^2 + \ldots]\mathbf{x} \\ &= \sum_{k=0}^{\infty} \frac{\epsilon^k}{k!}\mathrm{X}^k\mathbf{x} \qquad (2.26)\end{aligned}$$

where the operator $\mathrm{X} = \mathrm{X}(\mathbf{x})$ *is defined by* (2.24) *and the operator* $\mathrm{X}^k = \mathrm{X}\mathrm{X}^{k-1}$, $k = 1, 2, \ldots$; *in particular* $\mathrm{X}^k F(\mathbf{x})$ *is the function obtained by applying the operator* X *to the function* $\mathrm{X}^{k-1}F(\mathbf{x})$, $k = 1, 2, \ldots$, *with* $\mathrm{X}^0 F(\mathbf{x}) \equiv F(\mathbf{x})$.

Proof. Let

$$\mathrm{X} = \mathrm{X}(\mathbf{x}) = \sum_{i=1}^{n} \xi_i(\mathbf{x}) \frac{\partial}{\partial x_i} \qquad (2.27a)$$

and

$$\mathrm{X}(\mathbf{x}^*) = \sum_{i=1}^{n} \xi_i(\mathbf{x}^*) \frac{\partial}{\partial x_i^*} \qquad (2.27b)$$

where

$$\mathbf{x}^* = \mathbf{X}(\mathbf{x}; \epsilon) \qquad (2.28)$$

is the Lie group of transformations (2.6). From Taylor's theorem, expanding (2.28) about $\epsilon = 0$, we have

$$\mathbf{x}^* = \sum_{k=0}^{\infty} \frac{\epsilon^k}{k!} \left(\left. \frac{\partial^k \mathbf{X}(\mathbf{x};\epsilon)}{\partial \epsilon^k} \right|_{\epsilon=0} \right) = \sum_{k=0}^{\infty} \frac{\epsilon^k}{k!} \left(\left. \frac{d^k \mathbf{x}^*}{d\epsilon^k} \right|_{\epsilon=0} \right). \qquad (2.29)$$

For any differentiable function $F(\mathbf{x})$,

$$\frac{d}{d\epsilon}F(\mathbf{x}^*) = \sum_{i=1}^{n} \frac{\partial F(\mathbf{x}^*)}{\partial x_i^*}\frac{dx_i^*}{d\epsilon} = \sum_{i=1}^{n} \xi_i(\mathbf{x}^*)\frac{\partial F(\mathbf{x}^*)}{\partial x_i^*} = \mathrm{X}(\mathbf{x}^*)F(\mathbf{x}^*). \qquad (2.30)$$

Hence it follows that

$$\frac{d\mathbf{x}^*}{d\epsilon} = \mathrm{X}(\mathbf{x}^*)\mathbf{x}^*, \qquad (2.31a)$$

$$\frac{d^2\mathbf{x}^*}{d\epsilon^2} = \frac{d}{d\epsilon}\left(\frac{d\mathbf{x}^*}{d\epsilon}\right) = \mathrm{X}(\mathbf{x}^*)\mathrm{X}(\mathbf{x}^*)\mathbf{x}^* = \mathrm{X}^2(\mathbf{x}^*)\mathbf{x}^*, \tag{2.31b}$$

and in general,

$$\frac{d^k\mathbf{x}^*}{d\epsilon^k} = \mathrm{X}^k(\mathbf{x}^*)\mathbf{x}^*, \quad k = 1, 2, \ldots . \tag{2.31c}$$

Consequently

$$\left.\frac{d^k\mathbf{x}^*}{d\epsilon^k}\right|_{\epsilon=0} = \mathrm{X}^k(\mathbf{x})\mathbf{x} = \mathrm{X}^k\mathbf{x}, \quad k = 1, 2, \ldots \tag{2.32}$$

which leads to (2.26). □

Thus the Taylor series expansion about $\epsilon = 0$ of a function $\mathbf{X}(\mathbf{x}; \epsilon)$ which defines a Lie group of transformations (2.6) (with law of composition $\phi(a, b) = a + b$) is determined by the coefficient of its $O(\epsilon)$ term, which is the infinitesimal

$$\left.\frac{\partial \mathbf{X}}{\partial \epsilon}(\mathbf{x}; \epsilon)\right|_{\epsilon=0} = \boldsymbol{\xi}(\mathbf{x}).$$

In summary there are two ways to find explicitly a one-parameter Lie group of transformations from its infinitesimal transformation: (i) Express the group in terms of a power series (2.26), called a *Lie series,* which is developed from the infinitesimal generator (2.24) corresponding to the infinitesimal transformation; (ii) Solve the initial value problem (2.23a,b). Here one first finds the explicit general solution of the system of first order differential equations (2.23a).

The following corollary is an immediate consequence of the proof of Theorem 2.2.3-1:

Corollary 2.2.3-1. *If* $F(\mathbf{x})$ *is infinitely differentiable, then for a Lie group of transformations* (2.6) *with infinitesimal generator* (2.27a),

$$F(\mathbf{x}^*) = F(e^{\epsilon \mathrm{X}}\mathbf{x}) = e^{\epsilon \mathrm{X}} F(\mathbf{x}). \tag{2.33}$$

Proof. See Exercise 2.2-5. □

As an example consider the rotation group

$$x^* = x \cos \epsilon + y \sin \epsilon, \tag{2.34a}$$

$$y^* = -x \sin \epsilon + y \cos \epsilon. \tag{2.34b}$$

$$\frac{dx^*}{d\epsilon} = -x \sin \epsilon + y \cos \epsilon; \quad \frac{dy^*}{d\epsilon} = -x \cos \epsilon - y \sin \epsilon.$$

The infinitesimal for (2.34a,b) is

$$\boldsymbol{\xi}(\mathbf{x}) = (\xi_1(x, y), \xi_2(x, y)) = (\xi(x, y), \eta(x, y))$$

$$= \left(\left.\frac{dx^*}{d\epsilon}\right|_{\epsilon=0}, \left.\frac{dy^*}{d\epsilon}\right|_{\epsilon=0} \right) = (y, -x). \tag{2.35}$$

The infinitesimal generator for (2.34a,b) is

$$\mathrm{X} = \xi(x,y)\frac{\partial}{\partial x} + \eta(x,y)\frac{\partial}{\partial y} = y\frac{\partial}{\partial x} - x\frac{\partial}{\partial y}. \tag{2.36}$$

The Lie series corresponding to (2.36) is

$$(x^*, y^*) = (e^{\epsilon \mathrm{X}}x, e^{\epsilon \mathrm{X}}y).$$

$$\mathrm{X}x = y; \quad \mathrm{X}^2 x = \mathrm{X}y = -x.$$

Hence

$$\mathrm{X}^{4n}x = x,\ \mathrm{X}^{4n-1}x = -y,\ \mathrm{X}^{4n-2}x = -x,\ \mathrm{X}^{4n-3}x = y,\ n = 1,2,\dots;$$

$$\mathrm{X}^{4n}y = y,\ \mathrm{X}^{4n-1}y = x,\ \mathrm{X}^{4n-2}y = -y,\ \mathrm{X}^{4n-3}y = -x,\ n = 1,2,\dots.$$

Consequently

$$\begin{aligned} x^* &= e^{\epsilon \mathrm{X}}x = \sum_{k=0}^{\infty} \frac{\epsilon^k}{k!}\mathrm{X}^k x \\ &= \left(1 - \frac{\epsilon^2}{2!} + \frac{\epsilon^4}{4!} + \cdots\right)x + \left(\epsilon - \frac{\epsilon^3}{3!} + \frac{\epsilon^5}{5!} + \cdots\right)y \\ &= x\cos\epsilon + y\sin\epsilon. \end{aligned}$$

Similarly

$$y^* = e^{\epsilon \mathrm{X}}y = -x\sin\epsilon + y\cos\epsilon.$$

2.2.4 Invariant Functions

Definition 2.2.4-1. An infinitely differentiable function $F(\mathbf{x})$ is an *invariant function* of the Lie group of transformations (2.6) if and only if for any group transformation (2.6)

$$F(\mathbf{x}^*) \equiv F(\mathbf{x}). \tag{2.37}$$

If $F(\mathbf{x})$ is an invariant function of (2.6), then $F(\mathbf{x})$ is called an *invariant* of (2.6) and $F(\mathbf{x})$ is said to be *invariant under* (2.6).

Theorem 2.2.4-1. $F(\mathbf{x})$ *is invariant under* (2.6) *if and only if*

$$\mathrm{X}F(\mathbf{x}) \equiv 0. \tag{2.38}$$

Proof.

$$F(\mathbf{x}^*) \equiv e^{\epsilon \mathrm{X}}F(\mathbf{x}) \equiv \sum_{k=0}^{\infty} \frac{\epsilon^k}{k!}\mathrm{X}^k F(\mathbf{x})$$

$$\equiv F(\mathbf{x}) + \epsilon \mathrm{X}F(\mathbf{x}) + \frac{\epsilon^2}{2!}\mathrm{X}^2F(\mathbf{x}) + \cdots. \tag{2.39}$$

Suppose $F(\mathbf{x}^*) \equiv F(\mathbf{x})$. Then $\mathrm{X}F(\mathbf{x}) \equiv 0$ follows from (2.39).

Conversely, let $F(\mathbf{x})$ be such that $\mathrm{X}F(\mathbf{x}) \equiv 0$. Then $\mathrm{X}^nF(\mathbf{x}) \equiv 0$, $n = 1, 2, \ldots$. Hence from (2.39), $F(\mathbf{x}^*) \equiv F(\mathbf{x})$. □

Theorem 2.2.4-2. *For a Lie group of transformations* (2.6), *the identity*

$$F(\mathbf{x}^*) \equiv F(\mathbf{x}) + \epsilon \tag{2.40}$$

holds if and only if $F(\mathbf{x})$ *is such that*

$$\mathrm{X}F(\mathbf{x}) \equiv 1. \tag{2.41}$$

Proof. Let $F(\mathbf{x})$ be such that (2.40) holds. Then

$$F(\mathbf{x}) + \epsilon \equiv F(\mathbf{x}) + \epsilon \mathrm{X}F(\mathbf{x}) + \frac{\epsilon^2}{2!}\mathrm{X}^2F(\mathbf{x}) + \cdots.$$

Hence $\mathrm{X}F(\mathbf{x}) \equiv 1$.

Conversely, let $F(\mathbf{x})$ be such that (2.41) holds. Then $\mathrm{X}^nF(\mathbf{x}) \equiv 0$, $n = 2, 3, \ldots$. Hence $F(\mathbf{x}^*) \equiv e^{\epsilon \mathrm{X}}F(\mathbf{x}) \equiv F(\mathbf{x}) + \epsilon \mathrm{X}F(\mathbf{x}) \equiv F(\mathbf{x}) + \epsilon$. □

2.2.5 Canonical Coordinates

Suppose one makes a change of coordinates (one-to-one and continuously differentiable in some appropriate domain)

$$\mathbf{y} = \mathbf{Y}(\mathbf{x}) = (y_1(\mathbf{x}), y_2(\mathbf{x}), \ldots, y_n(\mathbf{x})). \tag{2.42}$$

For the one-parameter Lie group of transformations (2.6) the infinitesimal generator $\mathrm{X} = \sum_{i=1}^n \xi_i(\mathbf{x})\frac{\partial}{\partial x_i}$ with respect to coordinates $\mathbf{x} = (x_1, x_2, \ldots, x_n)$ becomes the infinitesimal generator

$$\mathrm{Y} = \sum_{i=1}^n \eta_i(\mathbf{y})\frac{\partial}{\partial y_i} \tag{2.43}$$

with respect to coordinates $\mathbf{y} = (y_1, y_2, \ldots, y_n)$ defined by (2.42). Then $\mathrm{Y} = \mathrm{X}$ in order to have the same group action. The infinitesimal with respect to coordinates $\mathbf{y}$ is

$$\eta(\mathbf{y}) = (\eta_1(\mathbf{y}), \eta_2(\mathbf{y}), \ldots, \eta_n(\mathbf{y})) = \mathrm{Y}\mathbf{y}. \tag{2.44}$$

Theorem 2.2.5-1.

$$\eta(\mathbf{y}) = \mathrm{X}\mathbf{y}. \tag{2.45}$$

Proof. Using the chain rule, we have

$$\mathrm{X}=\sum_{i=1}^{n}\xi_i(\mathbf{x})\frac{\partial}{\partial x_i}=\sum_{i,j=1}^{n}\xi_i(\mathbf{x})\frac{\partial y_j(\mathbf{x})}{\partial x_i}\frac{\partial}{\partial y_j}$$

$$=\sum_{j=1}^{n}\eta_j(\mathbf{y})\frac{\partial}{\partial y_j}=\mathrm{Y}$$

where

$$\eta_j(\mathbf{y})=\sum_{i=1}^{n}\xi_i(\mathbf{x})\frac{\partial y_j(\mathbf{x})}{\partial x_i}$$
$$=\mathrm{X}y_j,\quad j=1,2,\ldots,n. \quad \square$$

Theorem 2.2.5-2. *With respect to coordinates* $\mathbf{y}$ *given by* (2.42), *the Lie group of transformations* (2.6) *is*

$$\mathbf{y}^*=e^{\epsilon \mathrm{Y}}\mathbf{y}. \tag{2.46}$$

Proof. From equations (2.33), (2.42), we obtain

$$\mathbf{y}^*=\mathbf{Y}(\mathbf{x}^*)=e^{\epsilon\mathrm{X}}\mathbf{Y}(\mathbf{x})=e^{\epsilon\mathrm{X}}\mathbf{Y}=e^{\epsilon\mathrm{Y}}\mathbf{y}. \quad \square$$

Definition 2.2.5-1. A change of coordinates (2.42) defines a set of *canonical coordinates* for the one-parameter Lie group of transformations (2.6) if in terms of such coordinates the group (2.6) becomes

$$y_i^*=y_i,\quad i=1,2,\ldots,n-1, \tag{2.47a}$$

$$y_n^*=y_n+\epsilon. \tag{2.47b}$$

Theorem 2.2.5-3. *For any Lie group of transformations* (2.6) *there exists a set of canonical coordinates* $\mathbf{y}=(y_1,y_2,\ldots,y_n)$ *such that* (2.6) *is equivalent to* (2.47a,b).

Proof. From Theorem 2.2.4-1 we have

$$y_i^*=y_i(\mathbf{x}^*)=y_i(\mathbf{x})$$

if and only if

$$\mathrm{X}y_i(\mathbf{x})\equiv 0, \tag{2.48}$$

$i=1,2,\ldots,n-1$. The homogeneous first order linear partial differential equation

$$\mathrm{X}u(\mathbf{x})=\xi_1(\mathbf{x})\frac{\partial u}{\partial x_1}+\xi_2(\mathbf{x})\frac{\partial u}{\partial x_2}+\cdots+\xi_n(\mathbf{x})\frac{\partial u}{\partial x_n}=0 \tag{2.49}$$

has $n-1$ functionally independent solutions for $u(\mathbf{x})$. These solutions are $n-1$ essential constants $(y_1(\mathbf{x}), y_2(\mathbf{x}), \ldots, y_{n-1}(\mathbf{x}))$ appearing in the general solution of the system of n first order ordinary differential equations

$$\frac{d\mathbf{x}}{dt} = \boldsymbol{\xi}(\mathbf{x}), \tag{2.50}$$

resulting from the characteristic equations

$$\frac{dx_1}{\xi_1(\mathbf{x})} = \frac{dx_2}{\xi_2(\mathbf{x})} = \cdots = \frac{dx_n}{\xi_n(\mathbf{x})}.$$

This yields the $n-1$ coordinates satisfying (2.47a).

The canonical coordinate $y_n(\mathbf{x})$ follows from Theorem 2.2.4-2:

$$y_n^* = y_n(\mathbf{x}^*) = y_n(\mathbf{x}) + \epsilon$$

if and only if

$$\mathrm{X}y_n(\mathbf{x}) \equiv 1. \tag{2.51}$$

Hence $v(\mathbf{x}) = y_n(\mathbf{x})$ is a particular solution of the nonhomogeneous first order linear partial differential equation

$$\mathrm{X}v(\mathbf{x}) = \xi_1(\mathbf{x})\frac{\partial v}{\partial x_1} + \xi_2(\mathbf{x})\frac{\partial v}{\partial x_2} + \cdots + \xi_n(\mathbf{x})\frac{\partial v}{\partial x_n} = 1, \tag{2.52}$$

and is found by determining a particular solution of the corresponding characteristic system of $n+1$ first order ordinary differential equations

$$\frac{dv}{dt} = 1, \tag{2.53a}$$

$$\frac{d\mathbf{x}}{dt} = \boldsymbol{\xi}(\mathbf{x}). \quad \square \tag{2.53b}$$

Theorem 2.2.5-4. *In terms of any set of canonical coordinates* $\mathbf{y} = (y_1, y_2, \ldots, y_n)$, *the infinitesimal generator of the one-parameter Lie group of transformations* (2.6) *is*

$$\mathrm{Y} = \frac{\partial}{\partial y_n}. \tag{2.54}$$

Proof. We have

$$\mathrm{Y} = \sum_{i=1}^{n} \eta_i(\mathbf{y})\frac{\partial}{\partial y_i}.$$

In terms of canonical coordinates, from (2.48) and (2.51) it follows that

$$\eta_i(\mathbf{y}) = \mathrm{X}y_i = 0, \quad i = 1, 2, \ldots, n-1;$$
$$\eta_n(\mathbf{y}) = \mathrm{X}y_n = 1.$$

Hence we get

$$\mathrm{Y} = \frac{\partial}{\partial y_n}. \quad \square$$

2.2.6 Examples of Sets of Canonical Coordinates

In $\mathbb{R}^2$ we set $x_1 = x$, $x_2 = y$ and label canonical coordinates as $y_1 = r$, $y_2 = s$, so that in terms of canonical coordinates a one-parameter Lie group of transformations becomes

$$r^* = r, \tag{2.55a}$$

$$s^* = s + \epsilon, \tag{2.55b}$$

with infinitesimal generator $\mathrm{Y} = \dfrac{\partial}{\partial s}$.

(1) *Group of Scalings*

For the group of scalings

$$x^* = e^{\epsilon} x, \tag{2.56a}$$

$$y^* = e^{2\epsilon} y, \tag{2.56b}$$

the infinitesimal generator is $\mathrm{X} = x\dfrac{\partial}{\partial x} + 2y\dfrac{\partial}{\partial y}$. The canonical coordinate $r(x, y)$ satisfies

$$\mathrm{X}r = x\frac{\partial r}{\partial x} + 2y\frac{\partial r}{\partial y} = 0. \tag{2.57}$$

The corresponding characteristic differential equations reduce to

$$\frac{dy}{dx} = \frac{2y}{x} \tag{2.58}$$

with solution

$$r(x, y) = \frac{y}{x^2} = \text{const}. \tag{2.59}$$

The canonical coordinate $s(x, y)$ satisfies

$$\mathrm{X}s = x\frac{\partial s}{\partial x} + 2y\frac{\partial s}{\partial y} = 1. \tag{2.60}$$

A particular solution of (2.60) is $s(x, y) = s(x)$ satisfying

$$\frac{ds}{dx} = \frac{1}{x}. \tag{2.61}$$

Thus

$$s(x, y) = \log x, \tag{2.62}$$

and (2.56a,b) has canonical coordinates $(r, s) = \left(\dfrac{y}{x^2}, \log x\right)$.

(2) *Group of Rotations*

For the group of rotations

$$x^* = x \cos\epsilon - y \sin\epsilon, \tag{2.63a}$$

$$y^* = x\sin\epsilon + y\cos\epsilon, \tag{2.63b}$$

the infinitesimal generator is $X = x\frac{\partial}{\partial y} - y\frac{\partial}{\partial x}$. Correspondingly

$$r(x, y) = \text{const.}$$

is the general solution of

$$\frac{dy}{dx} = -\frac{x}{y} \tag{2.64}$$

so that

$$r = \sqrt{x^2 + y^2}. \tag{2.65}$$

Then a candidate for $s(x, y)$ is a solution of

$$\frac{ds}{dy} = \frac{1}{x} = \frac{1}{\sqrt{r^2 - y^2}}. \tag{2.66}$$

Thus

$$s = \theta = \sin^{-1}\frac{y}{r}. \tag{2.67}$$

Canonical coordinates are polar coordinates

$$(r, s) = (r, \theta) = \left(\sqrt{x^2 + y^2}, \sin^{-1}\frac{y}{r}\right), \tag{2.68}$$

in terms of which the rotation group (2.63a,b) is expressed in the familiar form

$$r^* = r, \tag{2.69a}$$

$$\theta^* = \theta + \epsilon. \tag{2.69b}$$

2.2.7 Invariant Surfaces, Invariant Curves, Invariant Points

Definition 2.2.7-1. A surface $F(\mathbf{x}) = 0$ is an *invariant surface* for a one-parameter Lie group of transformations (2.6) if and only if $F(\mathbf{x}^*) = 0$ when $F(\mathbf{x}) = 0$.

Definition 2.2.7-2. A curve $F(x, y) = 0$ is an *invariant curve* for a one-parameter Lie group of transformations

$$x^* = X(x, y; \epsilon) = x + \epsilon\xi(x, y) + O(\epsilon^2), \tag{2.70a}$$

$$y^* = Y(x, y; \epsilon) = y + \epsilon\eta(x, y) + O(\epsilon^2), \tag{2.70b}$$

with infinitesimal generator

$$X = \xi(x, y)\frac{\partial}{\partial x} + \eta(x, y)\frac{\partial}{\partial y}, \tag{2.70c}$$

if and only if $F(x^*, y^*) = 0$ when $F(x, y) = 0$.

Theorem 2.2.7-1. (i) *A surface written in a solved form* $F(\mathbf{x}) = x_n - f(x_1, x_2, \ldots, x_{n-1}) = 0$, *is an invariant surface for* (2.6) *if and only if*

$$\mathrm{X}F(\mathbf{x}) = 0 \quad when \quad F(\mathbf{x}) = 0. \tag{2.71}$$

(ii) *A curve written in a solved form* $F(x, y) = y - f(x) = 0$, *is an invariant curve for* (2.70a,b) *if and only if*

$$\mathrm{X}F(x, y) = \eta(x, y) - \xi(x, y)f'(x) = 0$$

when $F(x, y) = y - f(x) = 0$, *i.e., if and only if*

$$\eta(x, f(x)) - \xi(x, f(x))f'(x) = 0. \tag{2.72}$$

The proof of Theorem 2.2.7-1 is left to Exercise 2.2-7. This theorem gives a means for finding the invariant surface of a given Lie group of transformations, namely by solving (2.71). As an example consider the scaling group

$$x^* = e^{\epsilon}x, \tag{2.73a}$$

$$y^* = e^{\epsilon}y. \tag{2.73b}$$

The corresponding infinitesimal generator is

$$\mathrm{X} = x\frac{\partial}{\partial x} + y\frac{\partial}{\partial y}. \tag{2.74}$$

A ray $y - \lambda x = 0$, $x > 0$, $\lambda = \text{const}$, is an invariant curve for (2.73a,b) since $\mathrm{X}(y - \lambda x) = y - \lambda x = 0$ when $y - \lambda x = 0$; a parabola $y - \lambda x^2 = 0$, $\lambda = \text{const}$, is not an invariant curve for (2.73a,b) since $\mathrm{X}(y - \lambda x^2) = y - 2\lambda x^2 \neq 0$ when $y - \lambda x^2 = 0$.

To find invariant curves $y - f(x) = 0$ for (2.73a,b) we first find the general solution $u(x, y)$ of the partial differential equation

$$x\frac{\partial u}{\partial x} + y\frac{\partial u}{\partial y} = 0$$

which is

$$u(x, y) = F\left(\frac{y}{x}\right)$$

where F is an arbitrary function of y/x. Invariant curves then include the curves

$$y - \lambda x = 0, \quad \lambda = \text{const}, \quad x > 0 \text{ or } x < 0.$$

Definition 2.2.7-3. A point $\mathbf{x}$ is an *invariant point* for the Lie group of transformations (2.6) if and only if $\mathbf{x}^* \equiv \mathbf{x}$ under (2.6).

Theorem 2.2.7-2. *A point* $\mathbf{x}$ *is an invariant point for the Lie group of transformations* (2.6) *if and only if*

$$\xi(\mathbf{x}) = 0. \tag{2.75}$$

The proof of Theorem 2.2.7-2 is left to Exercise 2.2-8. For the scaling group (2.73a,b), $\xi(x,y) = \eta(x,y) = 0$ if and only if $x = y = 0$, so that the only invariant point is the origin $(0,0)$.

The general solution of an ordinary differential equation is represented geometrically by a family of curves; the "general solution" of a partial differential equation is a family of surfaces. A group of transformations is said to be admitted by a differential equation if it maps any solution curve (surface) into another solution curve (surface); in particular a group of transformations admitted by a differential equation must leave a family of solution curves (surfaces) invariant. This leads us to consider the following definitions and theorems:

Definition 2.2.7-4. The family of surfaces

$$\omega(\mathbf{x}) = \text{const} = c$$

is an *invariant family of surfaces* for (2.6) if and only if

$$\omega(\mathbf{x}^*) = \text{const} = c^* \quad \text{when} \quad \omega(\mathbf{x}) = c.$$

Definition 2.2.7-5. The family of curves

$$\omega(x,y) = \text{const} = c$$

is an *invariant family of curves* for (2.70a,b) if and only if

$$\omega(x^*,y^*) = \text{const} = c^* \quad \text{when} \quad \omega(x,y) = c.$$

From these definitions it follows that

$$c^* = C(c;\epsilon) \tag{2.76}$$

for some function C of c and group parameter ϵ. Without loss of generality we assume $c^* \not\equiv c$, for otherwise each surface itself is an invariant surface (leading to a trivial invariant family of surfaces).

Theorem 2.2.7-3. (i) *A family of surfaces* $\omega(\mathbf{x}) = \text{const} = c$ *is an invariant family of surfaces for* (2.6) *if and only if*

$$\mathrm{X}\omega = \Omega(\omega) \tag{2.77}$$

for some infinitely differentiable function $\Omega(\omega)$.

(ii) *A family of curves* $\omega(x,y) = \text{const} = c$ *is an invariant family of curves for* (2.70a,b) *if and only if*

$$\mathrm{X}\omega = \xi(x,y)\frac{\partial\omega}{\partial x} + \eta(x,y)\frac{\partial\omega}{\partial y} = \Omega(\omega) \tag{2.78}$$

for some infinitely differentiable function $\Omega(\omega)$.

Proof. Let $\omega(\mathbf{x}) = c$ be an invariant family of surfaces for (2.6). Then

$$\begin{aligned}\omega(\mathbf{x}^*) &= e^{\epsilon \mathrm{X}}\omega(\mathbf{x}) \\ &= \omega(\mathbf{x}) + \epsilon \mathrm{X}\omega(\mathbf{x}) + \frac{\epsilon^2}{2}\mathrm{X}^2\omega(\mathbf{x}) + \cdots \\ &= c^* = C(c;\epsilon).\end{aligned}$$

Hence $\mathrm{X}\omega(\mathbf{x}) = \Omega(\omega)$ for some function $\Omega(\omega)$ when $\omega(\mathbf{x}) = c$. It follows that $\mathrm{X}^2\omega = \Omega'(\omega)\mathrm{X}\omega = \Omega'(\omega)\Omega(\omega)$, etc.

Conversely, suppose $\mathrm{X}\omega = \Omega(\omega)$ for some infinitely differentiable function $\Omega(\omega)$. Then $\mathrm{X}^2\omega = \Omega'(\omega)\Omega(\omega)$ and $\mathrm{X}^n\omega = f_n(\omega)$ for some function $f_n(\omega)$, $n = 1, 2, \ldots$. Consequently if $\omega(\mathbf{x}) = c$, then

$$\begin{aligned}\omega(\mathbf{x}^*) &= e^{\epsilon \mathrm{X}}\omega(\mathbf{x}) \\ &= \omega(\mathbf{x}) + \epsilon \mathrm{X}\omega(\mathbf{x}) + \frac{\epsilon^2}{2}\mathrm{X}^2\omega(\mathbf{x}) + \cdots \\ &= \omega(\mathbf{x}) + \sum_{n=1}^{\infty} \frac{\epsilon^n f_n(\omega(\mathbf{x}))}{n!} \\ &= c + \sum_{n=1}^{\infty} \frac{\epsilon^n f_n(c)}{n!} \\ &= \text{const} = c^*. \qquad \square\end{aligned}$$

In order to find the invariant family of surfaces (curves) for a Lie group of transformations, without loss of generality we can set $\Omega(\omega) \equiv 1$. This follows from the fact that if $\omega(\mathbf{x}) = c$ is an invariant family of surfaces then so is $F(\omega(\mathbf{x})) = F(c)$ for any function F; $\mathrm{X}F(\omega(\mathbf{x})) = F'(\omega)\mathrm{X}\omega = F'(\omega)\Omega(\omega)$, so that setting $F'(\omega) = \frac{1}{\Omega(\omega)}$, we have $\mathrm{X}F(\omega) \equiv 1$. [We assume that $\Omega(\omega) \not\equiv 0$, for otherwise each surface in the invariant family of surfaces is itself an invariant surface for (2.6).] As an example consider again the scaling group (2.73a,b). The invariant family of curves $\omega(x, y) = c$ solves

$$\mathrm{X}\omega = x\frac{\partial \omega}{\partial x} + y\frac{\partial \omega}{\partial y} = 1.$$

The corresponding characteristic equations are

$$\frac{d\omega}{1} = \frac{dx}{x} = \frac{dy}{y}$$

with general solution

$$\omega(x, y) = \log x + f\left(\frac{y}{x}\right) \quad \text{for arbitrary function } f.$$

Hence any family of curves

$$F(\omega) = F\left(\log x + f\left(\frac{y}{x}\right)\right) = \text{const} = c \tag{2.79}$$

is an invariant family of curves for (2.73a,b) for any choice of F, f. In particular the family of circles $x^2+y^2 = \text{const} = r^2$ is an invariant family of curves for (2.73a,b) found by choosing $F(\omega) = e^{2\omega}$ and $f(z) = \frac{1}{2}\log(1+z^2)$ in (2.79). The family of lines $x = \text{const}$ is invariant, corresponding to $F(\omega) = e^{\omega}$, $f(z) = 0$. The family of logarithmic spirals $r^2 e^{\theta} = \text{const}$ is invariant, corresponding to $F(\omega) = e^{2\omega}$, $f(z) = \frac{1}{2}(\log(1+z^2) + \arctan z)$.

Exercises 2.2

1. Consider the rotation group

$$x^* = \sqrt{1-\epsilon^2}x - \epsilon y, \tag{2.80a}$$

$$y^* = \epsilon x + \sqrt{1-\epsilon^2}y. \tag{2.80b}$$

   (a) Show that (2.80a,b) defines a one-parameter group of transformations in some neighborhood of $\epsilon = 0$. In particular find the law of composition $\phi(a,b)$ and ϵ^{-1}.

   (b) Determine $\Gamma(\epsilon)$ and the infinitesimal $\boldsymbol{\xi}(\mathbf{x})$ for (2.80a,b).

   (c) Integrate the initial value problem for the infinitesimal to obtain (2.80a,b).

   (d) Parameterize (2.80a,b) in terms of $\tau = \int_0^\epsilon \Gamma(\epsilon')d\epsilon'$.

2. Consider the group of transformations

$$x^* = x + \epsilon, \tag{2.81a}$$

$$y^* = \frac{xy}{x+\epsilon}. \tag{2.81b}$$

   (a) Determine $\Gamma(\epsilon)$, $\boldsymbol{\xi}(\mathbf{x})$ and explicitly integrate the initial value problem for the infinitesimal to obtain (2.81a,b).

   (b) Find canonical coordinates, invariant curves, invariant points, and invariant families of curves for (2.81a,b).

   (c) Determine (2.81a,b) in terms of its Lie series developed from $\boldsymbol{\xi}(\mathbf{x})$.

3. For the group of transformations (1.93) find the infinitesimal and explicitly integrate out the initial value problem for the infinitesimal, and find canonical coordinates, invariant curves (surfaces), invariant points, and invariant families of curves (surfaces)

(a) in (x,t)-space;

(b) in (x,t,u)-space.

4. Find the one-parameter groups of transformations and canonical coordinates corresponding to the infinitesimal generators:

(a) $X_1 = x\frac{\partial}{\partial x} + y\frac{\partial}{\partial y}$;

(b) $X_2 = x\frac{\partial}{\partial x} - y\frac{\partial}{\partial y}$;

(c) $X_3 = x^2\frac{\partial}{\partial x} + y^2\frac{\partial}{\partial y}$.

5. Derive (2.33).

6. Show that $X(\mathbf{x})\mathbf{x}^* = \xi(\mathbf{x}^*)$. Hence show that $X(\mathbf{x}^*) \equiv X(\mathbf{x}) \equiv X$.

7. Prove Theorem 2.2.7-1. Geometrically interpret (2.72).

8. Prove Theorem 2.2.7-2.

9. For the infinitesimal generator

$$X = x^2\frac{\partial}{\partial x} + xy\frac{\partial}{\partial y} - \left(\frac{y^2}{4} + \frac{x}{2}\right)z\frac{\partial}{\partial z}$$

find

(a) invariant functions, invariant points, and canonical coordinates;

(b) determine the corresponding one-parameter Lie group of transformations by

(i) integrating the appropriate initial value problem;

(ii) developing it in terms of a Lie series.

2.3 Extended Transformations (Prolongations)

In later chapters we will be interested in determining one-parameter Lie groups of transformations admitted by a given system S of differential equations. Such groups of transformations will be of the form

$$x^* = X(x, u; \epsilon), \tag{2.82a}$$

$$u^* = U(x, u; \epsilon), \tag{2.82b}$$

and act on the space of $n + m$ variables

$$x = (x_1, x_2, \ldots, x_n), \tag{2.83a}$$

$$u = (u^1, u^2, \dots, u^m), \tag{2.83b}$$

where x corresponds to the n *independent variables* and u corresponds to the m *dependent variables* appearing in S;

$$u = \Theta(x) = (\Theta^1(x), \Theta^2(x), \dots, \Theta^m(x)) \tag{2.84}$$

denotes a solution of S. A Lie group of transformations of the form (2.82a,b) admitted by S has the equivalent properties of (i) mapping any solution $u = \Theta(x)$ of S into another solution of S and (ii) leaving S invariant in the sense that S reads the same (is unchanged) in terms of the transformed variables (2.82a,b) for any solution $u = \Theta(x)$ of S. For a Lie group of transformations (2.82a,b) the derivatives of the dependent variables u with respect to the independent variables x, for any function $u = \Theta(x)$, are transformed "naturally" in order to preserve contact conditions.

Let $\underset{1}{u}$ denote the set of coordinates corresponding to all first order partial derivatives of u with respect to x:

$$\underset{1}{u} = \left(\frac{\partial u^1}{\partial x_1}, \frac{\partial u^1}{\partial x_2}, \dots, \frac{\partial u^1}{\partial x_n}, \frac{\partial u^2}{\partial x_1}, \frac{\partial u^2}{\partial x_2}, \dots, \frac{\partial u^2}{\partial x_n}, \dots, \right.$$
$$\left. \frac{\partial u^m}{\partial x_1}, \frac{\partial u^m}{\partial x_2}, \dots, \frac{\partial u^m}{\partial x_n}\right); \tag{2.85}$$

$\underset{1}{u}$ denotes nm coordinates.

In general for $k \geq 1$ let $\underset{k}{u}$ denote the set of coordinates corresponding to all kth order partial derivatives of u with respect to x; a coordinate in $\underset{k}{u}$ is denoted by $u^\mu_{i_1 i_2 \dots i_k} = \dfrac{\partial u^\mu}{\partial x_{i_1} \partial x_{i_2} \dots \partial x_{i_k}}$ with $\mu = 1, 2, \dots, m$ and $i_j = 1, 2, \dots, n$; $j = 1, 2, \dots, k$.

It turns out that the "natural" transformation of derivatives of the dependent variables leads successively to natural *extensions* (*prolongations*) of a one-parameter Lie group of transformations (2.82a,b) acting on (x, u)-space to one-parameter Lie groups of transformations acting on $(x, u, \underset{1}{u})$-space, $(x, u, \underset{1}{u}, \underset{2}{u})$-space, $\dots, (x, u, \underset{1}{u}, \underset{2}{u}, \dots, \underset{k}{u})$-space for any $k > 2$. [For a given system S of differential equations, k would be the order of the highest order derivative appearing in S.] Then the infinitesimal transformation of (2.82a,b) is naturally extended (prolonged) successively to infinitesimal transformations acting on $(x, u, \underset{1}{u}, \dots, \underset{k}{u})$-space, $k = 1, 2, \dots$.

In the following subsections we consider separately the cases of one dependent and one independent variable $[(m, n) = (1, 1)]$; one dependent variable and n independent variables $[(m, n) = (1, n)]$. Key results will be stated for the case of general (m, n) with proofs left to an exercise.

As a consequence, in later chapters we will see that we can formulate the problem of finding one-parameter Lie groups of transformations of the form (2.82a,b) admitted by a given system S of differential equations in terms of infinitesimal transformations admitted by S. This is shown to be an algorithmic procedure.

2.3.1 Extended Group Transformations—One Dependent and One Independent Variable

In studying the invariance properties of a kth order ordinary differential equation with independent variable x and dependent variable y we will aim to find admitted one-parameter Lie groups of transformations of the form

$$x^* = X(x, y; \epsilon), \tag{2.86a}$$

$$y^* = Y(x, y; \epsilon), \tag{2.86b}$$

where $y = y(x)$.

For a Lie group of transformations (2.86a,b) it is convenient to also use the notations

$$\mathbf{x}^* = (x^*, y^*) = \mathbf{X}(\mathbf{x}; \epsilon) = (X(x, y; \epsilon), Y(x, y; \epsilon))$$

$$= \mathbf{X}(x, y; \epsilon) = (X(\mathbf{x}; \epsilon), Y(\mathbf{x}; \epsilon)). \tag{2.87}$$

Let

$$y_k = \underset{k}{y} = \frac{d^k y}{dx^k}, \quad k = 1, 2, \ldots . \tag{2.88}$$

We naturally extend (2.86a,b) to $(x, y, \underset{1}{y}, \ldots, \underset{k}{y})$-space, $k = 1, 2, \ldots$, by demanding that (2.86a,b) preserve the contact conditions relating differentials dx, dy, dy_1, $dy_2, \ldots$:

$$dy = y_1 dx, \tag{2.89a}$$

and

$$dy_k = y_{k+1} dx, \quad k = 1, 2, \ldots . \tag{2.89b}$$

In particular under the action of the group of transformations (2.86a,b), the transformed derivatives $\{y_k^*\}$, $k = 1, 2, \ldots$, are defined successively by

$$dy^* = y_1^* dx^*, \tag{2.90a}$$

$$dy_k^* = y_{k+1}^* dx^*, \tag{2.90b}$$

where x^* is defined by (2.86a) and y^* by (2.86b). Then

$$dy^* = dY(\mathbf{x}; \epsilon) = \frac{\partial Y}{\partial x}(\mathbf{x}; \epsilon) dx + \frac{\partial Y}{\partial y}(\mathbf{x}; \epsilon) dy, \tag{2.91a}$$

$$dx^* = dX(\mathbf{x}; \epsilon) = \frac{\partial X}{\partial x}(\mathbf{x}; \epsilon) dx + \frac{\partial X}{\partial y}(\mathbf{x}; \epsilon) dy. \tag{2.91b}$$

Consequently from (2.90a), (2.91a,b), y_1^* satisfies

$$\frac{\partial Y}{\partial x}(\mathbf{x};\epsilon)dx + \frac{\partial Y}{\partial y}(\mathbf{x};\epsilon)dy = y_1^*\left[\frac{\partial X}{\partial x}(\mathbf{x};\epsilon)dx + \frac{\partial X}{\partial y}(\mathbf{x};\epsilon)dy\right]. \tag{2.92}$$

Substituting (2.89a) into (2.92), we see that

$$y_1^* = Y_1(x,y,y_1;\epsilon) = \frac{\frac{\partial Y}{\partial x}(\mathbf{x};\epsilon) + y_1\frac{\partial Y}{\partial y}(\mathbf{x};\epsilon)}{\frac{\partial X}{\partial x}(\mathbf{x};\epsilon) + y_1\frac{\partial X}{\partial y}(\mathbf{x};\epsilon)}. \tag{2.93}$$

Theorem 2.3.1-1. *The Lie group of transformations* (2.86a,b) *acting on* (x,y)*-space (naturally) extends to the following one-parameter Lie group of transformations acting on* (x,y,y_1)*-space:*

$$x^* = X(x,y;\epsilon), \tag{2.94a}$$

$$y^* = Y(x,y;\epsilon), \tag{2.94b}$$

$$y_1^* = Y_1(x,y,y_1;\epsilon), \tag{2.94c}$$

where $Y_1(x,y,y_1;\epsilon)$ *is given by* (2.93).

Proof. The proof is accomplished by showing that the closure property is preserved in this *first extension* of (2.86a,b) to (x,y,y_1)-space. The other properties of a one-parameter Lie group of transformations then follow immediately for this first extension.

Let $\phi(\epsilon,\delta)$ define the law of composition of parameters ϵ and δ. Let

$$\mathbf{x}^{**} = \mathbf{X}(\mathbf{x}^*;\delta).$$

Then from the closure property of the group (2.86a,b) it follows that $\mathbf{x}^{**} = \mathbf{X}(\mathbf{x};\phi(\epsilon,\delta))$. But y_1^{**} satisfies $dy^{**} = y_1^{**}dx^{**}$. Consequently

$$y_1^{**} = Y_1(x,y,y_1;\phi(\epsilon,\delta)) = \frac{\frac{\partial Y}{\partial x}(\mathbf{x};\phi(\epsilon,\delta)) + y_1\frac{\partial Y}{\partial y}(\mathbf{x};\phi(\epsilon,\delta))}{\frac{\partial X}{\partial x}(\mathbf{x};\phi(\epsilon,\delta)) + y_1\frac{\partial X}{\partial y}(\mathbf{x};\phi(\epsilon,\delta))}. \qquad \square$$

Theorem 2.3.1-2. *The Lie group of transformations* (2.86a,b) *extends to its second extension which is the following one-parameter Lie group of transformations acting on* (x,y,y_1,y_2)*-space:*

$$x^* = X(x,y;\epsilon), \tag{2.95a}$$

$$y^* = Y(x,y;\epsilon), \tag{2.95b}$$

$$y_1^* = Y_1(x,y,y_1;\epsilon), \tag{2.95c}$$

$$y_2^* = Y_2(x,y,y_1,y_2;\epsilon) = \frac{\frac{\partial Y_1}{\partial x} + y_1\frac{\partial Y_1}{\partial y} + y_2\frac{\partial Y_1}{\partial y_1}}{\frac{\partial X}{\partial x}(\mathbf{x};\epsilon) + y_1\frac{\partial X}{\partial y}(\mathbf{x};\epsilon)}, \tag{2.95d}$$

where $Y_1 = Y_1(x, y, y_1; \epsilon)$ *is defined by* (2.93).

The proof of Theorem 2.3.1-2 is left to Exercise 2.3-2.

The proof of the next theorem follows by induction:

Theorem 2.3.1-3. *The Lie group of transformations* (2.86a,b) *extends to its* k*th extension,* $k \geq 2$*, which is the following one-parameter Lie group of transformations acting on* $(x, y, y_1, \ldots, y_k)$*-space:*

$$x^* = X(x, y; \epsilon), \tag{2.96a}$$

$$y^* = Y(x, y; \epsilon), \tag{2.96b}$$

$$y_1^* = Y_1(x, y, y_1; \epsilon), \tag{2.96c}$$

$$\vdots$$

$$y_k^* = Y_k(x, y, y_1, \ldots, y_k; \epsilon) = \frac{\frac{\partial Y_{k-1}}{\partial x} + y_1 \frac{\partial Y_{k-1}}{\partial y} + \cdots + y_k \frac{\partial Y_{k-1}}{\partial y_{k-1}}}{\frac{\partial X}{\partial x}(\mathbf{x}; \epsilon) + y_1 \frac{\partial X}{\partial y}(\mathbf{x}; \epsilon)}, \tag{2.96d}$$

where $Y_1 = Y_1(x, y, y_1; \epsilon)$ *is defined by* (2.93), *and* $Y_{k-1} = Y_{k-1}(x, y, y_1, \ldots, y_{k-1}; \epsilon)$.

Note that we can extend any set of one-to-one transformations (not necessarily a group of transformations)

$$x^\dagger = X(x, y), \tag{2.97a}$$

$$y^\dagger = Y(x, y), \tag{2.97b}$$

from some domain D in (x, y)-space to another domain $D^\dagger$ in $(x^\dagger, y^\dagger)$-space where the functions $X(x, y)$ and $Y(x, y)$ are k times differentiable in D. One can (naturally) extend the transformations (2.97a,b) to $(x, y, y_1, \ldots, y_k)$-space so that the contact conditions (2.89a,b) are preserved:

$$dy^\dagger = y_1^\dagger dx^\dagger, \tag{2.98a}$$

$$dy_k^\dagger = y_{k+1}^\dagger dx^\dagger, \quad k = 1, 2, \ldots . \tag{2.98b}$$

Here the k times extended transformation from

$$(x, y, y_1, \ldots, y_k)\text{-space}$$

to

$$(x^\dagger, y^\dagger, y_1^\dagger, \ldots, y_k^\dagger)\text{-space}$$

is given by

$$x^\dagger = X(x, y), \tag{2.99a}$$

$$y^\dagger = Y(x, y), \tag{2.99b}$$

$$y_1^\dagger = Y_1(x, y, y_1), \tag{2.99c}$$

$$\vdots$$

$$y_k^{\dagger} = Y_k(x, y, y_1, \ldots, y_k) = \frac{\frac{\partial Y_{k-1}}{\partial x} + y_1 \frac{\partial Y_{k-1}}{\partial y} + \cdots + y_k \frac{\partial Y_{k-1}}{\partial y_{k-1}}}{\frac{\partial X}{\partial x}(x,y) + y_1 \frac{\partial X}{\partial y}(x,y)}, \qquad (2.99\text{d})$$

where

$$Y_1 = Y_1(x, y, y_1) = \frac{\frac{\partial Y}{\partial x}(x,y) + y_1 \frac{\partial Y}{\partial y}(x,y)}{\frac{\partial X}{\partial x}(x,y) + y_1 \frac{\partial X}{\partial y}(x,y)},$$

and $Y_{k-1} = Y_{k-1}(x, y, y_1, \ldots, y_{k-1})$.

Now consider examples of extended group transformations:

(1) *A Translation Group*

For the translation group

$$y^* = X = x + \epsilon, \qquad (2.100\text{a})$$

$$y^* = Y = y, \qquad (2.100\text{b})$$

we have

$$y_1^* = \left(\frac{dy}{dx}\right)^* = \frac{dy^*}{dx^*} = Y_1 = \frac{dy}{dx} = y_1,$$

and in general,

$$y_k^* = \left(\frac{d^k y}{dx^k}\right)^* = \frac{d^k y^*}{dx^{*k}} = Y_k = \frac{d^k y}{dx^k} = y_k, \quad k = 1, 2, \ldots .$$

Then the kth extended group for the translation group (2.100a,b) is given by

$$x^* = x + \epsilon, \qquad (2.101\text{a})$$

$$y^* = y, \qquad (2.101\text{b})$$

$$y_i^* = y_i, \quad i = 1, 2, \ldots, k. \qquad (2.101\text{c})$$

(2) *A Scaling Group*

For the scaling group

$$x^* = X = e^{\epsilon} x, \qquad (2.102\text{a})$$

$$y^* = Y = e^{2\epsilon} y, \qquad (2.102\text{b})$$

we have

$$y_1^* = \left(\frac{dy}{dx}\right)^* = \frac{dy^*}{dx^*} = Y_1 = \frac{y_1 \frac{\partial Y}{\partial y}}{\frac{\partial X}{\partial x}} = e^{\epsilon} y_1,$$

and in general,

$$y_k^* = \left(\frac{d^k y}{dx^k}\right)^* = \frac{d^k y^*}{dx^{*k}} = Y_k = \frac{y_k \frac{\partial Y_{k-1}}{\partial y_{k-1}}}{\frac{\partial X}{\partial x}} = e^{(2-k)\epsilon} y_k, \quad k = 1, 2, \ldots .$$

Here the kth extended Lie group of transformations is given by

$$x^* = e^{\epsilon} x, \tag{2.103a}$$

$$y^* = e^{2\epsilon} y, \tag{2.103b}$$

$$y_i^* = e^{(2-i)\epsilon} y_i, \quad i = 1, 2, \ldots, k. \tag{2.103c}$$

(3) *Rotation Group*

For the rotation group

$$x^* = X = x \cos \epsilon + y \sin \epsilon, \tag{2.104a}$$

$$y^* = Y = -x \sin \epsilon + y \cos \epsilon, \tag{2.104b}$$

we obtain

$$\frac{\partial X}{\partial x} = \cos \epsilon, \quad \frac{\partial X}{\partial y} = \sin \epsilon; \quad \frac{\partial Y}{\partial x} = -\sin \epsilon, \quad \frac{\partial Y}{\partial y} = \cos \epsilon.$$

Hence from (2.93) we obtain

$$Y_1 = \frac{-\sin \epsilon + y_1 \cos \epsilon}{\cos \epsilon + y_1 \sin \epsilon}.$$

Then

$$\frac{\partial Y_1}{\partial x} = \frac{\partial Y_1}{\partial y} = 0, \quad \frac{\partial Y_1}{\partial y_1} = \frac{1}{(\cos \epsilon + y_1 \sin \epsilon)^2}.$$

Consequently from (2.95d) we have

$$Y_2 = \frac{y_2}{(\cos \epsilon + y_1 \sin \epsilon)^3}.$$

Then

$$\frac{\partial Y_2}{\partial x} = \frac{\partial Y_2}{\partial y} = 0, \quad \frac{\partial Y_2}{\partial y_1} = \frac{-3 \sin \epsilon \, y_2}{(\cos \epsilon + y_1 \sin \epsilon)^4},$$

$$\frac{\partial Y_2}{\partial y_2} = \frac{1}{(\cos \epsilon + y_1 \sin \epsilon)^3}.$$

As a result from (2.96d), we get

$$Y_3 = \frac{(y_1 \sin \epsilon + \cos \epsilon) y_3 - 3 \sin \epsilon (y_2)^2}{(\cos \epsilon + y_1 \sin \epsilon)^5}.$$

Thus the thrice-extended Lie group of transformations corresponding to (2.104a,b) is given by

$$x^* = x \cos \epsilon + y \sin \epsilon, \tag{2.105a}$$

$$y^* = -x \sin \epsilon + y \cos \epsilon, \tag{2.105b}$$

$$y_1^* = \frac{-\sin\epsilon + y_1\cos\epsilon}{\cos\epsilon + y_1\sin\epsilon}, \tag{2.105c}$$

$$y_2^* = \frac{y_2}{(\cos\epsilon + y_1\sin\epsilon)^3}, \tag{2.105d}$$

$$y_3^* = \frac{(y_1\sin\epsilon + \cos\epsilon)y_3 - 3\sin\epsilon(y_2)^2}{(\cos\epsilon + y_1\sin\epsilon)^5}. \tag{2.105e}$$

This is a one-parameter Lie group of transformations acting on $(x, y, \underset{1}{y}, \underset{2}{y}, \underset{3}{y})$-space. Of course we can extend this Lie group of transformations indefinitely to $(x, y, \underset{1}{y}, \underset{2}{y}, \ldots, \underset{k}{y})$-space, $k = 4, 5, \ldots$, but the calculations get more and more horrendous as k increases!

From Section 2.2 we know that a one-parameter Lie group of transformations is characterized by its infinitesimal transformation. Since the kth extension of a one-parameter Lie group of transformations is also a one-parameter Lie group of transformations it follows that the study of extended Lie groups of transformations reduces to the study of extended infinitesimal transformations. In the next subsection we formulate Theorems 2.3.1-1, 2, 3 in terms of infinitesimal transformations. Consequently we will have an explicit algorithm to determine extended infinitesimal transformations (and corresponding infinitesimal generators) of an infinitesimal transformation.

Before proceeding further, the following notations are very convenient:

Let a subscript denote differentiation with respect to the corresponding coordinate, e.g., $F_x = \frac{\partial F}{\partial x}$, $F_y = \frac{\partial F}{\partial y}$.

Definition 2.3.1-1. The *total derivative operator* is defined by

$$\frac{D}{Dx} = \frac{\partial}{\partial x} + y_1\frac{\partial}{\partial y} + y_2\frac{\partial}{\partial y_1} + \cdots + y_{n+1}\frac{\partial}{\partial y_n} + \cdots; \tag{2.106}$$

given a differentiable function $F(x, y, y_1, y_2, \ldots, y_\ell)$,

$$\frac{D}{Dx}F(x, y, y_1, y_2, \ldots, y_\ell) = F_x + y_1F_y + y_2F_{y_1} + y_3F_{y_2} + \cdots + y_{\ell+1}F_{y_\ell}.$$

In terms of the total derivative operator (2.106) Theorem 2.3.1-3 can be restated as follows:

Theorem 2.3.1-4. *The kth extension of the one-parameter Lie group of transformations* (2.86a,b) *is given by*

$$x^* = X(x, y; \epsilon), \tag{2.107a}$$

$$y^* = Y(x, y; \epsilon), \tag{2.107b}$$

$$y_i^* = Y_i(x, y, y_1, \ldots, y_i; \epsilon) = \frac{\frac{DY_{i-1}}{Dx}}{\frac{DX(x,y;\epsilon)}{Dx}}, \quad i = 1, 2, \ldots, k, \tag{2.107c}$$

where

$$Y_0 = Y(x, y; \epsilon).$$

2.3.2 Extended Infinitesimal Transformations—One Dependent and One Independent Variable

The one-parameter Lie group of transformations

$$x^* = X(x, y; \epsilon) = x + \epsilon\xi(x, y) + O(\epsilon^2), \tag{2.108a}$$

$$y^* = Y(x, y; \epsilon) = y + \epsilon\eta(x, y) + O(\epsilon^2), \tag{2.108b}$$

acting on (x, y)-space, has as its infinitesimal

$$\boldsymbol{\xi}(\mathbf{x}) = (\xi(x, y), \eta(x, y)), \tag{2.108c}$$

with corresponding infinitesimal generator

$$\mathrm{X} = \xi(x, y)\frac{\partial}{\partial x} + \eta(x, y)\frac{\partial}{\partial y}. \tag{2.108d}$$

The kth extension of (2.108a,b), given by

$$x^* = X(x, y; \epsilon) = x + \epsilon\xi(x, y) + O(\epsilon^2), \tag{2.109a}$$

$$y^* = Y(x, y; \epsilon) = y + \epsilon\eta(x, y) + O(\epsilon^2), \tag{2.109b}$$

$$y_1^* = Y_1(x, y, y_1; \epsilon) = y_1 + \epsilon\eta^{(1)}(x, y, y_1) + O(\epsilon^2), \tag{2.109c}$$

$$\vdots \qquad \vdots \qquad \vdots$$

$$y_k^* = Y_k(x, y, y_1, \ldots, y_k; \epsilon) = y_k + \epsilon\eta^{(k)}(x, y, y_1, \ldots, y_k) + O(\epsilon^2), \tag{2.109d}$$

has as its (*kth extended*) *infinitesimal*

$$(\xi(x, y), \eta(x, y), \eta^{(1)}(x, y, y_1), \ldots, \eta^{(k)}(x, y, y_1, \ldots, y_k)), \tag{2.109e}$$

with corresponding (*kth extended*) *infinitesimal generator*

$$\mathrm{X}^{(k)} = \xi(x, y)\frac{\partial}{\partial x} + \eta(x, y)\frac{\partial}{\partial y} + \eta^{(1)}(x, y, y_1)\frac{\partial}{\partial y_1} + \cdots + \eta^{(k)}(x, y, y_1, \ldots, y_k)\frac{\partial}{\partial y_k}, \tag{2.109f}$$

$k = 1, 2, \ldots$. Explicit formulas for the extended infinitesimals $\{\eta^{(k)}\}$ result from the following theorem.

Theorem 2.3.2-1.

$$\eta^{(k)}(x, y, y_1, \ldots, y_k) = \frac{D\eta^{(k-1)}}{Dx} - y_k\frac{D\xi(x, y)}{Dx}, \quad k = 1, 2, \ldots, \tag{2.110}$$

where

$$\eta^{(0)} = \eta(x, y).$$

Proof. From (2.107c), (2.109a-d), and (2.106), we have

$$\begin{aligned} Y_k(x, y, y_1, \ldots, y_k; \epsilon) &= \frac{\frac{DY_{k-1}}{Dx}}{\frac{DX(x,y;\epsilon)}{Dx}} \\ &= \frac{\frac{D[y_{k-1}+\epsilon\eta^{(k-1)}+O(\epsilon^2)]}{Dx}}{\frac{D[x+\epsilon\xi(x,y)+O(\epsilon^2)]}{Dx}} \\ &= \frac{\left[y_k + \epsilon\frac{D\eta^{(k-1)}}{Dx}\right]}{\left[1 + \epsilon\frac{D\xi(x,y)}{Dx}\right]} + O(\epsilon^2) \\ &= y_k + \epsilon\left[\frac{D\eta^{(k-1)}}{Dx} - y_k\frac{D\xi(x,y)}{Dx}\right] + O(\epsilon^2) \\ &= y_k + \epsilon\eta^{(k)} + O(\epsilon^2), \end{aligned}$$

leading to (2.110). □

Explicit formulas for $\{\eta^{(k)}\}$ follow immediately from Theorem 2.3.2-1. In particular,

$$\eta^{(1)} = \eta_x + (\eta_y - \xi_x)y_1 - \xi_y(y_1)^2, \tag{2.111}$$

$$\begin{aligned} \eta^{(2)} = \eta_{xx} &+ (2\eta_{xy} - \xi_{xx})y_1 + (\eta_{yy} - 2\xi_{xy})(y_1)^2 \\ &- \xi_{yy}(y_1)^3 + (\eta_y - 2\xi_x)y_2 - 3\xi_y y_1 y_2, \end{aligned} \tag{2.112}$$

$$\begin{aligned} \eta^{(3)} = \eta_{xxx} &+ (3\eta_{xxy} - \xi_{xxx})y_1 + 3(\eta_{xyy} - \xi_{xxy})(y_1)^2 \\ &+ (\eta_{yyy} - 3\xi_{xyy})(y_1)^3 - \xi_{yyy}(y_1)^4 + 3(\eta_{xy} - \xi_{xx})y_2 \\ &+ 3(\eta_{yy} - 3\xi_{xy})y_1 y_2 - 6\xi_{yy}(y_1)^2 y_2 \\ &- 3\xi_y(y_2)^2 + (\eta_y - 3\xi_x)y_3 - 4\xi_y y_1 y_3. \end{aligned} \tag{2.113}$$

The following observations are important:

Theorem 2.3.2-2.

(i) $\eta^{(k)}$ *is linear in* y_k, $k = 2, 3, \ldots$.

(ii) $\eta^{(k)}$ *is a polynomial in* $y_1, y_2, \ldots, y_k$ *whose coefficients are linear homogeneous in* $(\xi(x,y), \eta(x,y))$ *up to their kth order partial derivatives.*

Proof. See Exercise 2.3-5. □

We now find extended infinitesimals $\{\eta^{(k)}\}$ for the examples of Section 2.3.1.

(1) *Translation Group* (2.100a,b)

Here $\eta^{(k)} = 0$, $k = 1, 2, \ldots$.

(2) *Scaling Group* (2.102a,b)

From the form of (2.103c) it is immediately obvious that

$$\eta^{(k)} = (2-k)y_k, \quad k = 1, 2, \ldots .$$

(3) *Rotation Group* (2.104a,b)

Here $\xi(x,y) = y$, $\eta(x,y) = -x$. So $\xi_y = 1$, $\eta_x = -1$, $\xi_x = \eta_y = 0$. From (2.111)–(2.113) we see that

$$\begin{aligned}\eta^{(1)} &= -[1 + (y_1)^2],\\ \eta^{(2)} &= -3y_1y_2,\\ \eta^{(3)} &= -[3(y_2)^2 + 4y_1y_3].\end{aligned}$$

From (2.110), for $k \geq 4$, $\eta^{(k)} = \frac{D\eta^{(k-1)}}{Dx} - y_k y_1$, so that

$$\begin{aligned}\eta^{(4)} &= -5[2y_2y_3 + y_1y_4],\\ \eta^{(5)} &= -[10(y_3)^2 + 15y_2y_4 + 6y_1y_5], \quad \text{etc.}\end{aligned}$$

2.3.3 Extended Transformations—One Dependent and n Independent Variables

In studying invariance properties of a kth order partial differential equation with dependent variable u and independent variables $x = (x_1, x_2, \ldots, x_n)$, with $u = u(x)$, we are naturally led to the problem of finding the extensions of transformations on (x, u)-space to $(x, u, \underset{1}{u}, \ldots, \underset{k}{u})$-space where $\underset{k}{u}$ represents all the kth order partial derivatives of u with respect to x.

First we consider extended transformations of a set of *point transformations*

$$x^\dagger = X(x, u), \tag{2.114a}$$

$$u^\dagger = U(x, u). \tag{2.114b}$$

The transformations (2.114a,b) are assumed to be one-to-one in some domain D in (x, u)-space with $(X(x, u), U(x, u))$ k times differentiable in D. The transformations (2.114a,b) preserve the contact conditions, i.e.,

$$du = \underset{1}{u}\, dx, \tag{2.115a}$$

$$\vdots$$

$$d\underset{k-1}{u} = \underset{k}{u}\, dx, \tag{2.115b}$$

in some domain D in $(x, u, \underset{1}{u}, \ldots, \underset{k}{u})$-space if and only if

$$du^\dagger = \underset{1}{u}^\dagger\, dx^\dagger, \tag{2.116a}$$

$$\vdots$$

$$d \underset{k-1}{u}^{\dagger} = \underset{k}{u}^{\dagger} dx^{\dagger}, \tag{2.116b}$$

in the corresponding domain $D^{\dagger}$ in $(x^{\dagger}, u^{\dagger}, \underset{1}{u}^{\dagger}, \ldots, \underset{k}{u}^{\dagger})$-space.

Let

$$u_i = \frac{\partial u}{\partial x_i}, \quad u_i^{\dagger} = \frac{\partial u^{\dagger}}{\partial x_i^{\dagger}} = \frac{\partial U}{\partial X_i}, \quad \text{etc.}$$

From now on we assume summation over a repeated index. In (2.115a), $du = \underset{1}{u}\, dx$ represents

$$du = u_j dx_j,$$

and in (2.115b), $d \underset{k-1}{u} = \underset{k}{u}\, dx$ represents a set of equations

$$du_{i_1 i_2 \cdots i_{k-1}} = u_{i_1 i_2 \cdots i_{k-1} j} dx_j, \quad i_\ell = 1, 2, \ldots, n \text{ for } \ell = 1, 2, \ldots, k-1.$$

Similar representations hold for (2.116a,b).

We introduce the total derivative operators

$$\mathrm{D}_i = \frac{D}{Dx_i} = \frac{\partial}{\partial x_i} + u_i \frac{\partial}{\partial u} + u_{ij} \frac{\partial}{\partial u_j} + \cdots + u_{i i_1 i_2 \cdots i_n} \frac{\partial}{\partial u_{i_1 i_2 \cdots i_n}} + \cdots, \tag{2.117}$$

$i = 1, 2, \ldots, n$. For a given differentiable function $F(x, u, \underset{1}{u}, \ldots, \underset{\ell}{u})$ we have:

$$\mathrm{D}_i F(x, u, \underset{1}{u}, \ldots, \underset{\ell}{u}) = \frac{\partial F}{\partial x_i} + u_i \frac{\partial F}{\partial u} + u_{ij} \frac{\partial F}{\partial u_j} + \cdots + u_{i i_1 i_2 \cdots i_\ell} \frac{\partial F}{\partial u_{i_1 i_2 \cdots i_\ell}},$$

$i = 1, 2, \ldots, n$.

Now consider the preserved contact condition (2.116a), $du^{\dagger} = u_j^{\dagger} dx_j^{\dagger}$, in order to determine the extended transformation

$$u_j^{\dagger} = U_j(x, u, \underset{1}{u}), \quad j = 1, 2, \ldots, n. \tag{2.118}$$

From (2.114a,b) we obtain

$$du^{\dagger} = (\mathrm{D}_i U) dx_i,$$

and

$$dx_j^{\dagger} = (\mathrm{D}_i X_j) dx_i, \quad j = 1, 2, \ldots, n,$$

where D_i is defined by (2.117), $i = 1, 2, \ldots, n$. Then

$$(\mathrm{D}_i X_j) u_j^{\dagger} = \mathrm{D}_i U, \quad i = 1, 2, \ldots, n.$$

Let the $n \times n$ matrix

$$A = \begin{bmatrix} \mathrm{D}_1 X_1 & \cdots & \mathrm{D}_1 X_n \\ \vdots & & \vdots \\ \mathrm{D}_n X_1 & \cdots & \mathrm{D}_n X_n \end{bmatrix} \tag{2.119}$$

and assume that A^{-1} exists. Then

$$\begin{bmatrix} u_1^\dagger \\ u_2^\dagger \\ \vdots \\ u_n^\dagger \end{bmatrix} = \begin{bmatrix} U_1 \\ U_2 \\ \vdots \\ U_n \end{bmatrix} = A^{-1} \begin{bmatrix} \mathrm{D}_1 U \\ \mathrm{D}_2 U \\ \vdots \\ \mathrm{D}_n U \end{bmatrix}. \tag{2.120}$$

This leads to the extended transformation in $(x, u, \underset{1}{u})$-space:

$$x^\dagger = X(x, u), \tag{2.121a}$$

$$u^\dagger = U(x, u), \tag{2.121b}$$

$$\underset{1}{u}^\dagger = \underset{1}{U}(x, u, \underset{1}{u}). \tag{2.121c}$$

It is easy to show that the extension to $(x, u, \underset{1}{u}, \ldots, \underset{k}{u})$-space is given by

$$x^\dagger = X(x, u), \tag{2.122a}$$

$$u^\dagger = U(x, u), \tag{2.122b}$$

$$\underset{1}{u}^\dagger = \underset{1}{U}(x, u, \underset{1}{u}), \tag{2.122c}$$

$$\vdots \qquad \vdots$$

$$\underset{k}{u}^\dagger = \underset{k}{U}(x, u, \underset{1}{u}, \ldots, \underset{k}{u}), \tag{2.122d}$$

where the components of $\underset{k}{u}^\dagger$ are determined by

$$\begin{bmatrix} u^\dagger_{i_1 i_2 \cdots i_{k-1} 1} \\ u^\dagger_{i_1 i_2 \cdots i_{k-1} 2} \\ \vdots \\ u^\dagger_{i_1 i_2 \cdots i_{k-1} n} \end{bmatrix} = \begin{bmatrix} U_{i_1 i_2 \cdots i_{k-1} 1} \\ U_{i_1 i_2 \cdots i_{k-1} 2} \\ \vdots \\ U_{i_1 i_2 \cdots i_{k-1} n} \end{bmatrix} = A^{-1} \begin{bmatrix} \mathrm{D}_1 U_{i_1 i_2 \cdots i_{k-1}} \\ \mathrm{D}_2 U_{i_1 i_2 \cdots i_{k-1}} \\ \vdots \\ \mathrm{D}_n U_{i_1 i_2 \cdots i_{k-1}} \end{bmatrix}, \tag{2.123}$$

$i_\ell = 1, 2, \ldots, n$ for $\ell = 1, 2, \ldots, k-1$ with $k = 2, 3, \ldots$; $\underset{1}{U}(x, u, \underset{1}{u})$ is determined by (2.120) and A is the matrix (2.119).

Now we specialize to the case where (2.114a,b) defines a Lie group of transformations.

If the transformations (2.114a,b) define a one-parameter Lie group of transformations

$$x^* = X(x, u; \epsilon), \tag{2.124a}$$

$$u^* = U(x, u; \epsilon), \tag{2.124b}$$

acting on (x, u)-space, then it is easy to show (following the proofs of Theorems 2.3.1-1 to 2.3.1-3) that its kth extension to $(x, u, \underset{1}{u}, \ldots, \underset{k}{u})$-space, given by

$$x^* = X(x, u; \epsilon), \tag{2.125a}$$

$$u^* = U(x, u; \epsilon), \tag{2.125b}$$

$$\underset{1}{u}^* = \underset{1}{U}(x, u, \underset{1}{u}; \epsilon), \tag{2.125c}$$

$$\vdots$$

$$\underset{k}{u}^* = \underset{k}{U}(x, u, \underset{1}{u}, \ldots, \underset{k}{u}; \epsilon), \tag{2.125d}$$

defines a k times extended one-parameter Lie group of transformations. In (2.125a–d),

$$\begin{bmatrix} u_1^* \\ u_2^* \\ \vdots \\ u_n^* \end{bmatrix} = \begin{bmatrix} U_1 \\ U_2 \\ \vdots \\ U_n \end{bmatrix} = A^{-1} \begin{bmatrix} \mathrm{D}_1 U \\ \mathrm{D}_2 U \\ \vdots \\ \mathrm{D}_n U \end{bmatrix}, \tag{2.126a}$$

$$\begin{bmatrix} u^*_{i_1 i_2 \cdots i_{k-1} 1} \\ u^*_{i_1 i_2 \cdots i_{k-1} 2} \\ \vdots \\ u^*_{i_1 i_2 \cdots i_{k-1} n} \end{bmatrix} = \begin{bmatrix} U_{i_1 i_2 \cdots i_{k-1} 1} \\ U_{i_1 i_2 \cdots i_{k-1} 2} \\ \vdots \\ U_{i_1 i_2 \cdots i_{k-1} n} \end{bmatrix} = A^{-1} \begin{bmatrix} \mathrm{D}_1 U_{i_1 i_2 \cdots i_{k-1}} \\ \mathrm{D}_2 U_{i_1 i_2 \cdots i_{k-1}} \\ \vdots \\ \mathrm{D}_n U_{i_1 i_2 \cdots i_{k-1}} \end{bmatrix}, \tag{2.126b}$$

where $\{u_i^* = U_i\}$ are the components of $\underset{1}{u}^* = \underset{1}{U}$ and $\{u^*_{i_1 i_2 \cdots i_{k-1} i} = U_{i_1 i_2 \cdots i_{k-1} i}\}$ are the components of $\underset{k}{u}^* = \underset{k}{U}$. In (2.126a,b) $i_\ell = 1, 2, \ldots, n$ for $\ell = 1, 2, \ldots, k-1$ with $k = 2, 3, \ldots$; the operators D_i are given by (2.117); A^{-1} is the inverse of the matrix A given by (2.119) for X and U given by (2.125a,b).

2.3.4 Extended Infinitesimal Transformations—One Dependent and n Independent Variables

The one-parameter Lie group of transformations

$$x_i^* = X_i(x, u; \epsilon) = x_i + \epsilon \xi_i(x, u) + O(\epsilon^2), \tag{2.127a}$$

$$u^* = U(x, u; \epsilon) = u + \epsilon \eta(x, u) + O(\epsilon^2), \tag{2.127b}$$

$i = 1, 2, \ldots, n$, acting on (x, u)-space has as its infinitesimal generator

$$\mathrm{X} = \xi_i(x, u) \frac{\partial}{\partial x_i} + \eta(x, u) \frac{\partial}{\partial u}. \tag{2.127c}$$

The *kth extension* of (2.127a,b), given by

$$x_i^* = X_i(x, u; \epsilon) = x_i + \epsilon \xi_i(x, u) + O(\epsilon^2), \tag{2.128a}$$

$$u^* = U(x,u;\epsilon) = u + \epsilon\eta(x,u) + O(\epsilon^2), \tag{2.128b}$$
$$u_i^* = U_i(x,u,\underset{1}{u};\epsilon) = u_i + \epsilon\eta_i^{(1)}(x,u,\underset{1}{u}) + O(\epsilon^2), \tag{2.128c}$$
$$\vdots$$
$$\begin{aligned} u^*_{i_1i_2\cdots i_k} &= U_{i_1i_2\cdots i_k}(x,u,\underset{1}{u},\ldots,\underset{k}{u};\epsilon) \\ &= u_{i_1i_2\cdots i_k} + \epsilon\eta^{(k)}_{i_1i_2\cdots i_k}(x,u,\underset{1}{u},\ldots,\underset{k}{u}) + O(\epsilon^2), \end{aligned} \tag{2.128d}$$

where $i = 1,2,\ldots,n$ and $i_\ell = 1,2,\ldots,n$ for $\ell = 1,2,\ldots,k$ with $k = 1,2,\ldots,$ has as its (*kth extended*) *infinitesimal*

$$(\xi(x,u), \eta^{(1)}(x,u,\underset{1}{u}),\ldots,\eta^{(k)}(x,u,\underset{1}{u},\ldots,\underset{k}{u})) \tag{2.128e}$$

with corresponding (*kth extended*) *infinitesimal generator*

$$\begin{aligned} \mathrm{X}^{(k)} = \xi_i(x,u)\frac{\partial}{\partial x_i} + \eta(x,u)\frac{\partial}{\partial u} + \eta_i^{(1)}(x,u,\underset{1}{u})\frac{\partial}{\partial u_i} + \cdots \\ + \eta^{(k)}_{i_1i_2\cdots i_k}\frac{\partial}{\partial u_{i_1i_2\cdots i_k}}, \quad k = 1,2,\ldots . \end{aligned} \tag{2.128f}$$

Explicit formulas for the extended infinitesimals $\{\eta^{(k)}\}$ result from the following theorem.

Theorem 2.3.4-1.

$$\eta_i^{(1)} = \mathrm{D}_i\eta - (\mathrm{D}_i\xi_j)u_j, \quad i = 1,2,\ldots,n; \tag{2.129a}$$

$$\eta^{(k)}_{i_1i_2\cdots i_k} = \mathrm{D}_{i_k}\eta^{(k-1)}_{i_1i_2\cdots i_{k-1}} - (\mathrm{D}_{i_k}\xi_j)u_{i_1i_2\cdots i_{k-1}j}, \tag{2.129b}$$

$i_\ell = 1,2,\ldots,n$ for $\ell = 1,2,\ldots,k$ with $k = 2,3,\ldots$.

Proof. From (2.119) and (2.128a) we have

$$\begin{aligned} A &= \begin{bmatrix} \mathrm{D}_1(x_1+\epsilon\xi_1) & \mathrm{D}_1(x_2+\epsilon\xi_2) & \cdots & \mathrm{D}_1(x_n+\epsilon\xi_n) \\ \mathrm{D}_2(x_1+\epsilon\xi_1) & \mathrm{D}_2(x_2+\epsilon\xi_2) & \cdots & \mathrm{D}_2(x_n+\epsilon\xi_n) \\ \vdots & \vdots & & \vdots \\ \mathrm{D}_n(x_1+\epsilon\xi_1) & \mathrm{D}_n(x_2+\epsilon\xi_2) & \cdots & \mathrm{D}_n(x_n+\epsilon\xi_n) \end{bmatrix} + O(\epsilon^2) \\ &= I + \epsilon B + O(\epsilon^2) \end{aligned}$$

where I is the $n \times n$ identity matrix and

$$B = \begin{bmatrix} \mathrm{D}_1\xi_1 & \mathrm{D}_1\xi_2 & \cdots & \mathrm{D}_1\xi_n \\ \mathrm{D}_2\xi_1 & \mathrm{D}_2\xi_2 & \cdots & \mathrm{D}_2\xi_n \\ \vdots & \vdots & & \vdots \\ \mathrm{D}_n\xi_1 & \mathrm{D}_n\xi_2 & \cdots & \mathrm{D}_n\xi_n \end{bmatrix}. \tag{2.130}$$

Then

$$A^{-1} = I - \epsilon B + O(\epsilon^2). \tag{2.131}$$

From (2.126a), (2.128b,c), (2.130), and (2.131) it follows that

$$\begin{bmatrix} u_1 + \epsilon\eta_1^{(1)} \\ u_2 + \epsilon\eta_2^{(1)} \\ \vdots \\ u_n + \epsilon\eta_n^{(1)} \end{bmatrix} = [I - \epsilon B] \cdot \begin{bmatrix} u_1 + \epsilon D_1\eta \\ u_2 + \epsilon D_2\eta \\ \vdots \\ u_n + \epsilon D_n\eta \end{bmatrix} + O(\epsilon^2),$$

and thus

$$\begin{bmatrix} \eta_1^{(1)} \\ \eta_2^{(1)} \\ \vdots \\ \eta_n^{(1)} \end{bmatrix} = \begin{bmatrix} D_1\eta \\ D_2\eta \\ \vdots \\ D_n\eta \end{bmatrix} - B \cdot \begin{bmatrix} u_1 \\ u_2 \\ \vdots \\ u_n \end{bmatrix},$$

leading to (2.129a). Then from (2.126b), (2.128c,d), (2.130), and (2.131) we get

$$\begin{bmatrix} u_{i_1i_2\cdots i_{k-1}1} + \epsilon\eta^{(k)}_{i_1i_2\cdots i_{k-1}1} \\ u_{i_1i_2\cdots i_{k-1}2} + \epsilon\eta^{(k)}_{i_1i_2\cdots i_{k-1}2} \\ \vdots \\ u_{i_1i_2\cdots i_{k-1}n} + \epsilon\eta^{(k)}_{i_1i_2\cdots i_{k-1}n} \end{bmatrix}$$

$$= [I - \epsilon B] \begin{bmatrix} u_{i_1i_2\cdots i_{k-1}1} + \epsilon D_1\eta^{(k-1)}_{i_1i_2\cdots i_{k-1}} \\ u_{i_1i_2\cdots i_{k-1}2} + \epsilon D_2\eta^{(k-1)}_{i_1i_2\cdots i_{k-1}} \\ \vdots \\ u_{i_1i_2\cdots i_{k-1}n} + \epsilon D_n\eta^{(k-1)}_{i_1i_2\cdots i_{k-1}} \end{bmatrix} + O(\epsilon^2),$$

and hence

$$\begin{bmatrix} \eta^{(k)}_{i_1i_2\cdots i_{k-1}1} \\ \eta^{(k)}_{i_1i_2\cdots i_{k-1}2} \\ \vdots \\ \eta^{(k)}_{i_1i_2\cdots i_{k-1}n} \end{bmatrix} = \begin{bmatrix} D_1\eta^{(k-1)}_{i_1i_2\cdots i_{k-1}} \\ D_2\eta^{(k-1)}_{i_1i_2\cdots i_{k-1}} \\ \vdots \\ D_n\eta^{(k-1)}_{i_1i_2\cdots i_{k-1}} \end{bmatrix} - B \cdot \begin{bmatrix} u_{i_1i_2\cdots i_{k-1}1} \\ u_{i_1i_2\cdots i_{k-1}2} \\ \vdots \\ u_{i_1i_2\cdots i_{k-1}n} \end{bmatrix},$$

$i_\ell = 1, 2, \ldots, n$ for $\ell = 1, 2, \ldots, k-1$ with $k = 2, 3, \ldots$, leading to (2.129b). □

Specializing Theorem 2.3.4-1 to the case of one dependent variable and two independent variables x_1 and x_2, we have, for the extended one-parameter Lie group of transformations given by

$$x_i^* = X_i(x_1, x_2, u; \epsilon) = x_i + \epsilon\xi_i(x_1, x_2, u) + O(\epsilon^2), \quad i = 1, 2, \tag{2.132a}$$

$$u^* = U(x_1, x_2, u; \epsilon) = u + \epsilon\eta(x_1, x_2, u) + O(\epsilon^2), \tag{2.132b}$$

$$u_i^* = U_i(x_1, x_2, u, u_1, u_2; \epsilon) = U_i + \epsilon\eta_i^{(1)}(x_1, x_2, u, u_1, u_2) + O(\epsilon^2), \quad i = 1, 2, \tag{2.132c}$$

$$u_{ij}^* = U_{ij}(x_1, x_2, u, u_1, u_2, u_{11}, u_{12}, u_{22}; \epsilon) = u_{ij} + \epsilon\eta_{ij}^{(2)}(x_1, x_2, u, u_1, u_2, u_{11}, u_{12}, u_{22}) + O(\epsilon^2), \; i, j = 1, 2, \tag{2.132d}$$

etc., the following extended infinitesimals:

$$\eta_1^{(1)} = \frac{\partial\eta}{\partial x_1} + \left[\frac{\partial\eta}{\partial u} - \frac{\partial\xi_1}{\partial x_1}\right] u_1 - \frac{\partial\xi_2}{\partial x_1}u_2 - \frac{\partial\xi_1}{\partial u}(u_1)^2 - \frac{\partial\xi_2}{\partial u}u_1u_2, \tag{2.133}$$

$$\eta_2^{(1)} = \frac{\partial\eta}{\partial x_2} + \left[\frac{\partial\eta}{\partial u} - \frac{\partial\xi_2}{\partial x_2}\right] u_2 - \frac{\partial\xi_1}{\partial x_2}u_1 - \frac{\partial\xi_2}{\partial u}(u_2)^2 - \frac{\partial\xi_1}{\partial u}u_1u_2, \tag{2.134}$$

$$\begin{aligned}\eta_{11}^{(2)} = {} & \frac{\partial^2\eta}{\partial x_1^2} + \left[2\frac{\partial^2\eta}{\partial x_1\partial u} - \frac{\partial^2\xi_1}{\partial x_1^2}\right] u_1 - \frac{\partial^2\xi_2}{\partial x_1^2}u_2 + \left[\frac{\partial\eta}{\partial u} - 2\frac{\partial\xi_1}{\partial x_1}\right] u_{11} \\ & - 2\frac{\partial\xi_2}{\partial x_1}u_{12} + \left[\frac{\partial^2\eta}{\partial u^2} - 2\frac{\partial^2\xi_1}{\partial x_1\partial u}\right](u_1)^2 - 2\frac{\partial^2\xi_2}{\partial x_1\partial u}u_1u_2 \\ & - \frac{\partial^2\xi_1}{\partial u^2}(u_1)^3 - \frac{\partial^2\xi_2}{\partial u^2}(u_1)^2u_2 - 3\frac{\partial\xi_1}{\partial u}u_1u_{11} - \frac{\partial\xi_2}{\partial u}u_2u_{11} \\ & - 2\frac{\partial\xi_2}{\partial u}u_1u_{12},\end{aligned} \tag{2.135}$$

$$\begin{aligned}\eta_{12}^{(2)} = {} & \eta_{21}^{(2)} \\ = {} & \frac{\partial^2\eta}{\partial x_1\partial x_2} + \left[\frac{\partial^2\eta}{\partial x_1\partial u} - \frac{\partial^2\xi_2}{\partial x_1\partial x_2}\right] u_2 + \left[\frac{\partial^2\eta}{\partial x_2\partial u} - \frac{\partial^2\xi_1}{\partial x_1\partial x_2}\right] u_1 \\ & - \frac{\partial\xi_2}{\partial x_1}u_{22} + \left[\frac{\partial\eta}{\partial u} - \frac{\partial\xi_1}{\partial x_1} - \frac{\partial\xi_2}{\partial x_2}\right] u_{12} - \frac{\partial\xi_1}{\partial x_2}u_{11} - \frac{\partial^2\xi_2}{\partial x_1\partial u}(u_2)^2 \\ & + \left[\frac{\partial^2\eta}{\partial u^2} - \frac{\partial^2\xi_1}{\partial x_1\partial u} - \frac{\partial^2\xi_2}{\partial x_2\partial u}\right] u_1u_2 - \frac{\partial^2\xi_1}{\partial x_2\partial u}(u_1)^2 - \frac{\partial^2\xi_2}{\partial u^2}u_1(u_2)^2 \\ & - \frac{\partial^2\xi_1}{\partial u^2}(u_1)^2u_2 - 2\frac{\partial\xi_2}{\partial u}u_2u_{12} - 2\frac{\partial\xi_1}{\partial u}u_1u_{12} \\ & - \frac{\partial\xi_1}{\partial u}u_2u_{11} - \frac{\partial\xi_2}{\partial u}u_1u_{22},\end{aligned} \tag{2.136}$$

$$\begin{aligned}\eta_{22}^{(2)} = {} & \frac{\partial^2\eta}{\partial x_2^2} + \left[2\frac{\partial^2\eta}{\partial x_2\partial u} - \frac{\partial^2\xi_2}{\partial x_2^2}\right] u_2 - \frac{\partial^2\xi_1}{\partial x_2^2}u_1 + \left[\frac{\partial\eta}{\partial u} - 2\frac{\partial\xi_2}{\partial x_2}\right] u_{22} \\ & - 2\frac{\partial\xi_1}{\partial x_2}u_{12} + \left[\frac{\partial^2\eta}{\partial u^2} - 2\frac{\partial\xi_2}{\partial x_2\partial u}\right](u_2)^2 - 2\frac{\partial^2\xi_1}{\partial x_2\partial u}u_1u_2 - \frac{\partial^2\xi_2}{\partial u^2}(u_2)^3 \\ & - \frac{\partial^2\xi_1}{\partial u^2}u_1(u_2)^2 - 3\frac{\partial\xi_2}{\partial u}u_2u_{22} - \frac{\partial\xi_1}{\partial u}u_1u_{22} - 2\frac{\partial\xi_1}{\partial u}u_2u_{12},\end{aligned} \tag{2.137}$$

etc.

2.3.5 Extended Transformations and Extended Infinitesimal Transformations—m Dependent and n Independent Variables

The situation of m dependent variables $u = (u^1, u^2, \ldots, u^m)$ and n independent variables $x = (x_1, x_2, \ldots, x_n)$, $u = u(x)$, with $m \geq 2$, arises in studying systems of differential equations. This leads us to consider extended transformations from (x, u)-space to $(x, u, \underset{1}{u}, \ldots, \underset{k}{u})$-space where $\underset{k}{u}$ denotes all kth order partial derivatives of u with respect to x. These extended transformations preserve the corresponding contact conditions.

We consider a point transformation

$$x^\dagger = X(x, u), \tag{2.138a}$$

$$u^\dagger = U(x, u). \tag{2.138b}$$

Let

$$u_i^\sigma = \frac{\partial u^\sigma}{\partial x_i}, \quad (u_i^\sigma)^\dagger = \frac{\partial (u^\sigma)^\dagger}{\partial x_i^\dagger} = \frac{\partial U^\sigma}{\partial X_i}, \quad \text{etc.},$$

$$\mathrm{D}_i = \frac{\partial}{\partial x_i} + u_i^\mu \frac{\partial}{\partial u^\mu} + u_{ij}^\mu \frac{\partial}{\partial u_j^\mu} + \cdots + u_{ii_1i_2\cdots i_n}^\mu \frac{\partial}{\partial u_{i_1i_2\cdots i_n}^\mu} + \cdots,$$

with summation over a repeated index. The kth extended transformation of (2.138a,b) is given by [see Exercise 2.3-11]

$$x^\dagger = X(x, u), \tag{2.139a}$$

$$u^\dagger = U(x, u), \tag{2.139b}$$

$$\underset{1}{u}^\dagger = \underset{1}{U}(x, u, \underset{1}{u}), \tag{2.139c}$$

$$\vdots$$

$$\underset{k}{u}^\dagger = \underset{k}{U}(x, u, \underset{1}{u}, \ldots, \underset{k}{u}), \tag{2.139d}$$

where the components $\{(u_i^\mu)^\dagger\}$ of $\underset{1}{u}^\dagger$ are determined by

$$\begin{bmatrix} (u_1^\mu)^\dagger \\ (u_2^\mu)^\dagger \\ \vdots \\ (u_n^\mu)^\dagger \end{bmatrix} = \begin{bmatrix} U_1^\mu \\ U_2^\mu \\ \vdots \\ U_n^\mu \end{bmatrix} = A^{-1} \begin{bmatrix} \mathrm{D}_1 U^\mu \\ \mathrm{D}_2 U^\mu \\ \vdots \\ \mathrm{D}_n U^\mu \end{bmatrix}, \tag{2.140}$$

where A^{-1} is the inverse (assumed to exist) of the matrix

$$A = \begin{bmatrix} \mathrm{D}_1 X_1 & \mathrm{D}_1 X_2 & \cdots & \mathrm{D}_1 X_n \\ \mathrm{D}_2 X_1 & \mathrm{D}_2 X_2 & \cdots & \mathrm{D}_2 X_n \\ \vdots & \vdots & & \vdots \\ \mathrm{D}_n X_1 & \mathrm{D}_n X_2 & \cdots & \mathrm{D}_n X_n \end{bmatrix}, \tag{2.141}$$

and the components $\{(u^\mu_{i_1 i_2\cdots i_k})^\dagger\}$ of $\underset{k}{u}^\dagger$ are determined by

$$\begin{bmatrix}(u^\mu_{i_1 i_2\cdots i_{k-1}1})^\dagger\\(u^\mu_{i_1 i_2\cdots i_{k-1}2})^\dagger\\ \vdots\\(u^\mu_{i_1 i_2\cdots i_{k-1}n})^\dagger\end{bmatrix}=\begin{bmatrix}U^\mu_{i_1 i_2\cdots i_{k-1}1}\\U^\mu_{i_1 i_2\cdots i_{k-1}2}\\ \vdots\\U^\mu_{i_1 i_2\cdots i_{k-1}n}\end{bmatrix}$$

$$= A^{-1}\begin{bmatrix}\mathrm{D}_1 U^\mu_{i_1 i_2\cdots i_{k-1}}\\\mathrm{D}_2 U^\mu_{i_1 i_2\cdots i_{k-1}}\\ \vdots\\\mathrm{D}_n U^\mu_{i_1 i_2\cdots i_{k-1}}\end{bmatrix}, \quad k=2,3,\ldots,n. \tag{2.142}$$

Now specialize transformation (2.138a,b) to a one-parameter Lie group of transformations

$$x^* = X(x,u;\epsilon), \tag{2.143a}$$

$$u^* = U(x,u;\epsilon). \tag{2.143b}$$

Here the kth extended transformation (2.139)–(2.142), with $\dagger$ replaced by $*$, is a one-parameter Lie group of transformations acting on $(x,u,\underset{1}{u},\ldots,\underset{k}{u})$-space. Then we have

$$x_i^* = X_i(x,u;\epsilon) = x_i + \epsilon\xi_i(x,u) + O(\epsilon^2), \tag{2.144a}$$

$$(u^\mu)^* = U^\mu(x,u;\epsilon) = u^\mu + \epsilon\eta^\mu(x,u) + O(\epsilon^2), \tag{2.144b}$$

$$(u_i^\mu)^* = U_i^\mu(x,u,\underset{1}{u};\epsilon) = u_i^\mu + \epsilon\eta_i^{(1)\mu}(x,u,\underset{1}{u}) + O(\epsilon^2), \tag{2.144c}$$

$$\vdots$$

$$\begin{aligned}(u^\mu_{i_1 i_2\cdots i_k})^* &= U^\mu_{i_1 i_2\cdots i_k}(x,u,\underset{1}{u},\ldots,\underset{k}{u};\epsilon)\\ &= u^\mu_{i_1 i_2\cdots i_k} + \epsilon\eta^{(k)\mu}_{i_1 i_2\cdots i_k}(x,u,\underset{1}{u},\ldots,\underset{k}{u}) + O(\epsilon^2),\end{aligned} \tag{2.144d}$$

with extended infinitesimals $\{\eta^{(k)\mu}_{i_1 i_2\cdots i_k}\}$ given by

$$\eta_i^{(1)\mu} = \mathrm{D}_i\eta^\mu - (\mathrm{D}_i\xi_j)u_j^\mu, \tag{2.145}$$

and

$$\eta^{(k)\mu}_{i_1 i_2\cdots i_k} = \mathrm{D}_{i_k}\eta^{(k-1)\mu}_{i_1 i_2\cdots i_{k-1}} - (\mathrm{D}_{i_k}\xi_j)u^\mu_{i_1 i_2\cdots i_{k-1}j}, \tag{2.146}$$

$i_\ell = 1,2,\ldots,n$ for $\ell = 1,2,\ldots,k$ with $k=2,3,\ldots$. Here the kth extended infinitesimal generator is

$$\mathrm{X}^{(k)} = \xi_i(x,u)\frac{\partial}{\partial x_i} + \eta^\mu(x,u)\frac{\partial}{\partial u^\mu} + \eta_i^{(1)\mu}(x,u,\underset{1}{u})\frac{\partial}{\partial u_i^\mu} + \cdots$$

$$+\ \eta^{(k)\mu}_{i_1 i_2 \cdots i_k}(x, u, \underset{1}{u}, \underset{2}{u}, \ldots, \underset{k}{u}) \frac{\partial}{\partial u^{\mu}_{i_1 i_2 \cdots i_k}}, \qquad k = 1, 2, \ldots. \tag{2.147}$$

Exercises 2.3

1. In Theorem 2.3.1-3 show that Y_k, $k \geq 2$, defined by (2.96d), is
   (a) linear in y_k;
   (b) a polynomial in $y_2, y_3, \ldots, y_k$ whose coefficients are functions of (x, y, y_1).
2. Prove Theorem 2.3.1-2.
3. For the rotation group (2.104a,b), determine $y_4^* = Y_4$,
   (a) using Theorem 2.3.1-3;
   (b) from its extended infinitesimals, i.e. using Theorem 2.3.2-1.
4. (a) Derive (2.111)–(2.113).
   (b) Determine $\eta^{(4)}$.
5. Prove Theorem 2.3.2-2.
6. For the group

$$x^* = x + \epsilon, \quad y^* = \frac{xy}{x + \epsilon}, \quad \text{with } y = y(x),$$

   determine
   (a) $\{\xi(x, y), \eta(x, y), \eta^{(1)}, \eta^{(2)}, \eta^{(3)}\}$;
   (b) $y_1^* = Y_1$, $y_2^* = Y_2$, $y_3^* = Y_3$.
7. Explain the geometrical significance of preserving the contact conditions (2.89a,b).
8. Show that each component of $\underset{k}{U}$, $k \geq 2$, defined by (2.125d), (2.126a,b), is linear in the components of $\underset{k}{u}$ and is a polynomial in the components of $\underset{2}{u}, \underset{3}{u}, \ldots, \underset{k}{u}$ whose coefficients are functions of the components of $x, u, \underset{1}{u}$.
9. State and prove the analog of Theorem 2.3.2-2 for the extended infinitesimals $\{\eta^{(k)}_{i_1 i_2 \cdots i_k}\}$ determined by Theorem 2.3.4-1.
10. Derive (2.133)–(2.137).
11. Derive (2.139a–d), (2.140)–(2.142).

12. Derive (2.145), (2.146).

13. For the following two examples, involving two independent variables (x,t) and one dependent variable $u = u(x,t)$, (arising from studying the group properties of the heat equation) determine (i) the extended infinitesimal generators $X^{(1)}$ and $X^{(2)}$ and (ii) the extended one-parameter Lie groups of transformations acting on $(x, u, \underset{1}{u})$-space and $(x, u, \underset{1}{u}, \underset{2}{u})$-space with

(a) $X = 2t\dfrac{\partial}{\partial x} - xu\dfrac{\partial}{\partial u}$;

(b) $X = 4xt\dfrac{\partial}{\partial x} + 4t^2\dfrac{\partial}{\partial t} - (x^2 + 2t)u\dfrac{\partial}{\partial u}$.

14. Consider the case of one independent variable x and one dependent variable $y = y(x)$. Assume that the transformation

$$x^\dagger = X(x,y), \tag{2.148a}$$

$$y^\dagger = Y(x,y), \tag{2.148b}$$

preserves the contact conditions and can be inverted so that

$$x = X^\dagger(x^\dagger, y^\dagger),$$

$$y = Y^\dagger(x^\dagger, y^\dagger),$$

where $(X^\dagger, Y^\dagger)$ are known explicitly as functions of $(x^\dagger, y^\dagger)$. Express y_1 and y_2 as functions of $(x^\dagger, y^\dagger, y_1^\dagger, y_2^\dagger)$. Show how this simplifies in the case when (2.148a,b) is a one-parameter Lie group of transformations. Illustrate for the rotation group (2.104a,b).

15. Consider the case of two independent variables (x,t) and one dependent variable $u = u(x,t)$. Assume that the transformation

$$x^\dagger = X(x,t,u), \tag{2.149a}$$

$$t^\dagger = T(x,t,u), \tag{2.149b}$$

$$u^\dagger = U(x,t,u), \tag{2.149c}$$

preserves the contact conditions and can be inverted so that

$$x = X^\dagger(x^\dagger, t^\dagger, u^\dagger),$$

$$t = T^\dagger(x^\dagger, t^\dagger, u^\dagger),$$

$$u = U^\dagger(x^\dagger, t^\dagger, u^\dagger),$$

where $(X^\dagger, T^\dagger, U^\dagger)$ are known explicitly as functions of $(x^\dagger, t^\dagger, u^\dagger)$. Express the components of $\underset{1}{u}$ and $\underset{2}{u}$ as functions of $(x^\dagger, t^\dagger, u^\dagger)$ and the components of $\underset{1}{u}^\dagger$ and $\underset{2}{u}^\dagger$. Show how this simplifies in the case when (2.149a–c) is a one-parameter Lie group of transformations. Illustrate for the one-parameter Lie group of transformations with infinitesimal generator $\mathrm{X} = 2t\frac{\partial}{\partial x} - xu\frac{\partial}{\partial u}$.

16. Consider one-parameter (ϵ) derivative-dependent transformations which preserve the contact conditions and are described formally in terms of infinitesimals $\{\hat{\eta}^\mu\}$ depending on $(x, u, \underset{1}{u}, \ldots, \underset{\ell}{u})$ for some finite ℓ, in particular of the form

$$x_i^* = x_i, \quad i = 1, 2, \ldots, n,$$

$$(u^\mu)^* = u^\mu + \epsilon\hat{\eta}^\mu(x, u, \underset{1}{u}, \ldots, \underset{\ell}{u}) + O(\epsilon^2), \quad \mu = 1, 2, \ldots, m.$$

(a) Show that the extended transformations are of the form

$$(u^\mu_{i_1 i_2 \cdots i_k})^* = u^\mu_{i_1 i_2 \cdots i_k} + \epsilon\hat{\eta}^{(k)\mu}_{i_1 i_2 \cdots i_k}(x, u, \underset{1}{u}, \ldots, \underset{\ell+k}{u}) + O(\epsilon^2)$$

where

$$\hat{\eta}_i^{(1)\mu} = \mathrm{D}_i\hat{\eta}^\mu$$

and

$$\hat{\eta}^{(k)\mu}_{i_1 i_2 \cdots i_k} = \mathrm{D}_{i_k}\hat{\eta}^{(k-1)\mu}_{i_1 i_2 \cdots i_{k-1}}, \quad k = 2, 3, \ldots .$$

(b) Specialize (a) to the case when

$$\hat{\eta}^\mu = \hat{\eta}^\mu(x, u, \underset{1}{u}) = \eta^\mu(x, u) - \sum_{j=1}^{n}\xi_j(x, u)u_j^\mu$$

for some functions $\{\eta^\mu(x, u), \xi_j(x, u)\}$, $\mu = 1, 2, \ldots, m$; $j = 1, 2, \ldots, n$. Show that here

$$\hat{\eta}_i^{(1)\mu} = \eta_i^{(1)\mu} - \sum_{j=1}^{n}\xi_j u^\mu_{ij}$$

and

$$\hat{\eta}^{(k)\mu}_{i_1 i_2 \cdots i_k} = \eta^{(k)\mu}_{i_1 i_2 \cdots i_k} - \sum_{j=1}^{n}\xi_j u^\mu_{i_1 i_2 \cdots i_k j},$$

where $\{\eta^{(k)\mu}_{i_1 i_2 \cdots i_k}\}$ are given by (2.145), (2.146).

17. For $X(x, u)$ defined by (2.114a) [(2.139a)] give criteria so that the corresponding matrix A defined by (2.119) [(2.141)] has an inverse.

2.4 Multi-Parameter Lie Groups of Transformations; Lie Algebras

So far in this chapter we have only considered one-parameter Lie groups of transformations. In the first chapter on dimensional analysis we encountered invariance under multi-parameter families of scalings. These are examples of multi-parameter Lie groups of transformations. In this section we summarize some key results pertaining to multi-parameter Lie groups of transformations. We assume a finite number r of parameters but we will encounter examples of infinite-parameter Lie groups in later chapters.

Each parameter of an r-parameter Lie group of transformations leads to an infinitesimal generator. The infinitesimal generators belong to an r-dimensional linear vector space on which there is an additional structure, called the *commutator.* This special vector space is called a *Lie algebra* (*r-dimensional Lie algebra*).

For our purposes the study of an r-parameter Lie group of transformations is equivalent to the study of its infinitesimal generators and the structure of the corresponding Lie algebra. The exponentiation of any infinitesimal generator is a one-parameter Lie group of transformations which is a subgroup of the r-parameter Lie group of transformations. Most importantly the discovery of multi-parameter Lie groups of transformations admitted by differential equations requires one to consider only invariance under one-parameter Lie groups of transformations.

Special Lie algebras called *solvable Lie algebras* play an important role in later chapters, especially in the study of invariance of ordinary differential equations of at least third order under multi-parameter Lie groups of transformations.

For details of the material of this section, including more precise statements of the key results and their proofs, the reader is referred to the books of Cohn (1965), Eisenhart (1933), Gilmore (1974), Olver (1986) and Ovsiannikov (1962, 1982).

2.4.1 r-Parameter Lie Groups of Transformations

For an *r-parameter Lie group of transformations,*

$$\mathbf{x}^* = \mathbf{X}(\mathbf{x};\boldsymbol{\epsilon}), \tag{2.150}$$

let $\mathbf{x} = (x_1, x_2 \ldots, x_n)$, and let the parameters be denoted by $\boldsymbol{\epsilon} = (\epsilon_1, \epsilon_2, \ldots, \epsilon_r)$. Let the law of composition of parameters be denoted by

$$\boldsymbol{\phi}(\boldsymbol{\epsilon},\boldsymbol{\delta}) = (\phi_1(\boldsymbol{\epsilon},\boldsymbol{\delta}), \phi_2(\boldsymbol{\epsilon},\boldsymbol{\delta}), \ldots, \phi_r(\boldsymbol{\epsilon},\boldsymbol{\delta}))$$

where $\boldsymbol{\delta} = (\delta_1, \delta_2, \ldots, \delta_r)$; $\boldsymbol{\phi}(\boldsymbol{\epsilon},\boldsymbol{\delta})$ satisfies the group axioms with $\boldsymbol{\epsilon} = 0$ corresponding to the identity $\epsilon_1 = \epsilon_2 = \cdots = \epsilon_r = 0$; $\boldsymbol{\phi}(\boldsymbol{\epsilon},\boldsymbol{\delta})$ is assumed to be analytic in its domain of definition.

The *infinitesimal matrix* $\Xi(\mathbf{x})$ is the $r \times n$ matrix with entries

$$\xi_{\alpha j}(\mathbf{x}) = \left.\frac{\partial x_j^*}{\partial \epsilon_\alpha}\right|_{\epsilon=0} = \left.\frac{\partial X_j(\mathbf{x};\epsilon)}{\partial \epsilon_\alpha}\right|_{\epsilon=0}, \tag{2.151}$$

$\alpha = 1, 2, \ldots, r$; $j = 1, 2, \ldots, n$. Let $\Theta(\epsilon)$ be the $r \times r$ matrix with entries

$$\Theta_{\alpha\beta}(\epsilon) = \left.\frac{\partial \phi_\beta(\epsilon,\delta)}{\partial \delta_\alpha}\right|_{\delta=0}, \tag{2.152}$$

and let

$$\Psi(\epsilon) = \Theta^{-1}(\epsilon), \tag{2.153}$$

the inverse of matrix $\Theta(\epsilon)$. Then *Lie's First Fundamental Theorem* for an r-parameter Lie group of transformations states that essentially (2.150) is equivalent to the solution of the IVP for the system of nr first order partial differential equations (in some neighborhood of $\epsilon = 0$):

$$\begin{bmatrix} \frac{\partial x_1^*}{\partial \epsilon_1} & \frac{\partial x_2^*}{\partial \epsilon_1} & \cdots & \frac{\partial x_n^*}{\partial \epsilon_1} \\ \frac{\partial x_1^*}{\partial \epsilon_2} & \frac{\partial x_2^*}{\partial \epsilon_2} & \cdots & \frac{\partial x_n^*}{\partial \epsilon_2} \\ \vdots & \vdots & \cdots & \vdots \\ \frac{\partial x_1^*}{\partial \epsilon_r} & \frac{\partial x_2^*}{\partial \epsilon_r} & \cdots & \frac{\partial x_n^*}{\partial \epsilon_r} \end{bmatrix} = \Psi(\epsilon)\Xi(\mathbf{x}^*), \tag{2.154a}$$

with

$$\mathbf{x}^* = \mathbf{x} \quad \text{at} \quad \epsilon = 0. \tag{2.154b}$$

Definition 2.4.1-1. The *infinitesimal generator* X_α, corresponding to the parameter ϵ_α of the r-parameter Lie group of transformations (2.150), is

$$\mathrm{X}_\alpha = \sum_{j=1}^{n} \xi_{\alpha j}(\mathbf{x}) \frac{\partial}{\partial x_j}, \quad \alpha = 1, 2, \ldots, r. \tag{2.155}$$

One can show that the r-parameter Lie group of transformations (2.150) is essentially equivalent to both

(i)
$$\mathbf{x}^* = e^{\epsilon \sum_{\alpha=1}^{r} \lambda_\alpha \mathrm{X}_\alpha} \mathbf{x}, \tag{2.156}$$

where $\epsilon, \lambda_1, \lambda_2, \ldots, \lambda_r$ are arbitrary real constants representing the r parameters; and

(ii)
$$\begin{aligned} \mathbf{x}^* &= \textstyle\prod_{\alpha=1}^{r} e^{\mu_\alpha \mathrm{X}_\alpha} \mathbf{x}, \\ &= e^{\mu_1 \mathrm{X}_1} e^{\mu_2 \mathrm{X}_2} \cdots e^{\mu_r \mathrm{X}_r} \mathbf{x}, \end{aligned} \tag{2.157}$$

where $\mu_1, \mu_2, \ldots, \mu_r$ are arbitrary real constants. [The order of the operations in (2.157) can be rearranged by renumbering the infinitesimal generators even though it is not necessarily true that

$$e^{\mu_\alpha X_\alpha} e^{\mu_\beta X_\beta} = e^{\mu_\beta X_\beta} e^{\mu_\alpha X_\alpha} \quad \text{for} \quad \alpha \neq \beta.]$$

One can also show that the one-parameter (ϵ) Lie group of transformations

$$\mathbf{x}^* = e^{\epsilon X}\mathbf{x} = e^{\epsilon \sum_{\alpha=1}^{r} \sigma_\alpha X_\alpha}\mathbf{x}, \tag{2.158}$$

obtained by exponentiating the infinitesimal generator

$$X = \sum_{\alpha=1}^{r} \sigma_\alpha X_\alpha = \sum_{j=1}^{n} \zeta_j(\mathbf{x}) \frac{\partial}{\partial x_j}, \tag{2.159}$$

where $\sigma_1, \sigma_2, \ldots, \sigma_r$ are any fixed real constants and

$$\zeta_j(\mathbf{x}) = \sum_{\alpha=1}^{r} \sigma_\alpha \xi_{\alpha j}(\mathbf{x}), \quad j = 1, 2, \ldots, n, \tag{2.160}$$

defines a one-parameter (ϵ) subgroup of the r-parameter Lie group of transformations (2.150).

As an example consider the two-parameter $\boldsymbol{\epsilon} = (\epsilon_1, \epsilon_2)$ Lie group of transformations $[(x_1, x_2) = (x, y)]$ given by

$$x^* = e^{\epsilon_1} x + \epsilon_2, \tag{2.161a}$$

$$y^* = e^{2\epsilon_1} y. \tag{2.161b}$$

Then

$$\begin{aligned} x^{**} &= e^{\delta_1} x^* + \delta_2 = e^{\phi_1(\boldsymbol{\epsilon}, \boldsymbol{\delta})} x + \phi_2(\boldsymbol{\epsilon}, \boldsymbol{\delta}), \\ y^{**} &= e^{2\delta_1} y^* = e^{2\phi_1(\boldsymbol{\epsilon}, \boldsymbol{\delta})} y, \end{aligned}$$

with law of composition given by

$$\phi(\boldsymbol{\epsilon}, \boldsymbol{\delta}) = (\phi_1(\boldsymbol{\epsilon}, \boldsymbol{\delta}), \phi_2(\boldsymbol{\epsilon}, \boldsymbol{\delta})) = (\epsilon_1 + \delta_1, e^{\delta_1}\epsilon_2 + \delta_2). \tag{2.162}$$

One can easily check that the two-parameter family of transformations (2.161a,b) with law of composition (2.162) defines a two-parameter Lie group of transformations with $x^* = x$, $y^* = y$, when $\boldsymbol{\epsilon} = (\epsilon_1, \epsilon_2) = 0$.

We now check that (2.154a,b) holds:

$$\frac{\partial x^*}{\partial \epsilon_1} = e^{\epsilon_1} x = x^* - \epsilon_2; \quad \frac{\partial y^*}{\partial \epsilon_1} = 2e^{2\epsilon_1} y = 2y^*;$$

$$\frac{\partial x^*}{\partial \epsilon_2} = 1; \quad \frac{\partial y^*}{\partial \epsilon_2} = 0.$$

Hence

$$\begin{bmatrix} \dfrac{\partial x^*}{\partial \epsilon_1} & \dfrac{\partial y^*}{\partial \epsilon_1} \\ \dfrac{\partial x^*}{\partial \epsilon_2} & \dfrac{\partial y^*}{\partial \epsilon_2} \end{bmatrix} = \begin{bmatrix} x^* - \epsilon_2 & 2y^* \\ 1 & 0 \end{bmatrix}. \tag{2.163}$$

Then

$$\xi_{11}(\mathbf{x}) = \left.\frac{\partial x^*}{\partial \epsilon_1}\right|_{\epsilon=0} = x; \quad \xi_{12}(\mathbf{x}) = \left.\frac{\partial y^*}{\partial \epsilon_1}\right|_{\epsilon=0} = 2y;$$

$$\xi_{21}(\mathbf{x}) = \left.\frac{\partial x^*}{\partial \epsilon_2}\right|_{\epsilon=0} = 1; \quad \xi_{22}(\mathbf{x}) = \left.\frac{\partial y^*}{\partial \epsilon_2}\right|_{\epsilon=0} = 0.$$

Consequently the infinitesimal matrix is given by

$$\Xi(\mathbf{x}) = \begin{bmatrix} x & 2y \\ 1 & 0 \end{bmatrix}. \tag{2.164}$$

To determine $\psi(\epsilon)$, we have

$$\frac{\partial \phi_1}{\partial \delta_1} = 1; \quad \frac{\partial \phi_2}{\partial \delta_1} = e^{\delta_1}\epsilon_2; \quad \frac{\partial \phi_1}{\partial \delta_2} = 0; \quad \frac{\partial \phi_2}{\partial \delta_2} = 1,$$

and hence

$$\Theta_{11} = \left.\frac{\partial \phi_1}{\partial \delta_1}\right|_{\delta=0} = 1; \quad \Theta_{12} = \left.\frac{\partial \phi_2}{\partial \delta_1}\right|_{\delta=0} = \epsilon_2;$$

$$\Theta_{21} = \left.\frac{\partial \phi_1}{\partial \delta_2}\right|_{\delta=0} = 0; \quad \Theta_{22} = \left.\frac{\partial \phi_2}{\partial \delta_2}\right|_{\delta=0} = 1.$$

Thus we get

$$\Theta(\epsilon) = \begin{bmatrix} 1 & \epsilon_2 \\ 0 & 1 \end{bmatrix}, \tag{2.165}$$

and

$$\Psi(\epsilon) = \Theta^{-1}(\epsilon) = \begin{bmatrix} 1 & -\epsilon_2 \\ 0 & 1 \end{bmatrix}. \tag{2.166}$$

Then it is easily seen that

$$\Psi(\epsilon)\Xi(\mathbf{x}^*) = \begin{bmatrix} x^* - \epsilon_2 & 2y^* \\ 1 & 0 \end{bmatrix},$$

which is matrix (2.163), checking out (2.154a,b). It is left to Exercise 2.4-1 to solve the IVP for the system of partial differential equations

$$\frac{\partial x^*}{\partial \epsilon_1} = x^* - \epsilon_2, \tag{2.167a}$$

$$\frac{\partial y^*}{\partial \epsilon_1} = 2y^*, \tag{2.167b}$$

$$\frac{\partial x^*}{\partial \epsilon_2} = 1, \tag{2.167c}$$

$$\frac{\partial y^*}{\partial \epsilon_2} = 0, \tag{2.167d}$$

with

$$x^* = x, \ y^* = y, \ \text{when} \quad \epsilon_1 = 0, \ \ \epsilon_2 = 0, \tag{2.167e}$$

to recover (2.161a,b).

For the two-parameter Lie group of transformations (2.161a,b) the corresponding infinitesimal generators are

$$\mathrm{X}_1 = x\frac{\partial}{\partial x} + 2y\frac{\partial}{\partial y}, \tag{2.168a}$$

$$\mathrm{X}_2 = \frac{\partial}{\partial x}. \tag{2.168b}$$

For any differentiable function $F(x, y)$ we have:

$$e^{\epsilon \mathrm{X}_1} F(x,y) = F(e^{\epsilon}x, e^{2\epsilon}y), \tag{2.169a}$$

$$e^{\epsilon \mathrm{X}_2} F(x,y) = F(x+\epsilon, y). \tag{2.169b}$$

We now check that the representations (2.157) and (2.156) lead to (2.161a,b).

From (2.169a,b) it follows that for any real constants μ_1, μ_2:

$$e^{\mu_1 \mathrm{X}_1} e^{\mu_2 \mathrm{X}_2}(x,y) = e^{\mu_1 \mathrm{X}_1}(x+\mu_2, y) = (e^{\mu_1}x + \mu_2, e^{2\mu_1}y), \tag{2.170}$$

and

$$e^{\mu_2 \mathrm{X}_2} e^{\mu_1 \mathrm{X}_1}(x,y) = e^{\mu_2 \mathrm{X}_2}(e^{\mu_1}x, e^{2\mu_1}y) = (e^{\mu_1}(x+\mu_2), e^{2\mu_1}y). \tag{2.171}$$

Let $\tilde{x} = \lambda_1 x + \lambda_2$. Then

$$e^{\epsilon(\lambda_1 \mathrm{X}_1 + \lambda_2 \mathrm{X}_2)}(x,y) = e^{\epsilon\lambda_1 \tilde{x}(\partial/\partial\tilde{x}) + 2\epsilon\lambda_1 y(\partial/\partial y)}\left(\frac{\tilde{x}-\lambda_2}{\lambda_1}, y\right)$$

$$= \left(\frac{e^{\epsilon\lambda_1}\tilde{x}-\lambda_2}{\lambda_1}, e^{2\epsilon\lambda_1}y\right) = \left(e^{\epsilon\lambda_1}x + \lambda_2\left[\frac{e^{\epsilon\lambda_1}-1}{\lambda_1}\right], e^{2\epsilon\lambda_1}y\right). \tag{2.172}$$

Thus (2.170) is identical to (2.161a,b), with the same law of composition (2.162); (2.171) is equivalent to (2.161a,b) with law of composition $\phi(\epsilon, \delta) = (\epsilon_1 + \delta_1, \epsilon_2 + e^{-\epsilon_1}\delta_2)$; (2.172) is equivalent to (2.161a,b) with law of composition

$$\phi(\epsilon,\delta) = \left(\epsilon_1+\delta_1, \frac{\epsilon_1+\delta_1}{e^{\epsilon_1+\delta_1}-1}\left\{e^{\delta_1}\left[\frac{\epsilon_2}{\epsilon_1}(e^{\epsilon_1}-1)+\frac{\delta_2}{\delta_1}\right]-\frac{\delta_2}{\delta_1}\right\}\right).$$

2.4.2 Lie Algebras

Definition 2.4.2-1. Consider an r-parameter Lie group of transformations (2.150) with infinitesimal generators $\{\mathrm{X}_\alpha\}$, $\alpha = 1, 2, \ldots, r$, defined by (2.151) and (2.155). The *commutator* of X_α and X_β is another first order operator

$$[\mathrm{X}_\alpha, \mathrm{X}_\beta] = \mathrm{X}_\alpha \mathrm{X}_\beta - \mathrm{X}_\beta \mathrm{X}_\alpha = \sum_{i,j=1}^{n} \left[\left(\xi_{\alpha i}(\mathbf{x}) \frac{\partial}{\partial x_i} \right) \left(\xi_{\beta j}(\mathbf{x}) \frac{\partial}{\partial x_j} \right) \right.$$

$$\left. - \left(\xi_{\beta i}(\mathbf{x}) \frac{\partial}{\partial x_i} \right) \left(\xi_{\alpha j}(\mathbf{x}) \frac{\partial}{\partial x_j} \right) \right] = \sum_{j=1}^{n} \eta_j(\mathbf{x}) \frac{\partial}{\partial x_j}, \tag{2.173a}$$

where

$$\eta_j(\mathbf{x}) = \sum_{i=1}^{n} \left[\xi_{\alpha i}(\mathbf{x}) \frac{\partial \xi_{\beta j}(\mathbf{x})}{\partial x_i} - \xi_{\beta i}(\mathbf{x}) \frac{\partial \xi_{\alpha j}(\mathbf{x})}{\partial x_i} \right]. \tag{2.173b}$$

It immediately follows that

$$[\mathrm{X}_\alpha, \mathrm{X}_\beta] = -[\mathrm{X}_\beta, \mathrm{X}_\alpha]. \tag{2.174}$$

Theorem 2.4.2-1 (Second Fundamental Theorem of Lie). *The commutator of any two infinitesimal generators of an r-parameter Lie group of transformations is also an infinitesimal generator, in particular*

$$[\mathrm{X}_\alpha, \mathrm{X}_\beta] = C^{\gamma}_{\alpha\beta} \mathrm{X}_\gamma, \tag{2.175}$$

where the coefficients $C^{\gamma}_{\alpha\beta}$ *are constants called* **structure constants**, $\alpha, \beta, \gamma = 1, 2, \ldots, r$. [*In* (2.175) *we assume the usual convention of summation over a repeated index.*]

Proof. The proof of this theorem essentially depends on the integrability conditions

$$\frac{\partial^2 x_i^*}{\partial \epsilon_\alpha \partial \epsilon_\beta} = \frac{\partial^2 x_i^*}{\partial \epsilon_\beta \partial \epsilon_\alpha}, \quad i = 1, 2, \ldots, n; \quad \alpha, \beta = 1, 2, \ldots, r, \tag{2.176}$$

applied to equations (2.154a). For complete details see any of the earlier mentioned references of this section. □

Definition 2.4.2-2. Equations (2.175) are called the *commutation relations* of the r-parameter Lie group of transformations (2.150) with infinitesimal generators (2.155).

For any three infinitesimal generators X_α, X_β, X_γ, one can show by direct computation that *Jacobi's identity* holds:

$$[\mathrm{X}_\alpha, [\mathrm{X}_\beta, \mathrm{X}_\gamma]] + [\mathrm{X}_\beta, [\mathrm{X}_\gamma, \mathrm{X}_\alpha]] + [\mathrm{X}_\gamma, [\mathrm{X}_\alpha, \mathrm{X}_\beta]] = 0. \tag{2.177}$$

From (2.174), (2.175), and (2.177), the following theorem relating the structure constants is easily proved:

Theorem 2.4.2-2 (Third Fundamental Theorem of Lie). *The structure constants, defined by the commutation relations* (2.175), *satisfy the relations*

$$C^{\gamma}_{\alpha\beta} = -C^{\gamma}_{\beta\alpha}, \tag{2.178a}$$

$$C^{\rho}_{\alpha\beta}\, C^{\delta}_{\rho\gamma} + C^{\rho}_{\beta\gamma}\, C^{\delta}_{\rho\alpha} + C^{\rho}_{\gamma\alpha}\, C^{\delta}_{\rho\beta} = 0. \tag{2.178b}$$

Definition 2.4.2-3. A *Lie algebra* $\mathcal{L}$ is a vector space over some field $\mathcal{F}$ with an additional law of combination of elements in $\mathcal{L}$ (*the commutator*) satisfying the properties (2.174) and (2.177) with, most importantly, closure with respect to commutation. In particular the infinitesimal generators $\{X_\alpha\}$, $\alpha = 1, 2, \ldots, r$, of an r-parameter Lie group of transformations (2.150) form an r-dimensional Lie algebra $\mathcal{L}^r$ over (the field) $\mathbb{R}$ since for any $X_\alpha, X_\beta, X_\gamma \in \mathcal{L}^r$, $a, b \in \mathbb{R}$:

(i) $aX_\alpha + bX_\beta \in \mathcal{L}^r$;

(ii) $X_\alpha + X_\beta = X_\beta + X_\alpha$;

(iii) $X_\alpha + (X_\beta + X_\gamma) = (X_\alpha + X_\beta) + X_\gamma$;

(iv) $[X_\alpha, X_\beta] \in \mathcal{L}^r$;

(v) $[X_\alpha, X_\beta] = -[X_\beta, X_\alpha]$;

(vi) $[X_\alpha, [X_\beta, X_\gamma]] + [X_\beta, [X_\gamma, X_\alpha]] + [X_\gamma, [X_\alpha, X_\beta]] = 0$;

(vii) $[aX_\alpha + bX_\beta, X_\gamma] = a[X_\alpha, X_\gamma] + b[X_\beta, X_\gamma]$.

One can motivate the existence of the commutator $[X_\alpha, X_\beta]$ in $\mathcal{L}^r$ by the following argument:

Let G^r denote the r-parameter Lie group of transformations (2.150). Any one-parameter (ϵ) subgroup of G^r has a corresponding infinitesimal generator in $\mathcal{L}^r$. For example $X_\alpha \in \mathcal{L}^r$ corresponds to $e^{\epsilon X_\alpha}\mathbf{x} \in G^r$, $\alpha = 1, 2, \ldots, r$; $aX_\alpha + bX_\beta \in \mathcal{L}^r$ corresponds to both $e^{\epsilon(aX_\alpha + bX_\beta)}\mathbf{x} \in G^r$ and $e^{\epsilon a X_\alpha} e^{\epsilon b X_\beta}\mathbf{x} \in G^r$. If $X_\alpha, X_\beta \in \mathcal{L}^r$, then $e^{\epsilon X_\alpha}\mathbf{x} \in G^r$ and $e^{\delta X_\beta}\mathbf{x} \in G^r$ for any $\epsilon, \delta \in \mathbb{R}$. It follows that the one-parameter (ϵ) commutator group transformations

$$e^{-\epsilon X_\alpha} e^{-\epsilon X_\beta} e^{\epsilon X_\alpha} e^{\epsilon X_\beta}\mathbf{x} = \left[e^{\epsilon X_\alpha}\right]^{-1} \left[e^{\epsilon X_\beta}\right]^{-1} e^{\epsilon X_\alpha} e^{\epsilon X_\beta}\mathbf{x} \in G^r.$$

But

$$
\begin{aligned}
&e^{-\epsilon X_\alpha} e^{-\epsilon X_\beta} e^{\epsilon X_\alpha} e^{\epsilon X_\beta} \\
&= \left(1-\epsilon X_\alpha + \frac{\epsilon^2}{2}(X_\alpha)^2\right)\left(1-\epsilon X_\beta + \frac{\epsilon^2}{2}(X_\beta)^2\right) \\
&\cdot \left(1+\epsilon X_\alpha + \frac{\epsilon^2}{2}(X_\alpha)^2\right)\left(1+\epsilon X_\beta + \frac{\epsilon^2}{2}(X_\beta)^2\right) + O(\epsilon^3) \\
&= \left(1-\epsilon(X_\alpha + X_\beta) + \epsilon^2(X_\alpha X_\beta + \frac{1}{2}(X_\alpha)^2 + \frac{1}{2}(X_\beta)^2)\right) \\
&\cdot \left(1+\epsilon(X_\alpha + X_\beta) + \epsilon^2(X_\alpha X_\beta + \frac{1}{2}(X_\alpha)^2 + \frac{1}{2}(X_\beta)^2)\right) + O(\epsilon^3) \\
&= 1 + \epsilon^2(2X_\alpha X_\beta + (X_\alpha)^2 + (X_\beta)^2 - (X_\alpha + X_\beta)^2) + O(\epsilon^3) \\
&= 1 + \epsilon^2(2X_\alpha X_\beta - X_\alpha X_\beta - X_\beta X_\alpha) + O(\epsilon^3) \\
&= 1 + \epsilon^2(X_\alpha X_\beta - X_\beta X_\alpha) + O(\epsilon^3) \\
&= 1 + \epsilon^2[X_\alpha, X_\beta] + O(\epsilon^3).
\end{aligned}
$$

Hence $[X_\alpha, X_\beta] \in \mathcal{L}^r$.

One can show that $e^{\epsilon X_\alpha} e^{\delta X_\beta} = e^{\delta X_\beta} e^{\epsilon X_\alpha} = e^{\epsilon X_\alpha + \delta X_\beta}$ if and only if $[X_\alpha, X_\beta] = 0$ [see Exercise 2.4-10].

Theorem 2.4.2-3. *Let* $X_\alpha^{(k)}$, $X_\beta^{(k)}$ *be the kth extended infinitesimal generators of the infinitesimal generators* X_α, X_β *and let* $[X_\alpha, X_\beta]^{(k)}$ *be the kth extended infinitesimal generator of the commutator* $[X_\alpha, X_\beta]$. *Then* $[X_\alpha, X_\beta]^{(k)} = [X_\alpha^{(k)}, X_\beta^{(k)}]$, $k = 1, 2, \ldots$. *Hence if* $[X_\alpha, X_\beta] = X_\gamma$, *then* $[X_\alpha^{(k)}, X_\beta^{(k)}] = X_\gamma^{(k)}$, $k = 1, 2, \ldots$.

Proof. See Exercise 2.4-11 [cf. Ovsiannikov (1962, 1982), Olver (1986)]. □

Definition 2.4.2-4. A subspace $\mathcal{J} \subset \mathcal{L}$ is called a *subalgebra* of the Lie algebra $\mathcal{L}$ if for any $X_\alpha, X_\beta \in \mathcal{J}$, $[X_\alpha, X_\beta] \in \mathcal{J}$.

2.4.3 Examples of Lie Algebras

(1) *Eight-Parameter Lie Group of Projective Transformations in* $\mathbb{R}^2$

Projective transformations in $\mathbb{R}^2$ map straight lines into straight lines. In particular they are defined by the eight-parameter Lie group of transformations

$$x^* = \frac{(1+\epsilon_3)x + \epsilon_4 y + \epsilon_5}{\epsilon_1 x + \epsilon_2 y + 1}, \tag{2.179a}$$

$$y^* = \frac{\epsilon_6 x + (1+\epsilon_7)y + \epsilon_8}{\epsilon_1 x + \epsilon_2 y + 1}, \tag{2.179b}$$

where the parameters $\epsilon_\alpha \in \mathbb{R}$, $\alpha = 1, 2, \ldots, 8$. The infinitesimal generators of the corresponding Lie algebra $\mathcal{L}^8$ are

$$\mathrm{X}_1 = x^2\frac{\partial}{\partial x} + xy\frac{\partial}{\partial y}, \quad \mathrm{X}_2 = xy\frac{\partial}{\partial x} + y^2\frac{\partial}{\partial y}, \quad \mathrm{X}_3 = x\frac{\partial}{\partial x}, \quad \mathrm{X}_4 = y\frac{\partial}{\partial x},$$

$$\mathrm{X}_5 = \frac{\partial}{\partial x}, \quad \mathrm{X}_6 = x\frac{\partial}{\partial y}, \quad \mathrm{X}_7 = y\frac{\partial}{\partial y}, \quad \mathrm{X}_8 = \frac{\partial}{\partial y}. \tag{2.180}$$

It is convenient to display the commutators of a Lie algebra through its *commutator table* whose (i, j)-th entry is $[\mathrm{X}_i, \mathrm{X}_j]$. From (2.174) it follows that the table is antisymmetric with its diagonal elements all zero. The structure constants are easily read off from the commutator table.

For the infinitesimal generators (2.180) we have the following commutator table:

	X_1	X_2	X_3	X_4
X_1	0	0	$-\mathrm{X}_1$	$-\mathrm{X}_2$
X_2	0	0	0	0
X_3	X_1	0	0	$-\mathrm{X}_4$
X_4	X_2	0	X_4	0
X_5	$2\mathrm{X}_3 + \mathrm{X}_7$	X_4	X_5	0
X_6	0	X_1	$-\mathrm{X}_6$	$\mathrm{X}_3 - \mathrm{X}_7$
X_7	0	X_2	0	X_4
X_8	X_6	$\mathrm{X}_3 + 2\mathrm{X}_7$	0	X_5

	X_5	X_6	X_7	X_8
X_1	$-2\mathrm{X}_3 - \mathrm{X}_7$	0	0	$-\mathrm{X}_6$
X_2	$-\mathrm{X}_4$	$-\mathrm{X}_1$	$-\mathrm{X}_2$	$-\mathrm{X}_3 - 2\mathrm{X}_7$
X_3	$-\mathrm{X}_5$	X_6	0	0
X_4	0	$\mathrm{X}_7 - \mathrm{X}_3$	$-\mathrm{X}_4$	$-\mathrm{X}_5$
X_5	0	X_8	0	0
X_6	$-\mathrm{X}_8$	0	X_6	0
X_7	0	$-\mathrm{X}_6$	0	$-\mathrm{X}_8$
X_8	0	0	X_8	0

(2) *Group of Rigid Motions in* $\mathbb{R}^2$

The group of rigid motions in $\mathbb{R}^2$ preserves distances between any two points in $\mathbb{R}^2$. It is the three-parameter Lie group of transformations of rotations and translations in $\mathbb{R}^2$ given by

$$x^* = x\cos\epsilon_1 - y\sin\epsilon_1 + \epsilon_2, \tag{2.181a}$$

$$y^* = x\sin\epsilon_1 + y\cos\epsilon_1 + \epsilon_3. \tag{2.181b}$$

The corresponding infinitesimal generators are

$$X_1 = -y\frac{\partial}{\partial x} + x\frac{\partial}{\partial y}, \quad X_2 = \frac{\partial}{\partial x}, \quad X_3 = \frac{\partial}{\partial y}. \tag{2.182}$$

The commutator table of its Lie algebra follows:

	X_1	X_2	X_3
X_1	0	$-X_3$	X_2
X_2	X_3	0	0
X_3	$-X_2$	0	0

(3) *Similitude Group in* $\mathbb{R}^2$

The similitude group in $\mathbb{R}^2$ consists of uniform scalings and rigid motions in $\mathbb{R}^2$. It is the four-parameter Lie group of transformations given by

$$x^* = e^{\epsilon_4}(x\cos\epsilon_1 - y\sin\epsilon_1) + \epsilon_2, \tag{2.183a}$$

$$y^* = e^{\epsilon_4}(x\sin\epsilon_1 + y\cos\epsilon_1) + \epsilon_3. \tag{2.183b}$$

The infinitesimal generators are X_1, X_2, X_3 given by (2.182) and

$$X_4 = x\frac{\partial}{\partial x} + y\frac{\partial}{\partial y}. \tag{2.184}$$

The corresponding commutator table is:

	X_1	X_2	X_3	X_4
X_1	0	$-X_3$	X_2	0
X_2	X_3	0	0	X_2
X_3	$-X_2$	0	0	X_3
X_4	0	$-X_2$	$-X_3$	0

The group of rigid motions in $\mathbb{R}^2$ [(2.181a,b)] is a three-parameter subgroup of the similitude group in $\mathbb{R}^2$ [(2.183a,b)]. This also follows from seeing that the Lie algebra with infinitesimal generators (2.182) is a three-dimensional subalgebra of the four-dimensional Lie algebra with infinitesimal generators (2.182) and (2.184).

By comparing the infinitesimal generators of the Lie algebra for the projective group (2.179a,b) and those for the similitude group (2.183a,b), one can see that the similitude group is a four-parameter subgroup of the eight-parameter projective group.

The commutator table can be most useful as an aid for finding additional symmetries. For example $[X_1, X_2] = -X_3$ for the infinitesimal generators (2.182) tells us that in $\mathbb{R}^2$, if a problem has rotational symmetry ($X_1 = -y\frac{\partial}{\partial x} + x\frac{\partial}{\partial y}$) and translational symmetry in the x-direction ($X_2 = \frac{\partial}{\partial x}$) then it also has translational symmetry in the y-direction ($X_3 = \frac{\partial}{\partial y}$).

2.4.4 Solvable Lie Algebras

In the next chapter we will consider nth order ordinary differential equations admitting r-parameter Lie groups of transformations. We will show that if $r = 1$ then the order can be reduced constructively by one; if $n \geq 2$ and $r = 2$ the order can be reduced constructively by two; if $n \geq 3$ and $r \geq 3$ it will not necessarily follow that the order can be reduced by more than two. However if the r-dimensional Lie algebra of infinitesimal generators of the admitted r-parameter group has a q-dimensional *solvable subalgebra* then the order of the differential equation can be reduced constructively by q.

Definition 2.4.4-1. A subalgebra $\mathcal{J} \subset \mathcal{L}$ is called an *ideal* or *normal subalgebra* of $\mathcal{L}$ if for any $\mathrm{X} \in \mathcal{J}$, $\mathrm{Y} \in \mathcal{L}$, $[\mathrm{X}, \mathrm{Y}] \in \mathcal{J}$.

Definition 2.4.4-2. $\mathcal{L}^q$ is a q-dimensional *solvable Lie algebra* if there exists a chain of subalgebras

$$\mathcal{L}^{(1)} \subset \mathcal{L}^{(2)} \subset \cdots \subset \mathcal{L}^{(q-1)} \subset \mathcal{L}^{(q)} = \mathcal{L}^q, \tag{2.185}$$

such that $\mathcal{L}^{(k)}$ is a k-dimensional Lie algebra and $\mathcal{L}^{(k-1)}$ is an ideal of $\mathcal{L}^{(k)}$, $k = 1, 2, \ldots, q$. [$\mathcal{L}^{(0)}$ is the null ideal which has no nonzero vectors.]

Definition 2.4.4-3. $\mathcal{L}$ is called an *Abelian Lie algebra* if for any $\mathrm{X}_\alpha, \mathrm{X}_\beta \in \mathcal{L}$, $[\mathrm{X}_\alpha, \mathrm{X}_\beta] = 0$.

The proof of the following theorem is obvious and left to Exercise 2.4-12:

Theorem 2.4.4-1. *Every Abelian Lie algebra is a solvable Lie algebra.*

The following theorem holds for any two-dimensional Lie algebra:

Theorem 2.4.4-2. *Every two-dimensional Lie algebra is solvable.*

Proof. Let $\mathcal{L}$ be a two-dimensional Lie algebra with infinitesimal generators X_1 and X_2 as basis vectors. Suppose $[\mathrm{X}_1, \mathrm{X}_2] = a\mathrm{X}_1 + b\mathrm{X}_2 = \mathrm{Y}$. If $c_1\mathrm{X}_1 + c_2\mathrm{X}_2 \in \mathcal{L}$, then

$$\begin{aligned} [\mathrm{Y}, c_1\mathrm{X}_1 + c_2\mathrm{X}_2] &= c_1[\mathrm{Y}, \mathrm{X}_1] + c_2[\mathrm{Y}, \mathrm{X}_2] \\ &= c_1 b[\mathrm{X}_2, \mathrm{X}_1] + c_2 a[\mathrm{X}_1, \mathrm{X}_2] \\ &= (c_2 a - c_1 b)\mathrm{Y}. \end{aligned}$$

Hence Y is a one-dimensional ideal of $\mathcal{L}$. □

It turns out that a three-dimensional Lie algebra is not necessarily solvable. For example the three-dimensional Lie algebra with infinitesimal generators

$$\mathrm{X}_1 = \frac{\partial}{\partial x}, \quad \mathrm{X}_2 = x\frac{\partial}{\partial x}, \quad \mathrm{X}_3 = x^2\frac{\partial}{\partial x}, \tag{2.186}$$

is not solvable.

As an example of a solvable three-dimensional Lie algebra consider the Lie algebra for the group of rigid motions (2.181a,b). The solvability of its Lie algebra follows from the chain

$$\mathcal{L}^{(1)} \subset \mathcal{L}^{(2)} \subset \mathcal{L}^{(3)} = \mathcal{L}$$

where $\mathcal{L}^{(3)}$ has basis vectors X_1, X_2, X_3 given by (2.182), $\mathcal{L}^{(2)}$ has basis vectors X_2, X_3 and $\mathcal{L}^{(1)}$ is X_2.

Exercises 2.4

1. Solve the IVP (2.167a–e) and recover (2.161a,b).

2. In the case of a one-parameter Lie group of transformations [$r = 1$], show that the law of composition $\phi(a,b)$ satisfies

$$\Gamma(\epsilon) = \left.\frac{\partial \phi(a,b)}{\partial b}\right|_{(a,b)=(\epsilon^{-1},\epsilon)} = \frac{1}{\left.\frac{\partial \phi(\epsilon,\delta)}{\partial \delta}\right|_{\delta=0}}.$$

[Hint: Consider $\phi(\epsilon^{-1}, \phi(\epsilon,\delta))$ in some neighborhood of $\delta = 0$.]

3. Show that the set of conformal transformations

$$x^* = X(x,y),$$

$$y^* = Y(x,y),$$

where $F(z) = X(x,y) + iY(x,y)$ is analytic in domain D, forms an infinite-parameter Lie group of transformations. Characterize the infinitesimal generators of the group ($z = x + iy$).

4. Consider the set of all conformal transformations which are one-to-one on the extended plane, i.e. the *bilinear* (*Möbius*) *transformations*,

$$z^* = \frac{az+b}{cz+d}, \quad ad - bc \neq 0, \tag{2.187}$$

where $a, b, c, d \in \mathbb{C}$ and $z = x + iy$.

   (a) Show that (2.187) defines a six-parameter Lie group of transformations.

   (b) Find the infinitesimal generators of the group.

   (c) Establish the commutator table of the corresponding Lie algebra.

   (d) Find the subalgebra of largest dimension which is identical to a subalgebra of the Lie algebra of the projective group (2.179a,b).

(e) Determine the subgroup of (2.187) with the largest number of parameters which is in common with a subgroup of the projective group (2.179a,b).

5. (a) Show that the infinitesimal generators X_3, X_4, X_6, X_7 of (2.180) form a four-dimensional Lie algebra.

   (b) Find the corresponding four-parameter Lie group of transformations.

6. Show that the three-parameter family of transformations

$$x^* = ax + b,$$

$$y^* = cx + y,$$

does not form a three-parameter Lie group of transformations

   (a) from the definition of a Lie group of transformations;

   (b) from the algebra of its infinitesimal generators.

7. Consider the three-parameter family of transformations

$$x^* = ax + b, \tag{2.188a}$$

$$y^* = cy. \tag{2.188b}$$

   (a) Show that (2.188a,b) defines a three-parameter Lie group of transformations.

   (b) Establish the commutator table of the corresponding infinitesimal generators.

   (c) Show that the Lie algebra of (2.188a,b) is solvable.

8. In Chapter 1 we showed that problem (1.46a–c) is invariant under the two-parameter family of transformations

$$x^* = \alpha(x - \beta t), \tag{2.189a}$$

$$t^* = \alpha^2 t, \tag{2.189b}$$

$$u^* = \frac{1}{\alpha} u\, e^{\frac{1}{2}\beta x - \frac{1}{4}\beta^2 t}. \tag{2.189c}$$

   (a) Show that (2.189a–c) defines a two-parameter Lie group of transformations.

   (b) Establish the commutator table of its Lie algebra.

10. Show that $e^{\epsilon X_\alpha} e^{\delta X_\beta} = e^{\delta X_\beta} e^{\epsilon X_\alpha} = e^{\epsilon X_\alpha + \delta X_\beta}$ if and only if $[X_\alpha, X_\beta] = 0$.
11. Prove Theorem 2.4.2-3.
12. Prove Theorem 2.4.4-1.

2.5 Discussion

In this chapter we have considered one-parameter Lie groups of transformations which are completely determined by their infinitesimal transformations. Actually such groups are one-parameter ***connected local Lie groups of transformations*** [cf. Gilmore (1974), Olver (1986), Ovsiannikov (1962, 1982)]. The global properties of Lie groups turn out to be unimportant for our eventual purposes of constructing solutions for differential equations.

Using the infinitesimal generator of a one-parameter Lie group of transformations we can construct various kinds of invariants (invariant surfaces, invariant points, invariant families of surfaces). Moreover we can determine canonical coordinates in terms of which the one-parameter Lie group of transformations becomes a group of translations.

When applying groups to the study of invariance properties of a differential equation, the coordinates of the group are separated into independent and dependent variables. A one-parameter Lie group of transformations acting on the space of independent and dependent variables is naturally extended to a one-parameter Lie group of transformations acting on any enlarged space which includes all derivatives of the dependent variables up to a fixed finite order. This is accomplished by requiring, under the group action, preservation of derivative relations or, equivalently, preservation of contact conditions connecting higher order differentials. This requirement induces a unique extended group action in any enlarged space. Consequently one-parameter extended Lie groups of transformations are characterized completely by their infinitesimals. Moreover these extended infinitesimals are determined from the infinitesimals of the group action on the space of independent and dependent variables. This will allow us to establish an algorithm to determine the infinitesimal transformations admitted by a given differential equation.

The study of multi-parameter Lie groups of transformations reduces to the study of infinitesimal generators of one-parameter subgroups. The infinitesimal generators form a vector space called a Lie algebra which is closed under an additional operation (commutation). For our purposes of constructing solutions to differential equations, a multi-parameter Lie group of transformations is completely characterized by its Lie algebra. The structure (commutator table) of a multi-parameter group's Lie algebra will play an essential role in applying infinitesimal transformations to differential equations.

The Lie groups of transformations introduced in this chapter are commonly called ***Lie groups of point transformations*** or ***point symmetries*** since the group action is closed on the space of independent and dependent variables. In Chapter 5 we will consider the invariance of differential equations under transformations whose actions are not closed on any extended space of independent and dependent variables and their derivatives of finite order. Such ***Lie–Bäcklund transformations*** are characterized by infinitesimal transformations whose infinitesimals depend on a finite number of derivatives of the dependent variables.

3

Ordinary Differential Equations

3.1 Introduction — Invariance of an Ordinary Differential Equation

In this chapter we apply infinitesimal transformations (Lie groups of transformations) to the study of an nth order ordinary differential equation (ODE) written in a solved form

$$y_n = f(x, y, y_1, \ldots, y_{n-1}), \tag{3.1}$$

where

$$y_k = \frac{d^k y}{dx^k}, \quad k = 1, 2, \ldots, n.$$

ODE (3.1) defines a surface in $(x, y, y_1, \ldots, y_n)$-space.

3.1.1 Invariance of an ODE

Definition 3.1.1-1. The one-parameter Lie group of transformations

$$x^* = X(x, y; \epsilon), \tag{3.2a}$$

$$y^* = Y(x, y; \epsilon), \tag{3.2b}$$

leaves ODE (3.1) *invariant* (*is admitted by ODE* (3.1)) if and only if its nth extension, defined by (2.96a–d) for $k = n$, leaves the surface (3.1) invariant.

A solution curve $y = \Theta(x)$ of (3.1) satisfies $\Theta^{(n)}(x) = f(x, \Theta(x), \Theta'(x), \ldots, \Theta^{(n-1)}(x))$ and hence lies on the surface (3.1) with $y = \Theta(x)$, $y_k = \Theta^{(k)}(x)$, $k = 1, 2, \ldots, n$. Invariance of the surface (3.1) under the nth extension of (3.2a,b) means that any solution curve $y = \Theta(x)$ of (3.1) maps into some other solution curve $y = \phi(x; \epsilon)$ of (3.1) under the action of the group (3.2a,b). Moreover, if a transformation (3.2a,b) maps any solution curve $y = \Theta(x)$ of (3.1) into another solution curve $y = \phi(x; \epsilon)$ of (3.1), then the surface (3.1) is invariant under (3.2a,b) with

$$y_k = \frac{\partial^k \phi(x; \epsilon)}{\partial x^k}, \quad k = 1, 2, \ldots, n.$$

It immediately follows that *the family of all solution curves of* (3.1) *is invariant under* (3.2a,b) *if and only if* (3.1) *admits* (3.2a,b).

The following theorem results from Definition 3.1.1-1, Theorem 2.2.7-1 on the infinitesimal criterion for an invariant surface, and Theorem 2.3.2-1 on extended infinitesimals.

Theorem 3.1.1-1 (Infinitesimal Criterion for Invariance of an ODE). *Let*

$$\mathrm{X} = \xi(x,y)\frac{\partial}{\partial x} + \eta(x,y)\frac{\partial}{\partial y} \tag{3.3}$$

be the infinitesimal generator of (3.2a,b). *Let*

$$\mathrm{X}^{(n)} = \xi(x,y)\frac{\partial}{\partial x} + \eta(x,y)\frac{\partial}{\partial y} + \eta^{(1)}(x,y,y_1)\frac{\partial}{\partial y_1}$$

$$+ \cdots + \eta^{(n)}(x,y,y_1,\ldots,y_n)\frac{\partial}{\partial y_n} \tag{3.4}$$

be the nth extended infinitesimal generator of (3.3) *where* $\eta^{(k)}(x,y,y_1,\ldots,y_k)$ *is given by* (2.110) *in terms of* $(\xi(x,y),\eta(x,y))$ *for* $k = 1,2,\ldots,n$. *Then* (3.2a,b) *is admitted by* (3.1) *if and only if*

$$\mathrm{X}^{(n)}(y_n - f(x,y,y_1,\ldots,y_n)) = 0,$$

i.e.,

$$\eta^{(n)}(x,y,y_1,\ldots,y_n) = \mathrm{X}^{(n-1)}f(x,y,y_1,\ldots,y_{n-1}) \tag{3.5}$$

when $y_n = f(x,y,y_1,\ldots,y_{n-1})$.

Proof. See Exercise 3.1-7. □

[More generally an ODE $F(x,y,y_1,\ldots,y_n) = 0$ admits (3.2a,b) with infinitesimal generator (3.3) if and only if

$$\mathrm{X}^{(n)}F(x,y,y_1,\ldots,y_n) = 0 \quad \text{when} \quad F(x,y,y_1,\ldots,y_n) = 0.]$$

We will show that this infinitesimal criterion for invariance of an ODE leads directly to an algorithm to determine the infinitesimals $(\xi(x,y), \eta(x,y))$ admitted by a given ODE. More importantly it will be shown that if a one-parameter Lie group of transformations is admitted by an ODE then one can reduce constructively its order by one. For a first order ODE this corresponds to a reduction to quadrature. It turns out that this reduction of order can always be accomplished by using canonical coordinates associated with the group.

For higher order ODE's ($n \geq 2$) the reduction in order can also be accomplished by using *differential invariants* (invariants of the nth extended group). Moreover through differential invariants one can directly reduce the order of an ODE by r if it is invariant under an r-parameter Lie group of transformations whose Lie algebra is solvable.

In Section 3.5 we illustrate the application of group invariance to boundary value problems. In Section 3.6 we construct special solutions (*invariant solutions*) of ODE's which are invariant curves of admitted Lie groups of transformations. For a first order ODE, we show that invariant solutions are determined algebraically and include separatrices and singular envelope solutions. For higher order ODE's we will see that invariant solutions are determined either algebraically or by solving the first order ODE for the invariant curves of the group.

To illustrate reduction of order and the mapping of solution curves into other solution curves, from invariance under a one-parameter Lie group of transformations, we consider two elementary examples.

3.1.2 Elementary Examples

(1) *Group of Translations*

The first order ODE

$$y_1 = F(x) \quad \left[y_1 = \frac{dy}{dx}\right] \tag{3.6}$$

trivially reduces to quadrature

$$y = \int F(x)\,dx + C. \tag{3.7}$$

What characterizes (3.6)? Obviously its right-hand side does not depend on y. In particular the one-parameter (ϵ) Lie group of translations

$$x^* = x, \tag{3.8a}$$

$$y^* = y + \epsilon, \tag{3.8b}$$

is admitted by (3.6) since under (3.8a,b)

$$y_1^* = \frac{dy^*}{dx^*} = \frac{dy}{dx} = y_1 \quad \text{and} \quad F(x^*) = F(x),$$

so that under (3.8a,b) the surface $y_1 = F(x)$ in (x, y, y_1)-space is invariant.

Moreover it is easy to see that ODE

$$\frac{dy}{dx} = f(x, y) \tag{3.9}$$

is invariant under (3.8a,b) if and only if for any value of parameter ϵ

$$f(x^*, y^*) \equiv f(x, y + \epsilon) \equiv f(x, y),$$

i.e. $f(x, y)$ is independent of y or, equivalently, $f(x, y) \equiv F(x)$ for some function $F(x)$. Thus the reduction of (3.6) to quadrature (3.7) is equivalent to the invariance of (3.6) under (3.8a,b).

Under the action of (3.8a,b) a solution curve $y = \Theta(x)$ of (3.6) maps into a curve $y^* = \Theta(x^*)$ which corresponds to the solution curve $y = \Theta(x) - \epsilon$

of (3.6) [Figure 3.1.2-1]. Thus from invariance of (3.6) under (3.8a,b) we see that if $y = \Theta(x)$ is a particular solution of (3.6) then $y = \Theta(x) + C$ is the general solution of (3.6) for arbitrary constant C.

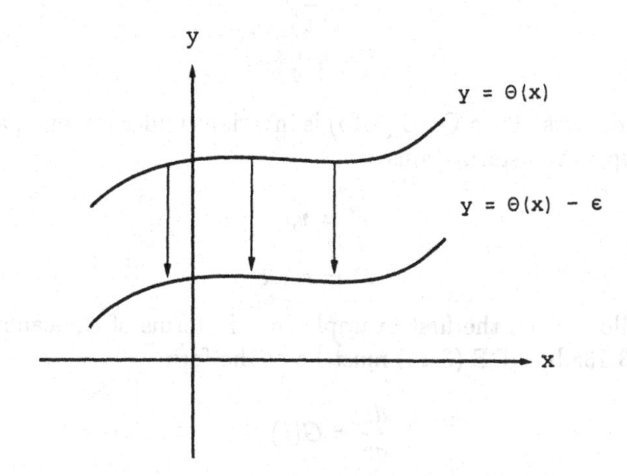

Figure 3.1.2-1

(2) *Group of Scalings*

The first order ODE

$$y_1 = F\left(\frac{y}{x}\right), \tag{3.10}$$

commonly called a *homogeneous equation,* admits the one-parameter (α) group of scalings

$$x^* = \alpha x, \tag{3.11a}$$

$$y^* = \alpha y, \tag{3.11b}$$

since

$$y_1^* = \frac{dy^*}{dx^*} = \frac{\alpha\, dy}{\alpha\, dx} = y_1, \quad \text{and} \quad F\left(\frac{y*}{x^*}\right) = F\left(\frac{y}{x}\right).$$

Under the action of (3.11a,b) a solution curve $y = \Theta(x)$ of (3.10) maps into a curve $y^* = \Theta(x^*)$ which corresponds to the solution curve

$$y = \frac{1}{\alpha}\Theta(\alpha x) \tag{3.12}$$

of (3.10). It follows that if $y = \Theta(x)$ is a particular solution of (3.10), and the curve $y - \Theta(x) = 0$ is not invariant under (3.11a,b) (i.e. $\Theta(x) \neq \lambda x$ for some fixed constant λ), then

$$y = \frac{1}{C}\Theta(Cx)$$

is the general solution of (3.10) for arbitrary constant C.

The reduction of order of (3.10) from invariance under (3.11a,b) is accomplished by choosing canonical coordinates

$$r = \frac{y}{x}, \tag{3.13a}$$

$$s = \log y, \tag{3.13b}$$

as new coordinates. Then ODE (3.10) is invariant under the one-parameter (ϵ) Lie group of transformations

$$r^* = r, \tag{3.14a}$$

$$s^* = s + \epsilon. \tag{3.14b}$$

Hence it follows from the first example that in terms of the canonical coordinates (3.13a,b) ODE (3.10) must be of the form

$$\frac{ds}{dr} = G(r) \tag{3.15}$$

for some function $G(r)$. Thus the general solution of ODE (3.10) is

$$s = \int G(r)\, dr + C, \tag{3.16}$$

or, in terms of coordinates (x, y),

$$y = \exp\left[\int^{\frac{y}{x}} G(r) dr\right] + C. \tag{3.17}$$

$G(r)$ is determined as follows:

$$ds = \frac{1}{y} dy, \quad dr = -\frac{y}{x^2} dx + \frac{1}{x} dy.$$

Hence

$$G(r) = \frac{ds}{dr} = \frac{y_1}{ry_1 - r^2} = \frac{F(r)}{rF(r) - r^2}, \tag{3.18}$$

where $F(r)$ is given by ODE (3.10).

3.1.3 Mapping of Solutions to Other Solutions from Group Invariance of an ODE

Under the action of a one-parameter Lie group of transformations admitted by an ODE a solution curve is mapped into a one-parameter family of solution curves if the solution curve is not an invariant curve of the group. We now derive a formula for this one-parameter family of solutions generated from a known solution. Without loss of generality we can assume that the one-parameter Lie group of transformations is parameterized so that it is of the form

$$x^* = X(x, y; \epsilon) = e^{\epsilon \mathrm{X}} x, \tag{3.19a}$$

$$y^* = Y(x, y; \epsilon) = e^{\epsilon \mathrm{X}} y, \tag{3.19b}$$

with infinitesimal generator

$$\mathrm{X} = \xi(x, y)\frac{\partial}{\partial x} + \eta(x, y)\frac{\partial}{\partial y}. \tag{3.19c}$$

Consider a solution $y = \Theta(x)$ of (3.1) which is not an invariant solution corresponding to (3.19c). Transformation (3.19a,b) maps a point $(x, \Theta(x))$ on the solution curve $y = \Theta(x)$ into the point (x^*, y^*) with

$$x^* = X(x, \Theta(x); \epsilon), \tag{3.20a}$$

$$y^* = Y(x, \Theta(x); \epsilon). \tag{3.20b}$$

For a fixed ϵ equations (3.20a,b) define a parametric representation of the new solution curve with x playing the role of a parameter [Figure 3.1.3-1]. One can eliminate x from (3.20a,b) by substituting the inverse transformation of (3.19a), i.e.,

$$x = X(x^*, y^*; -\epsilon) \tag{3.21}$$

into (3.20b):

$$\begin{aligned} y^* &= Y(X(x^*, y^*; -\epsilon), \Theta(X(x^*, y^*; -\epsilon)); \epsilon) \\ &= Y(e^{-\epsilon \mathrm{X}} x^*, \Theta(e^{-\epsilon \mathrm{X}} x^*); \epsilon). \end{aligned} \tag{3.22}$$

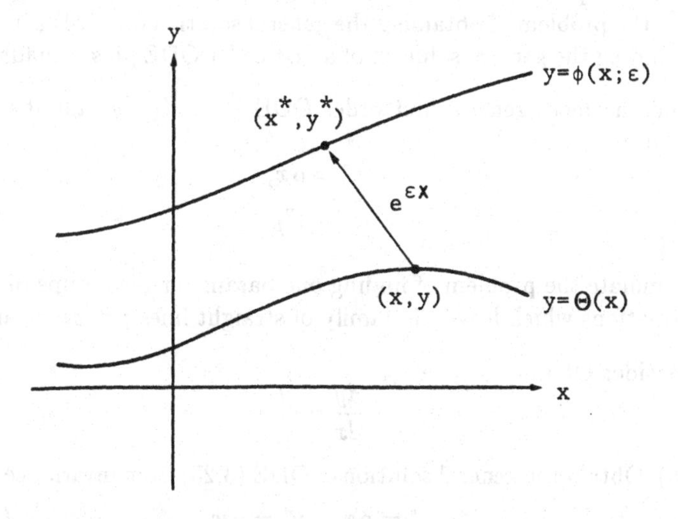

Figure 3.1.3-1. Mapping of a solution curve. A different solution curve $y = \phi(x; \epsilon)$ of (3.1) corresponds to each parameter value ϵ.

Equation (3.22) yields the relationship between the x and y coordinates of the new solution curve which we denote by $y = \phi(x;\epsilon)$. Substituting x and y for x^* and y^* and replacing ϵ by $-\epsilon$ in (3.22) we thus have:

Theorem 3.1.3-1. *Suppose*

(i) $y = \Theta(x)$ *is a solution (solution curve) of the nth order ODE* (3.1);

(ii) *ODE* (3.1) *admits* (3.19a,b);

(iii) $y = \Theta(x)$ *is not an invariant curve of* (3.19a,b).

Then

$$y = Y(e^{\epsilon X}x, \Theta(e^{\epsilon X}x); -\epsilon) \tag{3.23a}$$

$$= Y(X(x,y;\epsilon), \Theta(X(x,y;\epsilon)); -\epsilon) \tag{3.23b}$$

implicitly defines a one-parameter family of solutions $y = \phi(x;\epsilon)$ *of* (3.1).

Exercises 3.1

1. Consider ODE

$$\frac{d^2y}{dx^2} = F\left(x, \frac{dy}{dx}\right). \tag{3.24}$$

Assume that (3.24) is invariant under the one-parameter (α) Lie group of scalings

$$x^* = \alpha x,$$

$$y^* = \alpha y.$$

Find the special form of the function F for which this is true. Show how the problem of obtaining the general solution of (3.24) is reduced to finding the general solution of a first order ODE plus a quadrature.

2. Find the most general first order ODE $\frac{dy}{dx} = f(x,y)$ admitting the group

$$x^* = \alpha x,$$

$$y^* = \alpha^2 y.$$

3. Formulate the problem of finding one-parameter Lie groups of transformations which leave the family of straight lines $y = cx$ invariant.

4. Consider ODE

$$\frac{dy}{dx} = \frac{y}{x}. \tag{3.25}$$

(a) Obtain the general solution of ODE (3.25) from invariance under

(i) $$x^* = \alpha x, \quad y^* = \alpha y; \tag{3.26a}$$

(ii) $$x^* = x, \quad y^* = \beta y. \tag{3.26b}$$

(b) $y = \Theta(x) = x$ is a solution curve of (3.25). Apply Theorem 3.1.3-1 to this solution curve for each of the groups (3.26a,b). Explain your answer.

5. Show that if $y = \Theta(x)$ is an invariant curve of (3.19a,b) then (3.23a,b) yields

$$\phi(x;\epsilon) \equiv \Theta(x) \quad \text{for all} \quad \epsilon.$$

6. Let $y = \Theta(x)$ be a solution curve for an ODE invariant under the rotation group

$$x^* = x\cos\epsilon - y\sin\epsilon,$$

$$y^* = x\sin\epsilon + y\cos\epsilon.$$

Find the corresponding one-parameter family of solutions $y = \phi(x;\epsilon)$ for the ODE.

7. Prove Theorem 3.1.1-1.

3.2 First Order ODE's

We consider applications of infinitesimal transformations to the study of a first order ODE

$$y' = f(x,y) \quad \left[y' = \frac{dy}{dx}\right]. \tag{3.27}$$

We assume that ODE (3.27) admits a one-parameter Lie group of transformations

$$x^* = X(x,y;\epsilon) = x + \epsilon\xi(x,y) + O(\epsilon^2), \tag{3.28a}$$

$$y^* = Y(x,y;\epsilon) = y + \epsilon\eta(x,y) + O(\epsilon^2), \tag{3.28b}$$

with infinitesimal generator

$$\mathrm{X} = \xi(x,y)\frac{\partial}{\partial x} + \eta(x,y)\frac{\partial}{\partial y}. \tag{3.28c}$$

We show how to find the general solution of ODE (3.27) from the infinitesimal $(\xi(x,y), \eta(x,y))$ of an admitted group (3.28a,b) from two points of view:

(i) use of canonical coordinates;

(ii) determination of an integrating factor.

Alternatively, if a particular solution of (3.27) is known, and this particular solution is not an invariant curve of (3.28a,b), then (3.23a,b) yields the general solution of (3.27). The general solution of (3.27), obtained by using canonical coordinates or by determining an integrating factor, does not depend on knowledge of a particular solution of (3.27).

Having shown how to use the infinitesimal transformation arising from invariance, we then consider the problem of determining infinitesimal transformations admitted by a given first order ODE (3.27). We also show how to find all first order ODE's admitting a given one-parameter Lie group of transformations.

3.2.1 Canonical Coordinates

As discussed in Section 2.2.4, given any one-parameter Lie group of transformations (3.28a,b), there exist *canonical coordinates* $(r(x,y), s(x,y))$, determined by solving

$$\mathrm{X}r = 0,$$

$$\mathrm{X}s = 1,$$

such that (3.28a,b) becomes the translation group

$$r^* = r, \tag{3.29a}$$

$$s^* = s + \epsilon. \tag{3.29b}$$

In terms of canonical coordinates ODE (3.27) becomes

$$\frac{ds}{dr} = \frac{s_x + s_y y'}{r_x + r_y y'} = F(r,s) = \frac{s_x + s_y f(x,y)}{r_x + r_y f(x,y)}. \tag{3.30}$$

The invariance of ODE (3.27), and hence ODE (3.30), under (3.29a,b) means that $F(r,s)$ does not depend on s. Hence ODE (3.30) is of the form

$$\frac{ds}{dr} = G(r) = \frac{s_x + s_y f(x,y)}{r_x + r_y f(x,y)}. \tag{3.31}$$

Consequently the general solution of ODE (3.27) is given by

$$s(x,y) = \int^{r(x,y)} G(\rho)d\rho + C, \quad C = \text{const.} \tag{3.32}$$

In Section 3.1.2 we solved the homogeneous ODE (3.10) in terms of canonical coordinates arising from scaling invariance. Now consider two more familiar examples in terms of canonical coordinates:

(1) *Linear Homogeneous Equation*

The first order linear homogeneous ODE

$$y' + p(x)y = 0 \tag{3.33}$$

admits the one-parameter (α) Lie group of scaling transformations

$$x^* = x, \tag{3.34a}$$

$$y^* = \alpha y. \tag{3.34b}$$

In terms of corresponding canonical coordinates

$$r = x, \tag{3.35a}$$

$$s = \log y, \tag{3.35b}$$

ODE (3.33) becomes

$$\frac{ds}{dr} = \frac{y'}{y} = -p(r), \tag{3.36}$$

so that the general solution of (3.33) is given by

$$s(x, y) = \log y = -\int^x p(\rho)d\rho + C,$$

or

$$y = C \exp\left[-\int^x p(\rho)d\rho\right], \quad C = \text{const.}$$

(2) *Linear Nonhomogeneous Equation*

The first order linear nonhomogeneous ODE

$$y' + p(x)y = g(x) \tag{3.37}$$

admits the one-parameter (ϵ) Lie group of transformations

$$x^* = x, \tag{3.38a}$$

$$y^* = y + \epsilon\phi(x), \tag{3.38b}$$

where $u = \phi(x)$ is a particular solution of the associated homogeneous equation

$$u' + p(x)u = 0. \tag{3.39}$$

The infinitesimal generator corresponding to (3.38a,b) is

$$\mathrm{X} = \phi(x)\frac{\partial}{\partial y}$$

so that $\mathrm{X}s = 1$ has solution $s = \dfrac{y}{\phi(x)}$. In terms of canonical coordinates

$$r = x, \tag{3.40a}$$

$$s = \frac{y}{\phi(x)}, \tag{3.40b}$$

ODE (3.37) reduces to

$$\frac{ds}{dr} = \frac{g(r)}{\phi(r)},$$

leading to its general solution

$$s(x,y) = \frac{y}{\phi(x)} = \int^x \frac{g(\rho)}{\phi(\rho)} d\rho + C$$

or

$$y = \phi(x) \int^x \frac{g(\rho)}{\phi(\rho)} d\rho + C\phi(x), \quad C = \text{const.} \tag{3.41}$$

3.2.2 Integrating Factors

A first order ODE (3.27) can be written in a differential form

$$M(x,y)dx + N(x,y)dy = 0 \tag{3.42}$$

where $f(x,y) = -\frac{M(x,y)}{N(x,y)}$. If

$$\omega(x,y) = \text{const} \tag{3.43}$$

is the general solution (family of solution curves) of (3.27), then

$$N\frac{\partial \omega}{\partial x} - M\frac{\partial \omega}{\partial y} = 0. \tag{3.44}$$

We assume that ODE (3.27) admits a one-parameter Lie group of transformations (3.28a,b). Then (3.28a,b) leaves the family of solution curves (3.43) invariant. We assume that the group (3.28a,b) is such that the solution curves (3.43) of ODE (3.37) are not invariant curves of (3.28a,b). Then, without loss of generality, the family of solution curves (3.43) satisfies [cf. Section 2.2.7]

$$\mathrm{X}\omega = \xi(x,y)\frac{\partial \omega}{\partial x} + \eta(x,y)\frac{\partial \omega}{\partial y} = 1 \tag{3.45}$$

for infinitesimal generator (3.28c). The system given by (3.44) and (3.45) can be solved for the first partial derivatives

$$\frac{\partial \omega}{\partial x} = \frac{M}{M\xi + N\eta}, \quad \frac{\partial \omega}{\partial y} = \frac{N}{M\xi + N\eta}. \tag{3.46}$$

But $d\omega = \frac{\partial \omega}{\partial x}dx + \frac{\partial \omega}{\partial y}dy$ is an exact differential. Hence it follows that

$$\mu(x,y) = \frac{1}{M\xi + N\eta} \tag{3.47}$$

is an *integrating factor* for (3.42). Conversely we have the following theorem:

Theorem 3.2.2-1. *If $\mu(x,y)$ is an integrating factor of (3.42), then any $(\xi(x,y), \eta(x,y))$ satisfying (3.47) defines an infinitesimal generator $\mathrm{X} = \xi(x,y)\frac{\partial}{\partial x} + \eta(x,y)\frac{\partial}{\partial y}$ admitted by the first order ODE $y' = f(x,y)$.*

Proof. See Exercise 3.2-3. □

The following theorem concerns one-parameter Lie groups of transformations leaving solution curves of (3.27) invariant.

Theorem 3.2.2-2. *For any function $\xi(x,y)$, the one-parameter Lie group of transformations with infinitesimal generator*

$$X = \xi(x,y)\left[\frac{\partial}{\partial x} + f(x,y)\frac{\partial}{\partial y}\right] \tag{3.48}$$

leaves each solution curve of the first order ODE $y' = f(x,y)$ invariant.

Proof. Let $y = \Theta(x)$ be a solution curve of $y' = f(x,y)$. Then

$$y' = \Theta'(x) = f(x,\Theta(x)). \tag{3.49}$$

Consider infinitesimal generator X given by (3.48). Then

$$X(y-\Theta(x)) = \xi(x,y)[f(x,y) - \Theta'(x)].$$

Hence if $y = \Theta(x)$, then

$$\begin{aligned} X(y-\Theta(x)) &= \xi(x,\Theta(x))[f(x,\Theta(x)) - \Theta'(x)] \\ &= 0 \quad \text{from (3.49).} \end{aligned}$$

Consequently $y - \Theta(x) = 0$ is an invariant curve for the one-parameter Lie group of transformations with infinitesimal generator (3.48). □

From Theorems 3.2.2-1 and 3.2.2-2 we see that two types of one-parameter Lie groups of transformations are admitted by any first order ODE (3.27). Moreover (3.27) admits infinite-parameter Lie groups of transformations of both types:

(i) *Trivial one-parameter Lie groups* with infinitesimal generators $X = \xi(x,y)\frac{\partial}{\partial x} + \eta(x,y)\frac{\partial}{\partial y}$ are always admitted by $y' = f(x,y)$ if $\frac{\eta}{\xi} \equiv f(x,y)$. Here each solution curve of $y' = f(x,y)$ is an invariant curve. This type of group is useless for reducing $y' = f(x,y)$ to quadrature since in order to find the canonical coordinates of the group one must first find the general solution of $y' = f(x,y)$.

(ii) If the one-parameter Lie group with infinitesimal generator $X = \xi(x,y)\frac{\partial}{\partial x} + \eta(x,y)\frac{\partial}{\partial y}$ is admitted by $y' = f(x,y)$ and $\frac{\eta}{\xi} \not\equiv f(x,y)$ then the family of solution curves of $y' = f(x,y)$ is invariant but an arbitrary solution curve of $y' = f(x,y)$ is not invariant. This type of group defines a *nontrivial Lie group of transformations.* It is useful for reducing $y' = f(x,y)$ to quadrature provided one can solve the ODE $\frac{dy}{dx} = \frac{\eta}{\xi}$.

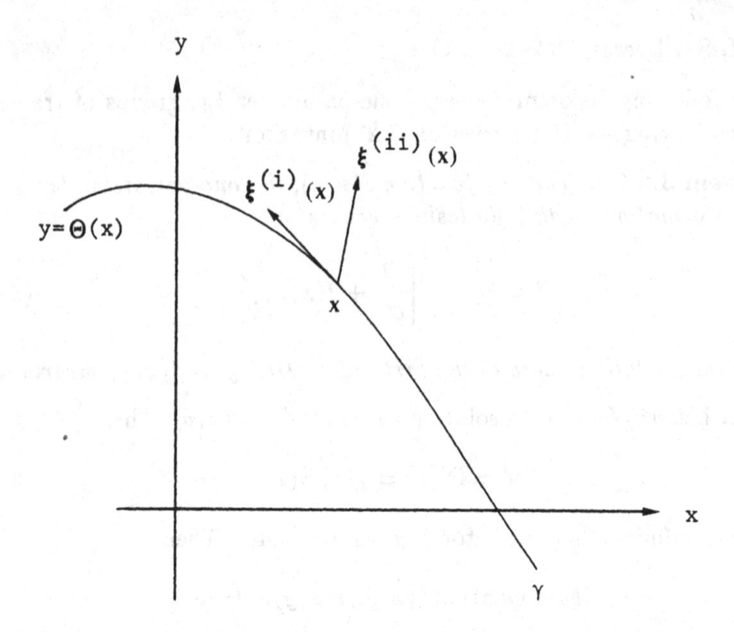

Figure 3.2.2-1. Illustration of groups of types (i) and (ii).

The geometrical situation is illustrated in Figure 3.2.2-1. Here $\boldsymbol{\xi}^{(i)}(\mathbf{x}) = (\xi^{(i)}(x,y), \eta^{(i)}(x,y))$ is the infinitesimal of a one-parameter Lie group $G^{(i)}$ of type (i) and $\boldsymbol{\xi}^{(ii)}(\mathbf{x}) = (\xi^{(ii)}(x,y), \eta^{(ii)}(x,y))$ is the infinitesimal of a one-parameter Lie group $G^{(ii)}$ of type (ii) admitted by $y' = f(x,y)$; γ is any solution curve $y = \Theta(x)$ of $y' = f(x,y)$. Along γ, $\boldsymbol{\xi}^{(i)}(\mathbf{x})$ is tangent to γ since $\frac{\eta}{\xi} \equiv f$, but $\boldsymbol{\xi}^{(ii)}(\mathbf{x})$ is not tangent to γ since $\frac{\eta}{\xi} \not\equiv f$. Consequently $G^{(i)}$ leaves γ invariant and $G^{(ii)}$ maps γ into a one-parameter family of solution curves given by (3.23a,b) with infinitesimal generator

$$\mathrm{X} = \xi^{(ii)}(x,y)\frac{\partial}{\partial x} + \eta^{(ii)}(x,y)\frac{\partial}{\partial y}.$$

3.2.3 Determining Equation for Infinitesimal Transformations of a First Order ODE

From Theorem 3.1.1-1 we see that the first order ODE

$$y' = f(x,y) \tag{3.50}$$

admits the one-parameter Lie group of transformations with infinitesimal generator

$$\mathrm{X} = \xi(x,y)\frac{\partial}{\partial x} + \eta(x,y)\frac{\partial}{\partial y} \tag{3.51}$$

if and only if

$$\eta^{(1)} = \xi f_x + \eta f_y \quad \text{when} \quad y' = f(x,y)$$

where [cf. (2.111)] the once-extended infinitesimal is given by

$$\eta^{(1)} = \eta_x + [\eta_y - \xi_x]y' - \xi_y(y')^2.$$

Thus the first order ODE (3.50) admits (3.51) if and only if $(\xi(x,y), \eta(x,y))$ satisfy

$$\eta_x + [\eta_y - \xi_x]f - \xi_y f^2 - \xi f_x - \eta f_y = 0. \tag{3.52}$$

Equation (3.52) is called the ***determining equation*** for the infinitesimal transformations admitted by (3.50).

It is easy to check that for *any* $\xi(x,y)$,

$$\eta(x,y) = \xi(x,y)f(x,y) \tag{3.53}$$

solves the determining equation (3.52). This represents the trivial infinite-parameter Lie group of transformations of type (i) which leaves each solution curve of (3.50) invariant.

For any $\xi(x,y)$,

$$\eta(x,y) = \xi(x,y)f(x,y) + \chi(x,y) \tag{3.54}$$

yields the general solution of (3.52) where $\chi(x,y)$ is the general solution of the first order linear partial differential equation

$$\chi_x + f\chi_y - f_y\chi = 0. \tag{3.55}$$

Thus (3.54) leads to the infinite-parameter Lie group of transformations of type (ii) admitted by (3.50) which maps solution curves into other solution curves of (3.50). [An infinite-parameter subgroup of type (ii) corresponds to $\eta = \chi$, $\xi \equiv 0$.] Moreover from (3.54) we see that the first order ODE (3.50) admits

$$\mathrm{X} = \xi(x,y)\frac{\partial}{\partial x} + \eta(x,y)\frac{\partial}{\partial y}$$

if and only if (3.50) admits

$$\mathrm{Y} = [\eta(x,y) - \xi(x,y)y_1]\frac{\partial}{\partial y} = [\eta(x,y) - \xi(x,y)f(x,y)]\frac{\partial}{\partial y}.$$

Consequently the problem of finding all Lie groups of transformations admitted by a given first order ODE (3.50) "reduces" to finding the general solution of (3.55). But in order to find the general solution of (3.55) we must know the general solution of ODE (3.50)! However any *particular solution* χ of (3.55) or, equivalently, (ξ, η) of (3.52) with $\eta \not\equiv \xi f$, leads to a one-parameter Lie group admitted by (3.50) and hence to the general solution of (3.50) through reduction by quadrature. [In turn this leads to determining the infinite-parameter Lie group admitted by (3.50).] Unfortunately a particular solution χ of (3.55) cannot be found by a deductive procedure.

Next we consider the converse problem of determining all first order ODE's which admit a given one-parameter Lie group of transformations.

3.2.4 Determination of First Order ODE's Invariant Under a Given Group

We show how to find all first order ODE's

$$y' = f(x,y) \tag{3.56}$$

which admit a given one-parameter Lie group of transformations with infinitesimal generator

$$\mathrm{X} = \xi(x,y)\frac{\partial}{\partial x} + \eta(x,y)\frac{\partial}{\partial y}. \tag{3.57}$$

This can be accomplished in two different ways:

(i) *Method of Canonical Coordinates.* Given (3.57) we compute *canonical coordinates* $(r(x,y), s(x,y))$ satisfying

$$\mathrm{X}r = 0, \quad \mathrm{X}s = 1, \tag{3.58}$$

so that the group corresponding to (3.57) becomes

$$r^* = r, \quad s^* = s + \epsilon. \tag{3.59}$$

Then

$$\frac{ds}{dr} = \frac{s_x + s_y y'}{r_x + r_y y'} \tag{3.60}$$

relates y' and $\frac{ds}{dr}$. Consequently the problem of finding (3.56) admitting (3.57) transforms to the problem of finding all first order ODE's

$$\frac{ds}{dr} = F(r,s) \tag{3.61}$$

which admit (3.59). Clearly $F(r,s)$ cannot depend on s. Hence the most general first order ODE admitting (3.57) is

$$\frac{ds}{dr} = G(r) \tag{3.62}$$

where $G(r)$ is an arbitrary function of r, which in terms of the given coordinates (x,y) becomes the ODE

$$\frac{s_x(x,y) + s_y(x,y)y'}{r_x(x,y) + r_y(x,y)y'} = G(r(x,y)). \tag{3.63}$$

(ii) *Method of (Differential) Invariants.* The first order ODE (3.56) admits (3.57) if and only if $f(x,y)$ satisfies the first order partial differential equation (3.52). The corresponding characteristic equations to determine f are

$$\frac{dx}{\xi(x,y)} = \frac{dy}{\eta(x,y)} = \frac{df}{\eta_x + (\eta_y - \xi_x)f - \xi_y f^2}. \tag{3.64}$$

The invariant

$$u(x,y) = r(x,y) = \text{const} = c_1 \tag{3.65}$$

is the integral of the first equation of (3.64). Eliminating y through (3.65) and setting

$$\frac{\eta(x,y)}{\xi(x,y)} = f_p(x;c_1), \tag{3.66}$$

one sees that

$$f = f_p(x;c_1), \tag{3.67}$$

is a particular solution of the second characteristic equation which now becomes

$$\frac{df}{dx} = A + Bf + Cf^2, \tag{3.68}$$

with

$$A = A(x;c_1) = \frac{\eta_x}{\xi}, \tag{3.69a}$$

$$B = B(x;c_1) = \frac{\eta_y - \xi_x}{\xi}, \tag{3.69b}$$

$$C = C(x;c_1) = -\frac{\xi_y}{\xi}. \tag{3.69c}$$

Equation (3.68) is a first order ODE of Riccati type. The general solution of (3.68) can be determined explicitly from knowledge of the particular solution (3.67) through its connection with a second order linear ODE. Specifically if z solves

$$z'' - \left(\frac{C'}{C} + B\right)z' + ACz = 0 \tag{3.70}$$

then

$$f = -\frac{1}{C}\frac{z'}{z} \tag{3.71}$$

solves (3.68). Hence the general solution of (3.70) leads to the general solution of (3.68) through the well-known Riccati transformation (3.71). A particular solution of (3.70) is given by

$$z = z_p = e^{-\int Cf_p dx} \tag{3.72}$$

where C is given by (3.69c), (3.65); f_p by (3.66), (3.65). The explicit general solution of (3.70) follows from (3.72) through the method of reduction of order (to be derived from group invariance in Section 3.3.3). Consequently one obtains the general solution of (3.68) which is given by

$$f = \phi(x;c_1,v) \tag{3.73}$$

where v is a constant of integration and ϕ is a known function of its arguments. Now in (3.73) replace f by y' and c_1 by $u(x,y)$. Then solving (3.73) for v, we obtain the *differential invariant* $v(x, y, y')$ of the first extension of (3.57) $[\mathrm{X}^{(1)}v = 0]$. The general solution of (3.52) is $v = \psi(c_1)$, i.e.,

$$v(x, y, f) = v(x, y, y') = \psi(u(x, y)) \tag{3.74}$$

where ψ is an arbitrary function of u. ODE (3.74) is the most general first order ODE which admits (3.51), written in terms of differential invariants.

Note that in terms of canonical coordinates $(r(x, y), s(x, y))$,

$$v(x, y, y') = \frac{ds}{dr} = \frac{s_x + s_y y'}{r_x + r_y y'}$$

satisfies $\mathrm{X}^{(1)}v = 0$ since $\frac{ds}{dr}$ is invariant under (3.57) $\left(\frac{ds^*}{dr^*} = \frac{ds}{dr}\right)$.

Consider examples for which we use both approaches to find first order ODE's admitting specific groups:

(1) *A Scaling Group*

Suppose (3.56) is invariant under the scaling group

$$x^* = e^{\epsilon} x, \tag{3.75a}$$

$$y^* = e^{\epsilon} y, \tag{3.75b}$$

with infinitesimal generator $\mathrm{X} = x\frac{\partial}{\partial x} + y\frac{\partial}{\partial y}$.

(i) Canonical coordinates for the Lie group (3.75a,b) are

$$r = \frac{y}{x}, \quad s = \log y.$$

Then:

$$s_x = 0, \quad s_y = \frac{1}{y}, \quad r_x = -\frac{y}{x^2}, \quad r_y = \frac{1}{x}.$$

Hence (3.63) becomes the ODE

$$\frac{y'}{ry' - r^2} = G(r). \tag{3.76}$$

Solving (3.76) for y', we find that the most general first order ODE admitting (3.75a,b) is of the form

$$y' = H\left(\frac{y}{x}\right) \tag{3.77}$$

where H is an arbitrary function of $\frac{y}{x}$.

(ii) In terms of the method of differential invariants, (3.64) becomes

$$\frac{dx}{x} = \frac{dy}{y} = \frac{df}{0}. \tag{3.78}$$

The integral of the first equation of (3.78) is given by

$$u(x,y) = \frac{y}{x} = \text{const} = c_1.$$

The second equation of (3.78) yields

$$f = \text{const} = c_2.$$

Then $f(x,y) = \psi(c_1) = \psi\left(\frac{y}{x}\right)$ yields again (3.77).

(2) *Rotation Group*

Let (3.56) admit the rotation group

$$x^* = x\cos\epsilon - y\sin\epsilon, \tag{3.79a}$$

$$y^* = x\sin\epsilon + y\cos\epsilon, \tag{3.79b}$$

with infinitesimal generator $\mathrm{X} = -y\frac{\partial}{\partial x} + x\frac{\partial}{\partial y}$.

(i) Canonical coordinates are polar coordinates

$$r = \sqrt{x^2+y^2}, \quad s = \theta = \sin^{-1}\frac{y}{r}.$$

Then $r_x = x/r$, $r_y = y/r$, $s_x = -y/r^2$, $s_y = x/r^2$. Thus (3.63) becomes the ODE

$$\frac{-y+xy'}{x+yy'} = rG(r).$$

Consequently the most general first order ODE admitting (3.79a,b) is given by

$$\frac{-y+xy'}{x+yy'} = H\left(\sqrt{x^2+y^2}\right) \tag{3.80}$$

where H is an arbitrary function of $\sqrt{x^2+y^2}$.

(ii) In terms of the method of differential invariants, (3.64) becomes

$$\frac{dx}{-y} = \frac{dy}{x} = \frac{df}{1+f^2}. \tag{3.81}$$

The integral of the first equation of (3.81) is given by

$$u(x,y) = \sqrt{x^2+y^2} = c_1. \tag{3.82}$$

The second characteristic equation of (3.81) then becomes

$$\frac{df}{1+f^2} = \frac{dy}{\sqrt{(c_1)^2-y^2}}. \tag{3.83}$$

Let $\alpha = \tan^{-1} f$, $\beta = \sin^{-1} \frac{y}{c_1} = \tan^{-1} \frac{y}{x}$. Then the integration of (3.83) yields

$$\alpha - \beta = c_2. \tag{3.84}$$

Then $\tan c_2 = \tan(\alpha - \beta) = \psi(c_1)$ leads to

$$\frac{f - \frac{y}{x}}{1 + \frac{fy}{x}} = \psi(c_1). \tag{3.85}$$

Replacing f by y' in (3.85) we again get (3.80).

Exercises 3.2

1. Let $y = \phi(x)$ be a particular solution of

$$y' + p(x)y = g(x). \tag{3.86}$$

   (a) Use this solution to find a one-parameter Lie group of transformations admitted by (3.86).

   (b) Find corresponding canonical coordinates and reduce (3.86) to quadrature.

   (c) Illustrate for the ODE $y' + y = x$.

2. Derive the integrating factor for (3.37) from its invariance under (3.38a,b).

3. Prove Theorem 3.2.2-1.

4. (a) Characterize the infinitesimal transformation of a one-parameter Lie group of transformations which leaves the family of straight lines $y =$ const invariant. Explicitly find all such groups for which $\xi(x, y) \equiv 1$.

   (b) Do the same for the family of straight lines $\frac{y}{x} =$ const.

5. Find all first order ODE's which admit $x^* = \alpha x$, $y^* = \alpha^k y$, $\alpha > 0$, with k a fixed constant.

6. For the first order ODE

$$y' = f(x, y) \tag{3.87}$$

   written in the differential form $Mdx + Ndy = 0$, let the operator $\mathcal{A} = N\frac{\partial}{\partial x} - M\frac{\partial}{\partial y}$.

   (a) Prove that the one-parameter Lie group with infinitesimal generator $\mathrm{X} = \xi\frac{\partial}{\partial x} + \eta\frac{\partial}{\partial y}$ is admitted by (3.87) if and only if the commutation relation $[\mathrm{X}, \mathcal{A}] = \lambda(x, y)\mathcal{A}$ is satisfied for some function $\lambda(x, y)$.

(b) What more can you say if $[\mathrm{X},\mathcal{A}]=0$?

(c) Illustrate for first order ODE's which admit

(i) $x\dfrac{\partial}{\partial x}+y\dfrac{\partial}{\partial y}$;

(ii) $-y\dfrac{\partial}{\partial x}+x\dfrac{\partial}{\partial y}$.

7. If a first order ODE $y'=f(x,y)$ admits two nontrivial groups with generators $\mathrm{X}_i=\xi_i\frac{\partial}{\partial x}+\eta_i\frac{\partial}{\partial y}$, $i=1,2$, show that

$$\Omega=\frac{\eta_2-f\xi_2}{\eta_1-f\xi_1}$$

is identically a constant or an integral of the ODE.

8. Find the most general first order ODE which admits the one-parameter (ϵ) Lie group of transformations

$$x^*=x+\epsilon,$$
$$y^*=\frac{xy}{x+\epsilon}.$$

9. Consider the first order ODE

$$\left(y-\frac{3}{2}x-3\right)y'+y=0.$$

(a) Find a nontrivial Lie group of transformations which is admitted by this ODE.

(b) Find the general solution of this ODE.

3.3 Second and Higher Order ODE's

Now consider the application of infinitesimal transformations to the study of second or higher order ODE's

$$y^{(n)}=f(x,y,y',\ldots,y^{(n-1)}),$$

or, equivalently,

$$y_n=f(x,y,y_1,\ldots,y_{n-1}), \tag{3.88}$$

$n\geq 2$, where we use the notations

$$y^{(k)}=y_k=\frac{d^ky}{dx^k},\quad k=1,2,\ldots,n;$$

$$y'=y^{(1)},\quad y''=y^{(2)},\text{ etc.}$$

Assume that ODE (3.88) admits a one-parameter Lie group of transformations

$$x^* = X(x, y; \epsilon) = x + \epsilon\xi(x, y) + O(\epsilon^2), \tag{3.89a}$$

$$y^* = Y(x, y; \epsilon) = y + \epsilon\eta(x, y) + O(\epsilon^2), \tag{3.89b}$$

with infinitesimal generator

$$\mathrm{X} = \xi(x, y)\frac{\partial}{\partial x} + \eta(x, y)\frac{\partial}{\partial y}. \tag{3.89c}$$

We show that the invariance of (3.88) under (3.89a,b) constructively reduces the nth order ODE (3.88) to an $(n-1)$th order ODE plus a quadrature. This can be done in two ways:

(i) reduction of order by canonical coordinates;

(ii) reduction of order by differential invariants.

[In Section 3.4 we show that reduction of order by differential invariants is the natural setting for reducing an nth order ODE which admits a multi-parameter Lie group of transformations.]

We then show how to find the infinitesimal transformation admitted by a given nth order ODE. Recall that for a given first order ODE we saw that admitted infinitesimals (ξ, η) satisfy a single linear partial differential equation whose general solution could not be found without knowing the general solution of the ODE itself; consequently we could not systematically deduce such symmetries. But when $n \geq 2$ it turns out that admitted infinitesimals satisfy a system of linear *determining equations* which have only a finite number of linearly independent solutions. In practice we are usually able to compute explicitly the admitted Lie group of transformations of an nth order ODE if $n \geq 2$.

We also consider the problem of finding all nth order ODE's which admit a given one-parameter Lie group of transformations.

3.3.1 Reduction of Order by Canonical Coordinates

The basic result is summarized in terms of the following theorem:

Theorem 3.3.1-1. *Assume that the nontrivial one-parameter Lie group of transformations* (3.89a,b) *with infinitesimal generator* (3.89c) *is admitted by the nth order ODE* (3.88), $n \geq 2$. *Let* $(r(x,y), s(x,y))$ *be corresponding canonical coordinates satisfying* $\mathrm{X}r = 0$, $\mathrm{X}s = 1$. *Then solving the nth order ODE* (3.88) *reduces to solving an* $(n-1)$*th order ODE*

$$\frac{d^{n-1}z}{dr^{n-1}} = G\left(r, z, \frac{dz}{dr}, \ldots, \frac{d^{n-2}z}{dr^{n-2}}\right) \tag{3.90}$$

where

$$\frac{ds}{dr} = z. \tag{3.91}$$

Proof. In terms of canonical coordinates (r, s), we have

$$\frac{ds}{dr} = \frac{s_x + s_y y'}{r_x + r_y y'}. \tag{3.92}$$

Solving (3.92) for y', we obtain

$$y' = F_0\left(r, s, \frac{ds}{dr}\right), \tag{3.93}$$

for some function $F_0(r, s, \frac{ds}{dr})$. Clearly $r_x + r_y y' \neq 0$ since we assume that (3.89a,b) is nontrivial (i.e. $\frac{\eta}{\xi} \neq y' = \Theta'(x)$ for an arbitrary solution curve $y = \Theta(x)$ of (3.88); in other words the group (3.89a,b) maps solution curves into other solution curves). Then

$$\frac{d^2 s}{dr^2} = \frac{d\left(\frac{s_x + s_y y'}{r_x + r_y y'}\right)}{dr} = y''\left[\frac{s_y r_x - s_x r_y}{(r_x + r_y y')^3}\right] + f_1\left(r, s, \frac{ds}{dr}\right), \tag{3.94}$$

for some function $f_1(r, s, \frac{ds}{dr})$. This leads to

$$y'' = \frac{d^2 s}{dr^2} F_1\left(r, s, \frac{ds}{dr}\right) + G_1\left(r, s, \frac{ds}{dr}\right), \tag{3.95}$$

where

$$F_1\left(r, s, \frac{ds}{dr}\right) = \frac{(r_x + r_y y')^3}{s_y r_x - s_x r_y},$$

and

$$G_1\left(r, s, \frac{ds}{dr}\right) = -f_1 F_1.$$

Since (r, s) are canonical coordinates, it follows that $s_y r_x - s_x r_y \neq 0$. Inductively one can show that

$$\frac{d^k s}{dr^k} = y_k\left[\frac{s_y r_x - s_x r_y}{(r_x + r_y y')^{k+1}}\right] + f_{k-1}\left(r, s, \frac{ds}{dr}, \ldots, \frac{d^{k-1}s}{dr^{k-1}}\right), \tag{3.96}$$

for some function

$$f_{k-1}\left(r, s, \frac{ds}{dr}, \ldots, \frac{d^{k-1}s}{dr^{k-1}}\right), \quad k = 2, 3, \ldots.$$

This leads to

$$y_k = \frac{d^k s}{dr^k} F_{k-1}\left(r, s, \frac{ds}{dr}\right) + G_{k-1}\left(r, s, \frac{ds}{dr}, \ldots, \frac{d^{k-1}s}{dr^{k-1}}\right), \tag{3.97}$$

where

$$F_{k-1}\left(r, s, \frac{ds}{dr}\right) = \frac{(r_x + r_y y')^{k+1}}{s_y r_x - s_x r_y},$$

and

$$G_{k-1}\left(r, s, \frac{ds}{dr}, \dots, \frac{d^{k-1}s}{dr^{k-1}}\right) = -f_{k-1}F_{k-1}, \quad k = 2, 3, \dots .$$

[Note that $F_{k-1} \neq 0, \infty$ for $k = 2, 3, \dots$.]

Thus it follows that in terms of canonical coordinates $(r(x,y), s(x,y))$ the nth order ODE (3.88) can be written as an nth order ODE in solved form

$$\frac{d^n s}{dr^n} = F\left(r, s, \frac{ds}{dr}, \dots, \frac{d^{n-1}s}{dr^{n-1}}\right), \tag{3.98}$$

for some function $F\left(r, s, \frac{ds}{dr}, \dots, \frac{d^{n-1}s}{dr^{n-1}}\right)$. But ODE (3.98) admits the group

$$r^* = r, \tag{3.99a}$$

$$s^* = s + \epsilon. \tag{3.99b}$$

Hence F is independent of s. Consequently ODE (3.88) reduces to (3.90), (3.91). □

Note that if

$$z = \phi(r; C_1, C_2, \dots, C_{n-1})$$

is the general solution of (3.90) then the general solution of ODE (3.88) is

$$s(x,y) = \int^{r(x,y)} \phi(\rho; C_1, C_2, \dots, C_{n-1})d\rho + C_n,$$

where $C_1, C_2, \dots, C_n$ are n essential constants of integration. Hence the invariance of the nth order ODE (3.88) under a one-parameter Lie group of transformations reduces (3.88) to an $(n-1)$th order ODE plus a quadrature.

3.3.2 Reduction of Order by Differential Invariants

The nth order ODE

$$\mathcal{F}(x, y, y_1, \dots, y_n) = y_n - f(x, y, y_1, \dots, y_{n-1}) = 0 \tag{3.100}$$

admits the group (3.89a,b) if and only if

$$\mathrm{X}^{(n)}\mathcal{F} = 0 \quad \text{when} \quad \mathcal{F} = 0$$

where $X^{(n)}$ is the nth extension of the infinitesimal generator X given by (3.89c). Hence it follows that $\mathcal{F}$ is some function of the invariants

$$u(x,y),\ v_1(x,y,y_1),\ldots,v_n(x,y,y_1,\ldots,y_n), \tag{3.101}$$

which satisfy

$$Xu(x,y)=0,$$

$$X^{(k)}v_k(x,y,y_1,\ldots,y_k)=0 \quad \text{with} \quad \frac{\partial v_k}{\partial y_k}\neq 0, \quad k=1,2,\ldots,n.$$

Clearly for the nth extension of the group (3.89a,b), we have $u^*=u$, $v_k^*=v_k$, $k=1,2,\ldots,n$; $v_k(x,y,y_1,\ldots,y_k)$ is a constant of integration of the characteristic equations

$$\frac{dx}{\xi(x,y)}=\frac{dy}{\eta(x,y)}=\frac{dy_1}{\eta^{(1)}(x,y,y_1)}=\cdots=\frac{dy_k}{\eta^{(k)}(x,y,y_1,\ldots,y_k)}$$

where $\eta^{(k)}$ is given by (2.110), $k=1,2,\ldots,n$.

For any set of invariants (3.101) the ODE (3.100) becomes

$$G(u,v_1,v_2,\ldots,v_n)=0 \tag{3.102}$$

for some function $G(u,v_1,v_2,\ldots,v_n)$. We now show that one can always choose the invariants (3.101), without the computation of a canonical coordinate $s(x,y)$, so that (3.102) becomes an $(n-1)$th order ODE. Moreover, the nth order ODE reduces to an $(n-1)$th order ODE plus a quadrature. [We have already accomplished this by using canonical coordinates $(r(x,y),s(x,y))$ with $u(x,y)=r(x,y)$ and

$$v_k(x,y,y_1,\ldots,y_k)=\frac{d^ks}{dr^k}, \quad k=1,2,\ldots,n.]$$

In Section 3.2.4 we showed how to find explicit invariants $u(x,y)$ and $v_1(x,y,y_1)=v(x,y,y_1)$ of the first extension $X^{(1)}$ corresponding to (3.89c). These were determined as constants of integration of the characteristic equations

$$\frac{dx}{\xi(x,y)}=\frac{dy}{\eta(x,y)}=\frac{dy_1}{\eta_x+(\eta_y-\xi_x)y_1-\xi_y(y_1)^2}. \tag{3.103}$$

Recall that if we could determine $u(x,y)$ explicitly then the computation of $v(x,y,y_1)$ reduced to quadrature.

Since $u(x,y)$, $v(x,y,y_1)$ are invariants under the action of the kth extended group of (3.89a,b), it follows that $\frac{dv}{du}$ is a group invariant under the action of the $(k+1)$th extended group of (3.89a,b) since

$$\left(\frac{dv}{du}\right)^*=\frac{dv^*}{du^*}=\frac{dv}{du}, \quad k\geq 1.$$

Continuing inductively, we see that

$$\frac{dv}{du}, \frac{d^2v}{du^2}, \dots, \frac{d^{n-1}v}{du^{n-1}}$$

are invariants of the nth extended group of (3.89a,b). These invariants are called *differential invariants* of the nth extended group of (3.89a,b). Moreover such differential invariants can be constructed for any choice of invariants $u(x,y)$, $v(x,y,y_1)$ $\left[\frac{\partial v}{\partial y_1} \neq 0\right]$ of the first extended group of (3.89a,b).

Then

$$\begin{aligned}
\frac{dv}{du} &= \frac{\frac{\partial v}{\partial x} + y_1 \frac{\partial v}{\partial y} + y_2 \frac{\partial v}{\partial y_1}}{\frac{\partial u}{\partial x} + y_1 \frac{\partial u}{\partial y}} \\
&= v_2(x, y, y_1, y_2) \\
&= y_2 \left[\frac{\frac{\partial v}{\partial y_1}}{\frac{\partial u}{\partial x} + y_1 \frac{\partial u}{\partial y}} \right] + g_1(x, y, y_1),
\end{aligned}$$

for some function $g_1(x,y,y_1)$. Inductively one can show that

$$\begin{aligned}
\frac{d^k v}{du^k} &= v_{k+1}(x, y, y_1, \dots, y_{k+1}) \\
&= y_{k+1} \left[\frac{\frac{\partial v}{\partial y_1}}{\left(\frac{\partial u}{\partial x} + y_1 \frac{\partial u}{\partial y} \right)^k} \right] + g_k(x, y, y_1, \dots, y_k)
\end{aligned}$$

for some function $g_k(x,y,y_1,\dots,y_k)$, $k = 1, 2, \dots, n-1$. Consequently the invariants $\{v_k(x,y,y_1,\dots,y_k)\}$, $k = 2, 3, \dots, n$, are constructed as differential invariants. Moreover it should be noted that

$$y_{k+1} = \frac{d^k v}{du^k} A_k(x, y, y_1) + B_k(x, y, y_1, \dots, y_k)$$

where

$$A_k(x, y, y_1) = \frac{\left(\frac{\partial u}{\partial x} + y_1 \frac{\partial u}{\partial y} \right)^k}{\frac{\partial v}{\partial y_1}},$$

$$B_k = -A_k g_k, \quad k = 1, 2, \dots, n-1.$$

[Note that $A_k \neq 0, \infty$ for $k = 1, 2, \dots, n-1$.]

Hence it follows that *constructively*, in terms of *differential invariants*, the reduced equation (3.102) is an $(n-1)$th order ODE

$$\frac{d^{n-1}v}{du^{n-1}} = H\left(u, v, \frac{dv}{du}, \dots, \frac{d^{n-2}v}{du^{n-2}}\right), \tag{3.104}$$

for some function $H\left(u, v, \frac{dv}{du}, \ldots, \frac{d^{n-2}v}{du^{n-2}}\right)$. Moreover if

$$v = \phi(u; C_1, C_2, \ldots, C_{n-1})$$

is the general solution of (3.104) where $C_1, C_2, \ldots, C_{n-1}$ are arbitrary constants, then the solution of the nth order ODE (3.100) is found by solving the first order ODE

$$v(x, y, y_1) = \phi(u(x, y); C_1, C_2, \ldots, C_{n-1})$$

which reduces to quadrature since it admits (3.89a,b).

3.3.3 Examples of Reduction of Order

(1) *Second Order Linear Homogeneous Equation (Invariance Under Scaling)*

Consider the second order linear homogeneous ODE

$$y'' + p(x)y' + q(x)y = 0$$

or, equivalently,

$$y_2 + p(x)y_1 + q(x)y = 0. \tag{3.105}$$

ODE (3.105) admits the one-parameter (α) Lie group of transformations

$$x^* = x, \tag{3.106a}$$

$$y^* = \alpha y. \tag{3.106b}$$

(i) *Reduction of Order by Canonical Coordinates.* Canonical coordinates corresponding to (3.106a,b) are

$$r(x, y) = x, \quad s(x, y) = \log y.$$

Then

$$\frac{ds}{dr} = \frac{y'}{y}, \tag{3.107}$$

so that

$$y' = y\frac{ds}{dr}.$$

Consequently

$$\begin{aligned} y'' &= y'\frac{ds}{dr} + y\frac{d^2s}{dr^2} \\ &= y\left(\frac{ds}{dr}\right)^2 + y\frac{d^2s}{dr^2}. \end{aligned}$$

Now let $z = \frac{ds}{dr}$. Then the second order ODE (3.105) reduces to the first order ODE of Riccati type

$$\frac{dz}{dr} + z^2 + p(r)z + q(r) = 0. \tag{3.108}$$

Note that (3.107) is the Riccati transformation relating (3.108) and (3.105).

(ii) *Reduction of Order by Differential Invariants.* It is obvious that invariants of the first extension of (3.106a,b) are

$$u(x,y) = x, \quad v(x,y,y_1) = \frac{y_1}{y}.$$

The corresponding differential invariant is

$$\frac{dv}{du} = \frac{y_2}{y} - \left(\frac{y_1}{y}\right)^2 = \frac{y_2}{y} - v^2.$$

Hence

$$y_2 = y\frac{dv}{du} + yv^2.$$

Consequently the second order ODE (3.105) reduces again to the Riccati equation

$$\frac{dv}{du} + v^2 + p(u)v + q(u) = 0. \tag{3.109}$$

If

$$v = \phi(u; C_1) \tag{3.110}$$

is the general solution of (3.109), then it follows from Section 3.3.2 that the first order ODE

$$v(x,y,y_1) = \frac{y_1}{y} = \phi(x; C_1)$$

admits (3.106a,b) which leads to its reduction to quadrature. In particular

$$\frac{dy}{y} = \phi(x; C_1)dx$$

or

$$y = C_2 e^{\int \phi(x;C_1)dx}.$$

(2) *Second Order Linear Homogeneous Equation* (*Reduction of Order from Knowledge of a Particular Solution*)

Suppose $y = \psi(x)$ is a particular solution of (3.105). Then the second order ODE (3.105) admits the one-parameter (ϵ) Lie group of transformations

$$x^* = x, \tag{3.111a}$$

$$y^* = y + \epsilon\psi(x). \tag{3.111b}$$

(i) *Reduction of Order by Canonical Coordinates.* Canonical coordinates corresponding to (3.111a,b) are

$$r(x,y) = x, \quad s(x,y) = \frac{y}{\psi(x)}.$$

Then

$$\frac{ds}{dr} = \frac{y'}{\psi(x)} - \frac{y\psi'(x)}{[\psi(x)]^2},$$

so that

$$y' = \psi(r)\frac{ds}{dr} + s\psi'(r).$$

Hence

$$y'' = 2\psi'(r)\frac{ds}{dr} + \psi(r)\frac{d^2s}{dr^2} + \psi''(r)s.$$

Let $z = \frac{ds}{dr}$. Then (3.105) reduces to the first order linear ODE

$$\psi(r)\frac{dz}{dr} + [2\psi'(r) + p(r)\psi(r)]z = 0.$$

(ii) *Reduction of Order by Differential Invariants.* Clearly $u(x,y) = x$. The invariant $v(x,y,y_1)$ of the first extension of (3.111a,b) $[\xi(x,y) = 0, \eta(x,y) = \psi(x)]$ is a constant of integration of the corresponding characteristic equations (3.103) which here become

$$\frac{dx}{0} = \frac{dy}{\psi(x)} = \frac{dy_1}{\psi'(x)}.$$

Hence

$$v(x,y,y_1) = y_1 - y\frac{\psi'(x)}{\psi(x)} = y_1 - y\frac{\psi'(u)}{\psi(u)}.$$

Then

$$\frac{dv}{du} = y_2 - y_1\frac{\psi'(u)}{\psi(u)} + y\left[\left(\frac{\psi'(u)}{\psi(u)}\right)^2 - \frac{\psi''(u)}{\psi(u)}\right].$$

Consequently

$$y_1 = y\frac{\psi'(u)}{\psi(u)} + v,$$

$$y_2 = \frac{dv}{du} + v\frac{\psi'(u)}{\psi(u)} + y\frac{\psi''(u)}{\psi(u)}.$$

Hence ODE (3.105) reduces to

$$\frac{dv}{du} + \left[\frac{\psi'(u)}{\psi(u)} + p(u)\right]v = 0. \tag{3.112}$$

If

$$v = \phi(u; C_1) \tag{3.113}$$

is the general solution of (3.112), then the first order ODE

$$v(x, y, y_1) = y_1 - \frac{\psi'(x)}{\psi(x)} y = \phi(x; C_1)$$

admits (3.111a,b) and accordingly can be reduced to quadrature. [See example (2) of Section 3.2.1.]

(3) *Blasius Equation*

The *Blasius equation* resulting from the Prandtl–Blasius problem for a flat plate [cf. Section 1.3.1] is

$$y''' + \frac{1}{2} y\, y'' = 0,$$

or, equivalently,

$$y_3 + \frac{1}{2} y\, y_2 = 0. \tag{3.114}$$

It is easy to see that this third order ODE admits the two-parameter (α, β) Lie group of transformations

$$x^* = \frac{1}{\alpha} x + \beta, \tag{3.115a}$$

$$y^* = \alpha y. \tag{3.115b}$$

We treat (3.115a,b) as two one-parameter groups by considering the parameters α and β separately to reduce (3.114) to two different second order ODE's through the use of differential invariants. Reduction by canonical coordinates is left to Exercise 3.3-2. In Section 3.4.2 we will show how to reduce (3.114) directly to a first order ODE plus two quadratures from invariance under (3.115a,b).

(i) *Reduction of* (3.114) *from Invariance Under Scalings* (α). Obvious invariants of the first extension of $x^* = \frac{1}{\alpha} x$, $y^* = \alpha y$, are

$$u(x, y) = xy, \quad v(x, y, y_1) = \frac{y_1}{y^2}.$$

Then $\frac{du}{dx} = y + x y_1 = y[1 + uv]$. Hence

$$\begin{aligned} y_2 &= \frac{d}{dx}(y^2 v) = 2 y y_1 v + y^2 \frac{dv}{du}\frac{du}{dx} \\ &= y^3 \left[2v^2 + (1 + uv)\frac{dv}{du} \right], \end{aligned}$$

and

$$y_3 = 3y^2y_1\left[2v^2+(1+uv)\frac{dv}{du}\right]+y^3\left\{\frac{d}{du}\left[2v^2+(1+uv)\frac{dv}{du}\right]\right\}\frac{du}{dx}$$

$$= y^4\left[(1+uv)^2\frac{d^2v}{du^2}+u(1+uv)\left(\frac{dv}{du}\right)^2+8v(1+uv)\frac{dv}{du}+6v^3\right].$$

Thus ODE (3.114) reduces to the second order ODE

$$(1+uv)^2\frac{d^2v}{du^2}+u(1+uv)\left(\frac{dv}{du}\right)^2+\left(8v+\frac{1}{2}\right)(1+uv)\frac{dv}{du}+6v^3+v^2=0, \tag{3.116}$$

plus a quadrature. In particular if

$$v=\phi(u;C_1,C_2) \tag{3.117}$$

is the general solution of (3.116), then the first order ODE

$$v=\frac{y_1}{y^2}=\phi(xy;C_1,C_2) \tag{3.118}$$

admits $x^*=\frac{1}{\alpha}x$, $y^*=\alpha y$. Choosing canonical coordinates $s=\log y$, $r=xy$, (3.118) becomes

$$\frac{ds}{dr}=\frac{\phi(r;C_1,C_2)}{r\phi(r;C_1,C_2)+1}.$$

Thus the general solution of the Blasius equation is

$$y=C_3\exp\left[\int^{xy}\frac{\phi(r;C_1,C_2)}{r\phi(r;C_1,C_2)+1}dr\right],$$

where C_1, C_2, C_3 are arbitrary constants.

Note that the second order ODE (3.116) does not admit any obvious one-parameter Lie group of transformations. In particular the group corresponding to the parameter β, namely, $x^*=x+\beta$, $y^*=y$ *does not* induce a group of point transformations admitted by (3.116). The reason for this will be explained in Section 3.4.

(ii) *Reduction of* (3.114) *from Invariance Under Translations* (β). Obvious invariants of the first extension of $x^*=x+\beta$, $y^*=y$, are

$$u(x,y)=y,\quad v(x,y,y_1)=y_1.$$

Then $\frac{du}{dx}=y_1=v$. Thus

$$y_2=\frac{dv}{du}\frac{du}{dx}=v\frac{dv}{du},$$

and

$$y_3 = v^2 \frac{d^2v}{du^2} + v\left(\frac{dv}{du}\right)^2.$$

Hence ODE (3.114) reduces to

$$v\frac{d^2v}{du^2} + \left(\frac{dv}{du}\right)^2 + \frac{1}{2}u\frac{dv}{du} = 0 \tag{3.119}$$

plus a quadrature. In particular if

$$v = \psi(u; C_1, C_2) \tag{3.120}$$

is the general solution of (3.119), then the first order ODE

$$v = y_1 = \psi(y; C_1, C_2) \tag{3.121}$$

admits $x^* = x + \epsilon$, $y^* = y$. Consequently the general solution of (3.114a) is given by

$$\int^y \frac{dz}{\psi(z; C_1, C_2)} = x + C_3,$$

where C_1, C_2, C_3 are arbitrary constants.

Note that the second order ODE (3.119) admits the obvious one-parameter (α) group $u^* = \alpha u$, $v^* = \alpha^2 v$. This group is induced by the invariance of (3.114) under $y^* = \alpha y$, $x^* = \frac{1}{\alpha}x$. Thus (3.119) can be reduced to a first order ODE. Why this is possible will be explained in Section 3.4.

3.3.4 Determining Equations for Infinitesimal Transformations of an nth Order ODE

From Theorem 3.1.1-1 we see that the nth order ODE

$$y_n = f(x, y, y_1, \ldots, y_{n-1}) \tag{3.122}$$

admits the one-parameter Lie group of transformations with infinitesimal generator

$$\mathrm{X} = \xi(x,y)\frac{\partial}{\partial x} + \eta(x,y)\frac{\partial}{\partial y} \tag{3.123}$$

if and only if

$$\mathrm{X}^{(n)}[y_n - f(x, y, y_1, \ldots, y_{n-1})] = 0 \tag{3.124}$$

when $y_n = f(x, y, y_1, \ldots, y_{n-1})$, where

$$\mathrm{X}^{(n)} = \xi(x,y)\frac{\partial}{\partial x} + \eta(x,y)\frac{\partial}{\partial y} + \eta^{(1)}(x,y,y_1)\frac{\partial}{\partial y_1}$$

$$+ \cdots + \eta^{(n)}(x, y, y_1, \ldots, y_n)\frac{\partial}{\partial y_n}$$

is the nth extended infinitesimal generator of (3.123). Recall that $\{\eta^{(k)}(x, y, y_1, \ldots, y_k)\}$ is given by (2.110) in terms of $(\xi(x,y), \eta(x,y))$. Then (3.124) becomes

$$\eta^{(n)} - \left[\xi\frac{\partial f}{\partial x} + \eta\frac{\partial f}{\partial y} + \eta^{(1)}\frac{\partial f}{\partial y_1} + \cdots + \eta^{(n-1)}\frac{\partial f}{\partial y_{n-1}}\right] = 0 \tag{3.125}$$

when $y_n = f(x, y, y_1, \ldots, y_{n-1})$.

From Theorem 2.3.2-2 it follows that if $f(x, y, y_1, \ldots, y_{n-1})$ is a polynomial in $y_1, y_2, \ldots, y_{n-1}$, then (3.125) is a polynomial equation in $y_1, y_2, \ldots, y_{n-1}$ whose coefficients are linear homogeneous in $(\xi(x,y),\ \eta(x,y))$ and their partial derivatives up to nth order. Since for any nth order ODE (3.122) we can assign arbitrary values to each of $y, y_1, y_2, \ldots, y_{n-1}$ at any fixed point x, it follows that the coefficient of each polynomial term in (3.125) must vanish. This leads to a system of *linear homogeneous* partial differential equations for $(\xi(x,y),\ \eta(x,y))$. This linear system defines the set of *determining equations* for the infinitesimals admitted by the nth order ODE (3.122). This set is *overdetermined* if $n \geq 2$ since the number of determining equations exceeds two (the number of unknowns ξ and η).

If $f(x, y, y_1, \ldots, y_{n-1})$ is not a polynomial in $y_1, y_2, \ldots, y_{n-1}$, we can still derive a corresponding set of determining equations from (3.125) based on the independence of the variables $y_1, y_2, \ldots, y_{n-1}$ in (3.125).

One can show that a second order ODE admits at most an eight-parameter Lie group of transformations; an nth order ODE ($n > 2$) admits at most an $(n+4)$-parameter Lie group of transformations [cf. Lie (1893, pp. 296–298), Dickson (1924), Ovsiannikov (1982)].

We state some theorems on the forms of infinitesimal generators which can be admitted by ODE's. These theorems cover many ODE's arising in applications and significantly simplify the tedious work involved in setting up and solving the determining equations for the infinitesimals $(\xi(x,y),\ \eta(x,y))$. They are concerned with the dependence of $(\xi(x,y),\ \eta(x,y))$ on y.

Theorem 3.3.4-1. *Consider an nth order ODE,* $n \geq 3$, *of the form*

$$y_n = g(x, y, y_1)y_{n-1} + h(x, y, y_1, \ldots, y_{n-2}). \tag{3.126}$$

If (3.126) *admits infinitesimal generator* (3.123), *then*

$$\xi_y = 0.$$

Theorem 3.3.4-2. *Consider an nth order ODE,* $n \geq 3$, *of the form*

$$y_n = g(x, y)y_{n-1} + h(x, y, y_1, \ldots, y_{n-2}). \tag{3.127}$$

If (3.127) *admits infinitesimal generator* (3.123), *then*

$$\xi_y = 0, \quad \eta_{yy} = 0.$$

Theorem 3.3.4-3. *Consider a second order ODE of the form*

$$y_2 = g(x,y)y_1 + h(x,y). \tag{3.128}$$

If (3.128) *admits infinitesimal generator* (3.123) *with* $\xi_y = 0$, *then*

$$\eta_{yy} = 0.$$

Theorems 3.3.4-1,2,3 are proved in Bluman (1989).
Consider two examples:

(1) *The Lie Group of Transformations Acting on* $\mathbb{R}^2$ *Which Maps Straight Lines into Straight Lines*

This is the Lie group of transformations admitted by the second order ODE

$$y_2 = 0. \tag{3.129}$$

The invariance criterion is

$$\eta^{(2)} = 0 \quad \text{when} \quad y_2 = 0,$$

where $\eta^{(2)}$ is given by (2.112), i.e., $(\xi(x,y), \eta(x,y))$ satisfy

$$\eta_{xx} + [2\eta_{xy} - \xi_{xx}]y_1 + [\eta_{yy} - 2\xi_{xy}](y_1)^2 - \xi_{yy}(y_1)^3 = 0. \tag{3.130}$$

Equation (3.130) is a polynomial equation of degree three in y_1. Equating to zero the coefficients of each polynomial term in (3.130), we obtain the set of *determining equations*

$$\xi_{yy} = 0, \tag{3.131a}$$

$$\eta_{xx} = 0, \tag{3.131b}$$

$$\eta_{yy} - 2\xi_{xy} = 0, \tag{3.131c}$$

$$\xi_{xx} - 2\eta_{xy} = 0. \tag{3.131d}$$

From (3.131a,b) we get

$$\xi = a(x)y + b(x),$$
$$\eta = c(y)x + d(y).$$

Then (3.131c,d) lead to

$$xc''(y) + d''(y) - 2a'(x) = 0, \tag{3.132a}$$

$$ya''(x) + b''(x) - 2c'(y) = 0. \tag{3.132b}$$

Taking $\frac{\partial}{\partial x}$ of (3.132a) and $\frac{\partial}{\partial y}$ of (3.132b), we see that $c''(y) = a''(x) = 0$. This leads to the eight-parameter Lie group of projective transformations

(2.179a,b) admitted by (3.126) since the solution of the set of determining equations (3.131a–d) is

$$\xi = \alpha_1 x^2 + \alpha_2 xy + \alpha_3 x + \alpha_4 y + \alpha_5,$$

$$\eta = \alpha_1 xy + \alpha_2 y^2 + \alpha_6 x + \alpha_7 y + \alpha_8,$$

where $\alpha_1, \alpha_2, \ldots, \alpha_8$ are arbitrary constants.

(2) *Blasius Equation*

The invariance criterion for the Blasius equation

$$y_3 + \frac{1}{2} y y_2 = 0 \tag{3.133}$$

is

$$\eta^{(3)} + \frac{1}{2} y_2 \eta + \frac{1}{2} y \eta^{(2)} = 0 \quad \text{when} \quad y_3 = -\frac{1}{2} y y_2, \tag{3.134}$$

where $\eta^{(2)}$, $\eta^{(3)}$ are given by (2.112), (2.113). Then (3.134) is the polynomial equation

$$\begin{aligned} &\left[\eta_{xxx} + \frac{1}{2} y \eta_{xx}\right] + \left[3\eta_{xxy} - \xi_{xxx} + y\eta_{xy} - \frac{1}{2} y \xi_{xx}\right] y_1 \\ &+ \left[3\eta_{xyy} - 3\xi_{xxy} + \frac{1}{2} y \eta_{yy} - y\xi_{xy}\right] (y_1)^2 - \left[\eta_{yyy} - 3\xi_{xyy} - \frac{1}{2} y \xi_{yy}\right] (y_1)^3 \\ &- \xi_{yyy}(y_1)^4 + \left[3\eta_{xy} - 3\xi_{xx} + \frac{1}{2} y \xi_x + \frac{1}{2}\eta\right] y_2 \\ &+ \left[3\eta_{yy} - 9\xi_{xy} + \frac{1}{2} y \xi_y\right] y_1 y_2 - 6\xi_{yy}(y_1)^2 y_2 - 3\xi_y (y_2)^2 = 0. \end{aligned} \tag{3.135}$$

The resulting set of determining equations for $(\xi(x,y), \eta(x,y))$ is:

$$\eta_{xxx} + \frac{1}{2} y \eta_{xx} = 0, \tag{3.136a}$$

$$3\eta_{xxy} - \xi_{xxx} + y\eta_{xy} - \frac{1}{2} y \xi_{xx} = 0, \tag{3.136b}$$

$$3\eta_{xyy} - 3\xi_{xxy} + \frac{1}{2} y \eta_{yy} - y \xi_{xy} = 0, \tag{3.136c}$$

$$3\xi_{xxy} + \frac{1}{2} y \xi_{yy} - \eta_{yyy} = 0, \tag{3.136d}$$

$$\xi_{yyy} = 0, \tag{3.136e}$$

$$3\eta_{xy} - 3\xi_{xx} + \frac{1}{2} y \xi_x + \frac{1}{2}\eta = 0, \tag{3.136f}$$

$$3\eta_{yy} - 9\xi_{xy} + \frac{1}{2}y\xi_y = 0, \tag{3.136g}$$

$$\xi_{yy} = 0, \tag{3.136h}$$

$$\xi_y = 0. \tag{3.136i}$$

Immediately one sees that if (3.136i) is satisfied $[\xi = \xi(x)]$ then so are (3.136h,e), and (3.136g) leads to

$$\eta_{yy} = 0, \tag{3.137}$$

so that (3.136c,d) are also satisfied. Taking $\frac{\partial}{\partial y}$ of (3.136b) and $\frac{\partial^2}{\partial x \partial y}$ of (3.136f), we are led to

$$\eta_{xy} = \xi''(x) = 0. \tag{3.138}$$

Moreover (3.136b) is satisfied. Then

$$\xi = \alpha + \beta x, \tag{3.139a}$$

$$\eta = \gamma y + a(x). \tag{3.139b}$$

Equation (3.136f) leads to $a(x) = 0$, $\gamma = -\beta$. No further restrictions arise from (3.136a). Hence the Blasius equation (3.133) only admits a two-parameter Lie group of transformations corresponding to the infinitesimals

$$\xi = \alpha + \beta x, \tag{3.140a}$$

$$\eta = -\beta y, \tag{3.140b}$$

where α, β are arbitrary constants.

If we use Theorem 3.3.4-2 the computations for finding $(\xi(x,y), \eta(x,y))$ simplify significantly. Since ODE (3.133) is of the form (3.127) it immediately follows that $\xi_y = 0$, $\eta_{yy} = 0$. Hence the set of determining equations resulting from (3.135) reduces to (3.136a,b,f).

3.3.5 Determination of nth Order ODE's Invariant Under a Given Group

Now we consider the problem of finding all nth order ODE's which admit a given one-parameter Lie group of transformations. This is accomplished by a simple extension of the procedure outlined in Section 3.2.4 for first order ODE's using either canonical coordinates or differential invariants.

Suppose (3.122) admits a given one-parameter Lie group of transformations with infinitesimal generator (3.123). We may proceed in two ways:

(i) *Method of Canonical Coordinates.* Corresponding to (3.123) we compute canonical coordinates $(r(x,y), s(x,y))$ satisfying $\mathrm{X}r = 0$, $\mathrm{X}s = 1$, so that the group admitted by (3.122) is now

$$r^* = r, \quad s^* = s + \epsilon. \tag{3.141}$$

Then invariants under (3.141) are

$$\frac{d^k s}{dr^k}, \quad k = 1, 2, \ldots, n.$$

These invariants can be expressed in terms of $x, y, y', \ldots, y^{(n)}$ through (3.92), (3.96). Consequently the most general nth order ODE, written in solved form, which admits (3.123), is given by

$$\frac{d^n s}{dr^n} = G\left(r, \frac{ds}{dr}, \frac{d^2 s}{dr^2}, \ldots, \frac{d^{n-1} s}{dr^{n-1}}\right) \tag{3.142}$$

where

$$G\left(r, \frac{ds}{dr}, \frac{d^2 s}{dr^2}, \ldots, \frac{d^{n-1} s}{dr^{n-1}}\right)$$

is an arbitrary function of its arguments. Note that ODE (3.142) is an $(n-1)$th order ODE for $\frac{ds}{dr}$.

(ii) *Method of Differential Invariants.* Here we first find invariants $u(x, y)$, $v(x, y, y_1)$ as we did for first order ODE's. Then we compute differential invariants

$$\frac{d^k v}{du^k}, \quad k = 1, 2, \ldots, n-1,$$

which can be expressed in terms of $x, y, y_1, \ldots, y_n$ as indicated in Section 3.3.2. Consequently the most general nth order ODE (in solved form) which admits (3.123) can be written in the form

$$\frac{d^{n-1} v}{du^{n-1}} = H\left(u, v, \frac{dv}{du}, \ldots, \frac{d^{n-2} v}{du^{n-2}}\right) \tag{3.143}$$

where

$$H\left(u, v, \frac{dv}{du}, \ldots, \frac{d^{n-2} v}{du^{n-2}}\right)$$

is an arbitrary function of its arguments. Note that ODE (3.143) is an $(n-1)$th order ODE for v.

If one wishes to find the most general nth order ODE which admits a particular *scaling* group it is simpler to directly compute $n+1$ invariants which respectively depend on (x, y), $(x, y, y_1), \ldots, (x, y, y_1, \ldots, y_n)$. The disadvantage of the direct approach is that reduction of order from n to $n-1$ is not automatic as is the case when the most general nth order ODE is obtained through canonical coordinates or differential invariants.

As an example we use these methods to find the most general second order ODE which admits the scaling group (3.75a,b). The reader is referred to the calculations in Section 3.2.4 where we found the most general first order ODE which admits (3.75a,b).

(1) *Method of Canonical Coordinates*

Canonical coordinates are

$$r = \frac{y}{x}, \quad s = \log y; \quad \frac{ds}{dr} = \frac{y'}{ry' - r^2};$$

$$\begin{aligned}\frac{d^2s}{dr^2} &= -\frac{r^2yy''}{(ry'-r^2)^3} + \frac{2ry' - (y')^2}{(ry'-r^2)^2} \\ &= -\frac{r^2yy''}{(ry'-r^2)^3} + \frac{2ry'}{(ry'-r^2)^2} - \left(\frac{ds}{dr}\right)^2.\end{aligned}$$

Hence the most general second order ODE which admits (3.75a,b) is given by

$$y'' = \frac{2(xy'-y)y'}{xy} + \frac{(xy'-y)^3}{x^4} G\left(\frac{y}{x}, \frac{x^2y'}{xyy'-y^2}\right) \tag{3.144}$$

where G is an arbitrary function of its arguments.

(2) *Method of Differential Invariants*

From (3.77) we see that $u = \frac{y}{x}$, $v = y'$. Then $\frac{dv}{du} = \frac{x^2y''}{xy'-y}$. Consequently the most general second order ODE which admits (3.75a,b) is given by

$$y'' = \frac{xy'-y}{x^2} H\left(\frac{y}{x}, y'\right) \tag{3.145}$$

where H is an arbitrary function of its arguments. Of course (3.144) and (3.145) must yield the *same* general second order ODE! In comparing (3.145) and (3.144) note that

$$\frac{x^2y'}{xyy'-y^2} = \frac{y'}{\left(\frac{y}{x}\right)y' - \left(\frac{y}{x}\right)^2}.$$

(3) *Direct Approach*

Obvious invariants of the scaling group (3.75a,b) are $\frac{y}{x}$, y', yy''. Hence the most general second order ODE which admits (3.75a,b) is given by

$$y'' = \frac{1}{y} I\left(\frac{y}{x}, y'\right) \tag{3.146}$$

where I is an arbitrary function of its arguments. In comparing (3.145) and (3.146) note that

$$\frac{xy'-y}{x^2} = \frac{1}{y}\left[\left(\frac{y}{x}\right)y' - \left(\frac{y}{x}\right)^2\right].$$

Exercises 3.3

1. Let $y = \phi(x)$ be a particular solution of the second order linear nonhomogeneous ODE

$$y'' + p(x)y' + q(x)y = g(x). \tag{3.147}$$

   (a) Find a one-parameter Lie group of transformations admitted by (3.147).

   (b) Use canonical coordinates to reduce (3.147) to a first order ODE plus a quadrature.

   (c) Use differential invariants to reduce (3.147) to a first order ODE plus a quadrature.

2. The third order Blasius equation (3.114) admits the two-parameter (α, β) Lie group of transformations (3.115a,b).

   (a) Use canonical coordinates associated with parameter β to reduce (3.114) to a second order ODE.

   (b) Find a symmetry of this second order ODE. How is this symmetry related to the parameter α of (3.115a,b)?

   (c) Find canonical coordinates for the symmetry of (b). Consequently reduce the Blasius equation to a first order ODE plus two quadratures.

3. Find the Lie group of transformations admitted by the ODE

$$y'' = \alpha(y')^N$$

   where $\alpha =$ const and N is a fixed integer; $N = 1, 2, 3, \ldots$. Investigate further the cases $N = 1, 2, 3$. Compare these cases with the group admitted by $y'' = 0$. [For related ODE's see Aguirre and Krause (1985).]

4. Consider the ODE

$$2\frac{d}{dx}\left(K(y)\frac{dy}{dx}\right) + x\frac{dy}{dx} = 0 \tag{3.148}$$

   which arises from studying the nonlinear diffusion equation.

   (a) Find an admitted Lie group of transformations when

      (i) $K(y) = \lambda(y+\kappa)^{\nu}$ where λ, κ, ν are arbitrary constants;

      (ii) $K(y) = \lambda e^{\nu y}$ where λ, ν are arbitrary constants.

   (b) Find all $K(y)$ for which (3.148) admits a Lie group of transformations.

5. Consider the ODE

$$K(y')y'' + xy' - y = 0. \tag{3.149}$$

(a) Find a Lie group of transformations admitted by (3.149) when

(i) $K(y') = \dfrac{1}{(y')^2+1}$;

(ii) $K(y') = \dfrac{1}{(y')^2+1}\exp[\lambda \arctan y']$ where $\lambda = \text{const.}$

(b) Solve (3.149) when $K(y') = \dfrac{1}{(y')^2+1}$.

6. Find the Lie group of transformations admitted by the family of curves $y = p(x)$ where $p(x)$ is an arbitrary polynomial of degree n.

7. Find the most general second order ODE which admits the scaling group $x^* = \alpha x$, $y^* = \alpha^k y$, k fixed. Reduce the ODE to a first order ODE plus a quadrature.

8. (a) Find the most general second order ODE which admits the rotation group.

(b) Interpret your answer geometrically.

(c) Find the general solution of the ODE

$$(x^2+y^2)y'' + 2(y - xy')(1 + (y')^2) = 0.$$

3.4 Invariance of ODE's Under Multi-Parameter Groups

Now we consider the invariance of an nth order ODE under an r-parameter Lie group of transformations, $2 \le r \le n$. If the corresponding r-dimensional Lie algebra is *solvable* [cf. Section 2.4.4] then we will show that the given nth order ODE can be reduced to an $(n-r)$th order ODE plus r quadratures. Theorem 2.4.2-2 showed that any two-dimensional Lie algebra is solvable. It turns out that every even-dimensional ($r = 2m$ for some integer m) Lie algebra acting on $\mathbb{R}^n$ contains a two-dimensional subalgebra. [Cf. Cohen (1911, p. 150), Dickson (1924). Both of these authors erroneously claim that this is true for *any* real Lie algebra acting on $\mathbb{R}^n$. Their proofs only hold for complex Lie algebras or real Lie algebras of even dimension. It is easy to show that the three-dimensional Lie algebra acting on $\mathbb{R}^3$, corresponding to the rotation group acting on $\mathbb{R}^3$, is not solvable.] By constructing all possible Lie algebras acting on $\mathbb{R}^2$, Lie (1893, pp. 292–362) [see also Campbell (1903, Chap. XXI, XXII)] implicitly showed that every

r-dimensional Lie algebra *acting on* $\mathbb{R}^2$ has a two-dimensional subalgebra. Since $\xi(x,y)\frac{\partial}{\partial x} + \eta(x,y)\frac{\partial}{\partial y}$ acts on $\mathbb{R}^2$, it follows that if an r-parameter Lie group is admitted by an nth order ODE, $2 \le r \le n$, then the order of the ODE can be reduced by at least two.

3.4.1 Invariance of a Second Order ODE Under a Two-Parameter Group

We show that if a second order ODE

$$y'' = f(x, y, y') \tag{3.150}$$

admits a two-parameter Lie group of transformations, then one can construct the general solution of (3.150) through a reduction to two quadratures.

Let X_1, X_2 be basis generators of the Lie algebra of the given two-parameter Lie group of transformations and let $X_i^{(k)}$ denote the kth extended infinitesimal generator of X_i, $i = 1, 2$. From Theorem 2.4.4-2, without loss of generality, we can assume that

$$[X_1, X_2] = \lambda X_1 \quad \text{for some constant } \lambda. \tag{3.151}$$

Let $u(x,y)$, $v(x,y,y_1)$ be invariants of $X_1^{(2)}$ such that

$$X_1 u = 0, \quad X_1^{(1)} v = 0.$$

Then the differential invariant $\frac{dv}{du}$ satisfies the equation [cf. Section 3.3.2]

$$X_1^{(2)} \frac{dv}{du} = 0,$$

and hence (3.150) reduces to

$$\frac{dv}{du} = H(u, v) \tag{3.152}$$

for some function $H(u,v)$. [Note that $\frac{\partial v}{\partial y_1} \neq 0$.] From the commutation relation (3.151) it follows that

$$X_1 X_2 u = X_2 X_1 u + \lambda X_1 u = 0.$$

Hence

$$X_2 u = \alpha(u) \tag{3.153a}$$

for some function $\alpha(u)$.

Similarly from Theorem 2.4.4-2 it follows that

$$X_1^{(1)} X_2^{(1)} v = 0, \quad X_1^{(2)} X_2^{(2)} \frac{dv}{du} = 0.$$

Hence

$$X_2^{(1)}v = \beta(u,v) \tag{3.153b}$$

for some function $\beta(u,v)$. Since (3.150) admits X_2 it follows that

$$X_2^{(2)}\left(\frac{dv}{du} - H(u,v)\right) = 0 \quad \text{when} \quad \frac{dv}{du} = H(u,v).$$

From (3.153a,b) it follows that in terms of (u,v) coordinates $X_2^{(1)}$ becomes

$$X_2^{(1)} = \alpha(u)\frac{\partial}{\partial u} + \beta(u,v)\frac{\partial}{\partial v},$$

and that this infinitesimal generator is admitted by (3.152). Let canonical coordinates $(R(u,v), S(u,v))$ be such that

$$X_2^{(1)}R = 0,$$

$$X_2^{(1)}S = 1.$$

Then $(R(u,v), S(u,v))$ satisfy

$$\alpha(u)\frac{\partial R}{\partial u} + \beta(u,v)\frac{\partial R}{\partial v} = 0,$$

$$\alpha(u)\frac{\partial S}{\partial u} + \beta(u,v)\frac{\partial S}{\partial v} = 1.$$

Thus the one-parameter Lie group of transformations

$$R^* = R, \tag{3.154a}$$

$$S^* = S + \epsilon, \tag{3.154b}$$

is admitted by (3.152). Hence (3.152) reduces to

$$\frac{dS}{dR} = I(R) \tag{3.155}$$

for some function $I(R)$. Then (3.155) integrates out to

$$S(u,v) = \int^{R(u,v)} I(R)dR + C_1, \tag{3.156}$$

where C_1 is an arbitrary constant. The differential equation

$$S(u(x,y), v(x,y,y')) = \int^{R(u(x,y),v(x,y,y'))} I(R)dR + C_1$$

admits X_1 and hence reduces to quadrature by the method of canonical coordinates after one determines $(r(x,y), s(x,y))$ such that

$$X_1 r = 0, \quad X_1 s = 1.$$

Consequently any second order ODE which admits a two-parameter Lie group of transformations reduces completely to quadratures.

As an example consider the second order linear nonhomogeneous ODE

$$y'' + p(x)y' + q(x)y = g(x). \tag{3.157}$$

Let $z = \phi_1(x)$, $z = \phi_2(x)$, be linearly independent solutions of the corresponding homogeneous equation

$$z'' + p(x)z' + q(x)z = 0. \tag{3.158}$$

Then (3.157) admits the two-parameter (ϵ_1, ϵ_2) Lie group of transformations

$$x^* = x, \tag{3.159a}$$

$$y^* = y + \epsilon_1\phi_1(x) + \epsilon_2\phi_2(x). \tag{3.159b}$$

The corresponding infinitesimal generators are

$$\mathrm{X}_1 = \phi_1(x)\frac{\partial}{\partial y}, \quad \mathrm{X}_2 = \phi_2(x)\frac{\partial}{\partial y}$$

with $[\mathrm{X}_1, \mathrm{X}_2] = 0$. Then

$$\mathrm{X}_i^{(1)} = \phi_i(x)\frac{\partial}{\partial y} + \phi_i'(x)\frac{\partial}{\partial y_1}, \quad i = 1, 2;$$

$$u = x, \quad v = \frac{y'}{\phi_1'(x)} - \frac{y}{\phi_1(x)};$$

$$\mathrm{X}_2 u = \mathrm{X}_2 x = 0, \quad \mathrm{X}_2^{(1)} v = \frac{\phi_2'(x)}{\phi_1'(x)} - \frac{\phi_2(x)}{\phi_1(x)} = \frac{W(x)}{\phi_1(x)\phi_1'(x)}$$

where $W(x)$ is the Wronskian $W(x) = \phi_1\phi_2' - \phi_2\phi_1'$. Now in terms of x and v,

$$\mathrm{X}_2^{(1)} = \frac{W(x)}{\phi_1(x)\phi_1'(x)}\frac{\partial}{\partial v}.$$

Canonical coordinates $(R(x,v), S(x,v))$ satisfy

$$\mathrm{X}_2^{(1)} R = \frac{W}{\phi_1\phi_1'}\frac{\partial R}{\partial v} = 0, \quad \mathrm{X}_2^{(1)} S = \frac{W}{\phi_1\phi_1'}\frac{\partial S}{\partial v} = 1,$$

and hence

$$R = x,$$

$$S = \frac{v\phi_1\phi_1'}{W}.$$

Consequently, by a simple calculation,

$$\frac{dS}{dx} = \frac{g(x)\phi_1(x)}{W(x)},$$

so that

$$S = \frac{y'\phi_1 - y\phi_1'}{W} = \int \frac{g\phi_1}{W} dx + C_1, \tag{3.160}$$

where C_1 is an arbitrary constant.

By construction the first order ODE (3.160) admits $X_1 = \phi_1(x)\frac{\partial}{\partial y}$. In terms of canonical coordinates $r = x$, $s = \frac{y}{\phi_1(x)}$, (3.160) reduces to

$$\frac{ds}{dx} = \frac{W}{(\phi_1)^2}\left[\int \frac{g\phi_1}{W} dx + C_1\right]. \tag{3.161}$$

But $\frac{W}{(\phi_1)^2} = \left(\frac{\phi_2}{\phi_1}\right)'$. Hence

$$\frac{W}{(\phi_1)^2}\int \frac{g\phi_1}{W} dx = \frac{d}{dx}\left[\frac{\phi_1}{\phi_1}\int \frac{g\phi_1}{W} dx\right] - \frac{g\phi_2}{W}.$$

Thus

$$s = C_1\frac{\phi_2}{\phi_1} + \frac{\phi_2}{\phi_1}\int \frac{g\phi_1}{W} dx - \int \frac{g\phi_2}{W} dx + C_2$$

which leads to the familiar general solution

$$y = C_1\phi_2 + C_2\phi_1 + \phi_2\int \frac{g\phi_1}{W} dx - \phi_1\int \frac{g\phi_2}{W} dx \tag{3.162}$$

of (3.157) (derived by variation of parameters in standard textbooks).

3.4.2 Invariance of an nth Order ODE Under a Two-Parameter Group

Now consider the nth order ODE

$$y_n = f(x, y, y_1, \ldots, y_{n-1}), \tag{3.163}$$

$n \geq 3$, assumed to be invariant under a two-parameter Lie group of transformations with infinitesimal generators X_1, X_2 such that $[X_1, X_2] = \lambda X_1$ for some constant λ.

As in Section 3.4.1 let $u(x,y)$, $v(x,y,y_1)$ be invariants of $X_1^{(2)}$. Then $X_1^{(2)}\dot{v} = 0$ where $\dot{v} = \frac{dv}{du}$, and (3.163) reduces to

$$\frac{d^{n-1}v}{du^{n-1}} = H\left(u, v, \frac{dv}{du}, \ldots, \frac{d^{n-2}v}{du^{n-2}}\right) \tag{3.164}$$

for some function

$$H\left(u, v, \frac{dv}{du}, \ldots, \frac{d^{n-2}v}{du^{n-2}}\right).$$

Since $[X_1^{(k)}, X_2^{(k)}] = \lambda X_1^{(k)}$, $k = 1, 2, \ldots$, it follows that

$$X_2 u = \alpha(u),$$

$$X_2^{(1)}v = \beta(u,v),$$
$$X_2^{(2)}\dot{v} = \gamma(u,v,\dot{v}),$$

for some functions $\alpha(u), \beta(u,v), \gamma(u,v,\dot{v})$. Then $X_2^{(1)} = \alpha(u)\frac{\partial}{\partial u} + \beta(u,v)\frac{\partial}{\partial v}$, with first extension given by

$$X_2^{(2)} = \alpha(u)\frac{\partial}{\partial u} + \beta(u,v)\frac{\partial}{\partial v} + \gamma(u,v,\dot{v})\frac{\partial}{\partial \dot{v}},$$

is admitted by (3.164). [Note that $\frac{\partial\gamma}{\partial\dot{v}} \neq 0$. Why?] Let $U(u,v)$, $V(u,v,\dot{v})$ be such that

$$X_2^{(1)}U = 0, \quad X_2^{(2)}V = 0.$$

Then

$$X_2^{(3)}\frac{dV}{dU} = 0.$$

Consequently (3.164), and hence (3.163), reduces to

$$\frac{d^{n-2}V}{dU^{n-2}} = I\left(U, V, \frac{dV}{dU}, \ldots, \frac{d^{n-3}V}{dU^{n-3}}\right) \tag{3.165}$$

for some function $I\left(U, V, \frac{dV}{dU}, \ldots, \frac{d^{n-3}V}{dU^{n-3}}\right)$. If

$$V = \phi(U; C_1, C_2, \ldots, C_{n-2})$$

is the general solution of (3.165), then the first order ODE

$$V\left(u, v, \frac{dv}{du}\right) = \phi(U(u,v); C_1, C_2, \ldots, C_{n-2}) \tag{3.166}$$

admits $X_2^{(1)} = \alpha(u)\frac{\partial}{\partial u} + \beta(u,v)\frac{\partial}{\partial v}$. Hence (3.166) reduces to quadrature

$$v = \psi(u; C_1, C_2, \ldots, C_{n-2}, C_{n-1}).$$

But the first order ODE

$$v(x, y, y') = \psi(u(x,y); C_1, C_2, \ldots, C_{n-2}, C_{n-1}) \tag{3.167}$$

admits X_1. Thus (3.167) reduces to quadrature which leads to the general solution of (3.163).

Hence we have shown that *if an nth order ODE* ($n \geq 3$) *admits a two-parameter Lie group of transformations then it can be reduced constructively to an* $(n-2)$*th order ODE plus two quadratures.* Note that the order of using the operators X_1 and X_2 is *crucial* if $\lambda \neq 0$.

As an example we again consider the Blasius equation

$$y''' + \frac{1}{2}yy'' = 0 \tag{3.168}$$

which admits the two-parameter Lie group of transformations with infinitesimal generators [cf. (3.140a,b)]

$$\mathrm{X}_1 = \frac{\partial}{\partial x}, \quad \mathrm{X}_2 = x\frac{\partial}{\partial x} - y\frac{\partial}{\partial y}.$$

Then

$$[\mathrm{X}_1, \mathrm{X}_2] = \mathrm{X}_1.$$

Invariants of $\mathrm{X}_1^{(2)}$ are

$$u = y, \quad v = y_1, \quad \dot{v} = \frac{dv}{du} = \frac{y_2}{y_1}.$$

It follows that

$$\mathrm{X}_2^{(2)} = x\frac{\partial}{\partial x} - y\frac{\partial}{\partial y} - 2y_1\frac{\partial}{\partial y_1} - 3y_2\frac{\partial}{\partial y_2};$$

$$\mathrm{X}_2 u = -y = -u; \quad \mathrm{X}_2^{(1)} v = -2y_1 = -2v;$$
$$\mathrm{X}_2^{(2)} \dot{v} = -\frac{y_2}{y_1} = -\dot{v}.$$

Without loss of generality we set

$$\mathrm{X}_2^{(2)} = u\frac{\partial}{\partial u} + 2v\frac{\partial}{\partial v} + \dot{v}\frac{\partial}{\partial \dot{v}}.$$

Then

$$\mathrm{X}_2^{(1)} U(u, v) = 0 \quad \text{leads to} \quad U = \frac{v}{u^2},$$

and

$$\mathrm{X}_2^{(2)} V(u, v, \dot{v}) = 0 \quad \text{leads to} \quad V = \frac{\dot{v}}{u}.$$

Then the third order Blasius equation (3.168) yields

$$\begin{aligned} \frac{dV}{dU} &= \frac{d\left(\frac{y_2}{yy_1}\right)}{d\left(\frac{y_1}{y^2}\right)} \\ &= \frac{y^2 y_1 y_3 - y^2 (y_2)^2 - y(y_1)^2 y_2}{y(y_1)^2 y_2 - 2(y_1)^4} \\ &= \frac{\frac{1}{2} y^3 y_1 y_2 + y^2 (y_2)^2 + y(y_1)^2 y_2}{2(y_1)^4 - y(y_1)^2 y_2}, \end{aligned}$$

which in terms of U and V becomes the first order ODE

$$\frac{dV}{dU} = \frac{V}{U}\left[\frac{\frac{1}{2} + V + U}{2U - V}\right]. \tag{3.169}$$

If $V = \phi(U; C_1)$ is the general solution of (3.169) then the first order ODE

$$\dot{v} = \frac{dv}{du} = u\phi\left(\frac{v}{u^2}; C_1\right) \tag{3.170}$$

admits $X_2^{(1)} = u\frac{\partial}{\partial u} + 2v\frac{\partial}{\partial v}$. In terms of corresponding canonical coordinates $s = \log v$, $r = \frac{v}{u^2}$, (3.170) becomes

$$\frac{ds}{dr} = \frac{\phi(r; C_1)}{r[\phi(r; C_1) - 2r]}. \tag{3.171}$$

This leads to the quadrature

$$v = C_2 \exp\left[\int^r \frac{\phi(\rho; C_1)}{\rho[\phi(\rho; C_1) - 2\rho]} d\rho\right], \tag{3.172}$$

where $v = y_1$, $r = \frac{y_1}{y^2}$. In principle (3.172) can be expressed in a solved form

$$y_1 = \psi(y; C_1, C_2)$$

which admits $X_1 = \frac{\partial}{\partial x}$, and hence reduces to quadrature

$$\int \frac{dy}{\psi(y; C_1, C_2)} = x + C_3. \tag{3.173}$$

Equation (3.173) represents a general solution of the Blasius equation.

3.4.3 Invariance of an nth Order ODE Under an r-Parameter Lie Group with a Solvable Lie Algebra

If an r-parameter Lie group ($r \geq 3$) is admitted by an nth order ODE ($n \geq r$) it does not always follow that we can have a reduction to an $(n-r)$th order ODE plus r quadratures. We show that such a reduction is always possible if the Lie algebra $\mathcal{L}^r$, formed by the infinitesimal generators of the group, is a *solvable Lie algebra* [cf. Section 2.4.4]. Then $\mathcal{L}^r$ has a basis set $\{X_1, X_2 \ldots, X_r\}$ satisfying commutation relations of the form

$$[X_i, X_j] = \sum_{k=1}^{j-1} C_{ij}^k X_k, \quad 1 \leq i < j, \quad j = 2, \ldots, r \tag{3.174}$$

for some real structure constants $\{C_{ij}^k\}$ [see Exercise 3.4-7]. For the same constants $\{C_{ij}^k\}$ the corresponding mth extended infinitesimal generators $\{X_j^{(m)}\}$ satisfy

$$[X_i^{(m)}, X_j^{(m)}] = \sum_{k=1}^{j-1} C_{ij}^k X_k^{(m)}, \quad 1 \leq i < j, \quad j = 2, \ldots, r. \tag{3.175}$$

Now consider the nth order ODE

$$y_n = F_n(x, y, y_1, \ldots, y_{n-1}), \tag{3.176}$$

where F_n is a given function of its arguments. We assume that (3.176) admits an r-parameter Lie group of transformations ($3 \leq r \leq n$) whose infinitesimal generators form a solvable Lie algebra. Without loss of generality we can assume that the infinitesimal generators $\{X_i\}$, $i = 1, 2, \ldots, r$, satisfy (3.174).

Let $x_{(1)}(x, y)$, $y_{(1)}(x, y, y_1)$ be such that

$$X_1 x_{(1)} = 0, \quad X_1^{(1)} y_{(1)} = 0.$$

Then

$$X_1^{(k+1)} \frac{d^k y_{(1)}}{dx_{(1)}^k} = 0, \quad k = 1, 2, \ldots, n-1.$$

Let

$$y_{(1)k} = \frac{d^k y_{(1)}}{dx_{(1)}^k}, \quad k = 1, 2, \ldots, n-1.$$

In terms of the invariants $x_{(1)}, y_{(1)}$, and the differential invariants $\{y_{(1)k}\}$, $k = 1, 2, \ldots, n-1$, of $X_1^{(n)}$, ODE (3.176) reduces to the $(n-1)$th order ODE

$$y_{(1)n-1} = F_{n-1}(x_{(1)}, y_{(1)}, y_{(1)1}, \ldots, y_{(1)n-2}), \tag{3.177}$$

for some function F_{n-1} of the indicated invariants of $X_1^{(n)}$.

From (3.174), (3.175) it follows that

$$\begin{aligned} X_2 x_{(1)} &= \alpha_1(x_{(1)}), \\ X_2^{(1)} y_{(1)} &= \beta_1(x_{(1)}, y_{(1)}), \\ X_2^{(2)} y_{(1)1} &= \gamma_1(x_{(1)}, y_{(1)}, y_{(1)1}), \end{aligned}$$

for some functions α_1, β_1, γ_1 of the indicated arguments. Hence

$$X_2^{(1)} = \alpha_1(x_{(1)}) \frac{\partial}{\partial x_{(1)}} + \beta_1(x_{(1)}, y_{(1)}) \frac{\partial}{\partial y_{(1)}},$$

with first extension given by

$$X_2^{(2)} = X_2^{(1)} + \gamma_1(x_{(1)}, y_{(1)}, y_{(1)1}) \frac{\partial}{\partial y_{(1)1}},$$

is admitted by (3.177).

Let $x_{(2)}(x_{(1)}, y_{(1)})$, $y_{(2)}(x_{(1)}, y_{(1)}, y_{(1)1})$ be such that

$$X_2^{(1)} x_{(2)} = 0, \quad X_2^{(2)} y_{(2)} = 0. \tag{3.178}$$

Then

$$X_2^{(2+k)} \frac{d^k y_{(2)}}{dx_{(2)}^k} = 0, \quad k = 1, 2, \ldots, n-2.$$

Let

$$y_{(2)k} = \frac{d^k y_{(2)}}{dx_{(2)}^k}, \quad k = 1, 2, \ldots, n-2.$$

In terms of the invariants $x_{(2)}, y_{(2)}, \{y_{(2)k}\}$, $k = 1, 2, \ldots, n-2$, of $X_2^{(n)}$ (which are also invariants of $X_1^{(n)}$), ODE (3.177), and hence ODE (3.176), reduces to the $(n-2)$th order ODE

$$y_{(2)n-2} = F_{n-2}(x_{(2)}, y_{(2)}, y_{(2)1}, \ldots, y_{(2)n-3}), \tag{3.179}$$

for some function F_{n-2} of invariants of $X_2^{(n)}$, $X_1^{(n)}$.

From (3.174), (3.175) it follows that

$$X_1^{(1)} X_3^{(1)} x_{(2)} = 0, \tag{3.180a}$$

$$X_2^{(1)} X_3^{(1)} x_{(2)} = 0. \tag{3.180b}$$

Then (3.180a) leads to

$$X_3^{(1)} x_{(2)} = A(x_{(1)}, y_{(1)}), \tag{3.181}$$

for some function $A(x_{(1)}, y_{(1)})$. From (3.180b),

$$X_2^{(1)} A(x_{(1)}, y_{(1)}) = 0. \tag{3.182}$$

Then (3.178) leads to

$$X_3^{(1)} x_{(2)} = A(x_{(1)}, y_{(1)}) = \alpha_2(x_{(2)}),$$

for some function $\alpha_2(x_{(2)})$. Similarly

$$\begin{aligned} X_3^{(2)} y_{(2)} &= \beta_2(x_{(2)}, y_{(2)}), \\ X_3^{(3)} y_{(2)1} &= \gamma_2(x_{(2)}, y_{(2)}, y_{(2)1}), \end{aligned}$$

for some functions β_2, γ_2 of the indicated arguments. Hence

$$X_3^{(2)} = \alpha_2(x_{(2)}) \frac{\partial}{\partial x_{(2)}} + \beta_2(x_{(2)}, y_{(2)}) \frac{\partial}{\partial y_{(2)}},$$

with first extension given by

$$X_3^{(3)} = X_3^{(2)} + \gamma_2(x_{(2)}, y_{(2)}, y_{(2)1}) \frac{\partial}{\partial y_{(2)1}},$$

is admitted by (3.179).

Then let $x_{(3)}(x_{(2)}, y_{(2)}), y_{(3)}(x_{(2)}, y_{(2)}, y_{(2)1})$ be such that

$$X_3^{(2)} x_{(3)} = 0, \quad X_3^{(3)} y_{(3)} = 0. \tag{3.183}$$

Consequently

$$X_3^{(3+k)} \frac{d^k y_{(3)}}{dx_{(3)}^k} = 0, \quad k = 1, 2, \ldots, n-3.$$

Let

$$y_{(3)k} = \frac{d^k y_{(3)}}{dx_{(3)}^k}, \quad k = 1, 2, \ldots, n-3.$$

In terms of the invariants $x_{(3)}, y_{(3)}, \{y_{(3)k}\}$, $k = 1, 2, \ldots, n-3$, of $X_3^{(n)}$ (which are also invariants of $X_1^{(n)}$, $X_2^{(n)}$), ODE (3.179), and hence ODE (3.176), reduces to the $(n-3)$th order ODE

$$y_{(3)n-3} = F_{n-3}(x_{(3)}, y_{(3)}, y_{(3)1}, \ldots, y_{(3)n-4}), \tag{3.184}$$

for some function F_{n-3} of the indicated invariants.

Continue inductively and suppose for $q = 3, \ldots, m$, $m < r$,

$$x_{(q)}(x_{(q-1)}, y_{(q-1)}), \quad y_{(q)}(x_{(q-1)}, y_{(q-1)}, y_{(q-1)1})$$

are such that

$$X_p^{(q-1)} x_{(q)} = 0, \quad X_p^{(q)} y_{(q)} = 0, \quad p = 1, 2, \ldots, q,$$

$$X_p^{(q+k)} \frac{d^k y_{(q)}}{dx_{(q)}^k} = 0, \quad k = 1, 2, \ldots, n-q \text{ for } 1 \leq p \leq q,$$

with $y_{(q)k} = \frac{d^k y_{(q)}}{dx_{(q)}^k}$, $k = 1, 2, \ldots, n-q$, so that the nth order ODE (3.176) reduces to the $(n-m)$th order ODE

$$y_{(m)n-m} = F_{n-m}(x_{(m)}, y_{(m)}, y_{(m)1}, \ldots, y_{(m)n-m-1}) \tag{3.185}$$

for some function F_{n-m} of invariants of $X_m^{(n)}, X_{m-1}^{(n)}, \ldots, X_2^{(n)}, X_1^{(n)}$.

To go from step m to step $m+1$ we proceed as follows:

From (3.175) it follows that

$$X_j^{(m-1)} X_{m+1}^{(m-1)} x_{(m)} = 0, \quad j = 1, 2, \ldots, m.$$

The equation $X_1^{(m-1)} X_{m+1}^{(m-1)} x_{(m)} = 0$ leads to

$$X_{m+1}^{(m-1)} x_{(m)} = A_1(x_{(1)}, y_{(1)}, y_{(1)1}, \ldots, y_{(1)m-2})$$

for some function A_1 of the invariants of $X_1^{(m-1)}$; $X_2^{(m-1)}X_{m+1}^{(m-1)}x_{(m)} = 0$ leads to

$$A_1 = A_2(x_{(2)}, y_{(2)}, y_{(2)1}, \dots, y_{(2)m-3})$$

for some function A_2 of the invariants of $X_2^{(m-1)}$, $X_1^{(m-1)}$; $X_\ell^{(m-1)}X_{m+1}^{(m-1)}\ x_{(m)} = 0$ leads to

$$A_1 = A_\ell(x_{(\ell)}, y_{(\ell)}, y_{(\ell)1}, \dots, y_{(\ell)m-\ell-1})$$

for some function A_ℓ of the invariants of $X_\ell^{(m-1)}$, $X_{\ell-1}^{(m-1)}, \dots, X_1^{(m-1)}$, $1 \le \ell \le m-2$. Then the equation $X_{m-1}^{(m-1)}X_{m+1}^{(m-1)}x_{(m)} = 0$ leads to

$$A_1 = A_{m-1}(x_{(m-1)}, y_{(m-1)})$$

for some function $A_{m-1}(x_{(m-1)}, y_{(m-1)})$ of the invariants of $X_{m-1}^{(m-1)}, X_{m-2}^{(m-1)}$, $\dots, X_1^{(m-1)}$; finally $X_m^{(m-1)}X_{m+1}^{(m-1)}x_{(m)} = 0$ leads to

$$X_{m+1}^{(m-1)}x_{(m)} = A_1 = \alpha_m(x_{(m)})$$

for some function $\alpha_m(x_{(m)})$.

Similarly one can show that

$$\begin{aligned} X_{m+1}^{(m)}y_{(m)} &= \beta_m(x_{(m)}, y_{(m)}), \\ X_{m+1}^{(m+1)}y_{(m)1} &= \gamma_m(x_{(m)}, y_{(m)}, y_{(m)1}), \end{aligned}$$

for some functions β_m, γ_m of the indicated arguments. Hence

$$X_{m+1}^{(m)} = \alpha_m(x_{(m)})\frac{\partial}{\partial x_{(m)}} + \beta_m(x_{(m)}, y_{(m)})\frac{\partial}{\partial y_{(m)}},$$

with first extension given by

$$X_{m+1}^{(m+1)} = X_{m+1}^{(m)} + \gamma_m(x_{(m)}, y_{(m)}, y_{(m)1})\frac{\partial}{\partial y_{(m)1}},$$

is admitted by (3.185) since ODE (3.176) admits X_{m+1}. Now let $x_{(m+1)}(x_{(m)}, y_{(m)})$, $y_{(m+1)}(x_{(m)}, y_{(m)}, y_{(m)1})$ be such that

$$X_{m+1}^{(m)}x_{(m+1)} = 0, \quad X_{m+1}^{(m+1)}y_{(m+1)} = 0.$$

Then

$$X_{m+1}^{(m+1+k)}\frac{d^k y_{(m+1)}}{dx_{(m+1)}^k} = 0, \quad k = 1, 2, \dots, n-m-1.$$

Let

$$y_{(m+1)k} = \frac{d^k y_{(m+1)}}{dx_{(m+1)}^k}, \quad k = 1, 2, \dots, n-m-1.$$

In terms of the invariants $x_{(m+1)}, y_{(m+1)}, \{y_{(m+1)k}\}$, $k = 1, 2, \ldots, n-m-1$, of $\mathrm{X}_{m+1}^{(n)}$ (which are also invariants of $\mathrm{X}_1^{(n)}, \mathrm{X}_2^{(n)}, \ldots, \mathrm{X}_m^{(n)}$), ODE (3.185), and hence ODE (3.176), reduces to the $(n-m-1)$th order ODE

$$y_{(m+1)n-m-1} = F_{n-m-1}(x_{(m+1)}, y_{(m+1)}, y_{(m+1)1}, \ldots, y_{(m+1)n-m-2}), \tag{3.186}$$

for some function F_{n-m-1} of invariants of $\mathrm{X}_{m+1}^{(n)}$.

Finally two cases are distinguished:

Case I $(3 \leq r < n)$. Here ODE (3.176) reduces to an $(n-r)$th order ODE

$$y_{(r)n-r} = F_{n-r}(x_{(r)}, y_{(r)}, y_{(r)1}, \ldots, y_{(r)n-r-1}), \tag{3.187}$$

for some function F_{n-r} of invariants of $\mathrm{X}_r^{(n)}$ plus r quadratures. The quadratures arise as follows:

Suppose

$$y_{(r)} = \phi_r(x_{(r)}; C_1, C_2, \ldots, C_{n-r})$$

is the general solution of (3.187). Then the first order ODE

$$y_{(r)}(x_{(r-1)}, y_{(r-1)}, y_{(r-1)1}) = \phi_r(x_{(r)}(x_{(r-1)}, y_{(r-1)}); C_1, C_2, \ldots, C_{n-r})$$

admits

$$\mathrm{X}_r^{(r-1)} = \alpha_{r-1}(x_{(r-1)})\frac{\partial}{\partial x_{(r-1)}} + \beta_{r-1}(x_{(r-1)}, y_{(r-1)})\frac{\partial}{\partial y_{(r-1)}}$$

which leads to a quadrature

$$y_{(r-1)} = \phi_{r-1}(x_{(r-1)}; C_1, C_2, \ldots, C_{n-r+1})$$

for some function ϕ_{r-1} of the indicated arguments. Continuing inductively, assume we have obtained

$$y_{(k)} = \phi_k(x_{(k)}; C_1, C_2, \ldots, C_{n-k}).$$

Then the first order ODE

$$y_{(k)}(x_{(k-1)}, y_{(k-1)}, y_{(k-1)1}) = \phi_k(x_{(k)}(x_{(k-1)}, y_{(k-1)}); C_1, C_2, \ldots, C_{n-k})$$

admits

$$\mathrm{X}_k^{(k-1)} = \alpha_{k-1}(x_{(k-1)})\frac{\partial}{\partial x_{(k-1)}} + \beta_{k-1}(x_{(k-1)}, y_{(k-1)})\frac{\partial}{\partial y_{(k-1)}},$$

which leads to quadratures

$$y_{(k-1)} = \phi_{k-1}(x_{(k-1)}; C_1, C_2, \ldots, C_{n-k+1})$$

for some functions ϕ_{k-1} of the indicated arguments, $k = r, r-1, \ldots, 1$ $[y_{(0)} = y]$.

Case II $(3 \leq r = n)$. Here ODE (3.176) reduces to a first order ODE

$$y_{(n-1)1} = F_1(x_{(n-1)}, y_{(n-1)}) \tag{3.188}$$

for some function F_1 of the invariants $x_{(n-1)}$, $y_{(n-1)}$ of $X_{n-1}^{(n)}$ plus $n-1$ quadratures which are obtained as demonstrated for Case I. The first order ODE (3.188) reduces to quadrature since (3.188) admits

$$X_n^{(n-1)} = \alpha_{n-1}(x_{(n-1)})\frac{\partial}{\partial x_{(n-1)}} + \beta_{n-1}(x_{(n-1)}, y_{(n-1)})\frac{\partial}{\partial y_{(n-1)}}.$$

Thus the solution of ODE (3.176) is reduced to n quadratures.

Note that in reducing an nth order ODE to an $(n-r)$th order ODE plus r quadratures from invariance under an r-parameter Lie group whose infinitesimal generators form an r-dimensional solvable Lie algebra, *one does not need to determine the intermediate ODE's of orders* $n-1, n-2, \ldots, n-r+2$; *in Case I one does not need to determine the intermediate ODE of order* $n-r+1$.

As an example consider the fourth order ODE

$$\left[yy'\left(\frac{y}{y'}\right)''\right]' = 0 \tag{3.189}$$

which arises in studying the group properties of the linear wave equation in an inhomogeneous medium [see Section 4.3.4]. The ODE (3.189) obviously admits the three-parameter (a, b, c) Lie group of transformations

$$x^* = ax + b, \tag{3.190}$$

$$y^* = cy.$$

Corresponding infinitesimal generators $X_1 = \frac{\partial}{\partial x}$ (parameter b), $X_2 = x\frac{\partial}{\partial x}$ (parameter a), and $X_3 = y\frac{\partial}{\partial y}$ (parameter c) satisfy

$$[X_1, X_2] = X_1, \quad [X_2, X_3] = 0, \quad [X_1, X_3] = 0,$$

and thus commutation relations of the form (3.174). To carry out the reduction algorithm, we first need the following extended infinitesimal generators:

$$X_1^{(1)} = \frac{\partial}{\partial x}; \quad X_2^{(1)} = x\frac{\partial}{\partial x} - y_1\frac{\partial}{\partial y_1}, \quad X_2^{(2)} = x\frac{\partial}{\partial x} - y_1\frac{\partial}{\partial y_1} - 2y_2\frac{\partial}{\partial y_2};$$

$$X_3^{(1)} = y\frac{\partial}{\partial y} + y_1\frac{\partial}{\partial y_1}, \quad X_3^{(2)} = y\frac{\partial}{\partial y} + y_1\frac{\partial}{\partial y_1} + y_2\frac{\partial}{\partial y_2},$$

$$X_3^{(3)} = y\frac{\partial}{\partial y} + y_1\frac{\partial}{\partial y_1} + y_2\frac{\partial}{\partial y_2} + y_3\frac{\partial}{\partial y_3}.$$

From

$$X_1 x_{(1)} = 0, \quad X_1^{(1)} y_{(1)} = 0, \quad y_{(1)1} = \frac{dy_{(1)}}{dx_{(1)}},$$

we get:

$$x_{(1)} = y, \quad y_{(1)} = y_1, \quad y_{(1)1} = \frac{y_2}{y_1}.$$

Then

$$\alpha_1(x_{(1)}) = X_2 x_{(1)} = 0, \quad \beta_1(x_{(1)}, y_{(1)}) = X_2^{(1)} y_{(1)} = -y_{(1)},$$

$$\gamma_1(x_{(1)}, y_{(1)}, y_{(1)1}) = X_2^{(2)} y_{(1)1} = -\frac{y_2}{y_1} = -y_{(1)1}.$$

Thus in terms of $x_{(1)}, y_{(1)}, y_{(1)1}$, we have:

$$X_2^{(1)} = -y_{(1)}\frac{\partial}{\partial y_{(1)}}, \quad X_2^{(2)} = -y_{(1)}\frac{\partial}{\partial y_{(1)}} - y_{(1)1}\frac{\partial}{\partial y_{(1)1}}.$$

Now from

$$X_2^{(1)} x_{(2)} = 0, \quad X_2^{(2)} y_{(2)} = 0, \quad y_{(2)1} = \frac{dy_{(2)}}{dx_{(2)}},$$

we find

$$x_{(2)} = x_{(1)} = y, \quad y_{(2)} = \frac{y_{(1)1}}{y_{(1)}} = \frac{y_2}{(y_1)^2}, \quad y_{(2)1} = \frac{y_1 y_3 - 2(y_2)^2}{(y_1)^4}.$$

Then

$$\alpha_2 = X_3^{(1)} x_{(2)} = y = x_{(2)}, \quad \beta_2 = X_3^{(2)} y_{(2)} = -\frac{y_2}{(y_1)^2} = -y_{(2)},$$

$$\gamma_2 = X_3^{(3)} y_{(2)1} = \frac{4(y_2)^2 - 2y_1 y_3}{(y_1)^4} = -2y_{(2)1}.$$

Thus in terms of $x_{(2)}, y_{(2)}, y_{(2)1}$, we have

$$X_3^{(2)} = x_{(2)}\frac{\partial}{\partial x_{(2)}} - y_{(2)}\frac{\partial}{\partial y_{(2)}},$$

$$X_3^{(3)} = x_{(2)}\frac{\partial}{\partial x_{(2)}} - y_{(2)}\frac{\partial}{\partial y_{(2)}} - 2y_{(2)1}\frac{\partial}{\partial y_{(2)1}}.$$

Now from $X_3^{(2)} x_{(3)} = 0$, $X_3^{(3)} y_{(3)} = 0$, we get

$$x_{(3)} = x_{(2)} y_{(2)} = \frac{y y_2}{(y_1)^2}, \quad y_{(3)} = (x_{(2)})^2 y_{(2)1} = \frac{y^2[y_1 y_3 - 2(y_2)^2]}{(y_1)^4}.$$

It now must follow that ODE (3.189) reduces to

$$\frac{dy_{(3)}}{dx_{(3)}} = F(x_{(3)}, y_{(3)}) \tag{3.191}$$

for some function $F(x_{(3)}, y_{(3)})$. We now find $F(x_{(3)}, y_{(3)})$. We have

$$\frac{dy_{(3)}}{dx_{(3)}} = \frac{y}{(y_1)^2}\left[\frac{y(y_1)^2y_4 - 7yy_1y_2y_3 + 2(y_1)^3y_3 - 4(y_1)^2(y_2)^2 + 8y(y_2)^3}{(y_1)^2y_2 + yy_1y_3 - 2y(y_2)^2}\right]; \tag{3.192}$$

ODE (3.189) can be expressed as

$$\begin{aligned} y^2(y_1)^2y_4 = {} & 4y(y_1)^2(y_2)^2 + 5y^2y_1y_2y_3 - (y_1)^4y_2 \\ & - 3y(y_1)^3y_3 - 4y^2(y_2)^3. \end{aligned} \tag{3.193}$$

Replacing $y^2(y_1)^2y_4$ by the right-hand side of (3.193) in (3.192), we obtain

$$\frac{dy_{(3)}}{dx_{(3)}} = -\frac{(y_1)^2 + 2yy_2}{(y_1)^2} = -(1 + 2x_{(3)}) \tag{3.194}$$

which *fortunately* reduces to quadrature

$$y_{(3)} = -x_{(3)} - (x_{(3)})^2 - c_1. \tag{3.195}$$

Then

$$(x_{(2)})^2 y_{(2)1} = -x_{(2)}y_{(2)} - [x_{(2)}y_{(2)}]^2 - c_1 \tag{3.196}$$

admits

$$X_3^{(2)} = x_{(2)}\frac{\partial}{\partial x_{(2)}} - y_{(2)}\frac{\partial}{\partial y_{(2)}},$$

with corresponding canonical variables

$$R = x_{(2)}y_{(2)}, \quad S = \log y_{(2)}.$$

Then (3.196) becomes

$$\frac{dS}{dR} = \frac{1}{R} + \frac{1}{R^2 - c_1}. \tag{3.197}$$

Consider the case $c_1 > 0$, and let $c_1 = (C_1)^2$. Then

$$S = \log R + \log\left(\frac{R - C_1}{R + C_2}\right)^{1/2C_1} + c_2,$$

and consequently

$$y_{(2)} = \phi(x_{(2)}; C_1, C_2) = \frac{C_1}{x_{(2)}}\left(\frac{1 + A(x_2)}{1 - A(x_2)}\right), \tag{3.198}$$

where $A(x_2) = (C_2/x_{(2)})^{2C_1}$, with arbitrary constants C_1, C_2. Then the first order ODE resulting from (3.198), i.e.,

$$\frac{y_{(1)1}}{y_{(1)}} = \phi(x_{(1)}; C_1, C_2), \tag{3.199}$$

admits $X_2^{(1)} = -y_{(1)}\frac{\partial}{\partial y_{(1)}}$. Hence (3.199) reduces to

$$\frac{dy_{(1)}}{y_{(1)}} = \phi(x_{(1)}; C_1, C_2)dx_{(1)}$$

which integrates out to

$$y_{(1)} = \psi(y; C_1, C_2, C_3) = C_3 \exp\left[\int^y \phi(x_{(1)}; C_1, C_2)dx_{(1)}\right]. \tag{3.200}$$

Finally the first order ODE

$$y_1 = \frac{dy}{dx} = \psi(y; C_1, C_2, C_3)$$

admits $X_1 = \frac{\partial}{\partial x}$ and reduces to quadrature

$$\int^y \frac{dy}{\psi(y; C_1, C_2, C_3)} = x + C_4,$$

yielding a general solution of (3.189). The case $c_1 = -(C_1)^2$ substituted into (3.197), yields another general solution in solved form.

In using the reduction algorithm to reduce an nth order ODE to an $(n-r)$th order ODE plus r quadratures from invariance under an r-parameter Lie group whose infinitesimal generators form an r-dimensional solvable Lie algebra, we determine iteratively coordinates $\{x_{(i)}, y_{(i)}, y_{(i)1}\}$ and coefficients $\{\alpha_i(x_{(i)}), \beta_i(x_{(i)}, y_{(i)}), \gamma_i(x_{(i)}, y_{(i)}, y_{(i)1})\}$:

$$(x_{(1)}, y_{(1)}, y_{(1)1}) \to (\alpha_1, \beta_1, \gamma_1) \to (x_{(2)}, y_{(2)}, y_{(2)1}) \to (\alpha_2, \beta_2, \gamma_2)$$

$$\to \cdots \to (x_{(r-1)}, y_{(r-1)}, y_{(r-1)1}) \to (\alpha_{r-1}, \beta_{r-1}, \gamma_{r-1}) \to (x_{(r)}, y_{(r)}).$$

The nth order ODE reduces to an $(n-r)$th order ODE in coordinates $(x_{(r)}, y_{(r)})$. The quadratures follow from reversing the arrows of the iterative procedure.

Exercises 3.4

1. For each of the following second order ODE's find an admitted two-parameter Lie group of transformations and use appropriate differential invariants to find the general solution of the ODE:

(a) $y'' + Ay' + By = C$ where A, B, C are constants;

(b) $y'' + \frac{A}{x}y' + \frac{B}{x^2}y = 0$ where A, B are constants;

(c) $(x^2 + y^2)y'' + 2(y - xy')(1 + (y')^2) = 0$;

(d) $yy'' + (y')^2 = 1$.

2. Find all second and third order ODE's which admit the following two-parameter Lie groups of transformations whose Lie algebras are spanned by $\{X_1, X_2\}$:

(a) $X_1 = \frac{\partial}{\partial x}, \quad X_2 = \frac{\partial}{\partial y}$;

(b) $X_1 = x\frac{\partial}{\partial x}, \quad X_2 = y\frac{\partial}{\partial y}$;

(c) $X_1 = x\frac{\partial}{\partial x} - y\frac{\partial}{\partial y}, \quad X_2 = x\frac{\partial}{\partial x} + y\frac{\partial}{\partial y}$;

(d) $X_1 = \frac{\partial}{\partial x}, \quad X_2 = x\frac{\partial}{\partial x}$;

(e) $X_1 = \frac{\partial}{\partial x}, \quad X_2 = x\frac{\partial}{\partial x} + y\frac{\partial}{\partial y}$.

3. For the fourth order ODE (3.189) invariant under the three-parameter Lie group of transformations (3.190):

(a) Show that if $X_1 = y\frac{\partial}{\partial y}$, $X_2 = \frac{\partial}{\partial x}$, $X_3 = x\frac{\partial}{\partial x}$, then commutation relations of the form (3.174) are satisfied. Accordingly solve (3.189) by the reduction algorithm.

(b) What happens in trying the reduction algorithm with

$$X_1 = x\frac{\partial}{\partial x}, \quad X_2 = y\frac{\partial}{\partial y}, \quad X_3 = \frac{\partial}{\partial x}?$$

4. Let $\phi_1(x), \phi_2(x), \ldots, \phi_n(x)$ be n linearly independent solutions of the nth order linear homogeneous ODE

$$y^{(n)} + p_1(x)y^{(n-1)} + \cdots + p_{n-1}(x)y' + p_n(x)y = 0.$$

(a) Find an n-parameter Lie group of transformations which is admitted by the nth order linear nonhomogeneous ODE

$$y^{(n)} + p_1(x)y^{(n-1)} + \cdots + p_{n-1}(x)y' + p_n(x)y = g(x). \quad (3.201)$$

Show that the corresponding Lie algebra is solvable.

(b) Use group invariance to obtain the general solution of (3.201). In particular find the well-known formula obtained by variation of parameters.

5. Use group invariance to show that to within the one-parameter family (a) of scalings $x^* = ax$, $y^* = ay$, a general solution of the third order ODE

$$\left[yy'\left(\frac{y}{y'}\right)''\right]^2 = 1$$

reduces to the general solution of one of the following three first order ODE's with ν an arbitrary constant [cf. Bluman and Kumei (1987)]:

   (a) $y' = \dfrac{\sinh(\nu \log y)}{\nu}$;

   (b) $y' = \dfrac{\cosh(\nu \log y)}{\nu}$;

   (c) $y' = \dfrac{\sin(\nu \log y)}{\nu}$.

6. The overdetermined system of second order ODE's

$$x^2y'' + xy' - y = 0, \tag{3.202a}$$

$$yy'' - 2(y')^2 = 0, \tag{3.202b}$$

admits $X_1 = x\frac{\partial}{\partial x}$, $X_2 = y\frac{\partial}{\partial y}$. Find the solution of (3.202a,b) by using differential invariants corresponding to

   (a) X_1;

   (b) X_2;

   (c) $\{X_1, X_2\}$.

   See Bluman and Kumei (1989a) for more details on using group analysis to solve overdetermined systems of ODE's.

7. Show that if an r-dimensional Lie algebra is solvable then a spanning (basis) set of infinitesimal generators $\{X_1, X_2, \ldots, X_r\}$ can be found so that commutation relations of the form (3.174) hold.

8. (a) Let a second order ODE admit a two-parameter Lie group of transformations with infinitesimal generators X_1, X_2 such that $[X_1, X_2] = 0$, i.e., the Lie algebra is Abelian. Suppose "canonical coordinates" $(R(x,y), S(x,y))$ can be found such that

$$X_1R = 1, \quad X_2R = 0, \quad X_1S = 0, \quad X_2S = 1. \tag{3.203}$$

   Transform the given ODE to (R, S) coordinates and reduce it to two quadratures.

   (b) Show that if the Lie algebra of a two-parameter Lie group of transformations acting on $\mathbb{R}^2$ is Abelian then it does not necessarily follow that one can find "canonical coordinates" $(R(x,y), S(x,y))$ satisfying (3.203). Explain geometrically and give a specific example.

3.5 Applications to Boundary Value Problems

We show how the reduction algorithm is applied to a specific boundary value problem for an ODE. As a typical example we again consider the *Prandtl-Blasius problem for a flat plate* discussed in Section 1.3.1. BVP (1.62a–e) reduces to solving the *Blasius equation*

$$y''' + \frac{1}{2}yy'' = 0, \quad 0 < x < \infty \tag{3.204a}$$

with boundary conditions

$$y(0) = y'(0) = 0, \tag{3.204b}$$

$$y'(\infty) = 1. \tag{3.204c}$$

We wish to determine the value of $\sigma = y''(0)$.

In Section 3.4.2 we saw that the invariance of (3.204a) under the two-parameter Lie group of transformations (3.140a,b) reduced it to the first order ODE

$$\frac{dV}{dU} = \frac{V}{U}\left[\frac{\frac{1}{2} + V + U}{2U - V}\right] \tag{3.205}$$

plus quadratures (3.172) and (3.173) where $V = \phi(U; C_1)$ is the general solution of (3.205) with

$$V = \frac{y''}{yy'}, \quad U = \frac{y'}{y^2}.$$

Consider the full phase plane diagram in the UV-plane associated with (3.205). At some point on the solution curve of (3.204a–c) $y' > 0$. Then from the phase plane diagram of (3.205) it follows that $U > 0$ along the whole solution curve of (3.204a–c). Consequently, $y > 0$, $y' > 0$ for $0 < x < \infty$. Then at some point on the solution curve of (3.204a–c) $y'' > 0$. Hence the solution curve of (3.204a–c) must lie totally in the first quadrant [Figure 3.5-1]. It then follows that along the solution curve of (3.204a–c) $y > 0$, $y' > 0$, $y'' > 0$ for $0 < x < \infty$. Thus $y''(0) = \sigma > 0$. Then

$$\frac{dU}{dx} = \frac{y'}{y}[V - 2U] \gtrless 0 \quad \text{if} \quad V \gtrless 2U,$$

leads to the direction of increasing x indicated by arrows in Figure 3.5-1. Thus as $x \to \infty$, $(U, V) \to (0, 0)$. As $x \to 0$ three cases could arise: $(U, V) \to (0, \infty)$; $U \to \infty$ with $V << U$; $(U, V) \to (\infty, \infty)$ with $V = O(U)$. We examine each of these three cases separately in terms of the boundary conditions (3.204c):

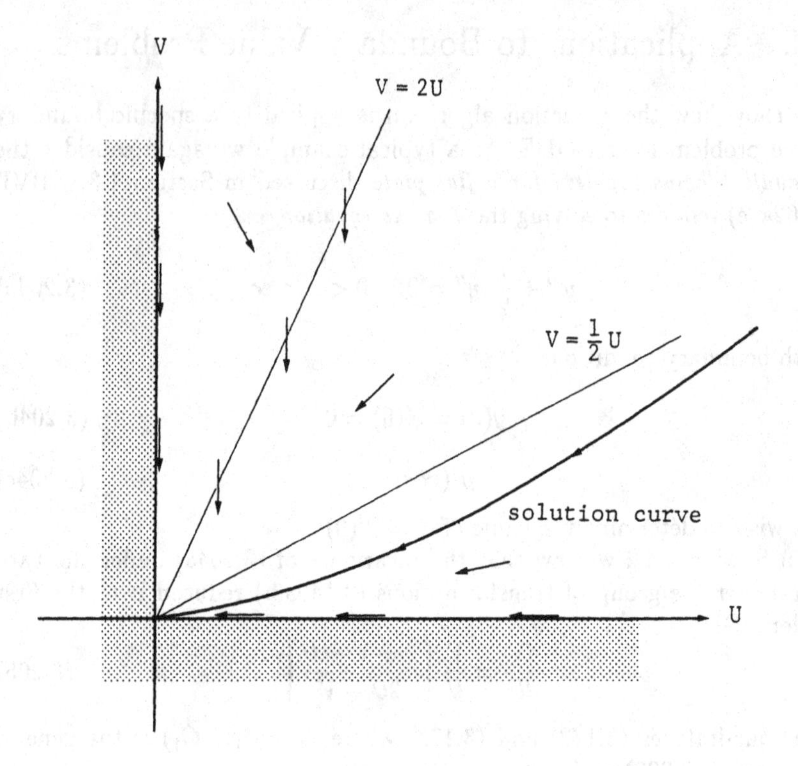

Figure 3.5-1

Case I. $(U, V) \to (0, \infty)$ as $x \to 0$: Here $\frac{dV}{dU} \sim -\frac{V}{U}$ as $x \to 0$. Then $\frac{y''}{y^3} = VU = C_1$ as $x \to 0$, which is impossible if $y''(0) = \sigma = \text{const}$.

Case II. $U \to \infty$ with $V << U$ as $x \to 0$: Here $\frac{dV}{dU} \sim \frac{V}{2U}$ as $x \to 0$. Then $\frac{y''}{(y')^{3/2}} = C_1$ as $x \to 0$ which is again impossible.

Thus Case III must hold:

Case III. $(U, V) \to (\infty, \infty)$ with $V = O(U)$ as $x \to 0$: From (3.205) it follows that the solution lies along an *exceptional path* (separatrix) [see Section 3.6]

$$V \sim \frac{1}{2}U \quad \text{as} \quad x \to 0,$$

i.e., $y'' \sim \frac{1}{2}\frac{(y')^2}{y}$ as $x \to 0$. Then (3.171) becomes

$$\frac{ds}{dr} \sim -\frac{1}{3r} \quad \text{as} \quad x \to 0,$$

which leads to

$$\frac{(y')^2}{y} \sim \text{const} = C_\infty \quad \text{as} \quad x \to 0^+.$$

Then $y''(0) = \frac{1}{2}C_\infty$.

As $x \to \infty$, the boundary condition (3.204c) leads to $(U, V) \to (0, 0)$ with $V << U$. Consequently as $x \to \infty$ we have

$$\frac{dV}{dU} \sim \frac{V}{4U^2},$$

so that

$$V \sim d_0 e^{-1/4U}$$

for some constant d_0. Thus from (3.171) we get $y' = s = \text{const} = C_0$ as $x \to \infty$. As shown in Section 1.3.1, it then follows that $\sigma = C_\infty/2(C_0)^{3/2}$.

The solution of (3.204a–c) is obtained by starting from the exceptional path as $U \to \infty$; integrating out to $V \sim d_0 e^{-1/4U}$ as $U \to 0$; determining constants C_0 and C_∞. Then $\sigma = C_\infty/2(C_0)^{3/2}$. See Dresner (1983) for further details.

Further examples of applications to boundary value problems are considered in Bluman and Cole (1974) and Dresner (1983).

Exercises 3.5

1. The nonlinear diffusion equation

$$u_t = (uu_x)_x, \quad 0 < x < \infty, \quad 0 < t < \infty,$$

with boundary conditions

$$u(x, 0) = 0, \quad x > 0,$$

$$u(0, t) = 1, \quad t > 0,$$

has an invariant solution of the form $u = y(\eta)$ with $\eta = x/\sqrt{t}$.

(a) Derive the second order ODE satisfied by $y(\eta)$ and corresponding boundary conditions.

(b) Show that this ODE admits a one-parameter Lie group of transformations. Reduce this ODE to a first order ODE plus a quadrature.

(c) Study the phase plane of this first order ODE and discuss which path gives the solution.

(d) Sketch $u(x, t) = y(\eta)$.

3.6 Invariant Solutions

Consider an nth order ODE

$$y^{(n)} = f(x, y, y', \ldots, y^{(n-1)}) \tag{3.206}$$

or, equivalently, $y_n = f(x, y, y_1, \ldots, y_{n-1})$, which admits a one-parameter Lie group of transformations with infinitesimal generator

$$X = \xi(x, y)\frac{\partial}{\partial x} + \eta(x, y)\frac{\partial}{\partial y}. \tag{3.207}$$

Definition 3.6-1. $y = \phi(x)$ is an *invariant solution* of (3.206) corresponding to infinitesimal generator (3.207) admitted by (3.206) if and only if

(i) $y = \phi(x)$ is an invariant curve of (3.207);

(ii) $y = \phi(x)$ solves (3.206).

It follows that $y = \phi(x)$ is an invariant solution of (3.206) resulting from invariance under (3.207) if and only if $y = \phi(x)$ satisfies

(i) $$\xi(x, \phi)\phi' = \eta(x, \phi); \tag{3.208a}$$

(ii) $$\phi^{(n)} = f(x, \phi, \phi', \ldots, \phi^{(n-1)}). \tag{3.208b}$$

More generally $\Phi(x, y) = 0$ defines an *invariant solution* of (3.206) corresponding to (3.206) being invariant under (3.207) if and only if

(i) $\Phi(x, y) = 0$ is an invariant curve of (3.207);

(ii) $\Phi(x, y) = 0$ solves (3.206).

From this more general definition of an invariant solution it follows that $\Phi(x, y) = 0$ is an invariant solution of (3.206) related to its invariance under (3.207) if and only if

(i) $\Phi(x, y) = 0$ is a solution of the first order ODE $y' = \frac{\eta(x,y)}{\xi(x,y)}$;

(ii) $\Phi(x, y) = 0$ solves $y^{(n)} = f(x, y, y', \ldots, y^{(n-1)})$.

The obvious way to find invariant solutions of ODE (3.206) related to its invariance under (3.207) is to first solve the ODE $y' = \frac{\eta(x,y)}{\xi(x,y)}$ and obtain its general solution $g(x, y; C) = 0$. The values of C, if any, are determined by substituting this general solution into (3.206). Any such value of $C = C^*$ determines an invariant solution $\Phi(x, y) = g(x, y; C^*) = 0$ of (3.206) related to its invariance under (3.207).

Alternatively, we now prove that usually *it is not necessary to solve the ODE* $y' = \frac{\eta(x,y)}{\xi(x,y)}$ *or any other ODE in order to find the invariant solutions of* (3.206) *related to the invariance of* (3.206) *under* (3.207):

Theorem 3.6-1. *Suppose* (3.206) *admits* (3.207). *Without loss of generality assume that* $\xi \not\equiv 0$. *Let* $Y = \frac{\partial}{\partial x} + \frac{\eta(x,y)}{\xi(x,y)}\frac{\partial}{\partial y}$ *and let* $\Psi(x, y) = \frac{\eta(x,y)}{\xi(x,y)}$. *With*

$$y_k = Y^{k-1}\Psi, \quad k = 1, 2, \ldots, n,$$

consider the algebraic expression $Q(x,y)$ *which is defined by*

$$Q(x,y) = y_n - f(x,y,y_1,\ldots,y_{n-1}) = \mathrm{Y}^{n-1}\Psi - f(x,y,\Psi,\mathrm{Y}\Psi,\ldots,\mathrm{Y}^{n-2}\Psi). \tag{3.209}$$

Three cases arise for the algebraic equation $Q(x,y) = 0$:

(i) $Q(x,y) = 0$ *defines no curves in the xy-plane;*

(ii) $Q(x,y) = 0$ *is identically satisfied for all* x, y;

(iii) $Q(x,y) = 0$ *defines curves in the xy-plane.*

In Case (i) *ODE* (3.206) *has no invariant solutions related to its invariance under* (3.207).

In Case (ii) **any** *solution of the ODE* $y' = \frac{\eta(x,y)}{\xi(x,y)}$ *is an invariant solution of ODE* (3.206) *related to its invariance under* (3.207).

In Case (iii) *an invariant solution of ODE* (3.206) *related to its invariance under* (3.207) *must satisfy* $Q(x,y) = 0$ *and conversely* **any curve** *satisfying* $Q(x,y) = 0$ *is an invariant solution of ODE* (3.206) *related to its invariance under* (3.207).

Proof. If $y_1 = y' = \frac{\eta}{\xi} = \Psi$, then $y_k = y^{(k)} = \mathrm{Y}^{k-1}\Psi$, $k = 1,2,\ldots,n$. Hence an invariant solution of ODE (3.206) related to its invariance under (3.207) must satisfy the algebraic equation

$$Q(x,y) = 0.$$

From this it immediately follows that (i) if $Q(x,y) = 0$ defines no curves in the xy-plane then ODE (3.206) has no invariant solutions related to its invariance under (3.207) and (ii) if $Q(x,y) \equiv 0$ for all x,y then *any* solution of $y' = \frac{\eta}{\xi}$ is an invariant solution of ODE (3.206).

In Case (iii) consider any curve satisfying

$$Q(x,y) = 0.$$

This curve is an invariant solution of ODE (3.206) related to its invariance under (3.207) if its differential consequence

$$Q_x + Q_y y' = 0$$

solves the ODE $y' = \frac{\eta(x,y)}{\xi(x,y)}$. This is true if $\mathrm{Y}Q = 0$ when $Q = 0$. We now prove that $\mathrm{Y}Q = 0$ when $Q = 0$:

Since ODE (3.206) admits (3.207) it follows that the invariance equation

$$\eta^{(n)} = \xi\frac{\partial f}{\partial y} + \eta\frac{\partial f}{\partial y} + \eta^{(1)}\frac{\partial f}{\partial y_1} + \cdots + \eta^{(n-1)}\frac{\partial f}{\partial y_{n-1}} \tag{3.210}$$

must hold for any values of $(x,y,y_1,\cdots,y_n)$ satisfying

$$y_n - f(x,y,y_1,\ldots,y_{n-1}) = 0 \tag{3.211}$$

where in terms of the total derivative operator $\frac{D}{Dx}$,

$$\eta^{(1)} = \frac{D\eta}{Dx} - y_1 \frac{D\xi}{Dx},$$

$$\eta^{(k)} = \frac{D\eta^{(k-1)}}{Dx} - y_k \frac{D\xi}{Dx}, \quad k = 2, 3, \ldots, n.$$

Let $y_k = \mathrm{Y}^{k-1}\Psi$, $k = 1, 2, \ldots, n$. Then (3.211) becomes $Q(x, y) = 0$ and the total derivative operator becomes $\frac{D}{Dx} \equiv \mathrm{Y}$. Then

$$\begin{aligned} \eta^{(1)} &= \mathrm{Y}\eta - \Psi\mathrm{Y}\xi \\ &= \mathrm{Y}(\xi\Psi) - \Psi\mathrm{Y}\xi \\ &= \xi\mathrm{Y}\Psi. \end{aligned}$$

We now show inductively that

$$\eta^{(k)} = \xi\mathrm{Y}^k\Psi, \quad k = 1, 2, \ldots, n.$$

If $\eta^{(k)} = \xi\mathrm{Y}^k\Psi$, then

$$\begin{aligned} \eta^{(k+1)} &= \mathrm{Y}\eta^{(k)} - (\mathrm{Y}^k\Psi)(\mathrm{Y}\xi) \\ &= \mathrm{Y}(\xi\mathrm{Y}^k\Psi) - (\mathrm{Y}^k\Psi)(\mathrm{Y}\xi) \\ &= \xi\mathrm{Y}^{k+1}\Psi. \end{aligned}$$

Consequently if ODE (3.206) admits (3.207) then from (3.210) it follows that for any curve satisfying $Q(x, y) = 0$, we have $[\xi \not\equiv 0]$

$$\mathrm{Y}^n\Psi = \frac{\partial f}{\partial x} + \frac{\eta}{\xi}\frac{\partial f}{\partial y} + (\mathrm{Y}\Psi)\frac{\partial f}{\partial y_1} + \cdots + (\mathrm{Y}^{n-1}\Psi)\frac{\partial f}{\partial y_{n-1}} \tag{3.212}$$

evaluted at $y_k = \mathrm{Y}^{k-1}\Psi$, $k = 1, 2, \ldots, n-1$. But from (3.209) we have

$$\begin{aligned} \mathrm{Y}Q \equiv Q_x + \frac{\eta}{\xi}Q_y \equiv \mathrm{Y}^n\Psi - \Bigg[&\frac{\partial f}{\partial x} + \frac{\eta}{\xi}\frac{\partial f}{\partial y} + (\mathrm{Y}\Psi)\frac{\partial f}{\partial y_1} \\ &+ \cdots + (\mathrm{Y}^{n-1}\Psi)\frac{\partial f}{\partial y_{n-1}}\Bigg] \end{aligned} \tag{3.213}$$

evaluted at $y_k = \mathrm{Y}^{k-1}\Psi$, $k = 1, 2, \ldots, n-1$. Then from (3.212) we get

$$\mathrm{Y}Q = 0 \quad \text{when} \quad Q = 0. \qquad \square$$

As a first example consider the nth order linear homogeneous ODE with constant coefficients

$$y^{(n)} + a_1 y^{(n-1)} + \cdots + a_{n-1}y' + a_n y = 0. \tag{3.214}$$

We find all invariant solutions of (3.214) corresponding to its invariance under translations in x and scalings in y generated by the infinitesimal generator

$$\mathrm{X} = \alpha\frac{\partial}{\partial x} + \beta y\frac{\partial}{\partial y}$$

with arbitrary constants α, β. Let $\lambda = \frac{\beta}{\alpha}$. Invariant solutions $y = \phi(x)$ satisfy

$$y' = \frac{\eta}{\xi} = \lambda y. \tag{3.215}$$

Substituting (3.215) into (3.214) ($\Psi = \lambda y$, $\mathrm{Y} = \lambda y\frac{\partial}{\partial y}$, $y^{(k)} = \mathrm{Y}^{k-1}\Psi + \lambda^k y$), we get

$$Q(x,y) = [\lambda^n + a_1\lambda^{n-1} + \cdots + a_{n-1}\lambda + a_n]y = 0,$$

and thus obtain the familiar characteristic polynomial equation $p(\lambda) = 0$ which λ must satisfy:

$$p(\lambda) = \lambda^n + a_1\lambda^{n-1} + \cdots + a_{n-1}\lambda + a_n = 0. \tag{3.216}$$

Any solution λ of (3.216) leads to an invariant solution of (3.214). In terms of Theorem 3.6-1, Case (ii) corresponds to λ being a root of $p(\lambda) = 0$; Case (iii) corresponds to λ not being a root of $p(\lambda) = 0$ and in this case $y = 0$ is the (trivial) invariant solution. In Case (ii) the invariant solution is any solution of (3.215), i.e. $y = Ce^{\lambda x}$.

As a second example consider the Blasius equation

$$y''' + \frac{1}{2}yy'' = 0 \tag{3.217}$$

which admits

$$\mathrm{X} = (\alpha x + \beta)\frac{\partial}{\partial x} - \alpha y\frac{\partial}{\partial y}$$

for arbitrary constants α, β. For $\alpha = 0$, the resulting invariant solution is $y = \text{const} = C$ for any constant C. For $\alpha \neq 0$, let $\lambda = \beta/\alpha$. We first consider the obvious approach. Then an invariant solution $y = \phi(x)$ satisfies

$$y' = \frac{\eta}{\xi} = -\frac{y}{x + \lambda} \tag{3.218a}$$

whose general solution is

$$y = \frac{C}{x + \lambda} \tag{3.218b}$$

where C and λ are arbitrary constants. Substituting (3.218b) into (3.217), we find that $C = 0$ or 6, which leads to invariant solutions

$$y = \phi_1(x) = 0, \quad y = \phi_2(x) = \frac{6}{x + \lambda}$$

of (3.217). Alternatively consider the approach arising from Theorem 3.6-1 for $\alpha \neq 0$. Here

$$\mathrm{Y} = \frac{\partial}{\partial x} - \frac{y}{x+\lambda}\frac{\partial}{\partial y}, \quad \Psi = -\frac{y}{x+\lambda};$$

$$\mathrm{Y}\Psi = \frac{2y}{(x+\lambda)^2}, \quad \mathrm{Y}^2\Psi = -\frac{6y}{(x+\lambda)^3}.$$

Then

$$Q(x,y) = \frac{y^2}{(x+\lambda)^2} - \frac{6y}{(x+\lambda)^3};$$

$Q(x,y) = 0$ leads to invariant solutions $y = 0$, $y = \frac{6}{x+\lambda}$.

For both examples there is little difference in the amount of computation between the two approaches. However if one is unable to obtain the general solution of ODE $y' = \eta/\xi$ then one must use the approach of Theorem 3.6-1.

Invariant solutions are especially interesting for first order ODE's.

3.6.1 Invariant Solutions for First Order ODE's—Separatrices and Envelopes

In the case of a first order ODE

$$y' = f(x,y) \tag{3.219}$$

it only makes sense to consider invariant solutions for a nontrivial infinitesimal generator (3.207) when $\frac{\eta(x,y)}{\xi(x,y)} \not\equiv f(x,y)$, i.e. $Q(x,y) \not\equiv 0$ for all x,y. Then from Theorem 3.6-1 it follows that any curve satisfying $Q(x,y) = f(x,y) - \frac{\eta(x,y)}{\xi(x,y)} = 0$ is an invariant solution of (3.219).

Consider the set of all solution curves of ODE (3.219) in the xy-plane (phase plane). This set may include *exceptional paths* (*separatrices*) which are solution curves which behave topologically "abnormally" in comparison with neighboring solutions curves [cf. Lefschetz (1963)]. Consequently a separatrix must be an invariant solution of ODE (3.219) for *all* Lie groups of transformations admitted by ODE (3.219). One can see this as follows: Two solutions of ODE (3.219) which are topologically different cannot be continuously deformed into each other and hence cannot be mapped into each other by any Lie group of transformations admitted by ODE (3.219) as the group parameters vary. Since a group of transformations admitted by ODE (3.219) maps any solution of ODE (3.219) into another solution of ODE (3.219) it follows that a separatrix is an invariant solution of ODE (3.219) for all Lie groups of transformations admitted by ODE (3.219).

By the same argument as that for separatrices it will follow that all *envelope solutions* (if any exist) for a first order ODE

$$F(x,y,y') = 0 \tag{3.220}$$

must be invariant solutions for any admitted Lie group of transformations. If ODE (3.220) admits a nontrivial one-parameter Lie group of transformations with infinitesimal generator $\mathrm{X} = \xi(x,y)\frac{\partial}{\partial x} + \eta(x,y)\frac{\partial}{\partial y}$, then

$$\mathrm{X}^{(1)}F = \xi F_x + \eta F_y + [\eta_x + (\eta_y - \xi_x)y' - \xi_y(y')^2]F_{y'} = 0$$

when

$$F(x,y,y') = 0,$$

and

$$F\left(x,y,\frac{\eta(x,y)}{\xi(x,y)}\right) \not\equiv 0 \quad \text{for all} \quad x,y.$$

By a simple extension of Theorem 3.6-1 it follows that the invariant solutions of ODE (3.220) related to its invariance under X are the curves defined by the algebraic equation

$$F\left(x,y,\frac{\eta(x,y)}{\xi(x,y)}\right) = 0. \tag{3.221}$$

Hence an envelope solution of ODE (3.220) satisfies (3.221) for *any* X admitted by ODE (3.220).

As a first example consider the first order ODE

$$y' = y^2, \tag{3.222}$$

which obviously admits

$$\mathrm{X}_1 = \frac{\partial}{\partial x}, \quad \mathrm{X}_2 = x\frac{\partial}{\partial x} - y\frac{\partial}{\partial y}.$$

From X_1 it follows that a separatrix solution $y = \phi(x)$ of (3.222) must satisfy

$$y' = \phi'(x) = 0 \quad \text{or} \quad \phi(x) = C_1 = \text{const.} \tag{3.223}$$

Substitution of (3.223) into (3.222) leads to the only possible candidate $C_1 = 0$, i.e. $y = 0$. [Alternatively, $Q(x,y) = f(x,y) - \frac{\eta(x,y)}{\xi(x,y)} = y^2 = 0$, which leads to $y = 0$.]

From X_2 it follows that a separatrix solution $y = \phi(x)$ of (3.222) must satisfy

$$y' = -\frac{y}{x}. \tag{3.224}$$

Substitution of (3.224) into (3.222) leads to $-y/x = y^2$ or

$$y = 0, \quad y = -\frac{1}{x}.$$

[Alternatively,

$$Q(x,y) = f(x,y) - \frac{\eta(x,y)}{\xi(x,y)} = y^2 + \frac{y}{x} = y\left(y + \frac{1}{x}\right) = 0,$$

which leads to $y = 0$, $y = -1/x$.] Since $y = -1/x$ is not an invariant solution of X_1, it cannot be a separatrix of (3.222). This solution is a particular solution of (3.222) associated with its general solution

$$y = -\frac{1}{x+C}, \quad C = \text{const}, \quad \text{with} \quad C = 0.$$

The solution curves are illustrated in Figure 3.6.1-1.

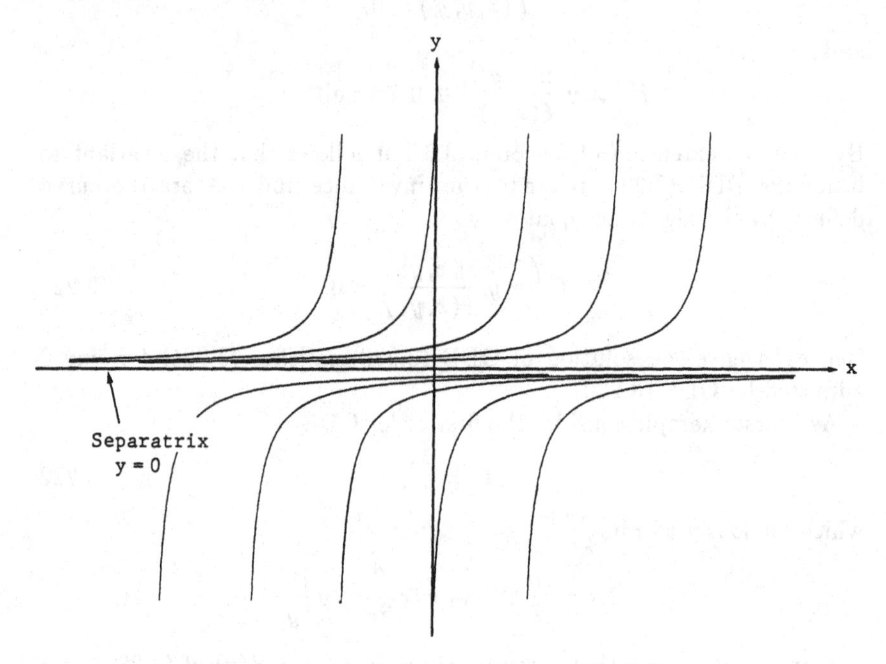

Figure 3.6.1-1. Solution curves of $y' = y^2$.

As a second example consider the first order ODE

$$y' = \frac{x\sqrt{x^2+y^2} + y(x^2+y^2-1)}{x(x^2+y^2-1) - y\sqrt{x^2+y^2}}, \tag{3.225}$$

which admits the rotation group

$$X = y\frac{\partial}{\partial x} - x\frac{\partial}{\partial y}. \tag{3.226}$$

The invariant curves of (3.226) are circles

$$x^2 + y^2 = c^2. \tag{3.227}$$

Then a separatrix must satisfy (3.227). After substituting (3.227) into ODE (3.225) we get

$$-\frac{x}{y} = \frac{xc + y(c^2-1)}{x(c^2-1) - yc},$$

so that $c = 1$. Hence the only possible separatrix is the circle

$$x^2 + y^2 = 1. \tag{3.228}$$

One can show that the circle (3.228) is a limit cycle of (3.225). Typical solution curves of (3.225) are illustrated in Figure 3.6.1-2.

As a third example consider Clairaut's equation

$$y = xy' + \frac{m}{y'} \tag{3.229}$$

where m is a constant

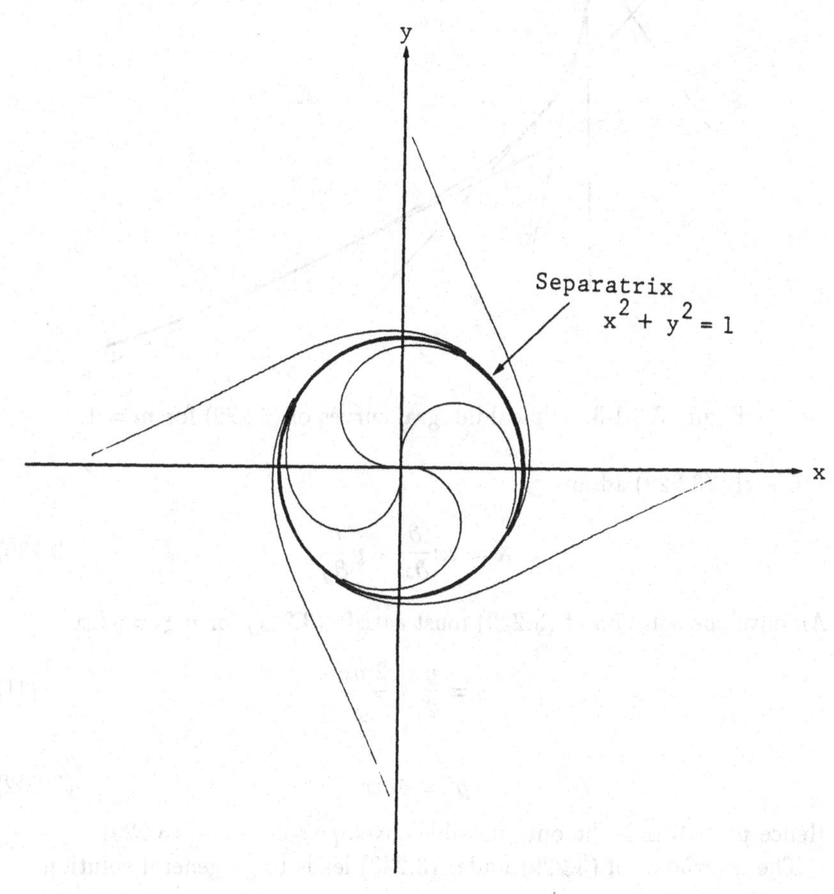

Figure 3.6.1-2. Solution curves of (3.225).

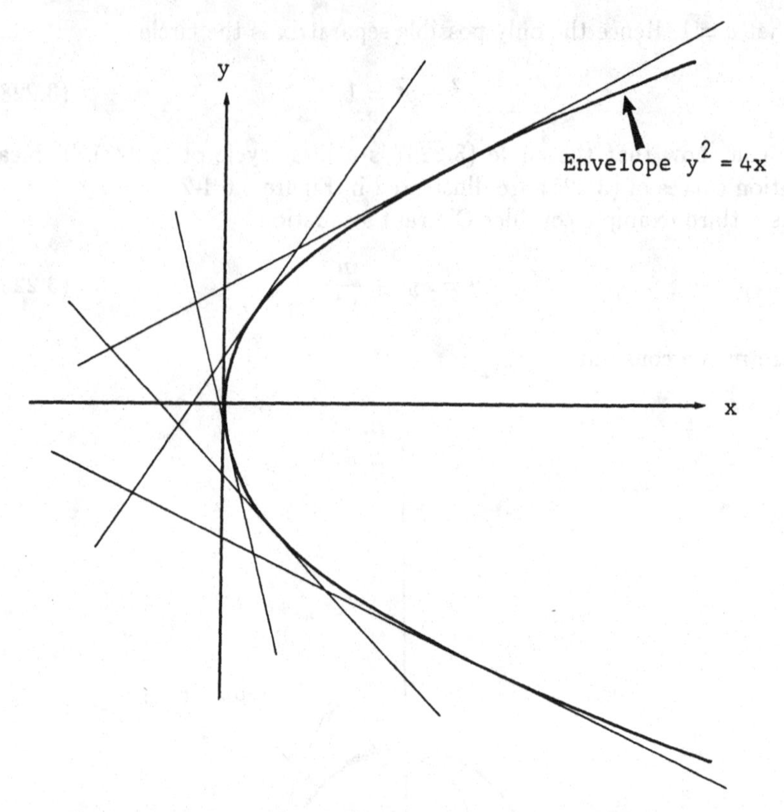

Figure 3.6.1-3. Typical integral curves of (3.229) for $m = 1$.

Clearly (3.229) admits

$$\mathrm{X} = 2x\frac{\partial}{\partial x} + y\frac{\partial}{\partial y}. \tag{3.230}$$

An envelope solution of (3.229) must satisfy (3.221) for $\eta/\xi = y/2x$:

$$y = \frac{y}{2} + \frac{2mx}{y} \tag{3.231}$$

or

$$y^2 = 4mx. \tag{3.232}$$

Hence $y^2 = 4mx$ is the only possible envelope solution of (3.229).

The invariance of (3.229) under (3.230) leads to its general solution

$$y = cx + \frac{m}{c} \tag{3.233}$$

where c is an arbitrary constant. Clearly the parabola equation (3.232) is the envelope of the family of straight lines (3.233). Typical integral curves of (3.229) are illustrated in Figure 3.6.1-3 for $m = 1$.

Exercises 3.6

1. (a) Suppose $\lambda = r$ is a double root of the characteristic polynomial equation (3.216).
   (i) Show that $X = \alpha\frac{\partial}{\partial x} + (\beta y + \gamma e^{rx})\frac{\partial}{\partial y}$ is admitted by (3.214) for arbitrary constants α, β, γ.
   (ii) Find corresponding invariant solutions of (3.214).

   (b) What can you say if $\lambda = r$ is a root of multiplicity k of (3.216)?

2. Consider the Euler equation

$$x^2y'' + Axy' + By = 0, \quad A = \text{const}, \quad B = \text{const}.$$

   Find its general solution in terms of invariant solutions. [Note that $X = \alpha x\frac{\partial}{\partial x} + \beta y\frac{\partial}{\partial y}$ is admitted by the Euler equation.]

3. Find all invariant solutions of (3.189) corresponding to its invariance under the three-parameter Lie group of transformations (3.190).

4. Find the general solution of (3.225) and sketch several integral curves.

5. Use the invariance of (3.229) under (3.230) to derive the general solution (3.233).

6. Find necessary conditions for a first order ODE $y' = f(x, y)$ so that it admits the rotation group $X = y\frac{\partial}{\partial x} - x\frac{\partial}{\partial y}$ and has the circle $x^2 + y^2 = 1$ as a limit cycle (separatrix).

7. Consider the ODE

$$y' = A\frac{y}{x}, \quad A = \text{const} \neq 0. \tag{3.234}$$

   (a) Find invariant solutions corresponding to the admitted infinitesimal generators
   (i) $X_1 = x\dfrac{\partial}{\partial x}$;
   (ii) $X_2 = y\dfrac{\partial}{\partial y}$;
   (iii) $X_3 = x\dfrac{\partial}{\partial x} + y\dfrac{\partial}{\partial y}$.

   (b) Determine the separatrices of (3.234).

   (c) Sketch typical solution curves of (3.234).

8. Use the invariance of $y' = x/y$ under $X = x\frac{\partial}{\partial y} + y\frac{\partial}{\partial y}$ to find its separatrices.

9. Find separatrices and sketch typical integral curves of the ODE

$$y' = \frac{y(y-2x)}{x(x-2y)}.$$

10. Without explicitly solving the ODE

$$y' = -\frac{x^3}{y},$$

show that it has no separatrices.

11. Consider the ODE

$$y + \frac{2}{3}(y')^3 - (x + (y')^2) = 0. \tag{3.235}$$

   (a) Find a one-parameter Lie group of transformations admitted by (3.235).
   (b) Sketch typical integral curves of (3.235).
   (c) Find the envelope of the solution curves of (3.235).

3.7 Discussion

In this chapter we have shown how to

(i) use infinitesimal transformations to construct solutions to ODE's;

(ii) find infinitesimal transformations admitted by a given ODE by means of Lie's algorithm; and

(iii) find the most general nth order ODE which admits a given Lie group of transformations.

If a given ODE admits a one-parameter Lie group of transformations then its order can be reduced constructively by one by the use of canonical coordinates or differential invariants. Moreover the solution of the given ODE is found by quadrature after solving the reduced ODE. Knowing the invariance of a first order ODE under a nontrivial one-parameter Lie group of transformations is equivalent to finding an integrating factor for the ODE.

Every first order ODE admits a nontrivial infinite-parameter Lie group of transformations. A second order ODE admits at most an eight-parameter Lie group of transformations whereas an nth order ODE ($n \geq 3$) admits at most an $(n+4)$-parameter group.

Olver (1986) showed that if an nth order ODE admits an r-parameter solvable Lie group of transformations ($r \leq n$) then it can be reduced to an

$(n-r)$th order ODE plus r quadratures. To our knowledge the reduction algorithm presented in this chapter to accomplish this has not appeared in the literature. The use of solvable Lie groups (called integrable groups in earlier literature) to reduce the order of a system of first order ODE's appears to have been first considered by Bianchi (1918, Sect. 167) [cf. Eisenhart (1933, Sect. 36)].

In the case of a boundary value problem for an nth order ODE, invariance of the ODE under an r-parameter solvable Lie group of transformations ($r \leq n$) reduces the given BVP to a BVP for an $(n-r)$th order ODE plus r quadratures. This is especially useful for obtaining qualitative results for the solution of the BVP. Moreover admittance of a Lie group of transformations by the ODE of a BVP may lead to an effective numerical method for solving the BVP. In particular if the boundary conditions are of the right type then the shooting method reduces to a single shooting combined with the use of group invariance to obtain a family of solutions from a given solution.

If an ODE admits a one-parameter Lie group of transformations then special solutions called invariant solutions (which are also invariant curves of the group) can be constructed. For a second or higher order ODE such invariant solutions can be found without explicitly solving the given ODE. If a first order ODE admits a nontrivial one-parameter Lie group of transformations then corresponding invariant solutions can be determined without solving any ODE. Moreover separatrices and envelope solutions (if they exist) are invariant solutions for any nontrivial one-parameter Lie group of transformations admitted by a given first order ODE. Consequently such solutions are constructed without determining the general solution of the given first order ODE. As far as we know the general results presented in Section 3.6 on the construction and geometrical significance of invariant solutions for ODE's have not appeared in the literature. Wulfman (1979) considers group aspects of separatrices which are limit cycles. The construction of separatrices in the case of scaling invariance is discussed by Dresner (1983). Discussions of envelope solutions from invariance considerations appear in Page (1896, 1897) and Cohen (1911).

Known results for reduction by symmetry for systems of ODE's are now summarized: Using ideas developed in Section 3.2 one can show that a system of first order ODE's always admits a nontrivial infinite-parameter Lie group of transformations and a trivial infinite-parameter Lie group of transformations but there is no deductive procedure for finding the group [see Ovsiannikov (1982, Sect. 8)]. Gonzalez–Gascon and Gonzalez–Lopez (1983) show that a system of m nth order ODE's can admit at most a $[2(m+1)^2]$-parameter Lie group of transformations if $n=2$ and at most a $(2m^2+mn+2)$-parameter group if $n>2$. If a system of nth order ODE's admits a one-parameter Lie group of transformations then the order of one of the ODE's of the system can be reduced by one. If a given system of m first order ODE's admits an r-parameter solvable Lie group of transformations

($r \leq m$) then it can be reduced to a system of $m - r$ first order ODE's plus r quadratures. The latter two results appear in Olver (1986).

In Chapter 4 we consider the invariance of partial differential equations (PDE's) under one-parameter Lie groups of transformations. Unlike the case for an ODE where invariance under a one-parameter Lie group of transformations leads to a reduced ODE whose solution includes all solutions of the given ODE, in general there is no known method for a PDE where invariance under a one-parameter Lie group of transformations leads to a reduced PDE which contains all solutions of the given PDE. However in precisely the same way as for ODE's we can define *invariant solutions* for PDE's resulting from invariance under one-parameter Lie groups of transformations. For ODE's such special solutions are obtained by solving reduced algebraic equations; for PDE's such special solutions are obtained by solving reduced PDE's with one less independent variable.

If a given ODE admits a multi-parameter Lie group with a solvable Lie algebra then each reduced ODE resulting from the reduction algorithm "inherits" symmetries of the given ODE. One can show that if a given PDE admits a multi-parameter Lie group with a solvable Lie algebra then each reduced PDE leading to invariant solutions inherits symmetries of the given PDE. In Chapter 7 we consider this for the especially interesting case where the reduced PDE is an ODE, i.e. the given PDE has two independent variables.

In Chapter 7 we will show how to extend Lie's theory for reducing the order of an ODE to include *potential symmetries.*

4

Partial Differential Equations

4.1 Introduction—Invariance of a Partial Differential Equation

In this chapter we apply infinitesimal transformations to the construction of solutions of partial differential equations (PDE's). We will consider scalar (single) PDE's and systems of PDE's.

As for ODE's we will show that the infinitesimal criterion for invariance of PDE's leads directly to an algorithm to determine infinitesimal generators X admitted by given PDE's. Invariant surfaces of the corresponding Lie group of point transformations lead to ***invariant solutions (similarity solutions)***. These solutions are obtained by solving PDE's with fewer independent variables than the given PDE's.

We will discuss how one can use infinitesimal transformations to solve boundary value problems (BVP's) for PDE's. If a one-parameter Lie group of transformations admitted by a PDE leaves the domain and boundary conditions of a BVP invariant, then the solution of the BVP is an invariant solution, and hence the given BVP is reduced to a BVP with one less independent variable. We will also consider the invariance of BVP's under multi-parameter Lie groups of transformations.

4.1.1 Invariance of a PDE

First we consider a scalar PDE. We represent a kth order PDE by

$$F(x, u, \underset{1}{u}, \underset{2}{u}, \ldots, \underset{k}{u}) = 0, \tag{4.1}$$

where $x = (x_1, x_2, \ldots, x_n)$ denotes n independent variables, u denotes the coordinate corresponding to the dependent variable, and $\underset{j}{u}$ denotes the set of coordinates corresponding to all jth order partial derivatives of u with respect to x; the coordinate of $\underset{j}{u}$ corresponding to $\dfrac{\partial^j u}{\partial x_{i_1} \partial x_{i_2} \cdots \partial x_{i_j}}$ is denoted by $u_{i_1 i_2 \cdots i_j}$, $i_j = 1, 2, \ldots, n$ for $j = 1, 2, \ldots, k$.

In terms of the coordinates of x, u, $\underset{1}{u}, \underset{2}{u}, \ldots, \underset{k}{u}$, equation (4.1) becomes an algebraic equation which defines a hypersurface in $(x, u, \underset{1}{u}, \underset{2}{u}, \ldots, \underset{k}{u})$-space. [$(x, u, \underset{1}{u})$-space is of dimension $2n + 1$, $(x, u, \underset{1}{u}, \underset{2}{u})$-space is of dimension

$\frac{1}{2}(n^2+5n+2), \ldots]$. For any solution $u=\Theta(x)$ of PDE (4.1) the equation

$$(x, u, \underset{1}{u}, \underset{2}{u}, \ldots, \underset{k}{u}) = (x, \Theta(x), \underset{1}{\Theta}(x), \underset{2}{\Theta}(x), \ldots, \underset{k}{\Theta}(x))$$

defines a solution surface which lies on the surface $F(x, u, \underset{1}{u}, \underset{2}{u}, \ldots, \underset{k}{u}) = 0$.

We also assume that PDE (4.1) can be written in solved form in terms of some ℓth order partial derivative of u:

$$F(x, u, \underset{1}{u}, \underset{2}{u}, \ldots, \underset{k}{u}) = u_{i_1 i_2 \cdots i_\ell} - f(x, u, \underset{1}{u}, \underset{2}{u}, \ldots, \underset{k}{u}) = 0 \tag{4.2}$$

where $f(x, u, \underset{1}{u}, \underset{2}{u}, \ldots, \underset{k}{u})$ does not depend on $u_{i_1 i_2 \cdots i_\ell}$.

Definition 4.1.1-1. The one-parameter Lie group of transformations

$$x^* = X(x, u; \epsilon), \tag{4.3a}$$

$$u^* = U(x, u; \epsilon), \tag{4.3b}$$

leaves PDE (4.1) *invariant* if and only if its kth extension, defined by (2.125a–d) and (2.126a,b), leaves the surface (4.1) invariant.

A solution $u = \Theta(x)$ of (4.1) satisfies (4.1) with

$$u_{i_1 i_2 \cdots i_j} = \frac{\partial^j \Theta(x)}{\partial x_{i_1} \partial x_{i_2} \cdots \partial x_{i_j}}, \quad i_j = 1, 2, \ldots, n \ \text{ for } \ j = 1, 2, \ldots, k.$$

Invariance of the surface (4.1) under the kth extension of (4.3a,b) means that any solution $u = \Theta(x)$ of (4.1) maps into some other solution $u = \phi(x; \epsilon)$ of (4.1) under the action of the group (4.3a,b). Moreover if a transformation (4.3a,b) maps any solution $u = \Theta(x)$ of (4.1) into another solution $u = \phi(x; \epsilon)$ of (4.1) then the surface (4.1) is invariant under (4.3a,b) where

$$u_{i_1 i_2 \cdots i_j} = \frac{\partial^j \phi(x; \epsilon)}{\partial x_{i_1} \partial x_{i_2} \cdots \partial x_{i_j}}, \quad i_j = 1, 2, \ldots, n \ \text{ for } \ j = 1, 2, \ldots, k.$$

Consequently *the family of all solutions of PDE* (4.1) *is invariant under* (4.3a,b) *if and only if* (4.1) *admits* (4.3a,b).

The following theorem results from Definition 4.1.1-1, Theorem 2.2.7-1 on the infinitesimal criterion for an invariant surface, and Theorem 2.3.4-1 on extended infinitesimals:

Theorem 4.1.1-1 (Infinitesimal Criterion for Invariance of a PDE). *Let*

$$\mathrm{X} = \xi_i(x, u) \frac{\partial}{\partial x_i} + \eta(x, u) \frac{\partial}{\partial u} \tag{4.4}$$

be the infinitesimal generator of (4.3a,b). *Let*

$$\mathrm{X}^{(k)} = \xi_i(x, u) \frac{\partial}{\partial x_i} + \eta(x, u) \frac{\partial}{\partial u} + \eta_i^{(1)}(x, u, \underset{1}{u}) \frac{\partial}{\partial u_i}$$

$$+\cdots+\eta^{(k)}_{i_1 i_2\cdots i_k}(x,u,\underset{1}{u},\underset{2}{u},\ldots,\underset{k}{u})\frac{\partial}{\partial u_{i_1 i_2\cdots i_k}} \tag{4.5}$$

be the kth extended infinitesimal generator of (4.4) *where* $\eta^{(1)}_i$ *is given by* (2.129a) *and* $\eta^{(j)}_{i_1 i_2\cdots i_j}$ *by* (2.129b), $i_j = 1,2,\ldots,n$ *for* $j = 1,2,\ldots,k$, *in terms of* $(\xi(x,u),\eta(x,u))$ $[\xi(x,u)$ *denotes* $(\xi_1(x,u),\xi_2(x,u),\ldots,\xi_n(x,u))]$. *Then* (4.3a,b) *is admitted by PDE* (4.1) *if and only if*

$$X^{(k)}F(x,u,\underset{1}{u},\underset{2}{u},\ldots,\underset{k}{u}) = 0$$

when

$$F(x,u,\underset{1}{u},\underset{2}{u},\ldots,\underset{k}{u}) = 0. \tag{4.6}$$

Proof. See Exercise 4.1-3. □

4.1.2 Elementary Examples

(1) *Group of Translations*

The second order PDE

$$u_{11} = f(u_{12},u_{22},u_1,u_2,u,x_1) \tag{4.7}$$

admits the one-parameter (ϵ) Lie group of translations

$$x_1^* = x_1, \tag{4.8a}$$

$$x_2^* = x_2+\epsilon, \tag{4.8b}$$

$$u^* = u, \tag{4.8c}$$

since under (4.8a–c) we have

$$u^*_{11} = u_{11},\quad u^*_{12} = u_{12},\quad u^*_{22} = u_{22},\quad u^*_1 = u_1,\quad u^*_2 = u_2$$

so that the surface defined by (4.7) is invariant in $(x,u,\underset{1}{u},\underset{2}{u})$-space. Then for any function $\Theta(x_1)$,

$$u = \Theta(x_1) \tag{4.9}$$

is invariant under (4.8a–c) and defines a solution (*invariant solution*) of (4.7) provided $\Theta(x_1)$ solves the second order ODE

$$\Theta''(x_1) = f(0,0,\Theta'(x_1),0,\Theta(x_1),x_1).$$

Note that

$$u = \Theta(x_1,x_2)$$

defines an *invariant surface* [cf. Theorem 2.2.7-1] of (4.8a–c) if and only if

$$X(u-\Theta(x_1,x_2)) = -\frac{\partial\Theta}{\partial x_2} = 0 \quad\text{when}\quad u = \Theta(x_1,x_2)$$

where $X = \frac{\partial}{\partial x_2}$ is the infinitesimal generator of (4.8a–c). This leads to the *invariant form* (*similarity form*) (4.9) for an invariant solution resulting from the invariance of (4.7) under (4.8a–c).

Under the action of (4.8a–c) a solution $u = \phi(x_1, x_2)$ of (4.7) is mapped into a *one-parameter family of solutions* $u = \phi(x_1, x_2 + \epsilon)$ of (4.7) provided $u = \phi(x_1, x_2)$ is not an invariant solution of (4.7) corresponding to its invariance under (4.8a–c).

(2) *Group of Scalings*

The wave equation

$$u_{11} = u_{22} \tag{4.10}$$

admits the one-parameter (α) Lie group of scalings

$$x_1^* = \alpha x_1, \tag{4.11a}$$

$$x_2^* = \alpha x_2, \tag{4.11b}$$

$$u^* = u, \tag{4.11c}$$

since $u_{11}^* = \frac{1}{\alpha^2} u_{11}$, $u_{22}^* = \frac{1}{\alpha^2} u_{22}$, and consequently $u_{11}^* = u_{22}^*$ when $u_{11} = u_{22}$.

If one chooses *canonical coordinates* $r = \frac{x_1}{x_2}$, $s = \log x_2$, u, so that (4.11a–c) becomes $r^* = r$, $s^* = s + \log \alpha$, $u^* = u$, then PDE (4.10) transforms to PDE

$$(1 - r^2) u_{rr} - u_{ss} + 2r u_{rs} + u_s - 2r u_r = 0. \tag{4.12}$$

Correspondingly

$$u = \Theta(r) = \Theta\left(\frac{x_1}{x_2}\right) \tag{4.13}$$

defines an invariant solution of (4.12), and hence of (4.10), provided that $\Theta(r)$ satisfies ODE

$$(1 - r^2)\Theta''(r) - 2r\Theta' = 0. \tag{4.14}$$

The infinitesimal generator of (4.11a–c) is given by $X = x_1 \frac{\partial}{\partial x_1} + x_2 \frac{\partial}{\partial x_2}$. Hence $u = \Theta(x_1, x_2)$ defines an invariant surface of (4.11a–c) if and only if

$$X(u - \Theta(x_1, x_2)) = 0 \quad \text{when} \quad u = \Theta(x_1, x_2),$$

i.e., if and only if

$$x_1 \frac{\partial \Theta}{\partial x_1} + x_2 \frac{\partial \Theta}{\partial x_2} = 0. \tag{4.15}$$

Corresponding characteristic equations

$$\frac{dx_1}{x_1} = \frac{dx_2}{x_2} = \frac{d\Theta}{0},$$

lead to the invariant form (4.13). Substitution of (4.13) into (4.10) leads to (4.14). We see that it is unnecessary to find the canonical coordinates of (4.11a–c) in order to find corresponding invariant solutions.

Moreover under the action of (4.11a–c) a solution $u = \phi(x_1, x_2)$ of (4.10) maps into the one-parameter family of solutions $u = \phi(\alpha x_1, \alpha x_2)$ of (4.10) provided $u = \phi(x_1, x_2)$ is not an invariant solution of (4.10) corresponding to its invariance under (4.11a–c).

(3) *Superposition of Invariant Solutions for Linear PDE's*

The wave equation (4.10) admits the one-parameter (ϵ) Lie group of transformations

$$x_1^* = x_1, \tag{4.16a}$$

$$x_2^* = x_2 + \epsilon, \tag{4.16b}$$

$$u^* = e^{\epsilon\lambda} u, \tag{4.16c}$$

for any constant $\lambda \in \mathbb{C}$. The infinitesimal generator of (4.16a–c) is

$$\mathrm{X} = \frac{\partial}{\partial x_2} + \lambda u \frac{\partial}{\partial u}.$$

Corresponding invariant solutions

$$u = \Theta(x_1, x_2)$$

satisfy

$$\lambda u = \frac{\partial \Theta}{\partial x_2} \quad \text{when} \quad u = \Theta(x_1, x_2),$$

i.e.,

$$u_2 = \lambda u. \tag{4.17}$$

As was the case of ODE's [cf. Section 3.6] we can find invariant solutions in two ways:

Method (I) (Invariant Form). The solution of (4.17) is the *invariant form*

$$u = \phi(x_1) e^{\lambda x_2}. \tag{4.18}$$

Substitution of (4.18) into (4.10) leads to

$$\phi''(x_1) = \lambda^2 \phi(x_1),$$

and hence to invariant solutions

$$u = \Theta(x_1, x_2) = Ce^{\lambda(x_2 \pm x_1)} \tag{4.19}$$

where C is an arbitrary complex constant.

Method (II) (Direct Substitution). Here we directly substitute (4.17) into (4.10) and avoid solving (4.17) explicitly. Then $u_{22} = \lambda u_2 = \lambda^2 u$. Hence $u_{11} = \lambda^2 u$ so that

$$u = \Psi(x_2)e^{\pm\lambda x_1}. \tag{4.20}$$

Then (4.17) leads to $\Psi'(x_2) = \lambda\Psi(x_2)$ and hence to (4.19).

Since (4.10) is a linear homogeneous PDE it follows that superpositions of invariant solutions

$$\sum_{\lambda} C(\lambda)e^{\lambda(x_2 \pm x_1)}, \quad \sum_{\lambda}\left[C_1(\lambda)e^{\lambda(x_2-x_1)} + C_2(\lambda)e^{\lambda(x_2+x_1)}\right],$$

$$\int_{\gamma} C(\lambda)e^{\lambda(x_2 \pm x_1)}d\lambda, \quad \text{etc.},$$

define solutions of (4.10) where $\lambda \in \mathbb{C}$ and γ defines a path in the complex λ-plane. Special superpositions correspond to Fourier series, Laplace transform, and Fourier transform representations of solutions.

Exercises 4.1

1. Find invariant solutions for the wave equation (4.10) corresponding to its invariance under the one-parameter (ϵ) group

$$x_1^* = x_1 + \epsilon,$$
$$x_2^* = x_2 + \epsilon\alpha,$$

   where $\alpha \in \mathbb{R}$ is constant. How do these solutions relate to the general solution of (4.10)?

2. (a) Show that the most general second order PDE which admits the two-parameter (ϵ_1, ϵ_2) Lie group of translations $x^* = x + \epsilon_1$, $t^* = t + \epsilon_2$, is of the form

$$F(u_{xx}, u_{xt}, u_{tt}, u_x, u_t, u) = 0.$$

   (b) Find invariant solutions of this PDE corresponding to its invariance under the one-parameter (ϵ) Lie group of translations $x^* = x + \epsilon c$, $t^* = t + \epsilon$ where $c \in \mathbb{R}$ is constant. Interpret these solutions.

3. Prove Theorem 4.1.1-1.

4.2 Invariance for Scalar PDE's

4.2.1 Invariant Solutions

Consider a kth order scalar PDE (4.1) ($k \geq 2$) which admits a one-parameter Lie group of transformations with infinitesimal generator (4.4). We assume that $\xi(x,u) \not\equiv 0$.

Definition 4.2.1-1. $u = \Theta(x)$ is an *invariant solution* of (4.1) corresponding to (4.4) admitted by PDE (4.1) if and only if

(i) $u = \Theta(x)$ is an invariant surface of (4.4);

(ii) $u = \Theta(x)$ solves (4.1).

It follows that $u = \Theta(x)$ is an invariant solution of (4.1) resulting from its invariance under (4.4) if and only if $u = \Theta(x)$ satisfies

(i) $\mathrm{X}(u - \Theta(x)) = 0$ when $u = \Theta(x)$, i.e.

$$\xi_i(x,\Theta(x))\frac{\partial \Theta}{\partial x_i} = \eta(x,\Theta(x)); \tag{4.21a}$$

(ii) $F(x,u,\underset{1}{u},\underset{2}{u},\ldots,\underset{k}{u}) = 0$, where $u_{i_1 i_2 \cdots i_j} = \dfrac{\partial^j \Theta(x)}{\partial x_{i_1}\partial x_{i_2}\cdots \partial x_{i_j}}$, (4.21b)

$i_j = 1,2,\ldots,n$ for $j = 1,2,\ldots,k$.

Equation (4.21a) is called the *invariant surface condition* for invariant solutions corresponding to (4.4). Invariant solutions were first considered by Lie (1881). They can be determined in two ways:

(I) Invariant Form Method. Here we first solve the invariant surface condition, i.e. the first order PDE (4.21a), by solving the corresponding characteristic equations for $u = \Theta(x)$:

$$\frac{dx_1}{\xi_1(x,u)} = \frac{dx_2}{\xi_2(x,u)} = \cdots = \frac{dx_n}{\xi_n(x,u)} = \frac{du}{\eta(x,u)}. \tag{4.22}$$

If $(X_1(x,u), X_2(x,u), \ldots, X_{n-1}(x,u))$, $v(x,u)$ are n independent invariants of (4.22) with $\frac{\partial v}{\partial u} \not\equiv 0$, then the solution $u = \Theta(x)$ of (4.21a) is given implicitly by the invariant form

$$v(x,u) = \phi(X_1(x,u), X_2(x,u), \ldots, X_{n-1}(x,u)), \tag{4.23}$$

where ϕ is an arbitrary function of $X_1, X_2, \ldots, X_{n-1}$. Note that $(X_1, X_2, \ldots,$ $X_{n-1}, v)$ are n independent group invariants of (4.4) and are canonical coordinates of (4.3a,b). Let $X_n(x,u)$ be the $(n+1)$th canonical coordinate satisfying

$$\mathrm{X}X_n = 1.$$

If PDE (4.1) is transformed to another kth order PDE in terms of independent variables $(X_1, X_2, \ldots, X_n)$ and dependent variable v, then the transformed PDE admits

$$X_i^* = X_i, \quad i = 1, 2, \ldots, n-1, \tag{4.24a}$$

$$X_n^* = X_n + \epsilon, \tag{4.24b}$$

$$v^* = v. \tag{4.24c}$$

Thus X_n does not appear explicitly in the transformed PDE, and hence the transformed PDE has solutions of the form $v = \phi(X_1, X_2, \ldots, X_{n-1})$. Accordingly PDE (4.1) has invariant solutions given implicitly by the form (4.23). These solutions are found by solving a reduced PDE with $n-1$ independent variables $(X_1, X_2, \ldots, X_{n-1})$ and dependent variable v. The variables $(X_1, X_2, \ldots, X_{n-1})$ are called *similarity variables.* This reduced PDE is obtained by substituting (4.23) into (4.1). We assume that this substitution does not lead to a singular differential equation for v. Note that if $\frac{\partial \xi}{\partial u} \equiv 0$, as is usually the case, then $X_i = X_i(x)$, $i = 1, 2, \ldots, n-1$; if $n = 2$ then the reduced PDE is an ODE and we denote the similarity variable by $\zeta = X_1$.

(II) Direct Substitution Method. This method is especially useful if one cannot explicitly solve the invariant surface condition (4.21a), i.e., the characteristic equations (4.22). Without loss of generality we assume that $\xi_n(x, u) \not\equiv 0$. Then (4.21a) becomes

$$u_n = -\sum_{i=1}^{n-1} \frac{\xi_i(x, u)}{\xi_n(x, u)} u_i + \frac{\eta(x, u)}{\xi_n(x, u)}. \tag{4.25}$$

It follows that any term involving derivatives with respect to x_n can be expressed in terms of x, u and derivatives $u_1, u_2, \ldots, u_{n-1}$. Hence after directly substituting (4.25) into PDE (4.1) for all terms involving derivatives of u with respect to x_n, we obtain a reduced PDE of order at most k with dependent variable u, $n-1$ independent variables $(x_1, x_2, \ldots, x_{n-1})$ and parameter x_n. Any solution of this reduced PDE defines an invariant solution of PDE (4.1) provided (4.25) or, equivalently, (4.1), is also satisfied. If $n = 2$ the reduced PDE is an ODE. The constants appearing in the general solution of this ODE are arbitrary functions of parameter x_2. These arbitrary functions are determined by substituting the general solution into (4.25) or (4.1).

4.2.2 Mapping of Solutions to Other Solutions from Group Invariance of a PDE

Under the action of a Lie group of transformations admitted by a PDE a solution of the PDE is mapped into a family of solutions if the solution is not an invariant solution of the group (i.e. the solution surface is not an invariant surface of the group). We derive a formula for this family of solutions generated from a known solution analogous to formula (3.23a,b) for ODE's. Without loss of generality we assume that a one-parameter Lie group of transformations is parameterized so that it is of the form

$$x^* = X(x, u; \epsilon) = e^{\epsilon \mathrm{X}} x, \tag{4.26a}$$

$$u^* = U(x, u; \epsilon) = e^{\epsilon \mathrm{X}} u, \tag{4.26b}$$

with infinitesimal generator

$$\mathrm{X} = \xi_i(x, u)\frac{\partial}{\partial x_i} + \eta(x, u)\frac{\partial}{\partial u}.$$

Consider a solution $u = \Theta(x)$ of (4.1) which is not an invariant solution corresponding to (4.26a,b). The transformation (4.26a,b) maps a point $(x, \Theta(x))$ on the solution surface $u = \Theta(x)$ into the point (x^*, u^*) with

$$x^* = X(x, \Theta(x); \epsilon), \tag{4.27a}$$

$$u^* = U(x, \Theta(x); \epsilon). \tag{4.27b}$$

For a fixed parameter value ϵ one can eliminate x from (4.27a,b) by substituting the inverse transformation of (4.26a), i.e.

$$x = X(x^*, u^*; -\epsilon)$$

into (4.27b):

$$\begin{aligned} u^* &= U(X(x^*, u^*; -\epsilon),\ \Theta(X(x^*, u^*; -\epsilon)); \epsilon) \\ &= U(e^{-\epsilon \mathrm{X}} x^*, \Theta(e^{-\epsilon \mathrm{X}} x^*); \epsilon) \end{aligned} \tag{4.28}$$

with

$$\mathrm{X} = \xi_i(x, u)\frac{\partial}{\partial x_i} + \eta(x, u)\frac{\partial}{\partial u} = \xi_i(x^*, u^*)\frac{\partial}{\partial x_i^*} + \eta(x^*, u^*)\frac{\partial}{\partial u^*}$$

[cf. Exercise 2.2-6]. Replacing $(x^*, u^*, -\epsilon)$ by (x, u, ϵ) in (4.28) we then have:

Theorem 4.2.2-1. *Suppose*

(i) $u = \Theta(x)$ *is a solution of PDE* (4.1);

(ii) *PDE* (4.1) *admits* (4.26a,b);

(iii) $u = \Theta(x)$ *is not an invariant surface of* (4.26a,b).

Then

$$u = U(e^{\epsilon X}x, \Theta(e^{\epsilon X}x); -\epsilon)$$
$$= U(X(x,u;\epsilon), \Theta(X(x,u;\epsilon)); -\epsilon)$$

implicitly defines a one-parameter family of solutions $u = \phi(x;\epsilon)$ *of PDE* (4.1).

4.2.3 Determining Equations for Infinitesimal Transformations of a kth Order PDE

From Theorem 4.1.1-1 we see that the kth order PDE ($k \geq 2$)

$$u_{i_1 i_2 \cdots i_\ell} = f(x, u, \underset{1}{u}, \underset{2}{u}, \ldots, \underset{k}{u}), \tag{4.29}$$

where $f(x, u, \underset{1}{u}, \underset{2}{u}, \ldots, \underset{k}{u})$ does not depend on $u_{i_1 i_2 \cdots i_\ell}$, admits

$$X = \xi_i(x,u)\frac{\partial}{\partial x_i} + \eta(x,u)\frac{\partial}{\partial u} \tag{4.30}$$

with kth extension given by (4.5), if and only if

$$\eta^{(\ell)}_{i_1 i_2 \cdots i_\ell} = \xi_j \frac{\partial f}{\partial x_j} + \eta \frac{\partial f}{\partial u} + \eta^{(1)}_j \frac{\partial f}{\partial u_j}$$
$$+ \cdots + \eta^{(k)}_{j_1 j_2 \cdots j_k} \frac{\partial f}{\partial u_{j_1 j_2 \cdots j_k}} \tag{4.31}$$

when $u_{i_1 i_2 \cdots i_\ell} = f(x, u, \underset{1}{u}, \underset{2}{u}, \ldots, \underset{k}{u})$.

It is easy to show that

(i) $\eta^{(p)}_{j_1 j_2 \cdots j_p}$ is linear in the components of $\underset{p}{u}$ if $p \geq 2$;

(ii) $\eta^{(p)}_{j_1 j_2 \cdots j_p}$ is a polynomial in the components of $\underset{1}{u}, \underset{2}{u}, \ldots, \underset{p}{u}$, whose coefficients are linear homogeneous in $(\xi(x,u), \eta(x,u))$ and in their partial derivatives with respect to (x,u) of order up to p.

From (i) and (ii) it follows that if $f(x, u, \underset{1}{u}, \underset{2}{u}, \ldots, \underset{k}{u})$ is a polynomial in the components of $\underset{1}{u}, \underset{2}{u}, \ldots, \underset{k}{u}$, then (4.31) is a polynomial equation in the components of $\underset{1}{u}, \underset{2}{u}, \ldots, \underset{k}{u}$ whose coefficients are linear homogeneous in $(\xi(x,u), \eta(x,u))$ and in their partial derivatives up to kth order. Observe that at any point x one can assign an arbitrary value to each component of $u, \underset{1}{u}, \underset{2}{u}, \ldots, \underset{k}{u}$ provided PDE (4.29) is satisfied; in particular, one can assign any values, except to the coordinate $u_{i_1 i_2 \cdots i_\ell}$, to the components of $u, \underset{1}{u}, \underset{2}{u}, \ldots, \underset{k}{u}$. Thus if we use (4.29) to eliminate $u_{i_1 i_2 \cdots i_\ell}$ from (4.31), then the resulting polynomial equation in the components of $\underset{1}{u}, \underset{2}{u}, \ldots, \underset{k}{u}$ must hold for arbitrary values of these components. Consequently the coefficients of the polynomial must vanish separately, resulting in a system of ***linear homogeneous*** partial differential equations for $n+1$ functions $(\xi(x,u), \eta(x,u))$. This system of linear PDE's is called the set of ***determining equations*** for the infinitesimal generator X admitted by (4.29). The set of determining equations is an ***overdetermined*** system of PDE's for $(\xi(x,u), \eta(x,u))$ since in general there are more than $n+1$ determining equations. [For PDE (4.29) with nonpolynomial function $f(x, u, \underset{1}{u}, \underset{2}{u}, \ldots, \underset{k}{u})$ one can still break up the equation (4.31) into a system of linear homogeneous PDE's for $(\xi(x,u), \eta(x,u))$ using the independence of the values of the components of $\underset{1}{u}, \underset{2}{u}, \ldots, \underset{k}{u}$ except for the value of $u_{i_1 i_2 \cdots i_\ell}$.]

Since the set of determining equations is an overdetermined system, it could happen that their only solution is the trivial solution $(\xi(x,u), \eta(x,u)) = (0,0)$. When the general solution of the determining equations is nontrivial two cases arise: If the general solution contains a finite number, say r, of essential arbitrary constants then it corresponds to an r-parameter Lie group of point transformations admitted by (4.29); if the general solution cannot be expressed in terms of a finite number of essential constants (for example when it contains an infinite number of essential constants or contains arbitrary functions of x and u) then it corresponds to an infinite-parameter Lie group of point transformations admitted by (4.29).

One can easily verify that a linear PDE, defined by a linear operator L,

$$\mathrm{L}u = g(x), \tag{4.32}$$

always admits a trivial infinite-parameter Lie group of transformations

$$x^* = x, \tag{4.33a}$$

$$u^* = u + \epsilon\omega(x), \tag{4.33b}$$

where $\omega(x)$ is any function satisfying

$$\mathrm{L}\omega = 0. \tag{4.34}$$

To within this trivial infinite-parameter Lie group of transformations, the Lie group of transformations admitted by a linear PDE usually has at most a finite number of parameters.

We now state some results on the form of admitted infinitesimal transformations which cover a wide class of scalar PDE's arising in applications. These results significantly simplify the tedious work involved in setting up and solving the determining equations for the infinitesimals $(\xi(x,u),$

$\eta(x,u))$. Suppose PDE (4.29) is such that $f(x,u,\underset{1}{u},\underset{2}{u},\ldots,\underset{k}{u})$ is linear in the components of $\underset{k}{u}$ and in addition the coefficients of components of $\underset{k}{u}$ depend only on (x,u). Then PDE (4.29) is of the form

$$A_{i_1 i_2 \cdots i_k}(x,u)u_{i_1 i_2 \cdots i_k} = g(x,u,\underset{1}{u},\ldots,\underset{k-1}{u}), \tag{4.35}$$

whose coefficients $\{A_{i_1 i_2 \cdots i_k}\}$ are symmetric with respect to their indices. The following theorems hold:

Theorem 4.2.3-1. *Suppose a kth order PDE ($k \geq 2$) of the form* (4.35) *admits a Lie group of transformations with infinitesimal generator* (4.30). *Let*

$$a_j = A_{11\cdots 1j}, \quad j = 1,2,\ldots,n.$$

(i) *If the coefficients $\{A_{i_1 i_2 \cdots i_k}\}$ do not satisfy*

$$A_{i_1 i_2 \cdots i_k} = (-1)^k a_{i_1} a_{i_2} \cdots a_{i_k} \tag{4.36}$$

then

$$\frac{\partial \xi_i}{\partial u} = 0, \quad i = 1,2,\ldots,n,$$

i.e., $\xi = \xi(x)$.

(ii) *If the coefficients $\{A_{i_1 i_2 \cdots i_k}\}$ satisfy* (4.36), *then*

$$a_i \frac{\partial \xi_j}{\partial u} = a_j \frac{\partial \xi_i}{\partial u}, \quad i,j = 1,2,\ldots,n.$$

Theorem 4.2.3-2. *Suppose PDE* (4.29) *is of the form* ($k \geq 2$)

$$B_{i_1 i_2 \cdots i_k}(x)u_{i_1 i_2 \cdots i_k} = g(x,u,\underset{1}{u},\ldots,\underset{k-1}{u}), \tag{4.37}$$

and admits a Lie group of transformations with infinitesimal generator (4.30). *If there does not exist a point transformation of its independent variables x such that PDE* (4.37) *is equivalent to*

$$\frac{\partial^k u}{\partial x_1^k} = G(x,u,\underset{1}{u},\ldots,\underset{k-1}{u}) \tag{4.38}$$

for some function $G(x,u,\underset{1}{u},\ldots,\underset{k-1}{u})$, then

$$\frac{\partial \xi_i}{\partial u} = 0, \quad i = 1,2,\ldots,n.$$

Theorem 4.2.3-3. *Suppose a PDE of the form* (4.38) ($k \geq 2$) *admits a Lie group of transformations with infinitesimal generator* (4.30). *Then*

$$\frac{\partial \xi_i}{\partial u} = 0, \quad i = 2,\ldots,n.$$

Theorem 4.2.3-4. *Suppose PDE* (4.29) *is of order* $k \geq 3$ *and is of the form*

$$A_{i_1 i_2 \cdots i_k}(x,u) u_{i_1 i_2 \cdots i_k} = B_{j_1 j_2 \cdots j_{k-1}}(x,u,\underset{1}{u}) u_{j_1 j_2 \cdots j_{k-1}} + h(x,u,\underset{1}{u},\ldots,\underset{k-2}{u}), \tag{4.39}$$

and admits infinitesimal generator (4.30). *Then*

$$\frac{\partial \xi_i}{\partial u} = 0, \quad i = 1,2,\ldots,n.$$

Theorem 4.2.3-5. *Suppose PDE* (4.29) ($k \geq 3$) *is of the form*

$$A_{i_1 i_2 \cdots i_k}(x,u) u_{i_1 i_2 \cdots i_k} = C_{j_1 j_2 \cdots j_{k-1}}(x,u) u_{j_1 j_2 \cdots i_{k-1}} + h(x,u,\underset{1}{u},\ldots,\underset{k-2}{u}), \tag{4.40}$$

and admits infinitesimal generator (4.30). *Then*

$$\frac{\partial \xi_i}{\partial u} = 0, \quad i = 1,2,\ldots,n,$$

and

$$\frac{\partial^2 \eta}{\partial u^2} = 0.$$

Theorem 4.2.3-6. *Suppose PDE* (4.40) *is of order two* ($k = 2$) *and admits infinitesimal generator* (4.30) *with*

$$\frac{\partial \xi_i}{\partial u} = 0, \quad i = 1,2,\ldots,n.$$

Then

$$\frac{\partial^2 \eta}{\partial u^2} = 0.$$

Theorem 4.2.3-7. *Suppose PDE* (4.29) ($k \geq 2$) *is a* **linear** *PDE which admits infinitesimal generator* (4.30). *Then*

$$\frac{\partial \xi_i}{\partial u} = 0, \quad i = 1,2,\ldots,n;$$

$$\frac{\partial^2 \eta}{\partial u^2} = 0.$$

Theorems 4.2.3-1 to 4.2.3-6 are proved in Bluman (1989). Theorem 4.2.3-7 is proved in Ovsiannikov (1962, Chapter 6; 1982, Sect. 27) for $k = 2$ and in Bluman (1989) for $k > 2$.

For $n = 2$, if $\frac{\partial \xi_1}{\partial u} = \frac{\partial \xi_2}{\partial u} = 0$, $\frac{\partial^2 \eta}{\partial u^2} = 0$, then an admitted infinitesimal generator is of the form

$$\mathrm{X} = \xi_1(x_1,x_2)\frac{\partial}{\partial x_1} + \xi_2(x_1,x_2)\frac{\partial}{\partial x_2} + [f(x_1,x_2)u + g(x_1,x_2)]\frac{\partial}{\partial u}. \tag{4.41}$$

From (2.133)–(2.137) it follows that for an infinitesimal generator of the form (4.41) we have

$$\eta_1^{(1)} = \frac{\partial g}{\partial x_1} + \frac{\partial f}{\partial x_1}u + \left[f - \frac{\partial \xi_1}{\partial x_1}\right]u_1 - \frac{\partial \xi_2}{\partial x_1}u_2; \tag{4.42}$$

$$\eta_2^{(1)} = \frac{\partial g}{\partial x_2} + \frac{\partial f}{\partial x_2}u - \frac{\partial \xi_1}{\partial x_2}u_1 + \left[f - \frac{\partial \xi_2}{\partial x_2}\right]u_2; \tag{4.43}$$

$$\begin{aligned}\eta_{11}^{(2)} = {} & \frac{\partial^2 g}{\partial x_1^2} + \frac{\partial^2 f}{\partial x_1^2}u + \left[2\frac{\partial f}{\partial x_1} - \frac{\partial^2 \xi_1}{\partial x_1^2}\right]u_1 \\ & - \frac{\partial^2 \xi_2}{\partial x_1^2}u_2 + \left[f - 2\frac{\partial \xi_1}{\partial x_1}\right]u_{11} - 2\frac{\partial \xi_2}{\partial x_1}u_{12};\end{aligned} \tag{4.44}$$

$$\begin{aligned}\eta_{12}^{(2)} = {} & \frac{\partial^2 g}{\partial x_1 \partial x_2} + \frac{\partial^2 f}{\partial x_1 \partial x_2}u + \left[\frac{\partial f}{\partial x_2} - \frac{\partial^2 \xi_1}{\partial x_1 \partial x_2}\right]u_1 \\ & + \left[\frac{\partial f}{\partial x_1} - \frac{\partial^2 \xi_2}{\partial x_1 \partial x_2}\right]u_2 - \frac{\partial \xi_1}{\partial x_2}u_{11} \\ & + \left[f - \frac{\partial \xi_1}{\partial x_1} - \frac{\partial \xi_2}{\partial x_2}\right]u_{12} - \frac{\partial \xi_2}{\partial x_1}u_{22};\end{aligned} \tag{4.45}$$

$$\begin{aligned}\eta_{22}^{(2)} = {} & \frac{\partial^2 g}{\partial x_2^2} + \frac{\partial^2 f}{\partial x_2^2}u - \frac{\partial^2 \xi_1}{\partial x_2^2}u_1 + \left[2\frac{\partial f}{\partial x_2} - \frac{\partial^2 \xi_2}{\partial x_2^2}\right]u_2 \\ & - 2\frac{\partial \xi_1}{\partial x_2}u_{12} + \left[f - 2\frac{\partial \xi_2}{\partial x_2}\right]u_{22}.\end{aligned} \tag{4.46}$$

4.2.4 EXAMPLES

(1) *Heat Equation*

Consider the heat equation

$$\frac{\partial^2 u}{\partial x_1^2} = \frac{\partial u}{\partial x_2}$$

or, equivalently,

$$u_{11} = u_2. \tag{4.47}$$

From Theorem 4.2.3-7 it follows that an infinitesimal generator admitted by (4.47) must be of the form (4.41). We now find all infinitesimal generators admitted by (4.47). For PDE (4.47) the invariance condition (4.31) is

$$\eta_{11}^{(2)} = \eta_2^{(1)} \tag{4.48}$$

when

$$u_{11} = u_2. \tag{4.49}$$

Substituting (4.43) and (4.44) into (4.48) and eliminating u_{11} by (4.49), we obtain

$$\left[\frac{\partial^2 g}{\partial x_1^2}-\frac{\partial g}{\partial x_2}\right]+\left[\frac{\partial^2 f}{\partial x_1^2}-\frac{\partial f}{\partial x_2}\right]u+\left[2\frac{\partial f}{\partial x_1}-\frac{\partial^2 \xi_1}{\partial x_1^2}+\frac{\partial \xi_1}{\partial x_2}\right]u_1$$
$$+\left[\frac{\partial \xi_2}{\partial x_2}-\frac{\partial^2 \xi_2}{\partial x_1^2}-2\frac{\partial \xi_1}{\partial x_1}\right]u_2-2\frac{\partial \xi_2}{\partial x_1}u_{12}=0. \tag{4.50}$$

Equation (4.50) must be an identity for all values of $(x_1, x_2, u, u_1, u_2, u_{12})$. Hence we obtain the following five *determining equations* for (ξ_1, ξ_2, f, g):

$$\frac{\partial \xi_2}{\partial x_1}=0; \tag{4.51a}$$

$$\frac{\partial \xi_2}{\partial x_2}-\frac{\partial^2 \xi_2}{\partial x_1^2}-2\frac{\partial \xi_1}{\partial x_1}=0; \tag{4.51b}$$

$$2\frac{\partial f}{\partial x_1}-\frac{\partial^2 \xi_1}{\partial x_1^2}+\frac{\partial \xi_1}{\partial x_2}=0; \tag{4.51c}$$

$$\frac{\partial^2 f}{\partial x_1^2}-\frac{\partial f}{\partial x_2}=0; \tag{4.51d}$$

$$\frac{\partial^2 g}{\partial x_1^2}-\frac{\partial g}{\partial x_2}=0. \tag{4.51e}$$

Hence $g(x_1, x_2)$ corresponds to the trivial infinite-parameter Lie group (4.33a,b) with $\omega(x)=g(x_1,x_2)$. Nontrivial infinitesimal generators arise from solving (4.51a–d). It is left to Exercise 4.2-4 to show that the solution of (4.51a–d) is

$$\xi_1(x_1,x_2)=\kappa+\delta x_1+\beta x_1+\gamma x_1 x_2, \tag{4.52a}$$

$$\xi_2(x_1,x_2)=\alpha+2\beta x_2+\gamma(x_2)^2, \tag{4.52b}$$

$$f(x_1,x_2)=-\gamma\left[\frac{(x_1)^2}{4}+\frac{x_2}{2}\right]-\frac{1}{2}\delta x_1+\lambda, \tag{4.52c}$$

where $\alpha,\beta,\gamma,\delta,\kappa,\lambda$ are six arbitrary parameters. Hence a nontrivial six-parameter Lie group of transformations acting on (x_1,x_2,u)-space is admitted by the heat equation (4.47) with infinitesimal generators given by

$$\mathrm{X}_1=\frac{\partial}{\partial x_1},\quad \mathrm{X}_2=\frac{\partial}{\partial x_2},\quad \mathrm{X}_3=x_1\frac{\partial}{\partial x_1}+2x_2\frac{\partial}{\partial x_2},$$

$$\mathrm{X}_4=x_1x_2\frac{\partial}{\partial x_1}+(x_2)^2\frac{\partial}{\partial x_2}-\left[\frac{(x_1)^2}{4}+\frac{x_2}{2}\right]u\frac{\partial}{\partial u},$$

$$\mathrm{X}_5 = x_2\frac{\partial}{\partial x_1} - \frac{1}{2}x_1 u\frac{\partial}{\partial u}, \quad \mathrm{X}_6 = u\frac{\partial}{\partial u}. \tag{4.53}$$

The commutator table for the Lie algebra arising from the infinitesimal generators (4.53) is:

	X_1	X_2	X_3	X_4	X_5	X_6
X_1	0	0	X_1	X_5	$-\frac{1}{2}\mathrm{X}_6$	0
X_2	0	0	$2\mathrm{X}_2$	$\mathrm{X}_3 - \frac{1}{2}\mathrm{X}_6$	X_1	0
X_3	$-\mathrm{X}_1$	$-2\mathrm{X}_2$	0	$2\mathrm{X}_4$	X_5	0
X_4	$-\mathrm{X}_5$	$-\mathrm{X}_3 + \frac{1}{2}\mathrm{X}_6$	$-2\mathrm{X}_4$	0	0	0
X_5	$\frac{1}{2}\mathrm{X}_6$	$-\mathrm{X}_1$	$-\mathrm{X}_5$	0	0	0
X_6	0	0	0	0	0	0

From the form of the infinitesimals (4.52a,b) we see that (4.52a,b) induces a five-parameter $(\alpha, \beta, \gamma, \delta, \kappa)$ Lie group of transformations acting on (x_1, x_2)-space with infinitesimal generators given by

$$\mathrm{Y}_1 = \frac{\partial}{\partial x_1}, \quad \mathrm{Y}_2 = \frac{\partial}{\partial x_2}, \quad \mathrm{Y}_3 = x_1\frac{\partial}{\partial x_1} + 2x_2\frac{\partial}{\partial x_2},$$

$$\mathrm{Y}_4 = x_1 x_2\frac{\partial}{\partial x_1} + (x_2)^2\frac{\partial}{\partial x_2}, \quad \mathrm{Y}_5 = x_2\frac{\partial}{\partial x_1}. \tag{4.54}$$

This five-parameter Lie group is a subgroup of the eight-parameter Lie group of projective transformations in $\mathbb{R}^2$ defined by (2.179a,b) with infinitesimal generators (2.180).

Consider infinitesimal generator X_4 (parameter γ). The corresponding one-parameter Lie group of transformations obtained by solving the IVP for the first order system of ODE's

$$\frac{dx_1^*}{d\epsilon} = x_1^* x_2^*,$$

$$\frac{dx_2^*}{d\epsilon} = (x_2^*)^2,$$

$$\frac{du^*}{d\epsilon} = -\left[\frac{(x_1^*)^2}{4} + \frac{x_2^*}{2}\right]u^*,$$

with $u^* = u$, $x_1^* = x_1$, $x_2^* = x_2$ at $\epsilon = 0$, is

$$x_1^* = X_1(x_1, x_2, u; \epsilon) = \frac{x_1}{1 - \epsilon x_2},$$

$$x_2^* = X_2(x_1, x_2, u; \epsilon) = \frac{x_2}{1 - \epsilon x_2},$$

$$u^* = U(x_1, x_2, u; \epsilon) = u\sqrt{1 - \epsilon x_2}\exp\left[-\frac{\epsilon (x_1)^2}{4(1 - \epsilon x_2)}\right].$$

Invariant solutions $u = \Theta(x_1, x_2)$ of PDE (4.47) corresponding to X_4 satisfy [cf. (4.21a)]

$$x_1 x_2 \frac{\partial \Theta}{\partial x_1} + (x_2)^2 \frac{\partial \Theta}{\partial x_2} = -\left[\frac{(x_1)^2}{4} + \frac{x_2}{2}\right] \Theta. \tag{4.55}$$

The solution of (4.55) is found by solving the characteristic equations [cf. (4.22)]

$$\frac{dx_1}{x_1 x_2} = \frac{dx_2}{(x_2)^2} = \frac{du}{-[\frac{(x_1)^2}{4} + \frac{x_2}{2}]u}$$

which have two invariants

$$X_1 = \frac{x_1}{x_2}, \quad v = \sqrt{x_2} e^{(x_1)^2/4x_2} u.$$

Thus the solution of (4.55) is defined by the *invariant form* [cf. (4.23)]

$$\sqrt{x_2} e^{(x_1)^2/4x_2} u = \phi\left(\frac{x_1}{x_2}\right)$$

or, solving for u,

$$u = \Theta(x_1, x_2) = \frac{1}{\sqrt{x_2}} e^{-(x_1)^2/4x_2} \phi(\zeta) \tag{4.56}$$

where the similarity variable is

$$\zeta = \frac{x_1}{x_2}.$$

Substitution of (4.56) into PDE (4.47) leads to $\phi(\zeta)$ satisfying

$$\phi''(\zeta) = 0,$$

so that invariant solutions of (4.47) resulting from X_4 are

$$u = \Theta(x_1, x_2) = \frac{1}{\sqrt{x_2}} e^{-(x_1)^2/4x_2} \left[C_1 + C_2 \frac{x_1}{x_2}\right] \tag{4.57}$$

where C_1 and C_2 are arbitrary constants.

For any solution $u = \Theta(x_1, x_2)$ of (4.47), which is not of the form (4.57), we find the one-parameter family of solutions $u = \phi(x_1, x_2; \epsilon)$ generated by X_4 [cf. (4.28a,b)]: Let

$$\hat{x}_1 = X_1(x_1, x_2, u; \epsilon) = \frac{x_1}{1 - \epsilon x_2},$$
$$\hat{x}_2 = X_2(x_1, x_2, u; \epsilon) = \frac{x_2}{1 - \epsilon x_2},$$
$$\hat{u} = \Theta(\hat{x}_1, \hat{x}_2).$$

Then

$$u = \phi(x_1, x_2; \epsilon) = U(\hat{x}_1, \hat{x}_2, \hat{u}; -\epsilon)$$

$$= \frac{1}{\sqrt{1-\epsilon x_2}} \exp\left[\frac{\epsilon (x_1)^2}{4(1-\epsilon x_2)}\right] \Theta\left(\frac{x_1}{1-\epsilon x_2}, \frac{x_2}{1-\epsilon x_2}\right).$$

Lie (1881) found the group of the heat equation. Bluman (1967, Chap. II) [see also Bluman and Cole (1969, 1974 (Sect. 2.7))] constructed all corresponding invariant solutions of the heat equation.

(2) *Nonlinear Heat Conduction Equation*

As a second example we consider a group classification problem. In particular we completely classify the group properties of the nonlinear heat conduction equation

$$\frac{\partial}{\partial x_1}\left(K(u)\frac{\partial u}{\partial x_1}\right) = \frac{\partial u}{\partial x_2}$$

or, equivalently,

$$u_2 = K(u)u_{11} + K'(u)(u_1)^2. \tag{4.58}$$

Suppose (4.58) admits

$$\mathrm{X} = \xi_1(x_1, x_2, u)\frac{\partial}{\partial x_1} + \xi_2(x_1, x_2, u)\frac{\partial}{\partial x_2} + \eta(x_1, x_2, u)\frac{\partial}{\partial u}.$$

The invariance condition (4.31) here is

$$\eta_2^{(1)} = K'(u)\eta u_{11} + K(u)\eta_{11}^{(2)} + K''(u)\eta(u_1)^2 + 2K'(u)u_1\eta_1^{(1)} \tag{4.59}$$

when

$$u_2 = K(u)u_{11} + K'(u)(u_1)^2,$$

where $\eta_1^{(1)}$, $\eta_2^{(1)}$, $\eta_{11}^{(2)}$ are given by (2.133)–(2.135). After using (4.58) to eliminate u_2 from (4.59) we obtain a polynomial equation in u_{11}, u_{12}, u_1 which must hold for arbitrary values of $x_1, x_2, u, u_1, u_{11}, u_{12}$; from the coefficients of u_{12}, $u_{12}u_1$, $u_{11}u_1$ we get

$$\frac{\partial \xi_1}{\partial u} = \frac{\partial \xi_2}{\partial u} = \frac{\partial \xi_2}{\partial x_1} = 0. \tag{4.60a}$$

Using (4.60a), we get from the coefficients of u_1, u_{11}, $(u_1)^2$:

$$\frac{\partial \xi_1}{\partial x_2} + 2K'(u)\frac{\partial \eta}{\partial x_1} + K(u)\left[2\frac{\partial^2 \eta}{\partial x_1 \partial u} - \frac{\partial^2 \xi_1}{\partial x_1^2}\right] = 0; \tag{4.60b}$$

$$K(u)\left[\frac{\partial \xi_2}{\partial x_2} - 2\frac{\partial \xi_1}{\partial x_1}\right] + K'(u)\eta = 0; \tag{4.60c}$$

$$K(u)\frac{\partial^2\eta}{\partial u^2}+K'(u)\left[\frac{\partial\xi_2}{\partial x_2}-2\frac{\partial\xi_1}{\partial x_1}+\frac{\partial\eta}{\partial u}\right]+K''(u)\eta=0. \tag{4.60d}$$

The terms not involving u_{11}, u_{12}, u_1 lead to

$$K(u)\frac{\partial^2\eta}{\partial x_1^2}-\frac{\partial\eta}{\partial x_2}=0. \tag{4.60e}$$

Hence $\xi_2=\xi_2(x_2)$. Solving (4.60c) for η and substituting this result into (4.60e) we find that

$$\xi_1=\frac{1}{2}x_1\xi_2'(x_2)+\rho(x_1)^2+\beta x_1+\gamma(x_2),\quad \eta=\frac{K(u)}{K'(u)}[4\rho x_1+2\beta], \tag{4.61}$$

where ρ,β are arbitrary constants and $\gamma(x_2)$ is an arbitrary function of x_2. Substituting (4.61) into (4.60d) we find that if one of ρ,β is nonzero then it is necessary that $K(u)$ satisfy the ODE

$$\left(\frac{K(u)}{K'(u)}\right)''=0$$

whose solution is $K(u)=\lambda(u+\kappa)^\nu$ (with limiting case $K(u)=\lambda e^{\nu u}$) where λ,κ,ν are arbitrary constants. Finally substituting (4.61) into (4.60b) we obtain

$$2\gamma'(x_2)+x_1\xi_2''(x_2)+4\rho K(u)\left[7-\frac{4K(u)K''(u)}{[K'(u)]^2}\right]=0.$$

Hence for any $K(u)$ it immediately follows that $\gamma'(x_2)=\xi_2''(x_2)=0$ so that $\gamma=\text{const}$, $\xi_2=\delta x_2+\sigma$. Consequently there are five parameters $\beta,\rho,\gamma,\delta,\sigma$. The parameters γ,δ,σ exist for arbitrary $K(u)$ but the existence of parameters β,ρ depends on the form of $K(u)$. Three cases arise:

Case I. $\boxed{K(u)\text{ arbitrary}}$ Here $\rho=\beta=0$ and PDE (4.58) admits a three-parameter Lie group with infinitesimal generators

$$X_1=\frac{\partial}{\partial x_1},\quad X_2=\frac{\partial}{\partial x_2},\quad X_3=x_1\frac{\partial}{\partial x_1}+2x_2\frac{\partial}{\partial x_2}. \tag{4.62}$$

Case II. $\boxed{K(u)=\lambda(u+\kappa)^\nu}$ Here $\rho=0$ and PDE (4.58) admits a four-parameter group with infinitesimal generators (4.62) and

$$X_4=x_1\frac{\partial}{\partial x_1}+\frac{2}{\nu}(u+\kappa)\frac{\partial}{\partial u}. \tag{4.63}$$

In the limiting case $K(u)=\lambda e^{\nu u}$, this infinitesimal generator becomes

$$X_4=x_1\frac{\partial}{\partial x_1}+\frac{2}{\nu}\frac{\partial}{\partial u}.$$

Case III. $K(u) = \lambda(u+\kappa)^{-4/3}$ Here PDE (4.58) admits a five-parameter group with infinitesimal generators (4.62), (4.63) $[\nu = -\frac{4}{3}]$, and

$$X_5 = (x_1)^2 \frac{\partial}{\partial x_1} - 3x_1(u+\kappa)\frac{\partial}{\partial u}.$$

Ovsiannikov (1959, 1962) derived the above results by considering PDE (4.58) as a sytem of PDE's $v = K(u)\frac{\partial u}{\partial x_1}$, $\frac{\partial v}{\partial x_1} = \frac{\partial u}{\partial x_2}$. The classification as presented here appears in Bluman (1967) [see also Bluman and Cole (1974), Ovsiannikov (1982)].

(3) *Wave Equation for an Inhomogeneous Medium*

As a third example consider the complete group classification of the wave equation for a variable wave speed $c(x_1)$:

$$\frac{\partial^2 u}{\partial x_2^2} = c^2(x_1)\frac{\partial^2 u}{\partial x_1^2}$$

or, equivalently,

$$u_{22} = c^2(x_1)u_{11}. \tag{4.64}$$

Since (4.64) is a second order linear PDE it follows that (4.64) only admits nontrivial infinitesimal generators of the form (4.41). The invariance condition (4.31) for PDE (4.64) is

$$\eta_{22}^{(2)} = c^2(x_1)\eta_{11}^{(2)} + 2c(x_1)c'(x_1)\xi_1 u_{11}$$

when

$$u_{22} = c^2(x_1)u_{11},$$

where $\eta_{11}^{(2)}$, $\eta_{22}^{(2)}$ are given by (4.44) and (4.46) (without loss of generality we can set $g = 0$). For arbitrary $c(x_1)$, after eliminating u_{22} and using the independence of u_{12}, u_{11}, u_2, u_1, u we obtain the following determining equations for (ξ_1, ξ_2, f):

$$\frac{\partial \xi_1}{\partial x_2} - c^2(x_1)\frac{\partial \xi_2}{\partial x_1} = 0; \tag{4.65a}$$

$$c(x_1)\left[\frac{\partial \xi_2}{\partial x_2} - \frac{\partial \xi_1}{\partial x_1}\right] + c'(x_1)\xi_1 = 0; \tag{4.65b}$$

$$\frac{\partial^2 \xi_2}{\partial x_2^2} - c^2(x_1)\frac{\partial^2 \xi_2}{\partial x_1^2} - 2\frac{\partial f}{\partial x_2} = 0; \tag{4.65c}$$

$$\frac{\partial^2 \xi_1}{\partial x_2^2} + c^2(x_1)\left[2\frac{\partial f}{\partial x_1} - \frac{\partial^2 \xi_1}{\partial x_1^2}\right] = 0; \tag{4.65d}$$

$$\frac{\partial^2 f}{\partial x_2^2} - c^2(x_1)\frac{\partial^2 f}{\partial x_1^2} = 0. \tag{4.65e}$$

Solving (4.65a) for $\frac{\partial \xi_2}{\partial x_1}$ and (4.65b) for $\frac{\partial \xi_2}{\partial x_2}$ and setting

$$\frac{\partial^2 \xi_2}{\partial x_1 \partial x_2} = \frac{\partial^2 \xi_2}{\partial x_2 \partial x_1},$$

one finds that

$$\frac{\partial^2 \xi_1}{\partial x_1^2} - c^{-2}(x_1)\frac{\partial^2 \xi_1}{\partial x_2^2} - \frac{\partial}{\partial x_1}[H(x_1)\xi_1] = 0 \tag{4.66}$$

where

$$H(x_1) = c'(x_1)/c(x_1).$$

The solution of (4.66), (4.65d) leads to

$$f(x_1, x_2) = \frac{1}{2}H(x_1)\xi_1(x_1, x_2) + S(x_2) \tag{4.67}$$

where $S(x_2)$ is an arbitrary function of x_2. Substituting (4.67) into (4.65c) and then solving (4.65a) for $\frac{\partial \xi_1}{\partial x_2}$ and (4.65b) for $\frac{\partial \xi_1}{\partial x_1}$ and setting

$$\frac{\partial^2 \xi_1}{\partial x_1 \partial x_2} = \frac{\partial^2 \xi_1}{\partial x_2 \partial x_1},$$

we find that $S(x_2) = \text{const} = s$, so that

$$f = \frac{1}{2}H\xi_1 + s. \tag{4.68}$$

Substituting (4.68) into (4.65e) and using (4.65d), we obtain

$$H''\xi_1 + 2H'\frac{\partial \xi_1}{\partial x_1} + H\frac{\partial}{\partial x_1}(H\xi_1) = 0$$

or, equivalently,

$$\frac{\partial}{\partial x_1}[(2H' + H^2)(\xi_1)^2] = 0. \tag{4.69}$$

Three cases can arise.

Case I. $2H' + H^2 = 0$: In this case it is easy to show that

$$c(x_1) = (Ax_1 + B)^2, \tag{4.70}$$

where A, B are arbitrary constants. Then $H(x_1) = \frac{2A}{Ax_1+B}$. For *any* solution $\xi_1(x_1, x_2)$ of the corresponding equation (4.66), one finds that $\xi_2(x_1, x_2)$, $f(x_1, x_2)$ solving (4.65a–e) are given by:

$$\xi_2(x_1, x_2) = \int \left[\frac{\partial \xi_1}{\partial x_1} - H\xi_1\right] dx_2, \tag{4.71a}$$

$$f(x_1, x_2) = \frac{A\xi_1(x_1, x_2)}{Ax_1 + B}. \tag{4.71b}$$

The set of functions $\{\xi_1, \xi_2, f\}$ determined by any solution $\xi_1(x_1, x_2)$ of PDE (4.66) and by (4.71a,b) corresponds to invariance of PDE (4.64) under a nontrivial infinite-parameter Lie group admitted when $c(x_1) = (Ax_1 + B)^2$. One can show that for $A \neq 0$ the wave equation

$$\frac{\partial^2 u}{\partial x_2^2} = (Ax_1 + B)^4 \frac{\partial^2 u}{\partial x_1^2} \tag{4.72}$$

can be transformed to the wave equation

$$\frac{\partial^2 w}{\partial z_1 \partial z_2} = 0$$

by the point transformation [Bluman (1983)]

$$\begin{aligned} z_1 &= [Ax_1 + B]^{-1} + Ax_2, \\ z_2 &= [Ax_1 + B]^{-1} - Ax_2, \\ w &= [Ax_1 + B]^{-1} u. \end{aligned}$$

Hence the general solution of PDE (4.72) is

$$u = (Ax_1 + B)\ [F(z_1) + G(z_2)]$$

where $F(z_1)$, $G(z_2)$ are arbitrary twice differentiable functions of their arguments.

Case II. $2H' + H^2 \neq 0$, $\xi_1 \neq 0$: In this case from (4.69) it follows that $\xi_1(x_1, x_2)$ can be expressed in the separable form

$$\xi_1(x_1, x_2) = \alpha(x_1)\beta(x_2), \tag{4.73}$$

where

$$\alpha^2(x_1) = \rho[2H' + H^2]^{-1} \tag{4.74}$$

for some constant ρ; $\beta(x_2)$ is to be determined. Substituting (4.68) and (4.73) into (4.65d), one finds that

$$\frac{\beta''(x_2)}{\beta(x_2)} = \frac{c^2[\alpha' - H\alpha]'}{\alpha} = \text{const} = \sigma^2, \tag{4.75}$$

where σ is a real or imaginary constant. We distinguish between the subcases $\sigma = 0$, $\sigma \neq 0$.

Case IIa. $\sigma = 0$: Here $c(x_1)$ must satisfy the fourth order ODE

$$[\alpha' - H\alpha]' = 0 \tag{4.76}$$

and correspondingly

$$\beta(x_2) = p + qx_2,$$

where p, q are arbitrary constants. The substitution of (4.68) and (4.73) into (4.65e) leads to

$$(\alpha H)'' = 0. \tag{4.77}$$

The wave speed $c(x_1)$ must satisfy (4.76), (4.77) and (4.74). This leads to

$$\alpha(x_1) = B(x_1)^2 + Cx_1 + D,$$

$$\alpha(x_1)H(x_1) = A + 2Bx_1,$$

$$\rho = 4BD + A^2 - 2AC,$$

where A, B, C, D are arbitrary constants. Consequently,

$$\frac{c'}{c} = H = \frac{A + 2Bx_1}{B(x_1)^2 + Cx_1 + D},$$

i.e.,

$$\text{(i)}\ \boxed{c(x_1) = [B(x_1)^2 + Cx_1 + D] \exp\left[(A - C) \int [B(x_1)^2 + Cx_1 + D]^{-1} dx_1\right]}$$

The corresponding ξ_1 and f are obtained respectively from (4.73) and (4.68); ξ_2 is obtained from (4.65a,b). They yield a four-parameter Lie group of point transformations with infinitesimal generators:

$$X_1 = \frac{\partial}{\partial x_2},$$

$$X_2 = [B(x_1)^2 + Cx_1 + D]\frac{\partial}{\partial x_1} + [C - A]x_2\frac{\partial}{\partial x_2} + \frac{1}{2}[A + 2Bx_1]u\frac{\partial}{\partial u},$$

$$X_3 = x_2[B(x_1)^2 + Cx_1 + D]\frac{\partial}{\partial x_1} + \left[\frac{1}{2}(C - A)(x_2)^2 + \int \frac{B(x_1)^2 + Cx_1 + D}{c^2(x_1)} dx_1\right] \frac{\partial}{\partial x_2} + \frac{1}{2}x_2[A + 2Bx_1]u\frac{\partial}{\partial u},$$

$$X_4 = u\frac{\partial}{\partial u}.$$

The nonzero commutators of the corresponding Lie algebra are: $[X_1, X_2] = (C-A)X_1$, $[X_1, X_3] = X_2$, $[X_2, X_3] = (C-A)X_3$. One can show that the Lie algebra formed by X_1, X_2, X_3 is isomorphic to the Lie algebra of SO(2, 1) when $A \neq C$. When $A = C$, we have $c(x_1) = B(x_1)^2 + Cx_1 + D$.

It is easy to show that to within arbitrary scalings and translations in x_1, a wave speed $c(x_1)$ given by (i) is equivalent to one of the following canonical forms:

$$c(x_1) = (x_1)^A \quad [B = D = 0,\ C = 1];$$
$$c(x_1) = e^{x_1} \quad [B = C = 0,\ A = D = 1];$$
$$c(x_1) = [1 + (x_1)^2]e^{A \arctan x_1} \quad [C = 0,\ B = D = 1];$$
$$c(x_1) = (1 + x_1)^{1+A/2}(1 - x_1)^{1-A/2} \quad [C = 0,\ B = -1,\ D = 1];$$
$$c(x_1) = (x_1)^2 e^{1/x_1} \quad [C = D = 0,\ A = -1,\ B = 1].$$

We list special cases of wave speed (i), together with corresponding infinitesimal generators (constants A, B, C, D are renamed):

(ii) $\boxed{c(x_1) = (Ax_1 + B)^C,\ C \neq 0, 1, 2}$

$$X_1 = \frac{\partial}{\partial x_2}, \quad X_2 = [Ax_1 + B]\frac{\partial}{\partial x_1} + A(1 - C)x_2\frac{\partial}{\partial x_2} + \frac{1}{2}ACu\frac{\partial}{\partial u},$$

$$X_3 = x_2[Ax_1 + B]\frac{\partial}{\partial x_1} + \frac{1}{2}\left[A(1 - C)(x_2)^2 + \frac{(Ax_1 + B)^{2-2C}}{A(1 - C)}\right]\frac{\partial}{\partial x_2}$$
$$+ \frac{1}{2}ACx_2u\frac{\partial}{\partial u}, \quad X_4 = u\frac{\partial}{\partial u}.$$

The commutator table is the same as for (i) with $(C - A)$ replaced by $A(1 - C)$.

(iii) $\boxed{c(x_1) = Ax_1 + B}$

$$X_1 = \frac{\partial}{\partial x_2}, \quad X_2 = [Ax_1 + B]\frac{\partial}{\partial x_1} + \frac{1}{2}Au\frac{\partial}{\partial u},$$

$$X_3 = x_2[Ax_1 + B]\frac{\partial}{\partial x_1} + \frac{1}{A}\log(Ax_1 + B)\frac{\partial}{\partial x_2} + \frac{1}{2}Ax_2u\frac{\partial}{\partial u},$$

$$X_4 = u\frac{\partial}{\partial u}.$$

The nonzero commutators of the corresponding Lie algebra are $[X_1, X_3] = X_2$, $[X_2, X_3] = X_1$.

(iv) $\boxed{c(x_1) = Ae^{Bx_1}}$

$$X_1 = \frac{\partial}{\partial x_2}, \quad X_2 = \frac{\partial}{\partial x_1} - Bx_2\frac{\partial}{\partial x_2} + \frac{1}{2}Bu\frac{\partial}{\partial u},$$

$$X_3 = Ax_2\frac{\partial}{\partial x_1} - \frac{1}{2}\left[AB(x_2)^2 + (AB)^{-1}e^{-2Bx_1}\right]\frac{\partial}{\partial x_2}$$

$$+\frac{1}{2}ABx_2u\frac{\partial}{\partial u}, \quad X_4 = u\frac{\partial}{\partial u}.$$

The commutator table is the same as for (i) with $C-A$ replaced by $-AB$.

Case IIb. $\sigma \neq 0$: In this case (4.75) leads to $c(x_1)$ solving the fourth order ODE

$$c^2[\alpha' - H\alpha]' = \sigma^2\alpha, \tag{4.78}$$

where $H = c'/c$, and α is given by (4.74). [Without loss of generality $\rho = 1$ in (4.74).] One can show that if $c(x_1)$ satisfies (4.78) then the corresponding wave equation (4.64) admits a four-parameter Lie group of transformations with infinitesimal generators given by

$$X_1 = \frac{\partial}{\partial x_2}, \quad X_2 = e^{\sigma x_2}\left[\alpha\frac{\partial}{\partial x_1} + \sigma^{-1}(\alpha' - H\alpha)\frac{\partial}{\partial x_2} + \frac{1}{2}\alpha Hu\frac{\partial}{\partial u}\right],$$

$$X_3 = e^{-\sigma x_2}\left[\alpha\frac{\partial}{\partial x_1} - \sigma^{-1}(\alpha' - H\alpha)\frac{\partial}{\partial x_2} + \frac{1}{2}\alpha Hu\frac{\partial}{\partial u}\right],$$

$$X_4 = u\frac{\partial}{\partial u}.$$

The nonzero commutators of the corresponding Lie algebra are

$$[X_1, X_2] = \sigma X_2, \quad [X_1, X_3] = -\sigma X_3,$$

$$[X_2, X_3] = 2\sigma^{-1}[(\alpha' - H\alpha)^2 - (\sigma\alpha/c)^2]X_1.$$

It immediately follows that

$$(\alpha' - H\alpha)^2 - (\sigma\alpha/c)^2 = \text{const} = K. \tag{4.79}$$

Hence (4.79) is an integration of ODE (4.78), i.e., the commutator $[X_2, X_3]$ leads to an integration of ODE (4.78)! The third order ODE (4.79) for $c(x_1)$ admits a two-parameter Lie group. Using the methods of Chapter 3 it can be reduced to a first order ODE. If the reduced ODE can be solved, then the general solution of (4.79) is obtained by three quadratures. [When σ is imaginary, appropriate linear combinations of X_2 and X_3 yield the corresponding real infinitesimal generators.] When $K \neq 0$, the Lie algebra formed by X_1, X_2, X_3 is isomorphic to the Lie algebra of SO(2, 1).

Case III. $\xi_1 = 0$: From the determining equations (4.65a–e) it immediately follows that $\xi_2 = \text{const} = r$, $f = \text{const} = s$, and hence (4.64) only admits translations in x_2 and scalings of u. In particular if the wave speed $c(x_1)$ does not satisfy (4.74), (4.79) for any constants ρ, σ, K, then wave equation (4.64) admits a two-parameter Lie group with infinitesimal generators $X_1 = \frac{\partial}{\partial x_2}$, $X_2 = u\frac{\partial}{\partial u}$.

In summary we have the following theorem:

Theorem 4.2.4-1. *The wave equation (4.64), whose wave speed is a solution of system (4.74), (4.79) for some constants ρ, σ, K, admits a four-parameter Lie group of transformations. The group becomes an infinite-parameter group if and only if $c(x_1) = (Ax_1 + B)^C$, $C = 0, 2$. All other wave speeds $c(x_1)$ only admit the two-parameter Lie group of translations in x_2 and scalings of u.*

The group classification of the wave equation (4.64) appears in Bluman and Kumei (1987). This paper includes corresponding invariant solutions.

(4) *Biharmonic Equation*

As a final example we find the group admitted by the fourth order biharmonic equation

$$\nabla^2\nabla^2 u = 0,$$

where $\nabla^2 = \frac{\partial^2}{\partial x_1^2} + \frac{\partial^2}{\partial x_2^2}$, or, equivalently, the equation

$$u_{2222} = -2u_{1122} - u_{1111}. \tag{4.80}$$

The invariance condition (4.31) for PDE (4.80) is given by

$$\eta_{2222}^{(4)} = -2\eta_{1122}^{(4)} - \eta_{1111}^{(4)} \tag{4.81}$$

when $u_{2222} = -2u_{1122} - u_{1111}$. From Theorem 4.2.3-7 we see that an admitted nontrivial group has an infinitesimal generator of the form

$$\mathrm{X} = \xi_1(x_1, x_2)\frac{\partial}{\partial x_1} + \xi_2(x_1, x_2)\frac{\partial}{\partial x_2} + f(x_1, x_2)u\frac{\partial}{\partial u}.$$

It is straightforward to derive the following determining equations for (ξ_1, ξ_2, f):

$$\frac{\partial \xi_2}{\partial x_1} + \frac{\partial \xi_1}{\partial x_2} = 0; \tag{4.82a}$$

$$\frac{\partial \xi_1}{\partial x_1} - \frac{\partial \xi_2}{\partial x_2} = 0; \tag{4.82b}$$

$$2\frac{\partial f}{\partial x_1} - 3\frac{\partial^2 \xi_1}{\partial x_1^2} - \frac{\partial^2 \xi_1}{\partial x_2^2} = 0; \tag{4.82c}$$

$$2\frac{\partial f}{\partial x_2} - 4\frac{\partial^2 \xi_1}{\partial x_1 \partial x_2} - 3\frac{\partial^2 \xi_2}{\partial x_1^2} - \frac{\partial^2 \xi_2}{\partial x_2^2} = 0; \tag{4.82d}$$

$$2\frac{\partial f}{\partial x_1} - \frac{\partial^2 \xi_1}{\partial x_1^2} - 3\frac{\partial^2 \xi_1}{\partial x_2^2} - 4\frac{\partial^2 \xi_2}{\partial x_1 \partial x_2} = 0; \tag{4.82e}$$

$$2\frac{\partial f}{\partial x_2} - \frac{\partial^2 \xi_2}{\partial x_1^2} - 3\frac{\partial^2 \xi_2}{\partial x_2^2} = 0; \tag{4.82f}$$

$$3\frac{\partial^2 f}{\partial x_1^2}+\frac{\partial^2 f}{\partial x_2^2}-2\left[\frac{\partial^3\xi_1}{\partial x_1^3}+\frac{\partial^3\xi_1}{\partial x_1\partial x_2^2}\right]=0; \tag{4.82g}$$

$$2\frac{\partial^2 f}{\partial x_1\partial x_2}-\frac{\partial^3\xi_1}{\partial x_2^3}-\frac{\partial^3\xi_1}{\partial x_1^2\partial x_2}-\frac{\partial^3\xi_2}{\partial x_1^3}-\frac{\partial^3\xi_2}{\partial x_1\partial x_2^2}=0; \tag{4.82h}$$

$$3\frac{\partial^2 f}{\partial x_2^2}+\frac{\partial^2 f}{\partial x_1^2}-2\left[\frac{\partial^3\xi_2}{\partial x_2^3}+\frac{\partial^3\xi_2}{\partial x_1^2\partial x_2}\right]=0; \tag{4.82i}$$

$$4\left[\frac{\partial^3 f}{\partial x_1^3}+\frac{\partial^3 f}{\partial x_1\partial x_2^2}\right]-\frac{\partial^4\xi_1}{\partial x_1^4}-2\frac{\partial^4\xi_1}{\partial x_1^2\partial x_2^2}-\frac{\partial^4\xi_1}{\partial x_2^4}=0; \tag{4.82j}$$

$$4\left[\frac{\partial^3 f}{\partial x_2^3}+\frac{\partial^3 f}{\partial x_1^2\partial x_2}\right]-\frac{\partial^4\xi_2}{\partial x_1^4}-2\frac{\partial^4\xi_2}{\partial x_1^2\partial x_2^2}-\frac{\partial^4\xi_2}{\partial x_2^4}=0; \tag{4.82k}$$

$$\nabla^2\nabla^2 f=0. \tag{4.82l}$$

From (4.82a,b) it follows that

$$\nabla^2\xi_1=\nabla^2\xi_2=0. \tag{4.83}$$

Substituting (4.83) into (4.82c,f), we find that

$$f=\frac{\partial\xi_1}{\partial x_1}+s \tag{4.84}$$

where s is an arbitrary constant. Then equations (4.82d,e) are satisfied. Substituting (4.84), (4.82a,b) into (4.82g–i) we find that all third order derivatives of ξ_1 and ξ_2 are zero. Then equations (4.82j–l) are automatically satisfied. Hence

$$\xi_1=\alpha_1(x_1)^2+\beta_1x_1x_2+\gamma_1(x_2)^2+\delta_1x_1+\kappa_1x_2+\rho_1, \tag{4.85a}$$

$$\xi_2=\alpha_2(x_1)^2+\beta_2x_1x_2+\gamma_2(x_2)^2+\delta_2x_1+\kappa_2x_2+\rho_2, \tag{4.85b}$$

and, renaming s in (4.84), we have

$$f=2\alpha_1x_1+\beta_1x_2+s, \tag{4.85c}$$

where the indicated constants are to be determined. From (4.82a,b) it follows that

$$2\alpha_2x_1+\beta_2x_2+\delta_2=-\beta_1x_1-2\gamma_1x_2-\kappa_1,$$
$$2\alpha_1x_1+\beta_1x_2+\delta_1=\beta_2x_1+2\gamma_2x_2+\kappa_2.$$

Hence

$$\beta_1=-2\alpha_2,\ \gamma_2=-\alpha_2,\ \beta_2=2\alpha_1,\ \gamma_1=-\alpha_1,\ \kappa_2=\delta_1,\ \kappa_1=-\delta_2.$$

Consequently, after renaming constants δ_1, δ_2, ρ_1, ρ_2, s, we get a seven-parameter $(\alpha_1,\alpha_2,\ldots,\alpha_7)$ group admitted by the biharmonic equation (4.80) with infinitesimals:

$$\xi_1=\alpha_1\left[(x_1)^2-(x_2)^2\right]-2\alpha_2x_1x_2+\alpha_3x_1-\alpha_4x_2+\alpha_5, \tag{4.86a}$$

$$\xi_2 = 2\alpha_1 x_1 x_2 + \alpha_2 \left[(x_1)^2 - (x_2)^2\right] + \alpha_3 x_2 + \alpha_4 x_1 + \alpha_6, \tag{4.86b}$$

$$f = 2\alpha_1 x_1 - 2\alpha_2 x_2 + \alpha_7. \tag{4.86c}$$

Let the complex variable $z = x_1 + ix_2$. It is left to Exercise 4.2-5 to show that a seven-parameter Lie group of transformations with infinitesimals (4.86a–c) is given by the transformations

$$z^* = \frac{az+b}{cz+d}, \tag{4.87a}$$

$$u^* = \lambda \left| \frac{dz^*}{dz} \right| u, \tag{4.87b}$$

where a, b, c, d are arbitrary complex constants such that $ad - bc \neq 0$ and λ is an arbitrary real constant. Equation (4.87a) is a general Möbius (bilinear) transformation. This example is considered in Bluman and Gregory (1985).

Exercises 4.2

1. For the heat equation (4.47)

   (a) find invariant solutions resulting from the infinitesimal generator X_5, using both methods in Section 4.2.1;

   (b) for any solution $u = \Theta(x_1, x_2)$ of (4.47), which is not an invariant solution for X_5, find the one-parameter family of solutions $u = \phi(x_1, x_2; \epsilon)$ of (4.47) generated by X_5.

2. (a) For which solutions $u = \Theta(x_1, x_2)$ of (4.47) do the infinitesimal generators $X_1, X_2, \ldots, X_6$ given by (4.53) yield a six-parameter family of solutions $u = \phi(x_1, x_2; \epsilon_1, \epsilon_2, \ldots, \epsilon_6)$ of (4.47)?

   (b) Determine $\phi(x_1, x_2; \epsilon_1, \epsilon_2, \ldots, \epsilon_6)$.

3. For the wave equation $u_{22} = e^{2x_1} u_{11}$:

   (a) Find invariant solutions for the infinitesimal generators
   (i) $X_2 + sX_1$, and
   (ii) $X_3 + sX_1$,
   where s is an arbitrary constant.

   (b) Given any solution $u = \Theta(x_1, x_2)$ of this wave equation, find the four-parameter family of solutions generated by X_1, X_2, X_3, X_4. What condition must $\Theta(x_1, x_2)$ satisfy?

4. Show that (4.52a–c) is the general solution of (4.51a–d).

5. Show that (4.87a,b) defines a seven-parameter Lie group of transformations with infinitesimals (4.86a–c).

6. Consider the heat equation in two, three, and n spatial dimensions.

   (a) Find the nine-parameter group admitted by

$$\frac{\partial^2 u}{\partial x_1^2} + \frac{\partial^2 u}{\partial x_2^2} = \frac{\partial u}{\partial x_3}.$$

   (b) Find the 13-parameter group admitted by

$$\frac{\partial^2 u}{\partial x_1^2} + \frac{\partial^2 u}{\partial x_2^2} + \frac{\partial^2 u}{\partial x_3^2} = \frac{\partial u}{\partial x_4}.$$

   (c) Generalize to the case of the n-dimensional heat equation

$$\sum_{j=1}^{n} \frac{\partial^2 u}{\partial x_j^2} = \frac{\partial u}{\partial x_{n+1}}.$$

7. Consider the axisymmetric wave equation

$$\frac{\partial^2 u}{\partial x_2^2} = \frac{\partial^2 u}{\partial x_1^2} + \frac{1}{x_1}\frac{\partial u}{\partial x_1}. \tag{4.88}$$

   (a) Show that the Lie group admitted by (4.88) has as infinitesimal generators

$$X_1 = x_1 \frac{\partial}{\partial x_1} + x_2 \frac{\partial}{\partial x_2},$$

$$X_2 = 2x_1 x_2 \frac{\partial}{\partial x_1} + \left[(x_1)^2 + (x_2)^2\right] \frac{\partial}{\partial x_2} - x_2 u \frac{\partial}{\partial u},$$

$$X_3 = u\frac{\partial}{\partial u}, \quad X_4 = \frac{\partial}{\partial x_2}.$$

   (b) Find invariant solutions of (4.88) for the infinitesimal generators

      (i) $X_1 + sX_3$, and

      (ii) $X_2 + sX_3$,

      where s is an arbitrary constant.

8. Consider the nonlinear wave equation

$$\frac{\partial^2 u}{\partial x_2^2} = c^2(u)\frac{\partial^2 u}{\partial x_1^2}, \tag{4.89}$$

   where $c(u) \neq$ const. In terms of infinitesimal generators show that the group classification of (4.89) is:

   (a) $c(u)$ arbitrary:

$$X_1 = x_1 \frac{\partial}{\partial x_1} + x_2 \frac{\partial}{\partial x_2}, \quad X_2 = \frac{\partial}{\partial x_1}, \quad X_3 = \frac{\partial}{\partial x_2}.$$

(b) $c(u) = A(u+B)^C$, where A, B, and C are arbitrary constants:

$$\mathrm{X}_1, \mathrm{X}_2, \mathrm{X}_3, \quad \text{and} \quad \mathrm{X}_4 = Cx_1\frac{\partial}{\partial x_1} + (u+B)\frac{\partial}{\partial u},$$

(c) $c(u) = A(u+B)^2$, where A and B are arbitrary constants:

$$\mathrm{X}_1, \mathrm{X}_2, \mathrm{X}_3, \mathrm{X}_4\ (C=2), \text{ and } \mathrm{X}_5 = (x_1)^2\frac{\partial}{\partial x_1} + x_1(u+B)\frac{\partial}{\partial u}.$$

9. Consider Laplace's equation in $n \geq 3$ dimensions:

$$\sum_{j=1}^{n} \frac{\partial^2 u}{\partial x_j^2} = 0. \tag{4.90}$$

(a) Show that the $[1+\frac{1}{2}(n+1)(n+2)]$-parameter Lie group admitted by (4.90) can be represented by the infinitesimal generator

$$\mathrm{X} = \sum_{j=1}^{n} \xi_j(x)\frac{\partial}{\partial x_j} + f(x)u\frac{\partial}{\partial u}$$

with infinitesimals

$$\xi_j(x) = \alpha_j + \sum_{k=1}^{n} \beta_{jk}x_k - \gamma_j\sum_{k=1}^{n}(x_k)^2 + 2x_j\sum_{k=1}^{n}\gamma_k x_k + \lambda x_j,$$

$j = 1, 2, \ldots, n,$

$$f(x) = (2-n)\sum_{k=1}^{n}\gamma_k x_k + \delta,$$

where

$$\alpha_j, \delta, \lambda, \gamma_j, \quad \text{and} \quad \beta_{jk} = -\beta_{kj},$$

$j, k = 1, 2, \ldots, n$ are $1+\frac{1}{2}(n+1)(n+2)$ arbitrary constants. The subgroup corresponding to $\delta = 0$ is called the conformal group. It can be shown that the conformal group is isomorphic to the group $\mathrm{SO}(n+1, 1)$ [Bluman (1967)].

(b) Find a corresponding global seven-parameter Lie group admitted by (4.90) when $n = 2$.

10. Consider the nonlinear diffusion equation

$$\frac{\partial^2 u}{\partial x_1^2} = \left(\frac{\partial u}{\partial x_1}\right)^2 \frac{\partial u}{\partial x_2}. \tag{4.91}$$

(a) Find the infinite-parameter Lie group admitted by (4.91).

(b) Compare the Lie algebra admitted by (4.91) with the Lie algebra admitted by the linear heat equation (4.47).

(c) Consequently find a mapping which transforms (4.91) to (4.47). [Generalizations of this example will be discussed in detail in Chapter 6.]

11. Find the five-parameter Lie group of transformations admitted by Burgers' equation

$$\frac{\partial^2 u}{\partial x_1^2} - u\frac{\partial u}{\partial x_1} = \frac{\partial u}{\partial x_2}.$$

12. Find the four-parameter Lie group of transformations admitted by the Korteweg–de Vries equation

$$\frac{\partial^3 u}{\partial x_1^3} + u\frac{\partial u}{\partial x_1} + \frac{\partial u}{\partial x_2} = 0. \tag{4.92}$$

13. Find the four-parameter Lie group of transformations admitted by the cylindrical Korteweg–de Vries equation

$$\frac{\partial^3 u}{\partial x_1^3} + u\frac{\partial u}{\partial x_1} + \frac{1}{2}\frac{u}{x_2} + \frac{\partial u}{\partial x_2} = 0. \tag{4.93}$$

[The use of the groups of (4.93) and (4.92) to relate these PDE's will be discussed in Chapter 6.]

14. The motion of an incompressible constant-property fluid in two dimensions is described by the stream-function equation

$$\nabla^2 \frac{\partial u}{\partial x_3} + \frac{\partial u}{\partial x_2}\nabla^2 \frac{\partial u}{\partial x_1} - \frac{\partial u}{\partial x_1}\nabla^2 \frac{\partial u}{\partial x_2} = \nu \nabla^4 u, \tag{4.94}$$

where u is the stream function for the flow, $\nu = \text{const}$ is the kinematic viscosity, and $\nabla^2 = \frac{\partial^2}{\partial x_1^2} + \frac{\partial^2}{\partial x_2^2}$.

(a) If $\nu = 0$ show that the infinite-parameter Lie group admitted by (4.94) can be represented by the infinitesimal generator

$$\mathrm{X} = \sum_{j=1}^{3} \xi_j(x)\frac{\partial}{\partial x_j} + \eta(x,u)\frac{\partial}{\partial u}$$

with infinitesimals given by

$$\begin{aligned}
\xi_1(x) &= ax_1 + bx_2 + cx_2x_3 + f_1(x_3),\\
\xi_2(x) &= ax_2 - bx_1 - cx_1x_3 + f_2(x_3),\\
\xi_3(x) &= hx_3 + k,\\
\eta(x,u) &= (2a-h)u + \frac{1}{2}c\left[(x_1)^2 + (x_2)^2\right]\\
&\quad + f_1'(x_3)x_2 - f_2'(x_3)x_1 + f_3(x_3)
\end{aligned}$$

where a, b, c, h, k are arbitrary constants; $f_1(x_3)$, $f_2(x_3)$ and $f_3(x_3)$ are arbitrary once-differentiable functions of the time coordinate x_3.

(b) If $\nu \neq 0$ show that the group admitted by (4.94) is the same as for (a) except that $h = 2a$ [cf. Cantwell (1978)].

15. The nonlinear reaction-diffusion equation

$$\frac{\partial^2 u}{\partial x_1^2} + F(u) = \frac{\partial u}{\partial x_2} \tag{4.95}$$

admits a trivial two-parameter Lie group of transformations with infinitesimal generators $X_1 = \frac{\partial}{\partial x_1}$, $X_2 = \frac{\partial}{\partial x_2}$ for arbitrary $F(u)$. Show that (4.95) admits a three-parameter Lie group only if $F(u)$ takes on one of three forms:

$$Au^B, \quad u(A + B\log u), \quad Ae^{Bu},$$

to within translations in u, where A and B are constants. [cf. Liu and Fang (1986). For generalizations see Galaktionov, Doronitsyn, Elenin, Kurdyumov, and Samarskii (1988).]

16. Consider the nonlinear wave equation

$$\frac{\partial^2 u}{\partial x_2^2} = \frac{\partial}{\partial x_1}\left(c^2(u)\frac{\partial u}{\partial x_1}\right) \tag{4.96}$$

when $c(u) \neq$ const. In terms of infinitesimal generators show that the group classification of (4.96) is:

(a) $c(u)$ arbitrary:

$$X_1 = x_1\frac{\partial}{\partial x_1} + x_2\frac{\partial}{\partial x_2}, \quad X_2 = \frac{\partial}{\partial x_1}, \quad X_3 = \frac{\partial}{\partial x_2}.$$

(b) $c(u) = A(u + B)^C$, where A, B, and C are arbitrary constants:

$$X_1, X_2, X_3, \quad \text{and} \quad X_4 = Cx_1\frac{\partial}{\partial x_1} + (u + B)\frac{\partial}{\partial u}.$$

(c) $c(u) = A(u + B)^{-2}$, where A and B are arbitrary constants:

$$X_1, X_2, X_3, X_4\ (C = -2), \quad \text{and} \quad X_5 = (x_2)^2\frac{\partial}{\partial x_2} + x_2(u + B)\frac{\partial}{\partial u}.$$

(d) $c(u) = A(u + B)^{-2/3}$, where A and B are arbitrary constants:

$$X_1, X_2, X_3, X_4\left(C = -\frac{2}{3}\right), \quad \text{and} \quad X_5 = (x_1)^2\frac{\partial}{\partial x_1} - 3x_1(u+B)\frac{\partial}{\partial u}.$$

[cf. Ames, Lohner, and Adams (1981). A group classification for the PDE where $c(u)$ is replaced by $c(u,x)$ in (4.96) has been investigated by Torrisi and Valenti (1985).]

17. Show that the two-dimensional nonlinear Schroedinger equation

$$\frac{\partial^2 u}{\partial x_1^2} + \frac{\partial^2 u}{\partial x_2^2} + r|u^2|u = i\frac{\partial u}{\partial x_3}, \quad r = \text{const},$$

admits an eight-parameter Lie group with infinitesimal generators

$$\mathrm{X}_1 = \frac{\partial}{\partial x_1}, \quad \mathrm{X}_2 = \frac{\partial}{\partial x_2}, \quad \mathrm{X}_3 = \frac{\partial}{\partial x_3},$$

$$\mathrm{X}_4 = x_1\frac{\partial}{\partial x_1} + x_2\frac{\partial}{\partial x_2} + 2x_3\frac{\partial}{\partial x_3}, \quad \mathrm{X}_5 = x_2\frac{\partial}{\partial x_1} - x_1\frac{\partial}{\partial x_2},$$

$$\mathrm{X}_6 = x_3\frac{\partial}{\partial x_1} - \frac{i}{2}x_1 u\frac{\partial}{\partial u}, \quad \mathrm{X}_7 = x_3\frac{\partial}{\partial x_2} - \frac{i}{2}x_2 u\frac{\partial}{\partial u},$$

$$\mathrm{X}_8 = x_1x_3\frac{\partial}{\partial x_1} + x_2x_3\frac{\partial}{\partial x_2} + (x_3)^2\frac{\partial}{\partial x_3} - \left[x_3 + \frac{i}{4}[(x_1)^2 + (x_2)^2]\right] u\frac{\partial}{\partial u}$$

[cf. Tajiri (1983)].

4.3 Invariance for Systems of PDE's

Now consider a system of m PDE's ($m > 1$) with n independent variables $x = (x_1, x_2, \ldots, x_n)$ and m dependent variables $u = (u^1, u^2, \ldots, u^m)$:

$$F^\mu(x, u, \underset{1}{u}, \underset{2}{u}, \ldots, \underset{k}{u}) = 0, \quad \mu = 1, 2, \ldots, m. \tag{4.97}$$

We assume that each PDE in system (4.97) can be written in solved form. In particular

$$F^\mu(x, u, \underset{1}{u}, \underset{2}{u}, \ldots, \underset{k}{u}) = u^{\nu_\mu}_{i_1 i_2 \cdots i_{\ell_\mu}} - f^\mu(x, u, \underset{1}{u}, \underset{2}{u}, \ldots, \underset{k}{u}) = 0 \tag{4.98}$$

in terms of some ℓ_μth order partial derivative of u^{ν_μ} for some $\nu_\mu = 1, 2, \ldots, m$ where $f^\mu(x, u, \underset{1}{u}, \underset{2}{u}, \ldots, \underset{k}{u})$ does not depend explicitly on any of $u^{\nu_\mu}_{i_1 i_2 \cdots i_{\ell_\mu}}$, $\mu = 1, 2, \ldots, m$. Moreover we assume that the isolated partial derivatives $\{u^{\nu_\mu}_{i_1 i_2 \cdots i_{\ell_\mu}}\}$ are m distinct quantities.

4.3.1 Invariance of a System of PDE's

Definition 4.3.1-1. The one-parameter Lie group of transformations

$$x^* = X(x, u; \epsilon), \tag{4.99a}$$

$$u^* = U(x, u; \epsilon), \tag{4.99b}$$

leaves the system of PDE's (4.97) ***invariant*** if and only if its kth extension, defined by (2.144a–d), (2.140)–(2.142), leaves the surfaces in $(x, u, \underset{1}{u}, \underset{2}{u}, \ldots, \underset{k}{u})$-space defined by (4.97) invariant. In analogy to the case for a scalar PDE it is easy to prove the following theorem:

Theorem 4.3.1-1 (Infinitesimal Criterion for Invariance of a System of PDE's). *Let*

$$\mathrm{X} = \xi_i(x, u)\frac{\partial}{\partial x_i} + \eta^\mu(x, u)\frac{\partial}{\partial u^\mu} \tag{4.100}$$

be the infinitesimal generator of (4.99a,b). *Let*

$$\mathrm{X}^{(k)} = \xi_i(x, u)\frac{\partial}{\partial x_i} + \eta^\mu(x, u)\frac{\partial}{\partial u^\mu} + \eta_i^{(1)\mu}(x, u, \underset{1}{u})\frac{\partial}{\partial u_i^\mu}$$
$$+ \cdots + \eta_{i_1 i_2 \cdots i_k}^{(k)\mu}(x, u, \underset{1}{u}, \underset{2}{u}, \ldots, \underset{k}{u})\frac{\partial}{\partial u_{i_1 i_2 \cdots i_k}^\mu} \tag{4.101}$$

be the kth extended infinitesimal generator of (4.100) *where* $\eta_i^{(1)\mu}$ *is given by* (2.145) *and* $\eta_{i_1 i_2 \cdots i_j}^{(j)\mu}$ *by* (2.146), $\mu = 1, 2, \ldots, m$, *and* $i_j = 1, 2, \ldots, n$ *for* $j = 1, 2, \ldots, k$, *in terms of* $(\xi(x, u), \eta(x, u))$ [$\eta(x, u)$ *denotes* $(\eta^1(x, u), \eta^2(x, u), \ldots, \eta^m(x, u))$]. *Then* (4.99a,b) *is admitted by the system of PDE's* (4.97) *if and only if for each* $\nu = 1, 2, \ldots, m$,

$$\mathrm{X}^{(k)} F^\nu(x, u, \underset{1}{u}, \underset{2}{u}, \ldots, \underset{k}{u}) = 0, \tag{4.102a}$$

when

$$F^\mu(x, u, \underset{1}{u}, \underset{2}{u}, \ldots, \underset{k}{u}) = 0, \quad \mu = 1, 2, \ldots, m. \tag{4.102b}$$

Proof. See Exercise 4.3-1. □

Note that the invariance criterion (4.102a,b) involves substitutions of the m PDE's (4.102b) and (possibly) their differential consequences into each of the m equations defined by (4.102a).

4.3.2 Invariant Solutions

Consider a system of PDE's (4.97) which admits a one-parameter Lie group of transformations with infinitesimal generator (4.100). We assume that $\xi(x, u) \not\equiv 0$.

Definition 4.3.2-1. $u = \Theta(x)$, with components $u^\nu = \Theta^\nu(x)$, $\nu = 1, 2, \ldots, m$, is an *invariant solution* of (4.97) corresponding to (4.100) admitted by the system of PDE's (4.97) if and only if

(i) $u^\nu = \Theta^\nu(x)$ is an invariant surface of (4.100) for each $\nu = 1, 2, \ldots, m$;

(ii) $u = \Theta(x)$ solves (4.97).

It follows that $u = \Theta(x)$ is an invariant solution of (4.97) resulting from its invariance under (4.99a,b) if and only if $u = \Theta(x)$ satisfies

(i) $\mathrm{X}(u^\nu - \Theta^\nu(x)) = 0$ when $u = \Theta(x)$, $\nu = 1, 2, \ldots, m$, i.e.,

$$\xi_i(x, \Theta(x))\frac{\partial \Theta^\nu}{\partial x_i} = \eta^\nu(x, \Theta(x)), \quad \nu = 1, 2, \ldots, m; \tag{4.103a}$$

$$\text{(ii)} \quad F^\mu(x, u, \underset{1}{u}, \underset{2}{u}, \ldots, \underset{k}{u}) = 0, \quad \mu = 1, 2, \ldots, m, \tag{4.103b}$$

where the components of $\underset{j}{u}$ appearing in (4.97) are replaced by

$$u^\nu_{i_1 i_2 \cdots i_j} = \frac{\partial^j \Theta^\nu(x)}{\partial x_{i_1} \partial x_{i_2} \cdots \partial x_{i_j}},$$

$\nu = 1, 2, \ldots, m$ and $i_j = 1, 2, \ldots, n$ for $j = 1, 2, \ldots, k$.

Equations (4.103a) are the *invariant surface conditions* for invariant solutions corresponding to (4.100). As is the case for a scalar PDE, invariant solutions can be determined in two ways:

(I) Invariant Form Method. Here we first solve the invariant surface conditions (4.103a). The corresponding characteristic equations for $u = \Theta(x)$ are given by

$$\frac{dx_1}{\xi_1(x, u)} = \frac{dx_2}{\xi_2(x, u)} = \cdots = \frac{dx_n}{\xi_n(x, u)} = \frac{du^1}{\eta^1(x, u)} = \frac{du^2}{\eta^2(x, u)}$$

$$= \cdots = \frac{du^m}{\eta^m(x, u)}. \tag{4.104}$$

If $X_1(x, u), X_2(x, u), \ldots, X_{n-1}(x, u)$, $v^1(x, u), v^2(x, u), \ldots, v^m(x, u)$, are $n + m - 1$ independent invariants of (4.104) with the Jacobian

$$\frac{\partial(v^1, v^2, \ldots, v^m)}{\partial(u^1, u^2, \ldots, u^m)} \neq 0,$$

then the solution $u = \Theta(x)$ of (4.103a) is given implicitly by the invariant form

$$v^\nu(x, u) = \phi^\nu(X_1(x, u), X_2(x, u), \ldots, X_{n-1}(x, u)), \tag{4.105}$$

where ϕ^ν is an arbitrary function of $X_1, X_2, \ldots, X_{n-1}$, for $\nu = 1, 2, \ldots, m$. Note that $(X_1, X_2, \ldots, X_{n-1}, v^1, v^2, \ldots, v^m)$ are $n+m-1$ independent invariants of (4.100) and are canonical coordinates of (4.99a,b). Let $X_n(x, u)$ be the $(n+m)$th canonical coordinate satisfying

$$\mathrm{X}X_n = 1.$$

If the system of PDE's (4.97) is transformed to another system of PDE's in terms of independent variables $(X_1, X_2, \ldots, X_n)$ and dependent variables $(v^1, v^2, \ldots, v^m)$, then the transformed system of PDE's admits

$$\begin{aligned} X_i^* &= X_i, \quad i = 1, 2, \ldots, n-1, \\ X_n^* &= X_n + \epsilon, \\ (v^\mu)^* &= v^\mu, \quad \mu = 1, 2, \ldots, m. \end{aligned}$$

Hence X_n does not appear explicitly in the transformed system of PDE's. Consequently the transformed system of PDE's and thus the given system of PDE's (4.97) has invariant solutions of the form (4.105). These are found by solving a reduced system of PDE's with $n-1$ independent (similarity) variables $(X_1, X_2, \ldots, X_{n-1})$ and dependent variables $(v^1, v^2, \ldots, v^m)$. This reduced system of PDE's arises from substituting (4.105) into (4.97). If $\frac{\partial \xi}{\partial u} \equiv 0$, as is often the case, then $X_i = X_i(x)$, $i = 1, 2, \ldots, n-1$; if $n = 2$, the reduced system of PDE's is a system of ODE's and we denote the similarity variable by $\zeta = X_1$.

If one is unable to determine an explicit set of invariants $(X_1, X_2, \ldots, X_{n-1}, v^1, v^2, \ldots, v^m)$, then invariant solutions could still be found by using a direct substitution method.

(II) Direct Substitution Method. Without loss of generality we assume that $\xi_n(x, u) \not\equiv 0$. Then (4.103a) becomes

$$u_n^\nu = -\sum_{i=1}^{n-1} \frac{\xi_i(x, u)}{\xi_n(x, u)} u_i^\nu + \frac{\eta^\nu(x, u)}{\xi_n(x, u)}, \quad \nu = 1, 2, \ldots, m. \tag{4.106}$$

Hence when finding invariant solutions of the system of PDE's (4.97) any term involving derivatives of u with respect to x_n can be expressed in terms of x, u and derivatives of u with respect to $x_1, x_2, \ldots, x_{n-1}$. Consequently we obtain a reduced system of PDE's with dependent variables $(u^1, u^2, \ldots, u^m)$, independent variables $(x_1, x_2, \ldots, x_{n-1})$, and parameter x_n. Any solution of this reduced system defines an invariant solution of system (4.97) if the solution also satisfies (4.106) or, equivalently, (4.97). Here the invariant surface conditions (4.104) are satisfied *after* a solution of system (4.97) is determined from the direct substitution (4.106). If $n = 2$, the reduced system of PDE's is a system of ODE's. As is the case for a scalar PDE, the constants appearing in the general solution of this reduced system of ODE's are arbitrary functions of parameter x_2. The arbitrary functions are determined by substituting this general solution into the invariant surface conditions (4.103a) or into the system of PDE's (4.97).

4.3.3 Determining Equations for Infinitesimal Transformations of a System of PDE's

From Theorem 4.3.1-1 we see that the system of PDE's

$$u^{\nu_\mu}_{i_1 i_2 \cdots i_{\ell_\mu}} = f^\mu(x, u, \underset{1}{u}, \underset{2}{u}, \ldots, \underset{k}{u}), \tag{4.107}$$

where f^μ does not depend on $u^{\nu_\sigma}_{i_1 i_2 \cdots i_{\ell_\sigma}}$, $\sigma = 1, 2, \ldots, m$, for each $\mu = 1, 2, \ldots, m$, admits

$$\mathrm{X} = \xi_i(x, u)\frac{\partial}{\partial x_i} + \eta^\mu(x, u)\frac{\partial}{\partial u^\mu} \tag{4.108}$$

with kth extension given by (4.101) if and only if

$$\eta^{(\ell_\mu)\nu_\mu}_{i_1 i_2 \cdots i_{\ell_\mu}} = \xi_j \frac{\partial f^\mu}{\partial x_j} + \eta^\nu \frac{\partial f^\mu}{\partial u^\nu} + \eta^{(1)\nu}_j \frac{\partial f^\mu}{\partial u^\nu_j}$$

$$+ \cdots + \eta^{(k)\nu}_{j_1 j_2 \cdots j_k} \frac{\partial f^\mu}{\partial u^\nu_{j_1 j_2 \cdots j_k}}, \quad \mu = 1, 2, \ldots, m, \tag{4.109a}$$

when

$$u^{\nu_\sigma}_{i_1 i_2 \cdots i_{\ell_\sigma}} = f^\sigma(x, u, \underset{1}{u}, \underset{2}{u}, \ldots, \underset{k}{u}), \quad \sigma = 1, 2, \ldots, m. \tag{4.109b}$$

It is easy to show that $\eta^{(p)\nu}_{j_1 j_2 \cdots j_p}$ is a polynomial in the components of $u, \underset{1}{u}, \underset{2}{u}, \ldots, \underset{p}{u}$, whose coefficients are linear homogeneous in the components of $\xi(x, u)$ and $\eta(x, u)$ and their partial derivatives up to pth order. Thus ξ and η appear linearly in (4.109a). As in the case of a scalar PDE, equations (4.109a,b) lead to a *system of linear homogeneous PDE's for ξ and η*: First we eliminate $u^{\nu_\sigma}_{i_1 i_2 \cdots i_{\ell_\sigma}}$, $\sigma = 1, 2, \ldots, m$, from (4.109a) by using (4.109b). Then the components of x, u and the remaining components of $\underset{1}{u}, \underset{2}{u}, \ldots, \underset{k}{u}$ which appear in (4.109a) are independent variables. The condition that (4.109a) holds for any values of these independent variables leads to a linear system of PDE's for ξ and η called the *determining equations* for infinitesimal generators admitted by (4.97). In particular if each $f^\mu(x, u, \underset{1}{u}, \underset{2}{u}, \ldots, \underset{k}{u})$, $\mu = 1, 2, \ldots, m$, is a polynomial in the components of $\underset{1}{u}, \underset{2}{u}, \ldots, \underset{k}{u}$, then equations (4.109a) become polynomial equations in the independent components of $\underset{1}{u}, \underset{2}{u}, \ldots, \underset{k}{u}$. Clearly the polynomial coefficients must vanish. Consequently we obtain a linear homogeneous system of PDE's for ξ and η. In general the number of determining equations is greater than $n + m$, which is the number of unknowns (ξ, η), and hence the determining equations are *overdetermined.*

A linear system of PDE's, defined by a linear operator L,

$$\mathrm{L}u = g(x), \tag{4.110}$$

always admits a "trivial" infinite-parameter Lie group of transformations

$$x^* = x, \tag{4.111a}$$

$$u^* = u + \epsilon\omega(x), \tag{4.111b}$$

for any $\omega(x)$ satisfying

$$\mathrm{L}\omega = 0.$$

To within this trivial infinite-parameter Lie group of transformations, the Lie group of transformations admitted by a linear system of PDE's usually has at most a finite number of parameters.

In Chapter 6 we will show that it is necessary for a nonlinear system of PDE's to admit an infinite-parameter Lie group of transformations in order for it to be transformed to a linear system of PDE's by some invertible transformation. The problem of the existence and construction of a mapping of a nonlinear system of PDE's to a linear system of PDE's is fully considered in Section 6.4.1.

Unlike the case for scalar PDE's, there is very little known about the forms of admitted infinitesimals ξ and η for systems of PDE's. It is conjectured that for a linear system of PDE's (4.110) an admitted infinitesimal generator of point transformations is such that ξ is independent of u and η is linear in u, i.e.,

$$\frac{\partial \xi_i}{\partial u^\mu} = 0, \quad i = 1, 2, \ldots, n, \quad \text{and} \quad \mu = 1, 2, \ldots, m; \tag{4.112a}$$

$$\eta^\nu = k^\nu_\sigma(x) u^\sigma, \tag{4.112b}$$

for some functions $k^\nu_\sigma(x)$, $\sigma = 1, 2, \ldots, m$ for $\nu = 1, 2, \ldots, m$ to within the admitted "trivial" infinite-parameter Lie group (4.111a,b). We will assume that (4.112a,b) holds for a Lie group admitted by a linear system of PDE's. [It is easy to check that (4.112a,b) holds for all examples of linear systems of PDE's considered in this book.] For $n = 2$ and $m = 2$, we let $u = u^1$, $v = u^2$, $f = k^1_1$, $g = k^1_2$, $k = k^2_2$, $\ell = k^2_1$, and $\eta^1 = f(x_1, x_2)u + g(x_1, x_2)v$, $\eta^2 = k(x_1, x_2)v + \ell(x_1, x_2)u$. In this case for an admitted infinitesimal generator of the form

$$\mathrm{X} = \xi_1(x_1, x_2)\frac{\partial}{\partial x_1} + \xi_2(x_1, x_2)\frac{\partial}{\partial x_2} + [f(x_1, x_2)u + g(x_1, x_2)v]\frac{\partial}{\partial u}$$

$$+ [k(x_1, x_2)v + \ell(x_1, x_2)u]\frac{\partial}{\partial v}, \tag{4.113}$$

the once-extended infinitesimals are

$$\eta_1^{(1)1} = \frac{\partial f}{\partial x_1}u + \frac{\partial g}{\partial x_1}v + \left[f - \frac{\partial \xi_1}{\partial x_1}\right]u_1 - \frac{\partial \xi_2}{\partial x_1}u_2 + gv_1; \tag{4.114}$$

$$\eta_1^{(1)2} = \frac{\partial \ell}{\partial x_1}u + \frac{\partial k}{\partial x_1}v + \ell u_1 + \left[k - \frac{\partial \xi_1}{\partial x_1}\right]v_1 - \frac{\partial \xi_2}{\partial x_1}v_2; \tag{4.115}$$

$$\eta_2^{(1)1} = \frac{\partial f}{\partial x_2}u + \frac{\partial g}{\partial x_2}v - \frac{\partial \xi_1}{\partial x_2}u_1 + \left[f - \frac{\partial \xi_2}{\partial x_2}\right]u_2 + gv_2; \tag{4.116}$$

$$\eta_2^{(1)2} = \frac{\partial \ell}{\partial x_2}u + \frac{\partial k}{\partial x_2}v + \ell u_2 - \frac{\partial \xi_1}{\partial x_2}v_1 + \left[k - \frac{\partial \xi_2}{\partial x_2}\right]v_2. \tag{4.117}$$

4.3.4 EXAMPLES

(1) *Wave Equations*

Consider the linear system of wave equations

$$\frac{\partial v}{\partial x_2} = \frac{\partial u}{\partial x_1}, \tag{4.118a}$$

$$\frac{\partial u}{\partial x_2} = (x_1)^4\frac{\partial v}{\partial x_1}. \tag{4.118b}$$

This system admits an infinitesimal generator of the form (4.113) provided the invariance conditions (4.102a,b) hold, which here become

$$\eta_2^{(1)2} = \eta_1^{(1)1}, \tag{4.119a}$$

$$\eta_2^{(1)1} = 4(x_1)^3 v_1\xi_1 + (x_1)^4\eta_1^{(1)2}, \tag{4.119b}$$

when $v_2 = u_1$, $u_2 = (x_1)^4 v_1$. Substituting (4.114)–(4.117) into (4.119a,b), and then eliminating v_2 and u_2 through substitution of (4.118a,b), we obtain

$$\left[\frac{\partial \ell}{\partial x_2} - \frac{\partial f}{\partial x_1}\right]u + \left[\frac{\partial k}{\partial x_2} - \frac{\partial g}{\partial x_1}\right]v + \left[(x_1)^4\left(\ell + \frac{\partial \xi_2}{\partial x_1}\right) - g - \frac{\partial \xi_1}{\partial x_2}\right]v_1$$
$$+ \left[k - f + \frac{\partial \xi_1}{\partial x_1} - \frac{\partial \xi_2}{\partial x_2}\right]u_1 = 0, \tag{4.120a}$$

$$\left[\frac{\partial f}{\partial x_2} - (x_1)^4\frac{\partial \ell}{\partial x_1}\right]u + \left[\frac{\partial g}{\partial x_2} - (x_1)^4\frac{\partial k}{\partial x_1}\right]v$$
$$+ \left[g - \frac{\partial \xi_1}{\partial x_2} + (x_1)^4\left(\frac{\partial \xi_2}{\partial x_1} - \ell\right)\right]u_1$$
$$+ \left[(x_1)^4\left(f - k + \frac{\partial \xi_1}{\partial x_1} - \frac{\partial \xi_2}{\partial x_2}\right) - 4(x_1)^3\xi_1\right]v_1 = 0. \tag{4.120b}$$

Each of the equations (4.120a,b) must be an identity for all values of $(x_1, x_2, u, v, u_1, v_1)$. Consequently we obtain eight *determining equations* for $(\xi_1, \xi_2, f, g, k, \ell)$ which simplify to:

$$\frac{\partial k}{\partial x_2} - \frac{\partial g}{\partial x_1} = 0; \tag{4.121a}$$

$$\frac{\partial \ell}{\partial x_2} - \frac{\partial f}{\partial x_1} = 0; \tag{4.121b}$$

$$(x_1)^4 \ell - g = 0; \tag{4.121c}$$

$$(x_1)^4 \frac{\partial \xi_2}{\partial x_1} - \frac{\partial \xi_1}{\partial x_2} = 0; \tag{4.121d}$$

$$(x_1)^4 \frac{\partial k}{\partial x_1} - \frac{\partial g}{\partial x_2} = 0; \tag{4.121e}$$

$$(x_1)^4 \frac{\partial \ell}{\partial x_1} - \frac{\partial f}{\partial x_2} = 0; \tag{4.121f}$$

$$x_1 \left[\frac{\partial \xi_2}{\partial x_2} - \frac{\partial \xi_1}{\partial x_1} \right] + 2\xi_1 = 0; \tag{4.121g}$$

$$k - f + \frac{\partial \xi_1}{\partial x_1} - \frac{\partial \xi_2}{\partial x_2} = 0; \tag{4.121h}$$

It is left to Exercise 4.3-3 to show that the solution of (4.121a–h) is

$$\xi_1(x_1, x_2) = \alpha x_1 + 2\beta x_1 x_2, \tag{4.122a}$$

$$\xi_2(x_1, x_2) = -\alpha x_2 - \beta \left[(x_1)^{-2} + (x_2)^2 \right] + \gamma, \tag{4.122b}$$

$$f(x_1, x_2) = 3\beta x_2 + \delta, \tag{4.122c}$$

$$g(x_1, x_2) = -\beta x_1, \tag{4.122d}$$

$$k(x_1, x_2) = -2\alpha - \beta x_2 + \delta, \tag{4.122e}$$

$$\ell(x_1, x_2) = -\beta (x_1)^{-3}, \tag{4.122f}$$

where $\alpha, \beta, \gamma, \delta$ are four arbitrary constants. Hence a nontrivial four-parameter Lie group of transformations acting on (x_1, x_2, u, v)-space is admitted by the system of wave equations (4.118a,b) with its infinitesimal generators given by

$$\mathrm{X}_1 = \frac{\partial}{\partial x_2}, \quad \mathrm{X}_2 = x_1 \frac{\partial}{\partial x_1} - x_2 \frac{\partial}{\partial x_2} - 2v \frac{\partial}{\partial v},$$

$$\mathrm{X}_3 = 2x_1 x_2 \frac{\partial}{\partial x_1} - \left[(x_1)^{-2} + (x_2)^2 \right] \frac{\partial}{\partial x_2} + (3x_2 u - x_1 v) \frac{\partial}{\partial u}$$
$$- \left[x_2 v + (x_1)^{-3} u \right] \frac{\partial}{\partial v},$$

$$\mathrm{X}_4 = u \frac{\partial}{\partial u} + v \frac{\partial}{\partial v}.$$

The nonzero commutators of the corresponding Lie algebra are:

$$[X_1, X_2] = -X_1, \quad [X_1, X_3] = 2X_2 + 3X_4, \quad [X_2, X_3] = -X_3.$$

One can show that the Lie algebra formed by $Y_1 = X_1$, $Y_2 = X_2 + \frac{3}{2}X_4$, $Y_3 = X_3$ is isomorphic to the Lie algebra of SO(2, 1).

Consider infinitesimal generator X_3 (parameter β). We find corresponding invariant solutions $(u, v) = (\Theta^1(x), \Theta^2(x))$ by both methods outlined in Section 4.3.2.

(i) *Invariant Form Method.* The characteristic equations (4.104) become

$$\frac{dx_1}{2x_1x_2} = \frac{dx_2}{-[(x_1)^{-2} + (x_2)^2]} = \frac{du}{3x_2u - x_1v} = \frac{dv}{-[x_2v + (x_1)^{-3}u]}. \tag{4.123}$$

The first equality in (4.123) leads to the similarity variable (invariant)

$$X_1 = \zeta = \text{const} = (x_1)^{-1} - x_1(x_2)^2. \tag{4.124}$$

In order to determine the other invariants of (4.123) we consider the corresponding characteristic DE's

$$\frac{dx_1}{d\epsilon} = 2x_1x_2, \tag{4.125a}$$

$$\frac{dx_2}{d\epsilon} = -\left[(x_1)^{-2} + (x_2)^2\right], \tag{4.125b}$$

$$\frac{du}{d\epsilon} = 3x_2u - x_1v, \tag{4.125c}$$

$$\frac{dv}{d\epsilon} = -\left[x_2v + (x_1)^{-3}u\right]. \tag{4.125d}$$

Using the solution (4.124) of (4.125a), we obtain from (4.125b)

$$\zeta^{-1}x_1x_2 + \epsilon = \text{const} = E. \tag{4.126}$$

The constant E is related to the invariance of (4.125a–d) under translations in ϵ. Without loss of generality one can set $E = 0$. From (4.125a–d) we get

$$\frac{d^2v}{d\epsilon^2} = -4x_2\frac{dv}{d\epsilon} + 2[(x_1)^{-2} - (x_2)^2]v.$$

Then using (4.124) one can show that this equation reduces to

$$\frac{d^2}{d\epsilon^2}(x_1v) = 0.$$

Hence

$$x_1 v = v^1 \epsilon + v^2 \tag{4.127a}$$

where v^1 and v^2 are constants. Equation (4.125d) leads to

$$u = (x_1)^2[x_2(v^1\epsilon + v^2) - v^1]. \tag{4.127b}$$

Using $\epsilon = -\zeta^{-1}x_1x_2$ [equation (4.126)], we can eliminate ϵ from (4.127a,b), and thus obtain

$$u = (x_1)^2[-x_1(x_2)^2\zeta^{-1}v^1 + x_2v^2 - v^1],$$
$$v = (x_1)^{-1}[-x_1x_2\zeta^{-1}v^1 + v^2].$$

(ζ, v^1, v^2) are independent invariants of (4.123); ζ is the similarity variable of X_3. The invariant solutions of X_3 are now found by letting v^1, v^2 be functions of ζ, say $v^1 = F(\zeta)$, $v^2 = G(\zeta)$ [F, G correspond to ϕ^1, ϕ^2 of (4.105)]:

$$u = (x_1)^2[-x_1(x_2)^2\zeta^{-1}F(\zeta) + x_2G(\zeta) - F(\zeta)], \tag{4.128a}$$

$$v = (x_1)^{-1}[-x_1x_2\zeta^{-1}F(\zeta) + G(\zeta)]. \tag{4.128b}$$

We now substitute (4.128a,b) into (4.118a,b) to determine $F(\zeta)$ and $G(\zeta)$. Equation (4.118a) leads to

$$x_1x_2[2G(\zeta) + \zeta G'(\zeta)] + [F'(\zeta) - \zeta^{-1}F(\zeta)] = 0,$$

whereas equation (4.118b) leads to

$$[2G(\zeta) + \zeta G'(\zeta)] + \frac{(x_1)^2x_2}{[1-(x_1x_2)^2]}[\zeta F'(\zeta) - F(\zeta)] = 0.$$

Consequently

$$2G(\zeta) + \zeta G'(\zeta) = 0,$$
$$\zeta F'(\zeta) - F(\zeta) = 0.$$

Thus

$$G(\zeta) = a\zeta^{-2}, \quad F(\zeta) = b\zeta,$$

where a, b are arbitrary constants. This leads to the pair of linearly independent solutions

$$(u, v) = (x_1, x_2), \tag{4.129a}$$

and

$$(u, v) = \left(\frac{(x_1)^4x_2}{[1-(x_1x_2)^2]^2}, \frac{x_1}{[1-(x_1x_2)^2]^2}\right), \tag{4.129b}$$

of system (4.118a,b).

(ii) *Direct Substitution Method.* The equations (4.106) become

$$u_1 = \frac{1}{2}\left[[(x_1)^{-3}(x_2)^{-1} + (x_1)^{-1}x_2]u_2 + 3(x_1)^{-1}u - (x_2)^{-1}v\right], \quad (4.130a)$$

$$v_1 = \frac{1}{2}\left[[(x_1)^{-3}(x_2)^{-1} + (x_1)^{-1}x_2]v_2 - (x_1)^{-1}v - (x_1)^{-4}(x_2)^{-1}u\right]. \quad (4.130b)$$

Using (4.130a,b), we eliminate derivatives of u and v with respect to x_1 from (4.118a,b) so that

$$v_2 = \frac{1}{2}\left[[(x_1)^{-3}(x_2)^{-1} + (x_1)^{-1}x_2]u_2 + 3(x_1)^{-1}u - (x_2)^{-1}v\right], \quad (4.131a)$$

$$u_2 = \frac{1}{2}\left[[x_1(x_2)^{-1} + (x_1)^3 x_2]v_2 - (x_1)^3 v - (x_2)^{-1}u\right], \quad (4.131b)$$

which is a system of first order ODE's with x_2 as independent variable and x_1 as a parameter. Next we replace u_2 in (4.131a) by the right-hand side of (4.131b) and solve the resulting equation in terms of u. Then

$$u = \frac{(x_1)^3 x_2[3 + (x_1x_2)^2]v - x_1[1 - (x_1x_2)^2]^2 v_2}{5(x_1x_2)^2 - 1}. \quad (4.132)$$

Substituting (4.132) into (4.131b), we obtain

$$u_2 = \frac{(x_1)^3[1 + 3(x_1x_2)^2]}{[5(x_1x_2)^2 - 1]}[x_2v_2 - v]. \quad (4.133)$$

Taking $\frac{\partial}{\partial x_2}$ of (4.132) and then replacing u_2 by the right-hand side of (4.133), we get (setting $\tau = x_1x_2$) the ODE

$$(1 - 5\tau^2)(\tau^2 - 1)\frac{\partial^2 v}{\partial \tau^2} - 4\tau(5\tau^2 + 1)\frac{\partial v}{\partial \tau} + 4(5\tau^2 + 1)v = 0. \quad (4.134)$$

Linearly independent solutions of (4.134) are

$$v = \tau \quad \text{and} \quad v = (1 - \tau^2)^{-2}.$$

Hence

$$v = A(x_1)\tau + B(x_1)(1 - \tau^2)^{-2} \quad (4.135a)$$

is the general solution of (4.134) where $A(x_1)$, $B(x_1)$ are arbitrary functions of x_1. From equation (4.132) we then get

$$u = (x_1)^2 A(x_1) + (x_1)^2 B(x_1)\tau(1 - \tau^2)^{-2}. \quad (4.135b)$$

Using the given PDE (4.118a), we find that

$$[A(x_1) + x_1A'(x_1)] = \frac{\tau}{(1 - \tau^2)^2}[B(x_1) - x_1B'(x_1)]. \quad (4.136)$$

Since equation (4.136) must hold for all values of x_1 and τ, we get

$$A(x_1) = ax_1^{-1}, \quad B(x_1) = bx_1,$$

where a and b are arbitrary constants. This leads to the solutions (4.129a,b).

Note that the direct substitution method avoids integrations of the characteristic equations (4.104) and hence is more amenable to being done automatically on a computer using symbolic manipulation.

(2) *Nonlinear Heat Conduction Equation*

Consider again the nonlinear heat conduction equation

$$\frac{\partial}{\partial x_1}\left(K(u)\frac{\partial u}{\partial x_1}\right) = \frac{\partial u}{\partial x_2}. \tag{4.137}$$

We form an associated system of PDE's

$$\frac{\partial v}{\partial x_2} = K(u)\frac{\partial u}{\partial x_1}, \tag{4.138a}$$

$$\frac{\partial v}{\partial x_1} = u. \tag{4.138b}$$

The Lie group of point transformations admitted by system (4.138a,b) can lead to a symmetry group of the scalar PDE (4.137) which is not a Lie group of point transformations admitted by (4.137). Full details of such symmetry groups (*potential symmetries*) will be discussed in Chapter 7. We completely classify the group properties of the nonlinear system of PDE's (4.138a,b), leaving many details to the reader. Suppose (4.138a,b) admits an infinitesimal generator of the form

$$\mathrm{X} = \xi_1\frac{\partial}{\partial x_1} + \xi_2\frac{\partial}{\partial x_2} + \eta^1\frac{\partial}{\partial u} + \eta^2\frac{\partial}{\partial v} \tag{4.139}$$

where $\xi_1, \xi_2, \eta^1, \eta^2$ are functions of (x_1, x_2, u, v). The invariance conditions (4.102) become

$$\eta_2^{(1)2} = K'(u)u_1\eta^1 + K(u)\eta_1^{(1)1}, \tag{4.140a}$$

$$\eta_1^{(1)2} = \eta^1, \tag{4.140b}$$

when $v_2 = K(u)u_1$, $v_1 = u$. Substituting (2.145) into (4.140a,b), and then eliminating v_1 and v_2 through substitution of (4.138a,b), we obtain

$$\left[\frac{\partial \eta^2}{\partial x_2} - u\frac{\partial \xi_2}{\partial x_2} - K(u)\left(\frac{\partial \eta^1}{\partial x_1} + u\frac{\partial \eta^1}{\partial v}\right)\right]$$

$$+ \left[K(u)\left(\frac{\partial \eta^2}{\partial v} - \frac{\partial \xi_2}{\partial x_2} - \frac{\partial \eta^1}{\partial u} + \frac{\partial \xi_1}{\partial x_1}\right) - K'(u)\eta^1\right]u_1$$

$$+\left[\frac{\partial\eta^2}{\partial u}-u\frac{\partial\xi_1}{\partial x_2}+K(u)\left(\frac{\partial\xi_2}{\partial x_1}+u\frac{\partial\xi_2}{\partial v}\right)\right]u_2$$

$$+K(u)\left[\frac{\partial\xi_1}{\partial u}-K(u)\frac{\partial\xi_2}{\partial v}\right](u_1)^2=0, \tag{4.141a}$$

$$\left[\frac{\partial\eta^2}{\partial x_1}+u\frac{\partial\eta^2}{\partial v}-u\frac{\partial\xi_1}{\partial x_1}-u^2\frac{\partial\xi_1}{\partial v}-\eta^1\right]+\left[\frac{\partial\eta^2}{\partial u}-u\frac{\partial\xi_1}{\partial u}\right.$$

$$\left.-K(u)\left(\frac{\partial\xi_2}{\partial x_1}+u\frac{\partial\xi_2}{\partial v}\right)\right]u_1-\left[K(u)\frac{\partial\xi_2}{\partial u}\right](u_1)^2=0. \tag{4.141b}$$

Each of the equations (4.141a,b) must be an identity for all values of $(x_1, x_2, u, v, u_1, u_2)$. Consequently we obtain seven determining equations for $(\xi_1, \xi_2, \eta^1, \eta^2)$ which simplify to:

$$\frac{\partial\xi_2}{\partial u}=0; \tag{4.142a}$$

$$\frac{\partial\xi_2}{\partial x_1}+u\frac{\partial\xi_2}{\partial v}=0; \tag{4.142b}$$

$$\frac{\partial\eta^2}{\partial u}-u\frac{\partial\xi_1}{\partial u}=0; \tag{4.142c}$$

$$\frac{\partial\xi_1}{\partial u}-K(u)\frac{\partial\xi_2}{\partial v}=0; \tag{4.142d}$$

$$\frac{\partial\eta^2}{\partial x_2}-u\frac{\partial\xi_1}{\partial x_2}-K(u)\left(\frac{\partial\eta^1}{\partial x_1}+u\frac{\partial\eta^1}{\partial v}\right)=0; \tag{4.142e}$$

$$\frac{\partial\eta^2}{\partial v}-\frac{\partial\xi_2}{\partial x_2}-\frac{\partial\eta^1}{\partial u}+\frac{\partial\xi_1}{\partial x_1}-\frac{K'(u)}{K(u)}\eta^1=0; \tag{4.142f}$$

$$\frac{\partial\eta^2}{\partial x_1}+u\frac{\partial\eta^2}{\partial v}-u\frac{\partial\xi_1}{\partial x_1}-u^2\frac{\partial\xi_1}{\partial v}-\eta^1=0. \tag{4.142g}$$

The solution of the determining equations (4.142a–g) is left to Exercise 4.3-5. The results can be summarized as follows [Bluman, Kumei, and Reid (1988)]:

Case I. $\boxed{K(u)\text{ arbitrary}}$ Here the system (4.138a,b) admits a four-parameter Lie group with infinitesimal generators

$$\mathrm{X}_1=\frac{\partial}{\partial x_1},\quad \mathrm{X}_2=\frac{\partial}{\partial x_2},$$

$$\mathrm{X}_3=x_1\frac{\partial}{\partial x_1}+2x_2\frac{\partial}{\partial x_2}+v\frac{\partial}{\partial v},\quad \mathrm{X}_4=\frac{\partial}{\partial v}. \tag{4.143}$$

Case II. $\boxed{K(u) = \lambda(u+\kappa)^\nu}$ Here (4.138a,b) admits a five-parameter group with infinitesimal generators (4.143) and

$$X_5 = x_1\frac{\partial}{\partial x_1} + \frac{2}{\nu}(u+\kappa)\frac{\partial}{\partial u} + \left[\left(1+\frac{2}{\nu}\right)v + \frac{2\kappa x_1}{\nu}\right]\frac{\partial}{\partial v}. \tag{4.144}$$

Case III. $\boxed{K(u) = \lambda(u+\kappa)^{-2}}$ Here (4.138a,b) admits an infinite-parameter group with infinitesimal generators (4.143), (4.144),

$$X_6 = -x_1(v+\kappa x_1)\frac{\partial}{\partial x_1} + (u+\kappa)[v + x_1(u+2\kappa)]\frac{\partial}{\partial u}$$
$$+ [2\lambda x_2 + \kappa x_1(v+\kappa x_1)]\frac{\partial}{\partial v},$$
$$X_7 = -x_1[(v+\kappa x_1)^2 + 2\lambda x_2]\frac{\partial}{\partial x_1} + 4\lambda(x_2)^2\frac{\partial}{\partial x_2}$$
$$+ (u+\kappa)[6\lambda x_2 + (v+\kappa x_1)^2 + 2x_1(u+\kappa)(v+\kappa x_1)]\frac{\partial}{\partial u}$$
$$+ [\kappa x_1(v+\kappa x_1)^2 + 2\lambda x_2(2v+3\kappa x_1)]\frac{\partial}{\partial v},$$
$$X_\infty = \phi(z,x_2)\frac{\partial}{\partial x_1} - (u+\kappa)^2\frac{\partial\phi(z,x_2)}{\partial z}\frac{\partial}{\partial u} - \kappa\phi(z,x_2)\frac{\partial}{\partial v}, \tag{4.145}$$

where $z = v + \kappa x_1$, and $w = \phi(z, x_2)$ is an arbitrary solution of the linear heat equation

$$\lambda\frac{\partial^2 w}{\partial z^2} - \frac{\partial w}{\partial x_2} = 0.$$

The use of the infinitesimal generator X_∞ to transform

$$\frac{\partial}{\partial x_1}\left[\lambda(u+\kappa)^{-2}\frac{\partial u}{\partial x_1}\right] - \frac{\partial u}{\partial x_2} = 0$$

to a linear PDE will be discussed in Chapter 6.

Case IV. $\boxed{K(u) = \dfrac{1}{u^2+pu+q}\exp\left[r\displaystyle\int\frac{du}{u^2+pu+q}\right]}$, where p, q, and r are arbitrary constants such that $p^2 - 4q - r^2 \neq 0$.

Here (4.138a,b) admits a five-parameter group with infinitesimal generators (4.143) and

$$X_5 = v\frac{\partial}{\partial x_1} + (r-p)x_2\frac{\partial}{\partial x_2} - (u^2+pu+q)\frac{\partial}{\partial u} - (qx_1+pv)\frac{\partial}{\partial v}. \tag{4.146}$$

(3) *Wave Equation for an Inhomogeneous Medium*

Consider again the wave equation for a variable wave speed $c(x_1)$:

$$\frac{\partial^2 u}{\partial x_2^2} = c^2(x_1)\frac{\partial^2 u}{\partial x_1^2}. \tag{4.147}$$

We form an associated system [see Chapter 7]

$$\frac{\partial v}{\partial x_2} = \frac{\partial u}{\partial x_1}, \tag{4.148a}$$

$$\frac{\partial u}{\partial x_2} = c^2(x_1)\frac{\partial v}{\partial x_1}, \tag{4.148b}$$

and give a complete group classification of this linear system of PDE's. Suppose (4.148a,b) admits an infinitesimal generator of the form (4.113). It is left to Exercise 4.3-6 to show that the determining equations for $(\xi_1, \xi_2, f, g, k, \ell)$ are:

$$\frac{\partial k}{\partial x_2} - \frac{\partial g}{\partial x_1} = 0; \tag{4.149a}$$

$$\frac{\partial \ell}{\partial x_2} - \frac{\partial f}{\partial x_1} = 0; \tag{4.149b}$$

$$c^2(x_1)\ell - g = 0; \tag{4.149c}$$

$$c^2(x_1)\frac{\partial \xi_2}{\partial x_1} - \frac{\partial \xi_1}{\partial x_2} = 0; \tag{4.149d}$$

$$c^2(x_1)\frac{\partial k}{\partial x_1} - \frac{\partial g}{\partial x_2} = 0; \tag{4.149e}$$

$$c^2(x_1)\frac{\partial \ell}{\partial x_1} - \frac{\partial f}{\partial x_2} = 0; \tag{4.149f}$$

$$c(x_1)\left[\frac{\partial \xi_2}{\partial x_2} - \frac{\partial \xi_1}{\partial x_1}\right] + c'(x_1)\xi_1 = 0; \tag{4.149g}$$

$$k - f + \frac{\partial \xi_1}{\partial x_1} - \frac{\partial \xi_2}{\partial x_2} = 0. \tag{4.149h}$$

The integrability conditions arising from equations (4.149b,c,f) lead to $g(x_1, x_2)$ satisfying

$$\frac{\partial g}{\partial x_1}H + gH' = 0, \tag{4.150}$$

where

$$H(x_1) = c'(x_1)/c(x_1). \tag{4.151}$$

Then

$$g(x_1, x_2) = -\frac{a(x_2)}{2H}, \tag{4.152}$$

with $a(x_2)$ an arbitrary function of x_2. Two cases arise depending on whether or not $c(x_1)$ satisfies the equation

$$cc'(c/c')'' = \text{const} = \mu. \tag{4.153}$$

If $c(x_1)$ satisfies (4.153), then $a(x_2)$ satisfies $a'' = \mu a$. If $c(x_1)$ does not satisfy (4.153) for any constant μ, then $a(x_2) \equiv 0$ and system (4.148a,b) only admits two infinitesimal generators

$$\mathrm{X}_1 = \frac{\partial}{\partial x_2}, \quad \mathrm{X}_2 = u\frac{\partial}{\partial u} + v\frac{\partial}{\partial v}. \tag{4.154}$$

If $c(x_1)$ satisfies (4.153), then system (4.148a,b) admits a four-parameter Lie group of transformations. Solutions of (4.153) can be classified as follows in terms of the value of μ [cf. Bluman and Kumei (1987, 1988)]:

(I) $\mu = 0$: In this case, to within arbitrary scalings and translations in x_1,

$$c(x_1) = e^{x_1} \quad \text{or} \quad (x_1)^C, \tag{4.155}$$

where C is an arbitrary constant.

(II) $\mu \neq 0$: Here (4.153) cannot be solved explicitly but reduces to one of the following ODE's:

$$c' = \nu^{-1}\sin(\nu \log c); \tag{4.156a}$$

$$c' = \nu^{-1}\sinh(\nu \log c); \tag{4.156b}$$

$$c' = \log c; \tag{4.156c}$$

$$c' = \nu^{-1}\cosh(\nu \log c); \tag{4.156d}$$

$\nu \neq 0$ is an arbitrary constant. If $c(x_1) = \phi(x_1, \nu)$ is a solution of any one of equations (4.156a–d) then the corresponding general solution of the differential equation for $c(x_1)$ in (4.153) is

$$c(x_1) = K\phi(Lx_1 + M, \nu)$$

where $K^2L^2 = |\mu|$ for arbitrary constants L, M, ν. The admitted infinitesimal generators for subcases include:

Case I. $\mu = 0$:

Case Ia. $\boxed{c(x_1) = (x_1)^C}$, $C \neq 0, 1$:

$$\mathrm{X}_1 = \frac{\partial}{\partial x_2}, \quad \mathrm{X}_2 = x_1\frac{\partial}{\partial x_1} + (1-C)x_2\frac{\partial}{\partial x_2} - Cv\frac{\partial}{\partial v},$$

$$X_3 = 2x_1x_2\frac{\partial}{\partial x_1} + \left[(1-C)(x_2)^2 + \frac{(x_1)^{2-2C}}{1-C}\right]\frac{\partial}{\partial x_2}$$
$$+ [(2C-1)x_2u - x_1v]\frac{\partial}{\partial u} - [x_2v + (x_1)^{1-2C}u]\frac{\partial}{\partial v},$$
$$X_4 = u\frac{\partial}{\partial u} + v\frac{\partial}{\partial v}; \tag{4.157}$$

Case Ib. $\boxed{c(x_1) = x_1}$

$$X_1 = \frac{\partial}{\partial x_2}, \quad X_2 = x_1\frac{\partial}{\partial x_1} - v\frac{\partial}{\partial v},$$
$$X_3 = 2x_1x_2\frac{\partial}{\partial x_1} + 2\log x_1\frac{\partial}{\partial x_2} + [x_2u - x_1v]\frac{\partial}{\partial u} - [x_2v + (x_1)^{-1}u]\frac{\partial}{\partial v},$$
$$X_4 = u\frac{\partial}{\partial u} + v\frac{\partial}{\partial v}; \tag{4.158}$$

Case Ic. $\boxed{c(x_1) = e^{x_1}}$

$$X_1 = \frac{\partial}{\partial x_2}, \quad X_2 = \frac{\partial}{\partial x_1} - x_2\frac{\partial}{\partial x_2} - v\frac{\partial}{\partial v},$$
$$X_3 = -4x_2\frac{\partial}{\partial x_1} + 2\left[(x_2)^2 + e^{-2x_1}\right]\frac{\partial}{\partial x_2} + 2[-2x_2u + v]\frac{\partial}{\partial u} + 2e^{-2x_1}u\frac{\partial}{\partial v},$$
$$X_4 = u\frac{\partial}{\partial u} + v\frac{\partial}{\partial v}; \tag{4.159}$$

Case II. $\mu \neq 0$: If $c(x_1)$ satisfies (4.156a) or (4.156b) then (4.148a,b) admits

$$X_1 = \frac{\partial}{\partial x_2},$$
$$X_2 = e^{x_2}\left\{\frac{2c}{c'}\frac{\partial}{\partial x_1} + 2\left[\left(\frac{c}{c'}\right)' - 1\right]\frac{\partial}{\partial x_2} + \left(\left[2 - \left(\frac{c}{c'}\right)'\right]u - \frac{c}{c'}v\right)\frac{\partial}{\partial u}\right.$$
$$\left. - \left[\left(\frac{c}{c'}\right)'v + \frac{1}{cc'}u\right]\frac{\partial}{\partial v}\right\},$$
$$X_3 = e^{-x_2}\left\{\frac{2c}{c'}\frac{\partial}{\partial x_1} + 2\left[1 - \left(\frac{c}{c'}\right)'\right]\frac{\partial}{\partial x_2} + \left(\left[2 - \left(\frac{c}{c'}\right)'\right]u + \frac{c}{c'}v\right)\frac{\partial}{\partial u}\right.$$
$$\left. - \left[\left(\frac{c}{c'}\right)'v - \frac{1}{cc'}u\right]\frac{\partial}{\partial v}\right\},$$
$$X_4 = u\frac{\partial}{\partial u} + v\frac{\partial}{\partial v}. \tag{4.160}$$

Corresponding invariant solutions appear in Bluman and Kumei (1987, 1988). Special classes of invariant solutions will be discussed in Section 4.4.3 and in Chapter 7.

Exercises 4.3

1. Prove Theorem 4.3.1-1.

2. Show that infinitesimal generators admitted by (4.118a,b) are of the form (4.113).

3. Show that (4.122a–f) gives the general solution of (4.121a–h).

4. The linear system of wave equations (4.118a,b) admits the infinitesimal generator

$$\mathrm{X} = \mathrm{X}_3 + s\mathrm{X}_4 = 2x_1x_2\frac{\partial}{\partial x_1} - \left[(x_1)^{-2} + (x_2)^2\right]\frac{\partial}{\partial x_2}$$

$$+ \left[(3x_2 + s)u - x_1v\right]\frac{\partial}{\partial u} + \left[(s - x_2)v - (x_1)^{-3}u\right]\frac{\partial}{\partial v}.$$

   (a) For invariant solutions corresponding to X show that the invariant form is

$$u = (x_1)^2 e^{-sx_1x_2\zeta^{-1}}\left[-x_1(x_2)^2\zeta^{-1}F(\zeta;s) + x_2G(\zeta;s) - F(\zeta;s)\right],$$
$$v = (x_1)^{-1}e^{-sx_1x_2\zeta^{-1}}\left[-x_1(x_2)^2\zeta^{-1}F(\zeta;s) + G(\zeta;s)\right],$$

   where $F(\zeta;s)$ and $G(\zeta;s)$ are arbitrary functions of ζ and s. The similarity variable ζ is given by (4.124).

   (b) Determine the coupled system of ODE's which are satisfied by $\{F(\zeta;s), G(\zeta;s)\}$. Simplify and express the solution in terms of special functions.

   (c) Derive these invariant solutions by the direct substitution method.

5. Complete the group classification of the system (4.138a,b) and derive (4.143)–(4.146).

6. Derive the determining equations (4.149a–h).

7. Consider the two-dimensional nonstationary boundary layer equations ($x_1 = x$, $x_2 = y$ are spatial variables; $x_3 = t$ is time; u, v are components of the velocity vector; p is pressure; without loss of generality the viscosity and density both equal one):

$$\frac{\partial u}{\partial x_3} + u\frac{\partial u}{\partial x_1} + v\frac{\partial u}{\partial x_2} + \frac{\partial p}{\partial x_1} = \frac{\partial^2 u}{\partial x_2^2},$$

$$\frac{\partial p}{\partial x_2} = 0, \quad \frac{\partial u}{\partial x_1} + \frac{\partial v}{\partial x_2} = 0. \tag{4.161}$$

Show that the admitted infinitesimal generators of (4.161) are

$$X_1 = \frac{\partial}{\partial x_3}, \quad X_2 = 2x_1\frac{\partial}{\partial x_1} + x_2\frac{\partial}{\partial x_2} + 2x_3\frac{\partial}{\partial x_3} - v\frac{\partial}{\partial v},$$

$$X_3 = x_1\frac{\partial}{\partial x_1} + u\frac{\partial}{\partial u} + 2p\frac{\partial}{\partial p},$$

$$X_{\infty_1} = \delta(x_3)\frac{\partial}{\partial x_1} + \delta'(x_3)\frac{\partial}{\partial u} - x_1\delta''(x_3)\frac{\partial}{\partial p},$$

$$X_{\infty_2} = \epsilon(x_3)\frac{\partial}{\partial x_2} + \epsilon'(x_3)\frac{\partial}{\partial v},$$

$$X_{\infty_3} = \pi(x_3)\frac{\partial}{\partial p},$$

where $\delta(x_3)$, $\epsilon(x_3)$, and $\pi(x_3)$ are arbitrary sufficiently smooth functions of x_3 [Ovsiannikov (1982)].

8. Show that the two-dimensional steady-state boundary layer equations [$\frac{\partial u}{\partial x_3}$ is replaced by 0 in (4.161)] admit

$$X_1 = \frac{\partial}{\partial x_1}, \quad X_2 = 2x_1\frac{\partial}{\partial x_1} + x_2\frac{\partial}{\partial x_2} - v\frac{\partial}{\partial v},$$

$$X_3 = x_1\frac{\partial}{\partial x_1} + u\frac{\partial}{\partial u} + 2p\frac{\partial}{\partial p}, \quad X_4 = \frac{\partial}{\partial p},$$

$$X_\infty = \phi(x_1)\frac{\partial}{\partial x_2} + u\phi'(x_1)\frac{\partial}{\partial v},$$

where $\phi(x_1)$ is an arbitrary differentiable function [Ovsiannikov (1982)].

9. Show that the incompressible three-dimensional Navier–Stokes equations (x_1, x_2, x_3 are spatial variables; x_4 is time; u^1, u^2, u^3 are components of the velocity vector; u^4 is pressure; $\nabla^2 = \sum_{i=1}^3 \frac{\partial^2}{\partial x_i^2}$; viscosity is set to equal one):

$$\sum_{i=1}^{3} u_i^i = 0, \quad u_4^j + \sum_{i=1}^{3} u^i u_i^j + u_j^4 = \nabla^2 u^j, \quad j = 1,2,3,$$

admit

$$X_1 = \frac{\partial}{\partial x_4};$$

$$X_2 = \sum_{i=1}^{3}\left[x_i\frac{\partial}{\partial x_i} - u^i\frac{\partial}{\partial u^i}\right] + 2\left[x_4\frac{\partial}{\partial x_4} - u^4\frac{\partial}{\partial u^4}\right];$$

$$X_3 = x_2\frac{\partial}{\partial x_1} - x_1\frac{\partial}{\partial x_2} + u^2\frac{\partial}{\partial u^1} - u^1\frac{\partial}{\partial u^2};$$

$$X_4 = x_3\frac{\partial}{\partial x_1} - x_1\frac{\partial}{\partial x_3} + u^3\frac{\partial}{\partial u^1} - u^1\frac{\partial}{\partial u^3};$$

$$X_5 = x_3\frac{\partial}{\partial x_2} - x_2\frac{\partial}{\partial x_3} + u^3\frac{\partial}{\partial u^2} - u^2\frac{\partial}{\partial u^3};$$

$$X_{\infty_j} = \alpha_j(x_4)\frac{\partial}{\partial x_j} + \alpha_j'(x_4)\frac{\partial}{\partial u^j} - x_j\alpha_j''(x_4)\frac{\partial}{\partial u^4}, \quad j = 1,2,3;$$

$$X_{\infty_4} = \beta(x_4)\frac{\partial}{\partial u^4},$$

where $\beta(x_4)$, $\alpha_j(x_4)$, $j = 1,2,3$, are arbitrary. [See Boisvert, Ames, and Srivastava (1983). In this paper various invariant solutions are found.]

10. If the complex-valued wave equation $\psi(x_1, x_2)$ satisfies the cubic nonlinear Schroedinger equation

$$i\frac{\partial \psi}{\partial x_2} = -\frac{\partial^2 \psi}{\partial x_1^2} + V(x_1)\psi + |\psi|^2\psi \tag{4.162}$$

for an external potential $V(x_1)$, then the canonical transformation

$$\psi(x_1, x_2) = \sqrt{v}e^{-iu/2},$$

where u and v are real-valued functions, transforms (4.162) into the nonlinear system of PDE's, representing a Madelung fluid, given by

$$\frac{\partial u}{\partial x_2} + \frac{1}{2}\left(\frac{\partial u}{\partial x_1}\right)^2 + 2v + 2V(x_1) = 2v^{-1/2}\frac{\partial^2(v^{1/2})}{\partial x_1^2}, \tag{4.163a}$$

$$\frac{\partial v}{\partial x_2} + \frac{\partial}{\partial x_1}\left(v\frac{\partial u}{\partial x_1}\right) = 0. \tag{4.163b}$$

Show that if $V(x_1) = -x_1$, then (4.163a,b) admits the infinitesimal generators

$$X_1 = \left[x_1 + 3(x_2)^2\right]\frac{\partial}{\partial x_1} + 2x_2\frac{\partial}{\partial x_2} + \left[6x_1x_2 + 2(x_2)^3\right]\frac{\partial}{\partial u} - 2v\frac{\partial}{\partial v},$$

$$X_2 = \frac{\partial}{\partial x_2}, \quad X_3 = x_2\frac{\partial}{\partial x_1} + \left[x_1 + (x_2)^2\right]\frac{\partial}{\partial u},$$

$$X_4 = \frac{\partial}{\partial u}, \quad X_5 = \frac{\partial}{\partial x_1} + 2x_2\frac{\partial}{\partial u}.$$

[See Baumann and Nonnenmacher (1987) where these infinitesimal generators and the corresponding invariant solutions are given.]

11. Show that the two-dimensional coupled nonlinear system of Schroedinger equations

$$i\frac{\partial u}{\partial x_3} - \frac{\partial^2 u}{\partial x_1^2} + \frac{\partial^2 u}{\partial x_2^2} + |u|^2 u - 2uv = 0,$$

$$\frac{\partial^2 v}{\partial x_1^2} + \frac{\partial^2 v}{\partial x_2^2} - \frac{\partial^2}{\partial x_1^2}(|u|^2) = 0,$$

where u and v are complex-valued functions, admits infinitesimal generators

$$\mathrm{X}_j = \frac{\partial}{\partial x_j}, \quad j = 1,2,3;$$

$$\mathrm{X}_4 = x_1\frac{\partial}{\partial x_1} + x_2\frac{\partial}{\partial x_2} + 2x_3\frac{\partial}{\partial x_3} - u\frac{\partial}{\partial u} - 2v\frac{\partial}{\partial v};$$

$$\mathrm{X}_5 = -x_3\frac{\partial}{\partial x_1} + \frac{1}{2}ix_1u\frac{\partial}{\partial u}; \quad \mathrm{X}_6 = x_3\frac{\partial}{\partial x_2} + \frac{1}{2}ix_2u\frac{\partial}{\partial u};$$

$$\mathrm{X}_7 = x_1x_3\frac{\partial}{\partial x_1} + x_2x_3\frac{\partial}{\partial x_2} + (x_3)^2\frac{\partial}{\partial x_3}$$

$$- \left[x_3 + \frac{1}{4}i[(x_1)^2 - (x_2)^2]\right] u\frac{\partial}{\partial u} - 2x_3v\frac{\partial}{\partial v};$$

$$\mathrm{X}_8 = iu\frac{\partial}{\partial u}$$

[cf. Tajiri and Hagiwara (1983)].

4.4 Applications to Boundary Value Problems

So far we have neglected problems with boundary conditions imposed on a given PDE, i.e. boundary value problems (BVP's) for PDE's. For an infinitesimal generator admitted by a given PDE we have computed corresponding invariant solutions without regard to solving a specific BVP posed for the PDE. The application of infinitesimal transformations to BVP's for PDE's is much more restrictive than is the case for ODE's. In the case of an ODE an admitted infinitesimal generator leads to a reduction in the order of the ODE. In terms of the corresponding differential invariants, any posed BVP for the ODE is *automatically* reduced to a BVP for a lower order ODE. In the case of a PDE an invariant solution arising from an admitted infinitesimal generator solves a given BVP if the infinitesimal generator leaves all boundary conditions invariant. This means that the domain of the BVP or, equivalently, its boundary is invariant as well as the imposed conditions on the boundary of the BVP.

In the case of linear PDE's the situation is not as restrictive. Here a BVP need not be invariant (*incomplete invariance*). In particular:

(i) For a linear ***nonhomogeneous*** PDE with linear homogeneous boundary conditions an infinitesimal generator $X \neq u\frac{\partial}{\partial u}$, admitted by the associated linear homogeneous PDE, is useful if X leaves the homogeneous boundary conditions invariant. The BVP is solved by a superposition (eigenfunction expansion and/or integral transform representation) of invariant form functions arising from the infinitesimal generator $X + \lambda u\frac{\partial}{\partial u}$, where λ is an arbitrary constant, since $X + \lambda u\frac{\partial}{\partial u}$ is admitted by the associated linear homogeneous PDE, and the homogeneous boundary conditions. [λ will play the role of an eigenvalue.]

(ii) For a linear ***homogeneous*** PDE with $p(\geq 1)$ linear homogeneous boundary conditions and one linear nonhomogeneous boundary condition, an infinitesimal generator $X \neq u\frac{\partial}{\partial u}$, admitted by the PDE, is useful if X leaves the p homogeneous boundary conditions invariant. Consequently for any complex constant λ, the infinitesimal generator $X+\lambda u\frac{\partial}{\partial u}$ is admitted by the PDE and its p homogeneous boundary conditions. One solves the BVP by first constructing invariant solutions of the PDE and its homogeneous boundary conditions which arise from $X + \lambda u\frac{\partial}{\partial u}$, and then determining a superposition of these invariant solutions to solve the nonhomogeneous boundary condition. Note that in this case X does not necessarily leave the domain of the BVP invariant.

The results in Sections 4.4.1 and 4.4.2 first appeared in a more rudimentary form in Bluman (1967, 1974a) and Bluman and Cole (1974).

4.4.1 Formulation of Invariance of a BVP for a Scalar PDE

Consider a BVP for a kth order scalar PDE (4.1) ($k \geq 2$) which can be written in a solved form (4.2):

$$F(x, u, \underset{1}{u}, \underset{2}{u}, \ldots, \underset{k}{u}) = 0 \tag{4.164a}$$

defined on a domain Ω_x in x-space [$x = (x_1, x_2, \ldots, x_n)$] with boundary conditions

$$B_\alpha(x, u, \underset{1}{u}, \ldots, \underset{k-1}{u}) = 0 \tag{4.164b}$$

prescribed on boundary surfaces

$$\omega_\alpha(x) = 0, \quad \alpha = 1, 2, \ldots, s. \tag{4.164c}$$

Assume that BVP (4.164a–c) has a unique solution. Consider an infinitesimal generator of the form

$$X = \xi_i(x)\frac{\partial}{\partial x_i} + \eta(x, u)\frac{\partial}{\partial u}, \tag{4.165}$$

which defines a one-parameter Lie group of transformations in x-space as well as in (x, u)-space.

Definition 4.4.1-1. X is *admitted by BVP* (4.164a–c) if and only if

(i) $\mathrm{X}^{(k)}F(x,u,\underset{1}{u},\underset{2}{u},\ldots,\underset{k}{u})=0$ when

$$F(x,u,\underset{1}{u},\underset{2}{u},\ldots,\underset{k}{u})=0, \tag{4.166a}$$

(ii) $\mathrm{X}\omega_\alpha(x)=0$ when $\omega_\alpha(x)=0,\quad \alpha=1,2,\ldots,s;$ (4.166b)

(iii) $\mathrm{X}^{(k-1)}B_\alpha(x,u,\underset{1}{u},\ldots,\underset{k-1}{u})=0$ when

$$B_\alpha(x,u,\underset{1}{u},\ldots,\underset{k-1}{u})=0 \text{ on } \omega_\alpha(x)=0,\ \alpha=1,2,\ldots,s. \tag{4.166c}$$

Theorem 4.4.1-1. *Let BVP* (4.164a–c) *admit* (4.165). *Let* $X=(X_1(x), X_2(x),\ldots,X_{n-1}(x))$ *be* $n-1$ *independent group invariants of* (4.165) *depending only on* x. *Let* $v(x,u)$ *be a group invariant of* (4.165) *such that* $\frac{\partial v}{\partial u}\neq 0$. *Then BVP* (4.164a–c) *reduces to*

$$G(X,v,\underset{1}{u},\underset{2}{v},\ldots,\underset{k}{v})=0 \tag{4.167a}$$

defined on some domain Ω_X *in* X*-space with boundary conditions*

$$C_\alpha(X,v,\underset{1}{v},\ldots,\underset{k-1}{v})=0 \tag{4.167b}$$

prescribed on boundary surfaces

$$\nu_\alpha(X)=0, \tag{4.167c}$$

for some G, C_α, ν_α, $\alpha=1,2,\ldots,s$.

Proof. See Exercise 4.4-1. □

Note that the surfaces $X_j(x)=0$, $j=1,2,\ldots,n-1$, are invariant surfaces of the group. The condition (4.166b) means that each boundary surface $\omega_\alpha(x)=0$ is an invariant surface $\nu_\alpha(X)=0$ of the infinitesimal generator

$$\xi_i(x)\frac{\partial}{\partial x_i} \tag{4.168}$$

which is the restriction of X to x-space. From invariance under (4.165) the number of independent variables in BVP (4.164a–c) is reduced by one. The solution of BVP (4.164a–c) is an invariant solution

$$v=\phi(X_1,X_2,\ldots,X_{n-1}) \tag{4.169}$$

of PDE (4.164a) corresponding to its invariance under (4.165). The invariant solution $u=\Theta(x)$, defined by (4.164a), satisfies

$$\mathrm{X}(u-\Theta(x))=0 \quad\text{when}\quad u=\Theta(x), \tag{4.170}$$

i.e.

$$\xi_i(x)\frac{\partial\Theta}{\partial x_i} = \eta(x, \Theta(x)). \tag{4.171}$$

Theorem 4.4.1-2. *If* X *is of the form*

$$\mathrm{X} = \xi_i(x)\frac{\partial}{\partial x_i} + f(x)u\frac{\partial}{\partial u}, \tag{4.172}$$

then v is of the form $v = \frac{u}{g(x)}$ for a known function $g(x)$ and hence an invariant solution arising from X *is of the separated form*

$$u = \Theta(x) = g(x)\phi(X), \tag{4.173}$$

for an arbitrary function $\phi(X)$ of $X = (X_1, X_2, \ldots, X_{n-1})$.

Proof. See Exercise 4.4-2. □

If BVP (4.164a–c) admits an r-parameter Lie group of transformations with infinitesimal generators of the form

$$\mathrm{X}_i = \xi_{ij}(x)\frac{\partial}{\partial x_j} + \eta_i(x, u)\frac{\partial}{\partial u}, \quad i = 1, 2, \ldots, r, \tag{4.174}$$

then the unique solution $u = \Theta(x)$ of (4.164a–c) is an invariant solution satisfying

$$\mathrm{X}_i(u - \Theta(x)) = 0, \quad i = 1, 2, \ldots, r.$$

The proof of the following theorem is left to Exercise 4.4-3:

Theorem 4.4.1-3 (Invariance of a BVP Under a Multi-Parameter Lie Group of Transformations). *Suppose BVP (4.164a–c) admits an r-parameter Lie group of transformations with infinitesimal generators of the form*

$$\mathrm{X}_i = \xi_{ij}(x)\frac{\partial}{\partial x_j} + f_i(x)u\frac{\partial}{\partial u}, \quad i = 1, 2, \ldots, r. \tag{4.175}$$

Let R be the rank of the $r \times n$ matrix

$$\Xi(x) = \begin{bmatrix} \xi_{11}(x) & \xi_{12}(x) & \cdots & \xi_{1n}(x) \\ \xi_{21}(x) & \xi_{22}(x) & \cdots & \xi_{2n}(x) \\ \vdots & \vdots & & \vdots \\ \xi_{r1}(x) & \xi_{r2}(x) & \cdots & \xi_{rn}(x) \end{bmatrix}. \tag{4.176}$$

Let $q = n - R$ and let $Y_1(x), Y_2(x), \ldots, Y_q(x)$ be a complete set of functionally independent invariants of (4.175) satisfying

$$\xi_{ij}(x)\frac{\partial Y_\ell(x)}{\partial x_j} = 0, \quad i = 1, 2, \ldots, r; \;\; \ell = 1, 2, \ldots, q. \tag{4.177}$$

Let

$$v = \frac{u}{g(x)} \tag{4.178}$$

be an invariant of (4.175) *satisfying*

$$\mathrm{X}_i v = 0, \quad i = 1, 2, \ldots, r.$$

Then BVP (4.164a–c) *reduces to a BVP with $q = n - R$ independent variables $Y = (Y_1, Y_2, \ldots, Y_q)$ and dependent variable* (4.178). *The solution of BVP* (4.164a–c) *is an invariant solution of separated form*

$$u = g(x)\phi(Y) \tag{4.179}$$

where the function $\phi(Y)$ is to be determined.

The following examples are illustrative:

(1) *Fundamental Solutions of the Heat Equation*

Consider the heat equation (4.47) defined on the domain $x_2 > 0$, $a < x_1 < b$. Recall that (4.47) admits a six-parameter $(\alpha, \beta, \gamma, \delta, \kappa, \lambda)$ Lie group of transformations with infinitesimal generators of the form

$$\mathrm{X} = \xi_1(x_1, x_2)\frac{\partial}{\partial x_1} + \xi_2(x_2)\frac{\partial}{\partial x_2} + f(x_1, x_2)u\frac{\partial}{\partial u}$$

with

$$\xi_1(x_1, x_2) = \kappa + \delta x_2 + \beta x_1 + \gamma x_1 x_2, \tag{4.180a}$$

$$\xi_2(x_2) = \alpha + 2\beta x_2 + \gamma(x_2)^2, \tag{4.180b}$$

$$f(x_1, x_2) = -\gamma\left[\frac{(x_1)^2}{4} + \frac{x_2}{2}\right] - \frac{1}{2}\delta x_1 + \lambda. \tag{4.180c}$$

The boundary surfaces are $x_2 = 0$, $x_1 = a$, $x_1 = b$. Invariance of $x_2 = 0$ leads to

$$\xi_2(0) = 0,$$

and hence $\alpha = 0$. If $a = -\infty$ and $b = \infty$ then there is no further parameter reduction resulting from invariance of the boundary surfaces. If $a \neq -\infty$, then invariance of $x_1 = a$ leads to

$$\xi_1(a, x_2) = 0$$

for any $x_2 > 0$, and hence

$$\kappa = -\beta a, \quad \delta = -\gamma a.$$

If $b \neq \infty$, then invariance of $x_1 = b$ yields

$$\kappa = -\beta b, \quad \delta = -\gamma b.$$

Consequently if $a \neq -\infty$ and $b \neq \infty$, then $\beta = \gamma = \delta = \kappa = 0$, and hence there is no nontrivial group admitted by the boundary of a BVP for the heat equation (4.47) defined on a domain $x_2 > 0$, $a < x_1 < b$. However since (4.47) is a linear PDE it is not necessary to leave all boundary surfaces of the domain invariant as mentioned in the introductory remarks of Section 4.4 [see also Section 4.4.2].

If $a = -\infty$ and $b = \infty$, then a four-parameter group is admitted by the boundary of a BVP posed for the heat equation (4.47), and a BVP admits at most a five-parameter $(\beta, \gamma, \delta, \kappa, \lambda)$ group.

If $a \neq -\infty$ (without loss of generality $a = 0$) and $b = \infty$, then a two-parameter group is admitted by the boundary of a posed BVP, and a BVP admits at most a three-parameter (β, γ, λ) group with infinitesimals

$$\xi_1(x_1, x_2) = \beta x_1 + \gamma x_1 x_2, \tag{4.181a}$$

$$\xi_2(x_2) = 2\beta x_2 + \gamma (x_2)^2, \tag{4.181b}$$

$$f(x_1, x_2) = -\gamma \left[\frac{(x_1)^2}{4} + \frac{x_2}{2}\right] + \lambda. \tag{4.181c}$$

We derive fundamental solutions for the heat equation (4.47) when

$$u(x_1, 0) = \delta(x_1 - \hat{x}_1),$$

where $\delta(x_1 - \hat{x}_1)$ is the Dirac delta function centered at $\hat{x}_1$, $a < \hat{x}_1 < b$, for an infinite domain ($a = -\infty$, $b = \infty$) and a semi-infinite domain ($a = 0$, $b = \infty$).

(i) *Infinite Domain* $(a, b) = (-\infty, \infty)$: Consider the BVP

$$\frac{\partial^2 u}{\partial x_1^2} = \frac{\partial u}{\partial x_2}, \tag{4.182a}$$

on the domain $x_2 > 0$, $-\infty < x_1 < \infty$, with boundary conditions

$$u(\pm\infty, x_2) = 0, \quad x_2 > 0,$$

and

$$u(x_1, 0) = \delta(x_1). \tag{4.182b}$$

[Without loss of generality one can set $\hat{x}_1 = 0$.]

The infinitesimal (4.180a–c) with $\alpha = 0$ is admitted by (4.182b) provided that

$$f(x_1, 0) u(x_1, 0) = \xi_1(x_1, 0) \delta'(x_1)$$

when $u(x_1, 0) = \delta(x_1)$, i.e.

$$f(x_1, 0) \delta(x_1) = \xi_1(x_1, 0) \delta'(x_1). \tag{4.183}$$

Equation (4.183) is satisfied if

$$\xi_1(0,0) = 0 \tag{4.184a}$$

and

$$f(0,0) = -\frac{\partial \xi_1}{\partial x_1}(0,0). \tag{4.184b}$$

Thus

$$\kappa = 0, \quad \lambda = -\beta,$$

and hence a three-parameter (β, γ, δ) Lie group of transformations leaves the BVP (4.182a,b) invariant on the domain $x_2 > 0$, $-\infty < x_1 < \infty$. Infinitesimal generators of this group are

$$\mathrm{X}_1 = x_1 \frac{\partial}{\partial x_1} + 2x_2 \frac{\partial}{\partial x_2} - u\frac{\partial}{\partial u}, \quad \mathrm{X}_2 = x_2 \frac{\partial}{\partial x_1} - \frac{1}{2} x_1 u \frac{\partial}{\partial u},$$

$$\mathrm{X}_3 = x_1 x_2 \frac{\partial}{\partial x_1} + (x_2)^2 \frac{\partial}{\partial x_2} - \left[\frac{(x_1)^2}{4} + \frac{x_2}{2}\right] u \frac{\partial}{\partial u}. \tag{4.185}$$

Correspondingly the matrix

$$\Xi(x) = \begin{bmatrix} x_1 & 2x_2 \\ x_2 & 0 \\ x_1 x_2 & (x_2)^2 \end{bmatrix}$$

has rank two, so that group invariance reduces the BVP completely, i.e. the number of independent variables reduces to zero. Note that

$$\mathrm{X}_3 = \frac{1}{2}[x_2 \mathrm{X}_1 + x_1 \mathrm{X}_2]. \tag{4.186}$$

Hence an invariant solution corresponding to X_1 and X_2 is also an invariant solution corresponding to X_3. Let $u = \Theta(x)$ be an invariant solution corresponding to X_1 and X_2. Then

$$\mathrm{X}_1(u - \Theta(x)) = 0$$

leads to the invariant form

$$u = \Theta(x) = \frac{1}{\sqrt{x_2}} F_1(\zeta_1) \tag{4.187}$$

with similarity variable

$$\zeta_1 = \frac{x_1}{\sqrt{x_2}}.$$

The equation

$$\mathrm{X}_2(u - \Theta(x)) = 0$$

leads to the invariant form

$$u = \Theta(x) = e^{-(x_1)^2/4x_2} F_2(\zeta_2) \tag{4.188}$$

with similarity variable

$$\zeta_2 = x_2.$$

From the uniqueness of the solution of the BVP (4.182a,b), we have

$$\frac{1}{\sqrt{x_2}} F_1(\zeta_1) = e^{-(x_1)^2/4x_2} F_2(\zeta_2),$$

i.e.

$$\sqrt{\zeta_2} F_2(\zeta_2) = e^{(\zeta_1)^2/4} F_1(\zeta_1) = \text{const} = c.$$

Hence the solution of the BVP (4.182a,b) is the familiar expression

$$u = \frac{c}{\sqrt{x_2}} e^{-(x_1)^2/4x_2}. \tag{4.189}$$

The initial condition (4.182b) leads to

$$c = \frac{1}{\sqrt{4\pi}}.$$

From (4.186) it automatically follows that

$$\mathbf{X}_3 \left(u - \frac{e^{-(x_1)^2/4x_2}}{\sqrt{4\pi x_2}} \right) = 0.$$

(ii) *Semi-Infinite Domain* $(a, b) = (0, \infty)$: Consider the BVP

$$\frac{\partial^2 u}{\partial x_1^2} = \frac{\partial u}{\partial x_2} \tag{4.190a}$$

on the domain $x_2 > 0$, $x_1 > 0$, with boundary conditions

$$u(x_1, 0) = \delta(x_1 - \hat{x}_1), \quad 0 < \hat{x}_1 < \infty, \tag{4.190b}$$

$$u(0, x_2) = 0. \tag{4.190c}$$

The three-parameter Lie group of transformations with infinitesimals (4.181a–c) is admitted by PDE (4.190a), boundary surfaces $x_2 = 0$, $x_1 = 0$, and boundary condition (4.190c). Invariance of (4.190b) means that

$$f(x_1, 0) u(x_1, 0) = \xi_1(x_1, 0) \delta'(x_1 - \hat{x}_1)$$

when $u(x_1, 0) = \delta(x_1 - \hat{x}_1)$, i.e.

$$f(x_1, 0) \delta(x_1 - \hat{x}_1) = \xi_1(x_1, 0) \delta'(x_1 - \hat{x}_1).$$

Hence

$$\xi_1(\hat{x}_1, 0) = 0,$$

$$f(\hat{x}_1, 0) = -\frac{\partial \xi_1}{\partial x_1}(\hat{x}_1, 0).$$

Consequently in (4.181a–c) we must have

$$\beta = 0, \quad \lambda = \frac{\gamma(\hat{x}_1)^2}{4}.$$

Thus BVP (4.190a–c) admits infinitesimal generator

$$\mathrm{X} = x_1 x_2 \frac{\partial}{\partial x_1} + (x_2)^2 \frac{\partial}{\partial x_2} + \left[\frac{(\hat{x}_1)^2}{4} - \left(\frac{(x_1)^2}{4} + \frac{x_2}{2}\right)\right] u \frac{\partial}{\partial u}.$$

The corresponding invariant solution has the invariant form

$$u = \Theta(x) = \frac{e^{-[(x_1)^2 + (\hat{x}_1)^2]/4x_2}}{\sqrt{x_2}} F(\zeta) \tag{4.191}$$

where $F(\zeta)$ is an arbitrary function of the similarity variable

$$\zeta = x_1/x_2.$$

Substituting (4.191) into (4.190a), we find that $F(\zeta)$ satisfies the ODE

$$\frac{d^2F}{d\zeta^2} = \frac{(\hat{x}_1)^2}{4} F,$$

and hence

$$u = \Theta(x) = \frac{1}{\sqrt{x_2}} \left[C e^{-(x_1 - \hat{x}_1)^2/4x_2} + D e^{-(x_1 + \hat{x}_1)^2/4x_2}\right]$$

where C, D are arbitrary constants. From (4.190c) it follows that $D = -C$, and from (4.190b) we get $C = \frac{1}{\sqrt{4\pi}}$, leading to the well-known solution of (4.190a–c) (usually obtained by the method of images)

$$u = \Theta(x) = G(x_1 - \hat{x}_1, x_2) - G(x_1 + \hat{x}_1, x_2)$$

with

$$G(x_1, x_2) = \frac{e^{-(x_1)^2/4x_2}}{\sqrt{4\pi x_2}}.$$

(2) *Fundamental Solution of the Axisymmetric Wave Equation*

The fundamental solution of the axisymmetric wave equation (4.88) is the solution of the BVP ($x_1 = r = \sqrt{x^2 + y^2}$, $x_2 = t$)

$$\mathrm{L}u = \frac{\partial^2 u}{\partial x_2^2} - \frac{\partial^2 u}{\partial x_1^2} - \frac{1}{x_1}\frac{\partial u}{\partial x_1} = \frac{1}{2\pi x_1}\delta(x_1)\delta(x_2),$$

i.e.

$$x_1 \mathrm{L}u = \frac{1}{2\pi}\delta(x_1)\delta(x_2), \tag{4.192a}$$

where

$$u \equiv 0 \quad \text{if} \quad x_1 > x_2. \tag{4.192b}$$

In Exercise 4.2-7 it is shown that

$$\mathrm{L}u = 0$$

admits a four-parameter $(\alpha, \beta, \gamma, \lambda)$ Lie group of transformations represented by the infinitesimal generator

$$\mathrm{X} = \xi_1(x_1, x_2)\frac{\partial}{\partial x_1} + \xi_2(x_1, x_2)\frac{\partial}{\partial x_2} + f(x_2)u\frac{\partial}{\partial u} \tag{4.193}$$

with infinitesimals

$$\xi_1(x_1, x_2) = \alpha x_1 + 2\beta x_1 x_2, \tag{4.194a}$$

$$\xi_2(x_1, x_2) = \alpha x_2 + \beta[(x_1)^2 + (x_2)^2] + \gamma, \tag{4.194b}$$

$$f(x_2) = -\beta x_2 + \lambda. \tag{4.194c}$$

Invariance of the wavefront $x_1 = x_2$ leads to

$$\xi_1(x_1, x_1) = \xi_2(x_1, x_1)$$

and hence $\gamma = 0$.

Under the action of (4.193) we have

$$x_1^* \mathrm{L}^* u^* = \left[1 + \epsilon\left[f(x_2) - 2\frac{\partial \xi_2}{\partial x_2}(x_1, x_2) + \frac{\xi_1(x_1, x_2)}{x_1}\right] + O(\epsilon^2)\right] x_1 \mathrm{L}u;$$

$$\begin{aligned}\delta(x_1^*)\delta(x_2^*) = \delta(x_1)\delta(x_2) + \epsilon[&\xi_1(x_1, x_2)\delta'(x_1)\delta(x_2)\\ &+ \xi_2(x_1, x_2)\delta(x_1)\delta'(x_2)] + O(\epsilon^2).\end{aligned}$$

Hence (4.194a–c) is admitted by (4.192a) if

$$\begin{aligned}&\left[f(x_2) - 2\frac{\partial \xi_2}{\partial x_2}(x_1, x_2) + \frac{\xi_1(x_1, x_2)}{x_1}\right]\delta(x_1)\delta(x_2)\\ &= \xi_1(x_1, x_2)\delta'(x_1)\delta(x_2) + \xi_2(x_1, x_2)\delta(x_1)\delta'(x_2).\end{aligned} \tag{4.195}$$

Since $x_i\delta'(x_i) = -\delta(x_i)$, $i = 1, 2$, it follows from (4.195) that (4.192a) admits (4.194a–c) if

$$\left[f(x_2) - 2\frac{\partial \xi_2}{\partial x_2}(x_1, x_2) + \frac{2}{x_1}\xi_1(x_1, x_2)\right.$$

$$+ \left. \frac{1}{x_2}\xi_2(x_1, x_2)\right] \delta(x_1)\delta(x_2) = 0. \tag{4.196}$$

Equation (4.196) reduces to

$$\left[(\lambda + \alpha) + \beta\frac{(x_1)^2}{x_2}\right] \delta(x_1)\delta(x_2) = 0. \tag{4.197}$$

Since (4.197) needs to be satisfied only on the wavefront $x_1 = x_2$ at $x_2 = 0$, it follows that β is arbitrary and $\lambda = -\alpha$. Thus (4.192a,b) admits a two-parameter Lie group of transformations with infinitesimal generators

$$X_1 = x_1\frac{\partial}{\partial x_2} + x_2\frac{\partial}{\partial x_2} - u\frac{\partial}{\partial u}, \tag{4.198a}$$

$$X_2 = 2x_1x_2\frac{\partial}{\partial x_1} + [(x_1)^2 + (x_2)^2]\frac{\partial}{\partial x_2} - x_2u\frac{\partial}{\partial u}. \tag{4.198b}$$

Let $u = \Theta(x)$ be the invariant solution corresponding to X_1 and X_2. Then

$$X_1(u - \Theta(x)) = 0$$

leads to the invariant form

$$u = \Theta(x) = \frac{1}{x_2}F_1(\zeta_1) \tag{4.199}$$

with similarity variable

$$\zeta_1 = \frac{x_1}{x_2},$$

and

$$X_2(u - \Theta(x)) = 0$$

leads to the invariant form

$$u = \Theta(x) = \frac{1}{\sqrt{x_1}}F_2(\zeta_2) \tag{4.200}$$

with similarity variable

$$\zeta_2 = \frac{1}{x_1}[(x_2)^2 - (x_1)^2].$$

Uniqueness of the solution of (4.192a,b) leads to

$$\frac{1}{\sqrt{x_1}}F_2(\zeta_2) = \frac{1}{x_2}F_1(\zeta_1). \tag{4.201}$$

Since

$$x_2 = \frac{\zeta_2\zeta_1}{1 - (\zeta_1)^2},$$

$$x_1 = \frac{\zeta_2(\zeta_1)^2}{1-(\zeta_1)^2},$$

(4.201) reduces to

$$\sqrt{\zeta_2}F_2(\zeta_2) = \sqrt{1-(\zeta_1)^2}F_1(\zeta_1) = \text{const},$$

and hence

$$u = \frac{c}{\sqrt{(x_2)^2-(x_1)^2}} \tag{4.202}$$

for some constant c. One can show that $c = \frac{1}{2\pi}$.

(3) *Fokker–Planck Equation*

As a third example we consider probability distributions which arise as fundamental solutions of the Fokker–Planck equation with drift $\phi(x_1)$:

$$\frac{\partial u}{\partial x_2} = \frac{\partial^2 u}{\partial x_1^2} + \frac{\partial}{\partial x_1}[\phi(x_1)u], \quad x_2 > 0, \quad a < x_1 < b, \tag{4.203a}$$

with initial condition

$$u(x_1, 0) = \delta(x_1 - \hat{x}_1), \quad a < \hat{x}_1 < b, \tag{4.203b}$$

and reflecting boundaries $x_1 = a$, $x_1 = b$ such that

$$\lim_{x_1 \to a^+, b^-} \left[\frac{\partial u}{\partial x_1} + \phi(x_1)u\right] = 0. \tag{4.203c}$$

Let $u(x_1, x_2; \hat{x}_1)$ be the solution of BVP (4.203a–c). It then follows that for any $\hat{x}_1$, $a < \hat{x}_1 < b$, one has

$$\int_a^b u(x_1, x_2; \hat{x}_1)dx_1 = 1. \tag{4.204}$$

We give a group analysis of (4.203a,b) for the case when $\phi(x_1)$ is an odd function of x_1. Complete details are given in Bluman (1967, 1971) and Bluman and Cole (1974).

One can show that (4.203a) admits

$$\xi_1(x_1, x_2)\frac{\partial}{\partial x_1} + \xi_2(x_1, x_2)\frac{\partial}{\partial x_2} + f(x_1, x_2)u\frac{\partial}{\partial u} \tag{4.205}$$

if and only if

$$\begin{aligned}
\xi_2(x_1, x_2) &= \tau(x_2), \\
\xi_1(x_1, x_2) &= \frac{1}{2}x_1\tau(x_2) + A(x_2), \\
f_1(x_1, x_2) &= -\frac{1}{4}x_1\phi(x_1)\tau'(x_2) - \frac{1}{8}(x_1)^2\tau''(x_2) - \frac{1}{2}\phi(x_1)A(x_2) \\
&\quad - \frac{1}{2}x_1A'(x_2) + B(x_2),
\end{aligned}$$

where for a given $\phi(x_1)$, the functions $A(x_2)$, $B(x_2)$, and $\tau(x_2)$ satisfy

$$N_1(x_1, x_2) + N_2(x_1, x_2) = 0$$

with

$$N_1(x_1, x_2) = \frac{1}{4}\tau'(x_2)[(\phi(x_1))^2 + x_1\phi(x_1)\phi'(x_1) - 2\phi'(x_1) - x_1\phi''(x_1)] + \frac{1}{4}\tau''(x_2) - \frac{1}{8}(x_1)^2\tau'''(x_2) + B'(x_2),$$

$$N_2(x_1, x_2) = \frac{1}{2}A(x_2)[\phi(x_1)\phi'(x_1) - \phi''(x_1)] - \frac{1}{2}x_1 A''(x_2).$$

Imposing the restriction that $\phi(x_1)$ is an odd function of x_1, we see that

$$N_1(x_1, x_2) \equiv 0, \tag{4.206a}$$

$$N_2(x_1, x_2) \equiv 0. \tag{4.206b}$$

From (4.206a) it follows that if $\tau(x_2) \not\equiv 0$, the drift $\phi(x_1)$ must satisfy the fifth order ODE

$$[(\phi(x_1))^2 + x_1\phi(x_1)\phi'(x_1) - 2\phi'(x_1) - x_1\phi''(x_1)]''' = 0 \tag{4.207}$$

in order that a nontrivial group is admitted by (4.203a,b).

One can show that the solution of (4.207) reduces to $\phi(x_1)$ satisfying the Riccati equation

$$2\phi'(x_1) - [\phi(x_1)]^2 + \beta^2(x_1)^2 - \gamma + \frac{16\nu^2 - 1}{(x_1)^2} = 0 \tag{4.208}$$

where β, γ, and ν are arbitrary constants. Then from (4.206a), $\tau(x_2)$ and $B(x_2)$ satisfy the equations

$$\tau'''(x_2) = 4\beta^2\tau'(x_2), \tag{4.209a}$$

$$B'(x_2) = \frac{1}{4}[\gamma\tau'(x_2) - \tau''(x_2)]. \tag{4.209b}$$

From (4.206b) it follows that if $A(x_2) \not\equiv 0$, then $\phi(x_1)$ must satisfy the Riccati equation

$$2\phi'(x_1) - [\phi(x_1)]^2 + \beta^2(x_1)^2 - \gamma = 0 \tag{4.210}$$

and $A(x_2)$ must satisfy

$$A''(x_2) = \beta^2 A(x_2).$$

Thus we see that if an odd drift $\phi(x_1)$ satisfies (4.210) then a six-parameter Lie group is admitted by PDE (4.203a); if an odd drift $\phi(x_1)$

satisfies (4.208) with $\nu^2 \neq \frac{1}{16}$ then PDE (4.203a) admits a four-parameter Lie group. Invariance of the boundary surface $x_2 = 0$ leads to $\tau(0) = 0$ and thus reduces the number of parameters by one. Invariance of the source condition leads to

$$\xi_1(\hat{x}_1, 0) = 0,$$

$$f(\hat{x}_1, 0) = -\frac{\partial \xi_1}{\partial x_1}(\hat{x}_1, 0).$$

Hence a three-parameter Lie group is admitted by (4.203a,b) if an odd drift $\phi(x_1)$ satisfies (4.210); a one-parameter Lie group is admitted by (4.203a,b) if an odd drift $\phi(x_1)$ satisfies (4.208) with $\nu^2 \neq \frac{1}{16}$.

The standard substitution

$$\phi(x_1) = -2\frac{V'(x_1)}{V(x_1)} \tag{4.211}$$

transforms (4.208) to the second order linear ODE

$$4V''(x_1) + \left[\gamma - \beta^2(x_1)^2 - \frac{16\nu^2 - 1}{(x_1)^2}\right] V(x_1) = 0. \tag{4.212}$$

The general solution of (4.212) is of the form

$$V(x_1) = c_1 V_1(x_1) + c_2 V_2(x_1)$$

where $V_1(x_1)$, $V_2(x_1)$ are respectively even and odd functions of x_1. Then $\phi(x_1)$ is an odd function of x_1 which satisfies (4.208) if and only if

$$\phi(x_1) = -2\frac{V_1'(x_1)}{V_1(x_1)}$$

or

$$\phi(x_1) = -2\frac{V_2'(x_1)}{V_2(x_1)}.$$

Only $V_1(x_1)$ leads to a reasonable drift. One can show that

$$V_1(x_1) = \left(\frac{1}{2}\beta(x_1)^2\right)^{(1/4)+\nu} e^{-\beta(x_1)^2/4} M\left(c, d, \frac{1}{2}\beta(x_1)^2\right), \tag{4.213}$$

where $M(c, d, z)$ denotes Kummer's hypergeometric function of the first kind with

$$c = \frac{1}{2} + \nu - \frac{\gamma}{8\beta}, \quad d = 1 + 2\nu,$$

$\nu > -\frac{1}{2}$, $\gamma\beta \leq 0$, $\beta \neq 0$. The properties of $M(c, d, z)$ are well known [Abramowitz and Stegun (1964, Chapter 13)]:

As $z \to 0$,

$$M(c, d, z) = 1 + \frac{c}{d}z + O(z^2). \tag{4.214a}$$

As $z \to \infty$,

$$M(c,d,z) = \frac{\Gamma(d)}{\Gamma(c)} e^z z^{c-d} \left[1 + O\left(\frac{1}{z}\right)\right]. \tag{4.214b}$$

From (4.214a,b) it follows that

$$\lim_{x_1 \to \infty} \frac{\phi(x_1)}{x_1} = -\beta, \quad \lim_{x_1 \to 0} x_1 \phi(x_1) = -(4\nu + 1).$$

The following cases arise:

Case I. $\nu^2 = \frac{1}{16}$: Here a three-parameter Lie group is admitted by (4.203a,b) with infinitesimal generators given by

$$\mathrm{X}_1 = 2\beta x_1 \sinh 2\beta x_2 \frac{\partial}{\partial x_1} + 4 \sinh^2 \beta x_2 \frac{\partial}{\partial x_2} + [\gamma \sinh^2 \beta x_2$$

$$-\beta \sinh 2\beta x_2 (1 + x_1 \phi(x_1)) + \beta^2((\hat{x}_1)^2 - (x_1)^2 \cosh 2\beta x_2)] u \frac{\partial}{\partial u}, \tag{4.215a}$$

$$\mathrm{X}_2 = 2 \sinh \beta x_2 \frac{\partial}{\partial x_1} + [\beta(\hat{x}_1 - x_1 \cosh \beta x_2) - \phi(x_1) \sinh \beta x_2] u \frac{\partial}{\partial u}, \tag{4.215b}$$

$$\begin{aligned} \mathrm{X}_3 = {} & 4\beta(\hat{x}_1 \cosh \beta x_2 - x_1 \cosh 2\beta x_2) \frac{\partial}{\partial x_1} - 4 \sinh 2\beta x_2 \frac{\partial}{\partial x_2} \\ & + [4\beta \cosh^2 \beta x_2 + 2\beta \phi(x_1)(x_1 \cosh 2\beta x_2 - \hat{x}_1 \cosh \beta x_2) \\ & + 2\beta^2 x_1 (x_1 \sinh 2\beta x_2 - \hat{x}_1 \sinh \beta x_2) - \gamma \sinh 2\beta x_2] u \frac{\partial}{\partial u}. \end{aligned} \tag{4.215c}$$

Note that

$$\mathrm{X}_3 = -2 \coth \beta x_2 \mathrm{X}_1 + 2\beta (x_1 \operatorname{cosech} \beta x_2 + \hat{x}_1 \coth \beta x_2) \mathrm{X}_2.$$

Let $u = \Theta(x)$ be the corresponding invariant solution. Then

$$\mathrm{X}_1(u - \Theta(x)) = 0$$

leads to the invariant form

$$u = g_1(x) F_1(\zeta_1) \tag{4.216}$$

where

$$g_1(x) = \frac{V_1(x_1)}{\sqrt{\sinh \beta x_2}} \exp\left[\frac{\gamma x_2}{4} + \frac{\beta(\hat{x}_1)^2}{2[1 - e^{2\beta x_2}]} - \frac{\beta (x_1)^2 \coth \beta x_2}{4}\right],$$

and $F_1(\zeta_1)$ is an arbitrary function of the similarity variable

$$\zeta_1 = \frac{x_1}{2 \sinh \beta x_2}.$$

Equation

$$X_2(u - \Theta(x)) = 0$$

leads to

$$u = g_2(x)F_2(\zeta_2) \tag{4.217}$$

where

$$g_2(x) = V_1(x_1)\exp\left[\frac{\beta x_1 \hat{x}_1}{2\sinh\beta x_2} - \frac{\beta(x_1)^2\coth\beta x_2}{4}\right],$$

and $F_2(\zeta_2)$ is an arbitrary function of the similarity variable

$$\zeta_2 = x_2.$$

Assuming uniqueness of the solution, we equate the invariant forms (4.216), (4.217) and obtain

$$F_2(\zeta_2) = F_2(x_2) = \frac{D}{\sqrt{\sinh\beta x_2}}\exp\left[\frac{\gamma x_2}{4} + \frac{\beta(\hat{x}_1)^2}{2[1 - e^{2\beta x_2}]}\right],$$

where D is an arbitrary constant.

Now consider separately the subcases $\nu = \pm\frac{1}{4}$:

Case I(a). $\nu = -\frac{1}{4}$: Here

$$V_1(x_1) = e^{-\beta(x_1)^2/4}M\left(c, \frac{1}{2}, \frac{1}{2}\beta(x_1)^2\right), \quad c = \frac{1}{4} - \frac{\gamma}{8\beta}.$$

If there is no reflecting boundary, i.e. $a = -\infty$, $b = \infty$, then the solution of (4.203a,b) is

$$u = G_1(x_1, x_2; \hat{x}_1) = \frac{DM(c, \frac{1}{2}, \frac{1}{2}\beta(x_1)^2)}{\sqrt{\sinh\beta x_2}}$$

$$\times\exp\left[\frac{\gamma x_2}{4} - \frac{\beta}{4}(1 + \coth\beta x_2)\left(x_1 - \hat{x}_1 e^{-\beta x_2}\right)^2\right], \tag{4.218a}$$

with

$$D = \sqrt{\frac{\beta}{4\pi}}\left[M\left(c, \frac{1}{2}, \frac{1}{2}\beta(\hat{x}_1)^2\right)\right]^{-1}, \tag{4.218b}$$

$x_2 > 0$, $-\infty < x_1 < \infty$.

If $a = 0$ is a reflecting boundary and $b = \infty$, then the solution of (4.203a–c) is

$$u = G_1(x_1, x_2; \hat{x}_1) + G_1(x_1, x_2; -\hat{x}_1), \quad x_2 > 0, \ \ 0 < x_1 < \infty, \tag{4.219}$$

which is an even function of x_1 if x_1 is extended to $-\infty < x_1 < \infty$. This solution represents the response to sources located at $\pm\hat{x}_1$ when $x_2 = 0$.

Note that in the limiting case of $c = 0$, we have $\phi(x_1) = \beta x_1$, and the solutions (4.218a,b) and (4.219) are probability distributions for a free particle in Brownian motion.

Case I(b). $\nu = \frac{1}{4}$: Here

$$V_1(x_1) = x_1 e^{-\beta(x_1)^2/4} M\left(c, \frac{3}{2}, \frac{1}{2}\beta(x_1)^2\right), \quad c = \frac{3}{4} - \frac{\gamma}{8\beta}, \quad x_1 > 0,$$

with $a = 0$ as a reflecting boundary, $b = \infty$. The resulting solution of (4.203a–c) is

$$u = G_2(x_1, x_2; \hat{x}_1) + G_2(x_1, x_2; -\hat{x}_1), \quad x_2 > 0, \quad 0 < x_1 < \infty,$$

where

$$G_2(x_1, x_2; \hat{x}_1) = \frac{E x_1 M(c, \frac{3}{2}, \frac{1}{2}\beta(x_1)^2)}{\sqrt{\sinh \beta x_2}}$$

$$\times \exp\left[\frac{\gamma x_2}{4} - \frac{\beta}{4}(1 + \coth \beta x_2)\left(x_1 - \hat{x}_1 e^{-\beta x_2}\right)^2\right],$$

with

$$E = \frac{1}{\hat{x}_1}\sqrt{\frac{\beta}{4\pi}}\left[M\left(c, \frac{3}{2}, \frac{1}{2}\beta(\hat{x}_1)^2\right)\right]^{-1}.$$

Case II. $\nu > -\frac{1}{2}$: Here only a one-parameter group is admitted by (4.203a,b) with infinitesimal generator given by (4.215a). The corresponding invariant solution arises from the invariant form (4.216). Substituting (4.216) into (4.203a) and letting $\zeta = \zeta_1$, $F(\zeta) = F_1(\zeta_1)$, we find that $F(\zeta)$ satisfies a second order linear ODE whose general solution can be expressed in terms of modified Bessel functions:

$$F(\zeta) = \zeta^{1/2}[A_1 I_{2\nu}(\rho\zeta) + A_2 I_{-2\nu}(\rho\zeta)] \quad \text{for} \quad x_1 > 0;$$

$$F(\zeta) = |\zeta|^{1/2}[B_1 K_{2\nu}(\rho|\zeta|) + B_2 I_{2\nu}(\rho|\zeta|)] \quad \text{for} \quad x_1 < 0,$$

where $\rho = \beta\hat{x}_1$ and A_1, A_2, B_1, B_2 are arbitrary constants to be determined from boundary and continuity conditions. As $x_2 \to 0, \zeta \to \infty$. From Watson (1922, Sect. 7.23) we find that as $z \to \infty$:

$$K_{2\nu}(z) = \left(\frac{\pi}{2z}\right)^{1/2} e^{-z}\left[1 + O\left(\frac{1}{z}\right)\right],$$

$$I_{2\nu}(z) = \left(\frac{1}{2\pi z}\right)^{1/2} e^{z}\left[1 + O\left(\frac{1}{z}\right)\right].$$

One can show that with $A_2 = B_1 = B_2 = 0$, we obtain a solution valid for $\nu \neq -\frac{1}{4}$, $x_1 > 0$, $x_2 > 0$ (i.e. $x_1 = 0$ is a reflecting boundary), with

$$A_1 = 2(\hat{x}_1)^{-2\nu}\left(\frac{1}{2}\beta\right)^{(3/4)-\nu}\left[M\left(c, d, \frac{1}{2}\beta(\hat{x}_1)^2\right)\right]^{-1}.$$

One can show that (4.215a) is admitted by (4.203c) when $x_1 = a$. An important consequence is that the equation

$$1 = \int_0^\infty u(x_1, x_2; \hat{x}_1) dx_1$$

is invariant under (4.215a), and hence moments of the probability distribution can be computed using invariance [cf. Bluman and Cole (1974, pp. 272–274)].

4.4.2 Incomplete Invariance for a Linear Scalar PDE

Consider a BVP for a kth order linear scalar PDE ($k \geq 2$)

$$\mathrm{L}u = h(x) \tag{4.220a}$$

defined on a domain Ω_x where L is a kth order linear operator, with linear boundary conditions

$$\mathrm{L}_\alpha u = h_\alpha(x) \tag{4.220b}$$

prescribed on boundary surfaces

$$\omega_\alpha(x) = 0, \tag{4.220c}$$

where L_α is a linear operator of order at most $k-1$, $\alpha = 1, 2, \ldots, s$. Assume that BVP (4.220a–c) has a unique solution. Formally the solution of (4.220a–c) can be represented as a superposition

$$u = u_0 + \sum_{\beta=1}^{s} u_\beta$$

where u_0 satisfies

$$\mathrm{L}u_0 = h(x), \quad x \in \Omega_x,$$
$$\mathrm{L}_\alpha u_0 = 0 \quad \text{on} \quad \omega_\alpha(x) = 0, \quad \alpha = 1, 2, \ldots, s,$$

and u_β satisfies

$$\mathrm{L}u_\beta = 0, \quad x \in \Omega_x,$$
$$\mathrm{L}_\alpha u_\beta = \delta_{\alpha\beta} h_\alpha(x) \quad \text{on} \quad \omega_\alpha(x) = 0, \quad \alpha, \beta = 1, 2, \ldots, s.$$

[$\delta_{\alpha\beta}$ is the Kronecker symbol.]

The solution of BVP (4.220a–c) reduces to the solution of two types of BVP's:

(i) a linear nonhomogeneous PDE with linear homogeneous boundary conditions:

$$\mathrm{L}u = h(x), \quad x \in \Omega_x, \tag{4.221a}$$

$$\mathrm{L}_\alpha u = 0 \quad \text{on} \quad \omega_\alpha(x) = 0, \quad \alpha = 1, 2, \ldots, s. \tag{4.221b}$$

(ii) a linear homogeneous PDE with $s-1$ linear homogeneous boundary conditions and one linear nonhomogeneous boundary condition:

$$\mathrm{L}u = 0, \quad x \in \Omega_x, \tag{4.222a}$$

$$\mathrm{L}_\alpha u = 0 \quad \text{on} \quad \omega_\alpha(x) = 0, \quad \alpha = 1, 2, \ldots, s-1, \tag{4.222b}$$

$$\mathrm{L}_s u = h(x) \quad \text{on} \quad \omega_s(x) = 0. \tag{4.222c}$$

The BVP

$$\mathrm{L}u = 0, \quad x \in \Omega_x, \tag{4.223a}$$

$$\mathrm{L}_\alpha u = 0 \quad \text{on} \quad \omega_\alpha(x) = 0, \quad \alpha = 1, 2, \ldots, s, \tag{4.223b}$$

is the associated homogeneous BVP of (4.220a–c). Suppose the nontrivial infinitesimal generator

$$\mathrm{X}_1 = \xi_i(x)\frac{\partial}{\partial x_i} + f(x)u\frac{\partial}{\partial u}, \tag{4.224}$$

is admitted by $\mathrm{L}u = 0$ [$\xi(x) \not\equiv 0$]. Clearly (4.223a,b) admits

$$\mathrm{X}_2 = u\frac{\partial}{\partial u}. \tag{4.225}$$

Let

$$u = \phi(x; \lambda) \tag{4.226}$$

be the *invariant form* corresponding to the infinitesimal generator $\mathrm{X}_\lambda = \mathrm{X}_1 + \lambda \mathrm{X}_2$ where λ is an arbitrary complex constant.

If X_1 is admitted by (4.223a,b) then the superposition of invariant forms

$$u = \sum_\lambda \phi(x; \lambda) \tag{4.227}$$

solves (4.221a,b) if

$$\sum_\lambda \mathrm{L}\phi(x; \lambda) = h(x), \tag{4.228}$$

$$\mathrm{L}_\alpha \phi(x; \lambda) = 0 \quad \text{on} \quad \omega_\alpha(x) = 0, \quad \alpha = 1, 2, \ldots, s, \tag{4.229}$$

for each λ in the sum of (4.227). In (4.227) the superposition $\sum_\lambda$ could also represent $\int_\Gamma d\lambda$ for some curve Γ in the complex λ-plane. Typically we solve for $h(x) = \delta(x - \hat{x})$, $\hat{x} \in \Omega_x$. Then superposition over the resulting Green's function is used to solve (4.221a,b) for arbitrary $h(x)$.

If X_1 is admitted by (4.222a,b) then so is $\mathrm{X}_\lambda = \mathrm{X}_1 + \lambda \mathrm{X}_2$ for arbitrary complex λ. Let

$$u = \Theta(x; \lambda) \tag{4.230}$$

be the most general *invariant solution* of (4.222a,b) corresponding to X_λ, which may exist for only certain eigenvalues λ. The superposition of invariant solutions

$$u = \sum_\lambda \Theta(x;\lambda) \tag{4.231}$$

solves (4.222a–c) provided

$$\sum_\lambda L_s\Theta(x;\lambda) = h(x) \quad \text{on} \quad \omega_s(x) = 0. \tag{4.232}$$

Again the superposition $\sum_\lambda$ in (4.231) could be replaced by $\int_\Gamma d\lambda$ for some curve Γ in the complex λ-plane. An appropriate superposition is picked to satisfy (4.232). Typically one solves for $h(x) = \delta(x - \hat{x})$, $\hat{x} \in \Omega_x$, and uses superposition over the resulting Green's function to solve (4.222a–c) for arbitrary $h(x)$.

The following examples are illustrative:

(1) *Fundamental Solutions of the Heat Equation for a Finite Domain*

(i) *Nonhomogeneous Heat Equation with Homogeneous Boundary Conditions.* Consider the BVP for the nonhomogeneous heat equation

$$Lu = \frac{\partial u}{\partial x_2} - \frac{\partial^2 u}{\partial x_1^2} = \delta(x_1 - \hat{x}_1)\delta(x_2), \tag{4.233a}$$

defined on the finite domain $x_2 > 0$, $0 < x_1 < 1$, where $0 < \hat{x}_1 < 1$, with homogeneous boundary conditions

$$u(0, x_2) = u(1, x_2) = 0. \tag{4.233b}$$

Clearly

$$X_1 = \frac{\partial}{\partial x_2}$$

is admitted by $Lu = 0$ and (4.233b). The invariant form corresponding to

$$\frac{\partial}{\partial x_2} + \lambda u \frac{\partial}{\partial u}$$

is

$$u = \phi(x;\lambda) = y(x_1;\lambda)e^{\lambda x_2}. \tag{4.234}$$

Consider

$$\sum_\lambda y(x_1;\lambda)e^{\lambda x_2}. \tag{4.235}$$

Substituting (4.235) into (4.233a), we formally find that

$$\sum_\lambda \left(\frac{\partial^2 y}{\partial x_1^2} - \lambda y\right) e^{\lambda x_2} = -\delta(x_1 - \hat{x}_1)\delta(x_2). \tag{4.236}$$

To satisfy (4.233b) we demand that for each λ

$$y(0;\lambda) = y(1;\lambda) = 0. \tag{4.237}$$

The natural superposition arising from (4.234) is the inverse Laplace transform representation of the solution of (4.233a,b) given by

$$u(x_1, x_2) = \frac{1}{2\pi i}\int_{\gamma - i\infty}^{\gamma + i\infty} y(x_1;\lambda)e^{\lambda x_2}d\lambda, \tag{4.238}$$

where $\gamma \in \mathbb{R}$ lies to the right of all singularities of $y(x_1;\lambda)$ in the complex λ-plane. Formally

$$\delta(x_2) = \frac{1}{2\pi i}\int_{\gamma - i\infty}^{\gamma + i\infty} e^{\lambda x_2}d\lambda.$$

Hence $y(x_1;\lambda)$ satisfies

$$\frac{\partial^2 y}{\partial x_1^2} - \lambda y = -\delta(x_1 - \hat{x}_1) \tag{4.239}$$

and boundary conditions (4.237). Consequently

$$y(x_1;\lambda) = \begin{cases} \dfrac{\sinh\sqrt{\lambda}x_1 \sinh\sqrt{\lambda}(\hat{x}_1 - 1)}{\sqrt{\lambda}\sinh\sqrt{\lambda}}, & 0 < x_1 < \hat{x}_1; \\[2ex] \dfrac{\sinh\sqrt{\lambda}\hat{x}_1 \sinh\sqrt{\lambda}(x_1 - 1)}{\sqrt{\lambda}\sinh\sqrt{\lambda}}, & \hat{x}_1 < x_1 < 1. \end{cases}$$

Using residues we obtain the following solution representation of BVP (4.233a,b) which is useful for large values of x_2:

$$u(x_1, x_2) = \sum_{n=1}^{\infty} 2\sin n\pi\hat{x}_1 \sin n\pi x_1 e^{-n^2\pi^2 x_2}. \tag{4.240}$$

Using the asymptotic expansion of $y(x_1;\lambda)$ valid for large values of $|\lambda|$ along the Bromwich contour of the inverse Laplace transform we obtain the following solution representation of BVP (4.233a,b) which is useful for small values of x_2:

$$u(x_1, x_2) = \sum_{n=-\infty}^{\infty} [G(x_1 - \hat{x}_1 - 2n, x_2) - G(x_1 + \hat{x}_1 + 2n, x_2)]$$

where

$$G(x_1, x_2) = \frac{1}{\sqrt{4\pi x_2}}e^{-(x_1)^2/4x_2}.$$

(ii) *Homogeneous Heat Equation with a Nonhomogeneous Boundary Condition.* Consider the following BVP for the heat equation

$$\frac{\partial^2 u}{\partial x_1^2} - \frac{\partial u}{\partial x_2} = 0, \quad 0 < x_1 < 1, \quad x_2 > 0; \tag{4.241a}$$

$$u(0, x_2) = u(1, x_2) = 0; \tag{4.241b}$$

$$u(x_1, 0) = h(x_1). \tag{4.241c}$$

Clearly

$$X_\lambda = \frac{\partial}{\partial x_2} + \lambda u \frac{\partial}{\partial u}$$

is admitted by (4.241a,b). The invariant form of the corresponding invariant solution is

$$u = \Theta(x; \lambda) = y(x_1; \lambda) e^{\lambda x_2}, \tag{4.242}$$

which satisfies (4.241a,b) if

$$\lambda = \lambda_n = -n^2 \pi^2,$$

with

$$y(x_1; \lambda_n) = a_n \sin n\pi x_1,$$

where a_n is an arbitrary constant, $n = 1, 2, \ldots$. If $h(x_1) = \delta(x_1 - \hat{x}_1)$, the superposition of invariant solutions

$$u(x_1, x_2) = \sum_{n=1}^{\infty} \Theta(x; \lambda_n)$$

satisfies (4.241c) if $a_n = 2 \sin n\pi \hat{x}_1$. Of course this is the solution representation (4.240) since (4.233a,b) and (4.241a–c) are equivalent problems when $h(x_1) = \delta(x_1 - \hat{x}_1)$. Let

$$K(x_1, x_2; \hat{x}_1) = \sum_{n=1}^{\infty} 2 \sin n\pi \hat{x}_1 \sin n\pi x_1 e^{-n^2 \pi^2 x_2}.$$

Then the solution of BVP (4.241a–c) is

$$u(x_1, x_2) = \int_0^1 h(\hat{x}_1) K(x_1, x_2; \hat{x}_1) d\hat{x}_1.$$

(2) *An Inverse Stefan Problem*

A nontrivial example is illustrated by the inverse Stefan problem which is given by the BVP

$$\frac{\partial u}{\partial x_2} = \frac{\partial^2 u}{\partial x_1^2}, \quad 0 < x_1 < X(x_2), \quad x_2 > 0; \tag{4.243a}$$

$$u(X(x_2), x_2) = 0, \quad x_2 > 0; \tag{4.243b}$$

$$\frac{\partial u}{\partial x_1}(0, x_2) = h_1(x_2), \quad x_2 > 0; \tag{4.243c}$$

$$u(x_1, 0) = h_2(x_1), \quad 0 < x_1 < 1; \tag{4.243d}$$

$$h_3(x_2) = k\frac{\partial u}{\partial x_1}(X(x_2), x_2) - \frac{dX}{dx_2}, \quad x_2 > 0, \tag{4.243e}$$

where for a prescribed moving boundary $X(x_2)$ with $X(0) = 1$, arbitrary initial distribution $h_2(x_1)$, fixed constant k, and arbitrary flux $h_1(x_2)$, the aim is to determine $u(x_1, x_2)$ and the flux $h_3(x_2)$ so that BVP (4.243a–e) is satisfied.

Our strategy is to first obtain a solution $u = \Theta_1(x_1, x_2)$ of (4.243a–c). Then we solve (4.243a–d) with $h_1(x_2) \equiv 0$ and $u(x_1, 0) = h_2(x_1) - \Theta_1(x_1, 0)$ to obtain a solution $u = \Theta_2(x_1, x_2)$. Consequently the solution of (4.243a–d) is $u = \Theta_1(x_1, x_2) + \Theta_2(x_1, x_2)$, and

$$h_3(x_2) = k\left[\frac{\partial\Theta_1}{\partial x_1}(X(x_2), x_2) + \frac{\partial\Theta_2}{\partial x_1}(X(x_2), x_2)\right] - \frac{dX}{dx_2}.$$

Details on this example appear in Bluman and Cole (1974, pp. 213–219, 235–245) and Bluman (1974a).

In order to leave the moving boundary curve $x_1 = X(x_2)$ invariant, with $X(0) = 1$, it is necessary that $\kappa = \delta = 0$ for the infinitesimals (4.52a–c) admitted by the heat equation (4.47). Thus if the solution of (4.243a–e) can be represented as a superposition of invariant solutions, it is necessary that $X(x_2)$ be of the form

$$X(x_2) = \sqrt{1 + 2\beta x_2 + \gamma(x_2)^2},$$

for arbitrary constants β and γ. We examine in detail the interesting subcase $\gamma = \beta^2$, where

$$X(x_2) = 1 - \frac{x_2}{T}, \quad T = -\frac{1}{\beta}. \tag{4.244}$$

If $X(x_2)$ is of the form (4.244), then (4.243a,b) admits

$$\mathrm{X}_1 = \beta x_1[1 + \beta x_2]\frac{\partial}{\partial x_1} + [1 + \beta x_2]^2\frac{\partial}{\partial x_2} - \beta^2\left[\frac{(x_1)^2}{4} + \frac{x_2}{2}\right]u\frac{\partial}{\partial u}. \tag{4.245}$$

We consider the situation where $T > 0$, so that $0 < x_2 < T$, i.e. $\frac{dX}{dx_2} < 0$. The corresponding similarity variable is

$$\zeta = \frac{x_1}{X(x_2)} = \frac{x_1}{1 - x_2/T}, \quad 0 \le \zeta \le 1, \tag{4.246}$$

with

$$\zeta = 0 \quad \text{corresponding to} \quad x_1 = 0;$$

$$\zeta = 1 \quad \text{corresponding to} \quad x_1 = X(x_2) = 1 - \frac{x_2}{T}.$$

The similarity curves (invariant curves) ζ = const are illustrated in Figure 4.4.2-1.

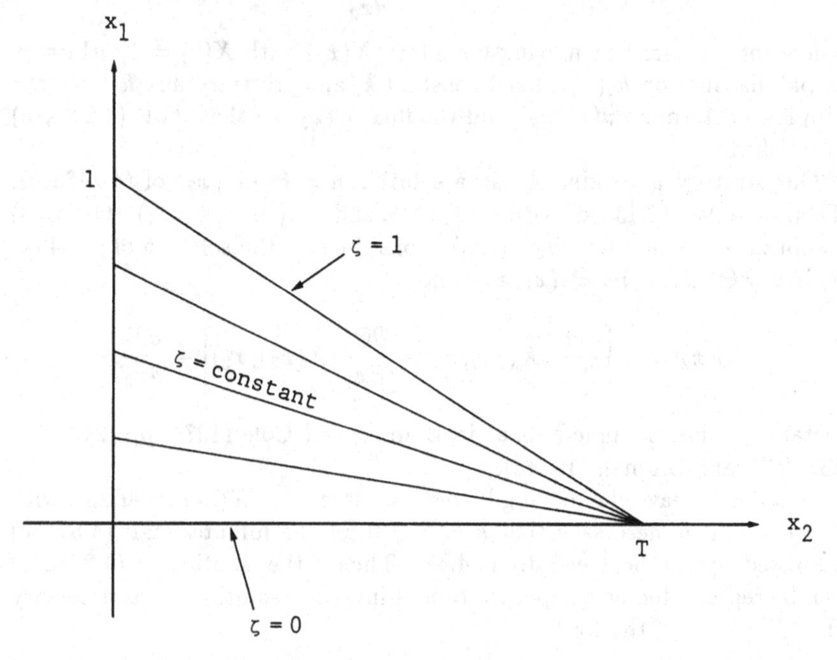

Figure 4.4.2-1. Invariance curves ζ = const.

The invariant form corresponding to the infinitesimal generator

$$\mathrm{X}_\lambda = \mathrm{X}_1 + \lambda u \frac{\partial}{\partial u}$$

is

$$u = \phi(x;\nu) = \frac{y(\zeta;\nu)\exp\left[-\frac{\nu^2 T}{X(x_2)} + \frac{\zeta^2 X(x_2)}{4T}\right]}{\sqrt{X(x_2)}}, \tag{4.247}$$

with

$$\nu^2 = \lambda - \frac{1}{2T}. \tag{4.248}$$

Substituting (4.247) into (4.243a) we find that $y(\zeta;\nu)$ satisfies

$$\frac{d^2y}{d\zeta^2} + \nu^2 y = 0. \tag{4.249}$$

The boundary condition (4.243b) leads to

$$y(1;\nu) = 0, \tag{4.250}$$

so that

$$y(\zeta;\nu) = A(\nu)\sin\nu(\zeta - 1), \tag{4.251}$$

for arbitrary constant $A(\nu)$. Hence any superposition of invariant solutions

$$u = \sum_{\nu} \frac{A(\nu)\sin\nu(\zeta-1)}{\sqrt{X(x_2)}} \exp\left[-\frac{\nu^2 T}{X(x_2)} + \frac{\zeta^2 X(x_2)}{4T}\right] \tag{4.252}$$

formally solves (4.243a,b).

Now let

$$t = \frac{x_2}{X(x_2)},$$

$$s = -\nu^2,$$

$$B(s) = \frac{2\pi i A(\nu) e^{-\nu^2 T}}{\sqrt{T}},$$

and replace $\sum_\nu$ by $\int_{\gamma-i\infty}^{\gamma+i\infty} ds$. Then formally we obtain the following solution representation of (4.243a–c) in terms of an inverse Laplace transform:

$$u = \Theta_1(x_1, x_2) = \sqrt{t+T}\, e^{\zeta^2/4(t+T)} \frac{1}{2\pi i} \int_{\gamma-i\infty}^{\gamma+i\infty} B(s) \sinh[\sqrt{s}(1-\zeta)] e^{st} ds. \tag{4.253}$$

Letting

$$H(t) = h_1(x_2) = h_1\left(\frac{tT}{t+T}\right),$$

and inverting (4.253) so that the boundary condition (4.243c) is satisfied, we find that

$$B(s) = -\frac{\beta}{\sqrt{s}\cosh\sqrt{s}} \int_0^\infty \frac{H(t)}{(t+T)^{3/2}} e^{-st} dt. \tag{4.254}$$

Now let

$$H_2(x_1) = h_2(x_1) - \Theta_1(x_1, 0),$$

and consider BVP (4.243a–d) with $h_1(x_2) \equiv 0$ and $h_2(x_1)$ replaced by $H_2(x_1)$, i.e. the BVP

$$\frac{\partial u}{\partial x_2} = \frac{\partial^2 u}{\partial x_1^2}, \quad 0 < x_1 < X(x_2), \quad x_2 > 0; \tag{4.255a}$$

$$u(X(x_2), x_2) = 0, \quad x_2 > 0; \tag{4.255b}$$

$$\frac{\partial u}{\partial x_1}(0, x_2) = 0, \quad x_2 > 0; \tag{4.255c}$$

$$u(x_1, 0) = H_2(x_1); \tag{4.255d}$$

with $X(x_2)$ given by (4.244). It is easy to see that the infinitesimal generator X_1 given by (4.245) is admitted by (4.255a–c). Consequently one obtains the similarity variable (4.246); the infinitesimal generator $\mathrm{X}_\lambda = \mathrm{X}_1 + \lambda u \frac{\partial}{\partial u}$ leads to the invariant form (4.247), (4.248). Substitution of (4.247) into (4.255a) leads to (4.249). Boundary conditions (4.255b,c) lead to (4.250) and

$$\frac{dy}{d\zeta}(0; \nu) = 0. \tag{4.256}$$

Thus

$$\nu = \nu_n = \left(n + \frac{1}{2}\right)\pi$$

with

$$y(\zeta; \nu_n) = A_n \cos \nu_n \zeta,$$

where A_n is an arbitrary constant, $n = 0, 1, 2, \ldots$. Formally the superposition of invariant solutions

$$u = \Theta_2(x_1, x_2) = \sum_{n=0}^{\infty} \frac{A_n \cos \nu_n \zeta}{\sqrt{X(x_2)}} \exp\left[-\frac{(\nu_n)^2 T}{X(x_2)} + \frac{\zeta^2 X(x_2)}{4T}\right] \tag{4.257}$$

satisfies (4.255a–c). The boundary condition (4.255d) is satisfied if

$$H_2(x_1) = \sum_{n=0}^{\infty} A_n \cos\left(n + \frac{1}{2}\right)\pi x_1 \exp\left[\frac{(x_1)^2}{4T} - \left(n + \frac{1}{2}\right)^2 \pi^2 T\right].$$

Let $\psi_n(x_1) = \cos(n + \frac{1}{2})\pi x_1$, $n = 0, 1, 2, \ldots$. Then the eigenfunctions $\{\psi_n(x_1)\}$ form a complete orthogonal set of functions on $[0, 1]$ with

$$\int_0^1 \psi_n(x_1)\psi_m(x_1)dx_1 = \frac{1}{2}\delta_{nm}.$$

Thus

$$A_n = 2e^{(n+\frac{1}{2})^2\pi^2 T} \int_0^1 H_2(x_1)\psi_n(x_1)e^{-x_1^2/4T}dx_1,$$

$n = 0, 1, 2, \ldots$.

The above solutions have been used to find numerical solutions for the nonlinear direct Stefan problem described by BVP (4.243a–e) where the aim is to find the unknown moving boundary $X(x_2)$ and $u(x_1, x_2)$ for arbitrary $h_1(x_2)$, $h_2(x_1)$, $h_3(x_2)$ and constant k [cf. Milinazzo (1974), Milinazzo and Bluman (1975)].

4.4.3 INCOMPLETE INVARIANCE FOR A LINEAR SYSTEM OF PDE'S

As an example consider an initial value problem for the system of wave equations

$$\frac{\partial v}{\partial x_2} = \frac{\partial u}{\partial x_1}, \tag{4.258a}$$

$$\frac{\partial u}{\partial x_2} = c^2(x_1)\frac{\partial v}{\partial x_1}, \tag{4.258b}$$

with

$$u(x_1, 0) = U(x_1), \tag{4.258c}$$

$$v(x_1, 0) = V(x_1), \tag{4.258d}$$

for some given functions $U(x_1)$ and $V(x_1)$ on the domain $-\infty < x_1 < \infty$, $-\infty < x_2 < \infty$. We consider the wave speed $c(x_1)$ found in Section 4.3.4 which satisfies the equation

$$c' = m\sin(\nu \log c). \tag{4.259}$$

This describes wave propagation in two-layered media with smooth transitions, with properties:

$$\lim_{x_1 \to -\infty} c(x_1) = 1, \tag{4.260a}$$

$$\lim_{x_1 \to \infty} c(x_1) = e^{\pi/\nu} = \gamma, \quad \gamma > 0, \tag{4.260b}$$

$$\max_{x_1 \in (-\infty,\infty)} c'(x_1) = m > 0, \tag{4.260c}$$

where γ, m are independent parameters with γ representing the ratio of asymptotic wave speeds. [One can easily adapt to the case where

$$\lim_{x_1 \to -\infty} c(x_1) = c_1^* > 0, \quad \lim_{x_1 \to \infty} c(x_1) = c_2^* > 0$$

by appropriate scalings.] A typical profile for $c(x_1)$ is shown in Figure 4.4.3-1. Without loss of generality $c'(0) = m$. The four-parameter group with infinitesimal generators (4.160) is admitted by (4.258a,b). The infinitesimal generator

$$\mathrm{X} = \mathrm{X}_2 + \mathrm{X}_3$$

leaves the curve $x_2 = 0$ invariant. One can show that the invariant solutions of (4.258a,b) corresponding to invariance under

$$\mathrm{X} + 4\nu n i\left[u\frac{\partial}{\partial u} + v\frac{\partial}{\partial v}\right] \tag{4.261}$$

are given by $(u, v) = (u_n(x_1, x_2), v_n(x_1, x_2))$ for $n = 0, 1, 2, \ldots$:

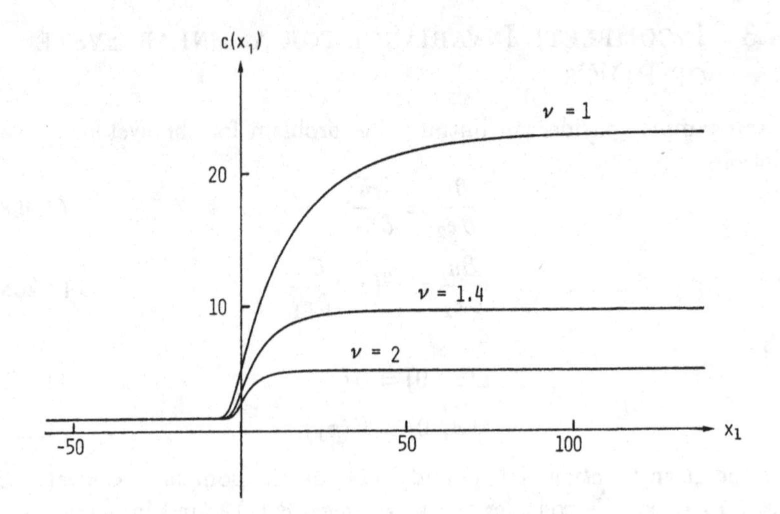

Figure 4.4.3-1. Profile of $c(x_1)$.

$$\begin{bmatrix} u_n(x_1, x_2) \\ v_n(x_1, x_2) \end{bmatrix} = [\sin y]^{1/2} e^{-i2n \arctan[\cot y \operatorname{sech} m\nu x_2]}$$

$$\times \begin{bmatrix} [c(x_1)]^{1/2} & 0 \\ 0 & [c(x_1)]^{-1/2} \end{bmatrix}$$

$$\times \begin{bmatrix} [\cosh m\nu x_2 + \sinh m\nu x_2 \cos y]^{1/2} & [\cosh m\nu x_2 - \sinh m\nu x_2 \cos y]^{1/2} \\ [\cosh m\nu x_2 + \sinh m\nu x_2 \cos y]^{1/2} & -[\cosh m\nu x_2 - \sinh m\nu x_2 \cos y]^{1/2} \end{bmatrix}$$

$$\times \begin{bmatrix} f_n(z) \\ g_n(z) \end{bmatrix}, \tag{4.262}$$

where

$$y = \nu \log c(x_1),$$

$$z = \sinh m\nu x_2 \sin y,$$

$$\begin{bmatrix} f_n(z) \\ g_n(z) \end{bmatrix} = M_n(z) \begin{bmatrix} f_0(z) \\ g_0(z) \end{bmatrix},$$

with

$$\begin{bmatrix} f_0(z) \\ g_0(z) \end{bmatrix} = (z^2+1)^{-1/2} \begin{bmatrix} \cos\psi(z) & \sin\psi(z) \\ -\sin\psi(z) & \cos\psi(z) \end{bmatrix} \begin{bmatrix} P_n \\ Q_n \end{bmatrix},$$

$$\psi(z) = \frac{1}{2\nu} \log\left(z + \sqrt{z^2+1}\right),$$

$$M_n(z) = R_n(z) \times R_{n-1}(z) \times \cdots \times R_1(z) \times R_0(z),$$

$$R_0(z) = \begin{bmatrix} 1 & 0 \\ 0 & 1 \end{bmatrix},$$

and for $n \geq 1$,

$$R_n(z) = \begin{bmatrix} \left(n^2 - \frac{1}{4}\right)\left(\frac{z-i}{z+i}\right) - \frac{1}{4\nu^2} & \frac{i-2nz}{2\nu\sqrt{z^2+1}} \\ \frac{2nz+i}{2\nu\sqrt{z^2+1}} & \left(n^2 - \frac{1}{4}\right)\left(\frac{z+i}{z-i}\right) - \frac{1}{4\nu^2} \end{bmatrix};$$

P_n, Q_n are arbitrary constants chosen separately for each invariant solution pair (u_n, v_n). For $n = -1, -2, \ldots$, it is convenient to define invariant solutions

$$\begin{bmatrix} u \\ v \end{bmatrix} = \begin{bmatrix} u_n(x_1, x_2) \\ v_n(x_1, x_2) \end{bmatrix} = \begin{bmatrix} \overline{u_{-n}}(x_1, x_2) \\ \overline{v_{-n}}(x_1, x_2) \end{bmatrix}$$

for system (4.258a,b) where a bar denotes the complex conjugate.

The solution of the initial value problem (4.258a–d) is represented formally in the form

$$\begin{bmatrix} u(x_1, x_2) \\ v(x_1, x_2) \end{bmatrix} = \sum_{n=-\infty}^{\infty} \begin{bmatrix} u_n(x_1, x_2) \\ v_n(x_1, x_2) \end{bmatrix}$$

$$= 2Re\left(\sum_{n=1}^{\infty} \begin{bmatrix} u_n(x_1, x_2) \\ v_n(x_1, x_2) \end{bmatrix}\right) + \begin{bmatrix} u_0(x_1, x_2) \\ v_0(x_1, x_2) \end{bmatrix}$$

with the constants P_n, Q_n determined from the initial condition (4.258c,d). Note that

$$u_n(x_1, 0) = (-1)^n [c(x_1) \sin y]^{1/2} [P_n + Q_n] e^{i2ny}, \tag{4.263a}$$

$$v_n(x_1, 0) = (-1)^n \left[\frac{\sin y}{c(x_1)}\right]^{1/2} [P_n - Q_n] e^{i2ny}, \tag{4.263b}$$

with $0 < y < \pi$. Consequently from the Fourier series representation (4.263a,b), we find that

$$\frac{P_n}{Q_n} = \frac{(-1)^n}{2\pi} \int_0^{\pi} e^{-i2ny} [\sin y]^{-1/2} \left[e^{-y/2\nu} U(x_1(y)) \pm e^{y/2\nu} V(x_1(y))\right] dy.$$

For a given initial value problem, after determining the constants $\{P_n, Q_n\}$, one can directly compute the solution for any time x_2, $-\infty < x_2 < \infty$. No marching is required as is the case for usual numerical procedures based on the method of characteristics.

Full details of the derivation of these solutions and their properties are given in Bluman and Kumei (1988).

Exercises 4.4

1. Prove Theorem 4.4.1-1.

2. Prove Theorem 4.4.1-2.

3. Prove Theorem 4.4.1-3.

4. Obtain the fundamental solution of the heat equation (4.47) for an infinite domain $(a,b)=(-\infty,\infty)$ using invariant forms arising from the following infinitesimal generators of (4.185):

   (a) X_1, X_3;

   (b) X_2, X_3.

5. The problem of finding the steady-state temperature distribution near the surface of the earth due to a periodic temperature variation at the earth's surface approximately reduces to finding the steady-state solution of the following BVP:

$$\frac{\partial u}{\partial x_2}=\frac{\partial^2 u}{\partial x_1^2}, \quad 0<x_1<\infty, \quad 0<x_2<\infty, \tag{4.264a}$$

$$u(x_1,0)=\Theta(x_1), \tag{4.264b}$$

$$u(0,x_2)=B\cos Ax_2, \tag{4.264c}$$

$$u(\infty,x_2)=0. \tag{4.264d}$$

   (a) Show that the steady-state solution of (4.264a–d) is independent of the initial distribution $\Theta(x_1)$.

   (b) Let $v(x_1,x_2)$ solve

$$\frac{\partial v}{\partial x_2}=\frac{\partial^2 v}{\partial x_1^2}, \quad 0<x_1<\infty, \quad 0<x_2<\infty, \tag{4.265a}$$

$$v(0,x_2)=Be^{iAt}, \tag{4.265b}$$

$$v(\infty,x_2)=0. \tag{4.265c}$$

   Find a one-parameter Lie group of transformations which is admitted by (4.265a–c). Find the corresponding invariant solution of (4.265a–c).

   (c) Find the steady-state solution of (4.264a–d).

6. Find the fundamental solution (Riemann function) for the Euler–Poisson–Darboux equation, i.e. solve

$$\frac{\partial^2 u}{\partial x_1^2} - \frac{\partial^2 u}{\partial x_2^2} + \frac{\lambda}{x_1}\frac{\partial u}{\partial x_1} = \delta(x_1 - \hat{x}_1)\delta(x_2 - \hat{x}_2). \tag{4.266}$$

This solution is the source solution for isentropic flow for a polytropic gas where

x_1 is the sound speed in the gas,

x_2 is the fluid velocity in some fixed direction,

u is the time variable,

and the constant λ is related to the ratio of specific heats of the gas.

(a) Show that (4.266) admits

$$\mathrm{X} = -\frac{2x_1}{\lambda}(x_2 - \hat{x}_2)\frac{\partial}{\partial x_1} + \left[\frac{(\hat{x}_1)^2 - (x_1)^2 - (x_2 - \hat{x}_2)^2}{\lambda}\right]\frac{\partial}{\partial x_2} + (x_2 - \hat{x}_2)u\frac{\partial}{\partial u}. \tag{4.267}$$

(b) Show that the similarity variable corresponding to (4.267) is

$$\zeta = \frac{(x_1 - \hat{x}_1)^2 - (x_2 - \hat{x}_2)^2}{x_1}. \tag{4.268}$$

(c) Derive the invariant form of the invariant solution.

(d) Derive the solution of (4.266):

$$u = u(x_1, x_2; \hat{x}_1, \hat{x}_2) = \frac{(2\hat{x}_1)^\lambda}{[(x_1 + \hat{x}_1)^2 - (x_2 - \hat{x}_2)^2]^{\lambda/2}} \times F\left(\frac{\lambda}{2}, \frac{\lambda}{2}; 1; \frac{(x_1 - \hat{x}_1)^2 - (x_2 - \hat{x}_2)^2}{(x_1 + \hat{x}_1)^2 - (x_2 - \hat{x}_2)^2}\right),$$

where $F(a, b; c; z)$ is the hypergeometric function [Bluman (1967)].

7. Consider the BVP for the response due to a unit impulse for a string with a nonlinear restoring force:

$$\frac{\partial^2 u}{\partial x_2^2} - \frac{\partial^2 u}{\partial x_1^2} + f(u) = \delta(x_1)\delta(x_2),$$

$$u \equiv 0 \quad \text{if} \quad x_1 > x_2,$$

with

$$f(u) = -f(-u); \quad f(u) > 0 \quad \text{if} \quad u > 0.$$

(a) Find an infinitesimal generator admitted by this BVP and the invariant form of the corresponding invariant solution.

(b) For the case $f(u) = ku^3$, study the solution in a suitable phase plane. What conditions apply at the wavefront $x_1 = x_2$?

8. Consider the Fokker–Planck equation

$$\frac{\partial u}{\partial x_2} = \frac{\partial^2 u}{\partial x_1^2} + \frac{\partial}{\partial x_1}[\phi(x_1)u], \quad x_2 > 0, \quad 0 < x_1 < \infty, \tag{4.269a}$$

with drift

$$\phi(x_1) = \frac{\alpha}{x_1} + \beta x_1, \quad \alpha < 1, \quad \beta > 0,$$

and initial condition

$$u(x_1, 0) = \delta(x_1 - \hat{x}_1), \quad 0 < \hat{x}_1 < \infty. \tag{4.269b}$$

(a) Find an infinitesimal generator X admitted by (4.269a,b).

(b) Let $u(x_1, x_2; \hat{x}_1)$ be the solution of (4.269a,b). Since

$$\int_0^\infty u(x_1, x_2; \hat{x}_1) dx_1 = 1 \tag{4.270}$$

for any values of x_2, $\hat{x}_1$, it follows that (4.270) admits X. Consequently show that the second moment of the solution of (4.269a,b) is

$$\begin{aligned}\langle (x_1)^2 \rangle &= \int_0^\infty (x_1)^2 u(x_1, x_2; \hat{x}_1) dx_1 \\ &= \left(\frac{1-\alpha}{\beta}\right)(1 - e^{-2\beta x_2}) + (\hat{x}_1)^2 e^{-2\beta x_2}\end{aligned} \tag{4.271}$$

without determining $u(x_1, x_2; \hat{x}_1)$ explicitly.

(c) Show that the solution of (4.269a,b) is the invariant solution arising from X:

$$\begin{aligned}u(x_1, x_2; \hat{x}_1) = \beta \sqrt{\frac{\hat{x}_1 \zeta}{2 \sinh \beta x_2}} \left(\frac{\hat{x}_1}{x_1}\right)^{\alpha/2} \Bigg(\exp\Bigg[\frac{1}{2}(1-\alpha)\beta x_2 \\ - \frac{1}{4}\beta(1 + \coth \beta x_2)\left(x_1 - \hat{x}_1 e^{-\beta x_2}\right)^2\Bigg]\Bigg) I_{(-\frac{1}{2}-\frac{1}{2}\alpha)}(\beta \hat{x}_1 \zeta),\end{aligned} \tag{4.272}$$

where

$$\zeta = \frac{x_1}{2 \sinh \beta x_2}$$

and $I_\nu(z)$ is a modified Bessel function.

(d) Use (4.270)–(4.272) to derive explicit expressions for integrals involving $I_\nu(z)$. Generalize [Bluman and Cole (1974)].

9. Use group methods to find the fundamental solution of the heat equation in two and three space dimensions, i.e. to solve the initial value problem

$$\frac{\partial u}{\partial x_n} = \sum_{i=1}^{n-1} \frac{\partial^2 u}{\partial x_i^2}, \quad x_n > 0, \ -\infty < x_i < \infty \text{ for } i = 1, 2, \ldots, n-1,$$

with

$$u(x_1, x_2, \ldots, x_{n-1}, 0) = \delta(x_1)\delta(x_2)\cdots\delta(x_{n-1}),$$

for $n = 2, 3$.

10. Consider the problem of finding the Green's function for an instantaneous line particle source diffusing in a gravitational field and under the influence of a linear shear wind [Neuringer (1968), Bluman and Cole (1974)] which reduces to solving

$$\frac{\partial u}{\partial x_3} + x_2\frac{\partial u}{\partial x_1} - \frac{\partial u}{\partial x_2} - d\left(\frac{\partial^2 u}{\partial x_1^2} + \frac{\partial^2 u}{\partial x_2^2}\right) = 0, \tag{4.273a}$$

$$u(x_1, x_2, 0) = \delta(x_1)\delta(x_2 - \hat{x}_2), \tag{4.273b}$$

$$-\infty < x_1 < \infty, \quad -\infty < x_2 < \infty, \quad -\infty < \hat{x}_2 < \infty, \quad x_3 > 0.$$

(a) Show that (4.273a,b) admits

$$\mathrm{X}_1 = (x_3)^2\frac{\partial}{\partial x_1} + 2x_3\frac{\partial}{\partial x_2} + \frac{1}{d}(\hat{x}_2 - x_2 - x_3)u\frac{\partial}{\partial u},$$

$$\mathrm{X}_2 = [(x_3)^3 - 6x_3]\frac{\partial}{\partial x_1} + 3(x_3)^2\frac{\partial}{\partial x_2} + \frac{1}{2d}[6x_1 - 3(x_3)^2 - 6x_2x_3]u\frac{\partial}{\partial u}. \tag{4.274}$$

(b) Find invariant forms for the solution of (4.273a,b) corresponding to each of the generators X_1 and X_2. Then show that the solution of (4.273a,b) reduces to solving a first order ODE.

(c) Show that the solution of (4.273a,b) is

$$u(x_1, x_2, x_3) = \frac{1}{4\pi d x_3\sqrt{1 + (x_3)^2/12}} \times \exp\left[-\frac{1}{16d}\left\{\frac{[2x_1 - (x_2 + \hat{x}_2)x_3]^2}{x_3[1 + (x_3)^2/12]} + \frac{4(x_2 - \hat{x}_2 + x_3)^2}{x_3}\right\}\right]. \tag{4.275}$$

(d) Use the infinitesimal generators (4.274) to compute the moments $\langle x_1\rangle$, $\langle x_2\rangle$, $\langle x_1x_2\rangle$, $\langle (x_1)^2\rangle$, and $\langle (x_2)^2\rangle$ directly *without* use of solution (4.275).

11. The Poisson kernel is the solution of

$$\nabla^2 u = \frac{\partial^2 u}{\partial x_1^2} + \frac{1}{x_1}\frac{\partial u}{\partial x_1} + \frac{1}{(x_1)^2}\frac{\partial^2 u}{\partial x_2^2} = 0, \quad 0 \le x_1 < 1, \quad 0 \le x_2 < 2\pi, \tag{4.276a}$$

with

$$u(1, x_2) = \delta(x_2). \tag{4.276b}$$

(a) Let $z = x_1 e^{ix_2}$. Show that (4.276a,b) admits the infinite-parameter Lie group corresponding to infinitesimal generator

$$\mathrm{X}_\infty = x_1 S(x_1, x_2)\frac{\partial}{\partial x_1} + T(x_1, x_2)\frac{\partial}{\partial x_2} + \lambda u \frac{\partial}{\partial u},$$

where

$$S(x_1, x_2) = \sum_{n=1}^{\infty}[a_n(z^n - (\overline{z})^{-n}) + b_n(z^{-n} - \overline{z}^n)],$$

$\overline{z} = x_1 e^{-ix_2}$, $T(x_1, x_2)$ is the harmonic conjugate of $S(x_1, x_2)$, $T(1,0) = 0$, $\lambda = -\frac{\partial T}{\partial x_2}(1,0)$, and $\{a_n, b_n\}$ are arbitrary complex parameters.

(b) Consider the subgroup for which $a_1 = a \neq 0$, $b_1 = b \neq 0$, $a_j = b_j = 0$ for $j \neq 1$, and let $\alpha = -i(a+b)$, $\beta = a - b$. Show that (4.276a,b) admits the two-parameter subgroup with infinitesimal generators

$$\mathrm{X}_1 = (1 - (x_1)^2)\sin x_2 \frac{\partial}{\partial x_1} + \left[\left(x_1 + \frac{1}{x_1}\right)\cos x_2 - 2\right]\frac{\partial}{\partial x_2},$$

$$\mathrm{X}_2 = ((x_1)^2 - 1)\cos x_2 \frac{\partial}{\partial x_1} + \left(x_1 + \frac{1}{x_1}\right)\sin x_2 \frac{\partial}{\partial x_2} - 2u\frac{\partial}{\partial u}.$$

(c) Show that the invariant solution corresponding to X_1 has the invariant form

$$u = A(\zeta)$$

with

$$\zeta = \frac{1 - (x_1)^2}{1 - 2x_1\cos x_2 + (x_1)^2}.$$

(d) Use the invariant surface condition

$$\mathrm{X}_2(u - A(\zeta)) = 0$$

to show that $-\zeta A'(\zeta) + A = 0$, and hence derive the Poisson kernel

$$u(x_1, x_2) = \frac{1}{2\pi} \frac{1-(x_1)^2}{1-2x_1\cos x_2 + (x_1)^2}$$

[Bluman and Cole (1974)].

12. Consider a well-posed BVP for a linear homogeneous PDE with independent variables $x = (x_1, x_2, \ldots, x_n)$:

$$\mathrm{L}u = 0, \quad x \in \Omega_x, \tag{4.277a}$$

with $k-1$ linear homogeneous boundary conditions

$$\mathrm{L}_\alpha u = 0 \text{ on } \omega_\alpha(x_1, x_2, \ldots, x_{n-1}) = 0, \;\; \alpha = 1, 2, \ldots, k-1, \tag{4.277b}$$

and one linear nonhomogeneous boundary condition (initial condition)

$$\mathrm{L}_k u = h(x_1, x_2, \ldots, x_{n-1}) \quad \text{when} \quad x_n = 0. \tag{4.277c}$$

If the associated homogeneous BVP admits $\mathrm{X} = \frac{\partial}{\partial x_n}$, show that the solution of (4.277a–c) has the inverse Laplace transform representation

$$u(x_1, x_2, \ldots, x_n) = \frac{1}{2\pi i}\int_{\gamma - i\infty}^{\gamma + i\infty} F(x_1, x_2, \ldots, x_{n-1}; s)e^{sx_n}\,ds, \tag{4.278}$$

where $F(x_1, x_2, \ldots, x_{n-1}; s)$ is determined by substituting (4.278) into (4.277a–c). What is the situation when the right-hand side of (4.277a) is $h(x_1, x_2, \ldots, x_{n-1})$ and the boundary condition (4.277c) becomes homogeneous?

4.5 Discussion

In this chapter we showed how to

(i) find infinitesimal transformations admitted by a given scalar PDE or a given system of PDE's;

(ii) use infinitesimal transformations to construct special solutions (invariant solutions) for PDE's; and

(iii) find infinitesimal transformations admitted by a boundary value problem posed for a PDE.

Invariant solutions for scalar PDE's were discovered by Lie (1881). Invariant solutions can be determined from an admitted infinitesimal transformation in two ways: Using the invariant form method, one must first explicitly solve the characteristic equations arising from the invariant surface conditions to obtain the invariant form for the invariant solution; the invariant solution is then determined by substituting the invariant form into the given PDE's. Using the direct substitution method, one first substitutes the invariant surface conditions and their necessary differential consequences into the given PDE's to eliminate all derivatives with respect to a particular independent variable; the invariant solution is then determined by solving the resulting reduced PDE's and then substituting the solution of the reduced PDE's into either the invariant surface conditions or the given PDE's. An important feature of the direct substitution method is that one could proceed to construct invariant solutions without first explicitly solving the invariant surface conditions. To our knowledge the direct substitution method has not previously appeared in the literature.

A one-parameter Lie group of transformations is admitted by a boundary value problem if it leaves the boundary, each boundary condition, and the governing PDE's invariant. If a well-posed BVP admits a one-parameter Lie group of transformations, then its solution is an invariant solution of the group. The construction of the solution of a BVP further simplifies if the BVP admits a multi-parameter group.

In applying infinitesimal transformations to a linear BVP it is not necessary to leave the BVP invariant since the associated homogeneous BVP always admits a uniform scaling of the dependent variable(s). In this case a superposition of invariant solutions or invariant forms, arising from invariance of the homogeneous part of the BVP, could be used to satisfy the nonhomogeneous part of the BVP.

Often the asymptotic solution of a boundary value problem for nonlinear PDE's is an invariant solution of self-similar type (*self-similar solution*) arising from scaling invariance of the governing PDE's. Comprehensive reviews of self-similar asymptotics appear in Newman (1984) and Galaktionov, Doronitsyn, Elenin, Kurdyumov, and Samarskii (1988). For applications of self-similar asymptotics to physical problems see Barenblatt and Zel'dovich (1972) and Barenblatt (1979). Barenblatt and Zel'dovich also consider examples of "intermediate asymptotics" where in an intermediate space-time domain the solution of a BVP is approximated by a similarity solution (invariant solution) which does not depend on the specific boundary conditions; in such examples the similarity solution is not an equilibrium state. Kamin (1973) rigorously justified the evolution of the solution of a porous medium equation to a self-similar solution. For other papers rigorously justifying self-similar asymptotics see Atkinson and Peletier (1974), Friedman and Kamin (1980), Galaktionov and Samarskii (1984), Kamin (1975).

In Chapter 5 we generalize the concept of infinitesimal transformations by allowing infinitesimals to depend on derivatives of dependent variables. The consideration of such symmetries (*Lie–Bäcklund symmetries*) arises naturally when seeking conservation laws through Noether's theorem.

In Chapter 6 we show that a nonlinear PDE must admit an infinite-parameter Lie group of transformations in order to have an invertible mapping to a linear PDE. The form of the infinitesimals shows when such a mapping exists and also leads to its construction.

In Chapter 7 we introduce nonlocal symmetries whose infinitesimal transformations can depend on integrals of dependent variables. In particular we consider ***potential symmetries*** which are nonlocal symmetries realized as local symmetries for a system of PDE's related to a given PDE which can be expressed in a conserved form.

5

Noether's Theorem and Lie–Bäcklund Symmetries

5.1 Introduction

In the preceding chapters we established the algorithm to determine Lie groups of point transformations of differential equations and developed methods to solve differential equations using such symmetries. In this chapter we study one of the most important applications of symmetries to physical problems, namely, the construction of conservation laws.

A *conservation law* of a physical system, which has independent variables t, x, y, z and dependent variables described by a state function $u(t, x, y, z)$ with components $u = (u^1, u^2, \dots, u^m)$, is an equation of the form

$$\operatorname{div} f = \mathrm{D}_t f^1 + \mathrm{D}_x f^2 + \mathrm{D}_y f^3 + \mathrm{D}_z f^4 = 0,$$

where the vector function $f = (f^1, f^2, f^3, f^4)$ can depend on t, x, y, z, u and derivatives $\underset{1}{u}, \underset{2}{u}, \dots, \underset{k}{u}$; $\mathrm{D}_t, \mathrm{D}_x, \mathrm{D}_y, \mathrm{D}_z$ are total derivative operators. Physically a conservation law means that the rate of change of f^1 inside any spatial domain must equal the flow (f^2, f^3, f^4) through the surface of the domain. For systems arising in classical mechanics, where time t is the only independent variable, the conservation law is simply $\mathrm{D}_t f^1 = 0$, and f^1 becomes a constant of the motion. Each constant of the motion restricts the motion of the system and can be used to reduce by one the number of degrees of freedom of the system. Indeed finding the conservation laws of a system is often the first step towards finding its solution: The more conservation laws one finds the closer one gets to the complete solution.

Construction of conservation laws for a given system is generally a nontrivial task. However for systems arising from a Lagrangian formulation there exists a fundamental theorem due to Emmy Noether [Noether (1918)]. Noether proved that for every infinitesimal transformation which is admitted by the action integral of a Lagrangian system one can constructively find a conservation law. For instance conservation of angular momentum is related to rotational invariance and conservation of energy to translational invariance in time. Noether's proof provides an algorithm to construct the conservation law for any admitted infinitesimal transformation. What is significant is that one can find all such transformations by examining the invariance properties of Euler–Lagrange equations, i.e., differential equations arising from a variational problem of the action integral. These aspects of conservation laws are discussed in Section 5.2.

The types of infinitesimal transformations considered by Noether go beyond those corresponding to Lie groups of point transformations (*point symmetries*) which we considered in earlier chapters. Noether allowed the infinitesimals ξ and η to depend, not only on x and u, but also on derivatives $\underset{1}{u}, \underset{2}{u}, \ldots$. Transformations of this type are commonly called *Lie–Bäcklund transformations* even though neither Lie nor Bäcklund considered such transformations. Not all of the important properties of point symmetries are shared by Lie–Bäcklund transformations. In particular an infinitesimal transformation of Lie–Bäcklund type cannot be integrated to a global transformation by the method of characteristics. But infinitesimal Lie–Bäcklund transformations can be used to construct conservation laws and invariant solutions. Moreover for a given system of differential equations one can find admitted infinitesimal Lie–Bäcklund transformations by a simple extension of Lie's algorithm to find admitted infinitesimal point symmetries. The well-known infinite sequences of conservation laws admitted by the Korteweg–de Vries (KdV) and sine-Gordon equations are directly related to admitted infinite sequences of Lie–Bäcklund symmetries via Noether's theorem. Basic properties of Lie–Bäcklund transformations and some of their applications are discussed in Sections 5.2 and 5.3.

5.2 Noether's Theorem

Many problems can be cast in a variational formulation:

Given a function $L(x, u, \underset{1}{u}, \underset{2}{u}, \ldots, \underset{k}{u})$ of x, $u(x)$, $\underset{1}{u}(x), \underset{2}{u}(x), \ldots, \underset{k}{u}(x)$ defined on a domain Ω in the space $x = (x_1, x_2, \ldots, x_n)$, find functions $u(x)$ which correspond to extrema of the integral

$$J[u] = \int_\Omega L(x, u, \underset{1}{u}, \underset{2}{u}, \ldots, \underset{k}{u})dx. \tag{5.1}$$

The function L is called a *Lagrangian* and the integral $J[u]$ of (5.1) an *action integral;* $u(x) = (u^1(x), u^2(x), \ldots, u^m(x))$ describes the state of the system and is usually subject to a set of conditions prescribed on the boundary $\partial\Omega$ of the domain Ω. If $u(x)$ is an extremum of (5.1), then any infinitesimal change $u(x) \to u(x) + \epsilon v(x)$, not altering the boundary conditions, should have no effect on $J[u]$ at least to order $O(\epsilon)$.

A *conservation law* of a system is, by definition, an equation in divergence-free form:

$$\operatorname{div} f = \mathrm{D}_i f^i = 0, \tag{5.2}$$

involving a vector function $f(x, u, \underset{1}{u}, \underset{2}{u}, \ldots, \underset{k}{u}) = (f^1, f^2, \ldots, f^n)$ depending on x, $u(x)$ and its derivatives to some order k. Equation (5.2) must hold for any extremal function $u(x)$ of (5.1). The vector f is called a *conserved flux* since (5.2) implies that a net flow of f through any closed surface in the space x is zero.

Noether's original paper considered transformations of the form

$$\begin{aligned} x^* &= x + \epsilon\xi(x, u, \underset{1}{u}, \underset{2}{u}, \dots, \underset{p}{u}) + O(\epsilon^2), \\ u^* &= u + \epsilon\eta(x, u, \underset{1}{u}, \underset{2}{u}, \dots, \underset{p}{u}) + O(\epsilon^2), \end{aligned} \tag{5.3}$$

which leave the action integral $J[u]$ invariant for *arbitrary* Ω. Noether established the explicit relationship between the infinitesimals ξ, η and the conserved flux f.

To establish Noether's theorem we first derive basic equations connected with the variational problem. Having established these equations we present a proof of Noether's theorem. The proof is based on a reformulation, due to Boyer (1967), and is considerably simpler than Noether's proof. We compare the relationship between these two formulations and give Noether's version which has some merit in spotting the existence of standard conservation laws.

5.2.1 Euler–Lagrange Equations

Consider an infinitesimal change of $u : u(x) \to u(x) + \epsilon v(x)$. The corresponding variation of the Lagrangian L is given by

$$\begin{aligned} \delta L = {} & L(x, u + \epsilon v, \underset{1}{u} + \epsilon \underset{1}{v}, \underset{2}{u} + \epsilon \underset{2}{v}, \dots, \underset{k}{u} + \epsilon \underset{k}{v}) \\ & - L(x, u, \underset{1}{u}, \underset{2}{u}, \dots, \underset{k}{u}) = \epsilon \left[\frac{\partial L}{\partial u^\gamma} v^\gamma + \frac{\partial L}{\partial u_i^\gamma} v_i^\gamma + \frac{\partial L}{\partial u_{ij}^\gamma} v_{ij}^\gamma \right. \\ & \left. + \dots + \frac{\partial L}{\partial u_{i_1 i_2 \cdots i_k}^\gamma} v_{i_1 i_2 \cdots i_k}^\gamma \right] + O(\epsilon^2), \end{aligned} \tag{5.4}$$

assuming summation over a repeated index. It is convenient to introduce the *Euler operator* E_γ defined by

$$\begin{aligned} \mathrm{E}_\gamma = {} & \frac{\partial}{\partial u^\gamma} - \mathrm{D}_i \frac{\partial}{\partial u_i^\gamma} + \mathrm{D}_i \mathrm{D}_j \frac{\partial}{\partial u_{ij}^\gamma} \\ & + \dots + (-1)^k \mathrm{D}_{i_1} \mathrm{D}_{i_2} \cdots \mathrm{D}_{i_k} \frac{\partial}{\partial u_{i_1 i_2 \cdots i_k}^\gamma} + \dots . \end{aligned} \tag{5.5}$$

Let

$$W^i[u, v] = v^\gamma \left[\frac{\partial L}{\partial u_i^\gamma} + \dots + (-1)^{k-1} \mathrm{D}_{i_1} \mathrm{D}_{i_2} \cdots \mathrm{D}_{i_{k-1}} \frac{\partial L}{\partial u_{i i_1 i_2 \cdots i_{k-1}}^\gamma} \right]$$

$$+ (\mathrm{D}_{i_1}v^\gamma)\left[\frac{\partial L}{\partial u^\gamma_{i_1 i}} + \cdots + (-1)^{k-2}\mathrm{D}_{i_2}\mathrm{D}_{i_3}\cdots\mathrm{D}_{i_{k-1}}\frac{\partial L}{\partial u^\gamma_{i_1 i i_2\cdots i_{k-1}}}\right]$$

$$+ \cdots + (\mathrm{D}_{i_1}\mathrm{D}_{i_2}\cdots\mathrm{D}_{i_{k-1}}v^\gamma)\frac{\partial L}{\partial u^\gamma_{i_1 i_2\cdots i_{k-1} i}}. \tag{5.6}$$

Then using integration by parts repeatedly one can show that

$$\delta L = \epsilon\left[\mathrm{E}_\gamma(L)v^\gamma + \mathrm{D}_i W^i[u,v]\right] + O(\epsilon^2), \tag{5.7}$$

where

$$\mathrm{E}_\gamma(L) = \frac{\partial L}{\partial u^\gamma} - \mathrm{D}_i\frac{\partial L}{\partial u^\gamma_i} + \mathrm{D}_i\mathrm{D}_j\frac{\partial L}{\partial u^\gamma_{ij}}$$

$$+ \cdots + (-1)^k\mathrm{D}_{i_1}\mathrm{D}_{i_2}\cdots\mathrm{D}_{i_k}\frac{\partial L}{\partial u^\gamma_{i_1 i_2\cdots i_k}}. \tag{5.8}$$

Note that in deriving (5.7) we imposed no boundary conditions on $u(x)$. Hence these equations are valid for arbitrary functions $u(x)$ and $v(x)$ as long as their derivatives in (5.6) and (5.8) exist.

Now we determine the conditions which must be satisfied by extremal functions $u(x)$. We evaluate the variation in $J[u]$ under a change $u(x) \to u(x) + \epsilon v(x)$. From (5.7) and the divergence theorem, we obtain the variation of $J[u]$:

$$\delta J[u] = J[u+\epsilon v] - J[u] = \int_\Omega \delta L\, dx$$

$$= \epsilon\int_\Omega \left[\mathrm{E}_\gamma(L)v^\gamma + \mathrm{D}_i W^i[u,v]\right] dx + O(\epsilon^2)$$

$$= \epsilon\left[\int_\Omega \mathrm{E}_\gamma(L)v^\gamma dx + \int_{\partial\Omega} W^i[u,v]n_i d\sigma\right] + O(\epsilon^2) \tag{5.9}$$

where $\int_{\partial\Omega}$ represents the surface integral over the boundary surface $\partial\Omega$ of Ω with $n = (n_1, n_2, \ldots, n_n)$ being the unit outward normal vector to $\partial\Omega$. For the function $u(x)$ to be an extremal function for $J[u]$ the $O(\epsilon)$ term of $\delta J[u]$ must vanish:

$$\int_\Omega \mathrm{E}_\gamma(L)v^\gamma dx + \int_{\partial\Omega} W^i[u,v]n_i d\sigma = 0. \tag{5.10}$$

Since a condition on $v(x)$ is that it does not alter boundary conditions imposed on $u(x)$, without loss of generality we assume that $v(x)$ and the derivatives of $v(x)$ appearing in $W^i[u,v]$ vanish on $\partial\Omega$. Then the surface integral in (5.10) also vanishes since the functions $W^i[u,v]$ are linear in v and its derivatives. Consequently the volume integral in (5.10) must vanish for any $v(x)$ satisfying the indicated homogeneous boundary conditions. Since the behaviour of $v(x)$ is arbitrary within Ω it follows that an extremum $u(x)$ satisfies

$$\mathrm{E}_\gamma(L) = \frac{\partial L}{\partial u^\gamma} - \mathrm{D}_i \frac{\partial L}{\partial u_i^\gamma} + \mathrm{D}_i \mathrm{D}_j \frac{\partial L}{\partial u_{ij}^\gamma} + \cdots$$

$$+ (-1)^k \mathrm{D}_{i_1} \mathrm{D}_{i_2} \cdots \mathrm{D}_{i_k} \frac{\partial L}{\partial u^\gamma_{i_1 i_2 \cdots i_k}} = 0, \quad \gamma = 1, 2, \ldots, m. \tag{5.11}$$

Equations (5.11) are called the *Euler–Lagrange equations* for an extremum $u(x)$ of $J[u]$. Thus the following theorem has been proved:

Theorem 5.2.1-1. *For a smooth function $u(x)$ to be an extremum of the action integral*

$$J[u] = \int_\Omega L(x, u, \underset{1}{u}, \underset{2}{u}, \ldots, \underset{k}{u}) dx$$

it is necessary that it satisfies the Euler–Lagrange equations (5.11).

Consider two examples:

(1) *sine-Gordon Equation:* Let

$$L = -\frac{1}{2} u_1 u_2 + \cos u. \tag{5.12}$$

The corresponding Euler–Lagrange equation

$$\frac{\partial L}{\partial u} - \mathrm{D}_1 \frac{\partial L}{\partial u_1} - \mathrm{D}_2 \frac{\partial L}{\partial u_2} = 0 \tag{5.13}$$

is the sine-Gordon equation

$$u_{12} - \sin u = 0. \tag{5.14}$$

(2) *Korteweg–de Vries Equation:* Let

$$L = \frac{1}{2} u_1^1 u_2^1 + \frac{1}{6} (u_1^1)^3 + u_1^1 u_1^2 + \frac{1}{2} (u^2)^2.$$

Corresponding Euler–Lagrange equations

$$\frac{\partial L}{\partial u^1} - \mathrm{D}_1 \frac{\partial L}{\partial u_1^1} - \mathrm{D}_2 \frac{\partial L}{\partial u_2^1} = 0,$$

$$\frac{\partial L}{\partial u^2} - \mathrm{D}_1 \frac{\partial L}{\partial u_1^2} - \mathrm{D}_2 \frac{\partial L}{\partial u_2^2} = 0,$$

are given by

$$u_{12}^1 + u_1^1 u_{11}^1 + u_{11}^2 = 0, \tag{5.15a}$$

$$u_{11}^1 - u^2 = 0. \tag{5.15b}$$

Eliminating u^2, we find that (5.15a,b) reduce to

$$u^1_{12} + u^1_1 u^1_{11} + u^1_{1111} = 0. \tag{5.16}$$

Let $x = x_1$, $t = x_2$, $w = u^1_1$. Then (5.16) becomes the Korteweg–de Vries (KdV) equation

$$\frac{\partial w}{\partial t} + w\frac{\partial w}{\partial x} + \frac{\partial^3 w}{\partial x^3} = 0.$$

The following theorem is useful:

Theorem 5.2.1-2. *For the Euler operator* E_γ, *defined by* (5.5), *the following identities hold for any twice continuously differentiable function* $F(x, u, \underset{1}{u}, \underset{2}{u}, \ldots, \underset{\ell}{u})$:

$$\mathrm{E}_\gamma \mathrm{D}_i F(x, u, \underset{1}{u}, \underset{2}{u}, \ldots, \underset{\ell}{u}) \equiv 0, \quad \gamma = 1, 2, \ldots, m; \quad i = 1, 2, \ldots, n. \tag{5.17}$$

Proof. See Exercise 5.2-2. □

From Theorem 5.2.1-2 two theorems follow immediately:

Theorem 5.2.1-3. *The Euler–Lagrange equations for a Lagrangian L are identically zero if L can be expressed in divergence form:*

$$L = \mathrm{D}_i F^i(x, u, \underset{1}{u}, \underset{2}{u}, \ldots, \underset{\ell}{u}).$$

Theorem 5.2.1-4. *Two Lagrangians L and L′ have the same set of Euler–Lagrange equations if* $L - L' = \operatorname{div} A$ *for some vector*

$$A(x, u, \underset{1}{u}, \underset{2}{u}, \ldots, \underset{\ell}{u}) = (A^1, A^2, \ldots, A^n).$$

The converses of Theorems 5.2.1-3,4 also hold [cf. Courant and Hilbert (1953, IV.3.5) and Olver (1986, Sect. 4.1)].

5.2.2 Variational Symmetries and Conservation Laws; Boyer's Formulation

Since Euler–Lagrange equations are the governing equations of many physical systems, one might expect that conservation laws would arise directly from properties of the Euler–Lagrange equations. However to find conservation laws Noether showed that it is more fruitful if one examines transformations which leave the action integral $J[u] = \int_\Omega L\,dx$ invariant. She established a direct relationship between such invariances and conservation laws. Transformations considered by Noether are of the general form

$$\begin{aligned} x^* &= x + \epsilon\xi(x, u, \underset{1}{u}, \underset{2}{u}, \ldots, \underset{p}{u}) + O(\epsilon^2), \\ u^* &= u + \epsilon\eta(x, u, \underset{1}{u}, \underset{2}{u}, \ldots, \underset{p}{u}) + O(\epsilon^2), \end{aligned} \tag{5.18}$$

which allow ξ and η to depend on derivatives of u. Boyer (1967) recognized that the generality of transformation (5.18) is superfluous. In particular one can get by with a simpler form of transformation in which x is invariant:

$$x^* = x, \quad u^* = u + \epsilon\eta(x, u, \underset{1}{u}, \underset{2}{u}, \ldots, \underset{p}{u}) + O(\epsilon^2). \tag{5.19}$$

This follows from the general result that any transformation of the form (5.18) can be represented by the form (5.19). The proof is given in Section 5.2.3. Here we take Boyer's view and derive Noether's theorem.

The extended transformations for (5.19) are obviously given by

$$u_i^* = u_i + \epsilon \mathrm{D}_i\eta + O(\epsilon^2), \quad u_{ij}^* = u_{ij} + \epsilon \mathrm{D}_i \mathrm{D}_j\eta + O(\epsilon^2), \ldots, \tag{5.20}$$

and the corresponding kth extended infinitesimal generator is denoted by

$$\mathrm{U}^{(k)} = \eta^\gamma \frac{\partial}{\partial u^\gamma} + \eta_i^\gamma \frac{\partial}{\partial u_i^\gamma} + \eta_{ij}^\gamma \frac{\partial}{\partial u_{ij}^\gamma} + \cdots + \eta_{i_1 i_2 \cdots i_k}^\gamma \frac{\partial}{\partial u_{i_1 i_2 \cdots i_k}^\gamma} \tag{5.21}$$

where $\eta_{ij\cdots s}^\gamma = \mathrm{D}_i \mathrm{D}_j \cdots \mathrm{D}_s \eta^\gamma$. Under a transformation (5.19), (5.20), the Lagrangian L varies by $\delta L = \epsilon \mathrm{U}^{(k)} L + O(\epsilon^2)$.

Definition 5.2.2-1. A transformation (5.19) is a *variational symmetry* of the action integral $J[u]$ (variational symmetry for L) if for *any* $u(x)$ there exists some vector function

$$A(x, u, \underset{1}{u}, \underset{2}{u}, \ldots, \underset{r}{u}) = (A^1, A^2, \ldots, A^n)$$

of x, u and its derivatives to some finite order r, such that

$$\mathrm{U}^{(k)} L = \mathrm{D}_i A^i. \tag{5.22a}$$

Variational symmetries are also called *Noether transformations.*

As an example consider

$$L = 3(u_{11})^2 - (u_1)^3 - 3u_1 u_2, \quad \mathrm{U} = (x_2 u_1 - x_1)\frac{\partial}{\partial u}.$$

The corresponding Euler–Lagrange equation is the KdV equation (5.16). Here

$$\mathrm{U}^{(2)} L = 3x_2[2u_{11}u_{111} - (u_1)^2 u_{11} - u_2 u_{11} - u_1 u_{12}] + 3u_2 = \mathrm{D}_1 A^1 + \mathrm{D}_2 A^2$$

with

$$A^1 = x_2[3(u_{11})^2 - (u_1)^3 - 3u_1 u_2], \quad A^2 = 3u.$$

Hence U is a variational symmetry for L.

Now consider identity (5.7) which holds for arbitrary u and v. Setting $v = \eta$, $v_i = \eta_i$, $v_{ij} = \eta_{ij}, \ldots$ in (5.7), we have

$$\mathrm{U}^{(k)} L = \mathrm{E}_\gamma(L)\eta^\gamma + \mathrm{D}_i W^i[u, \eta] \tag{5.22b}$$

since $\delta L = \epsilon \mathrm{U}^{(k)} L + O(\epsilon^2)$. Comparing (5.22a) and (5.22b), we obtain for any variational symmetry (5.19) for L:

$$\mathrm{E}_\gamma(L)\eta^\gamma + \mathrm{D}_i W^i[u,\eta] = \mathrm{D}_i A^i. \tag{5.23}$$

If $u(x)$ is a solution of the Euler–Lagrange equations, i.e., $\mathrm{E}_\gamma(L) = 0$, $\gamma = 1, 2, \ldots, m$, then (5.23) yields a conservation law $\mathrm{D}_i(W^i - A^i) = 0$. Thus we have:

Theorem 5.2.2-1 (Boyer's Formulation of Noether's Theorem). *Let* $\mathrm{U} = \eta^\gamma \dfrac{\partial}{\partial u^\gamma}$ *be the infinitesimal generator of transformations* (5.19) *and let* $\mathrm{U}^{(k)}$ *be its kth extension* (5.21). *If* U *is the infinitesimal generator of a variational symmetry of an action integral* (5.1) *so that* $\mathrm{U}^{(k)} L = \mathrm{D}_i A^i$ *holds for any* $u(x)$*, then the conservation law*

$$D_i(W^i[u,\eta] - A^i) = 0$$

holds for any solution $u(x)$ *of the Euler–Lagrange equations* $\mathrm{E}_\gamma(L) = 0$, $\gamma = 1, 2, \ldots, m$.

The construction of conservation laws via Noether's theorem requires explicit expressions for $W^i[u,\eta]$, $i = 1, 2, \ldots, n$. Sometimes these can be obtained more easily by directly re-expressing $\mathrm{U}^{(k)} L$ in the form (5.22b) rather than using (5.6). For example again consider $L = 3(u_{11})^2 - (u_1)^3 - 3u_1u_2$. Let $\mathrm{U}^{(2)}$ be the twice extended operator of $\mathrm{U} = \eta \dfrac{\partial}{\partial u}$. Then

$$\mathrm{U}^{(2)} L = 6u_{11}(\mathrm{D}_1)^2\eta - 3(u_1)^2 \mathrm{D}_1\eta - 3u_2\mathrm{D}_1\eta - 3u_1\mathrm{D}_2\eta.$$

Applying integration by parts to each term, we get

$$u_{11}(\mathrm{D}_1)^2\eta = \mathrm{D}_1(u_{11}\mathrm{D}_1\eta) - u_{111}\mathrm{D}_1\eta = \mathrm{D}_1(u_{11}\mathrm{D}_1\eta - u_{111}\eta) + u_{1111}\eta;$$

$$(u_1)^2\mathrm{D}_1\eta = \mathrm{D}_1((u_1)^2\eta) - 2u_1u_{11}\eta;$$

$$u_2\mathrm{D}_1\eta = \mathrm{D}_1(u_2\eta) - u_{12}\eta; \quad u_1\mathrm{D}_2\eta = \mathrm{D}_2(u_1\eta) - u_{12}\eta.$$

Consequently

$$\mathrm{U}^{(2)} L = \mathrm{E}(L)\eta + \mathrm{D}_1 W^1[u,\eta] + \mathrm{D}_2 W^2[u,\eta]$$

where

$$\mathrm{E}(L) = 6[u_{1111} + u_1u_{11} + u_{12}],$$
$$W^1[u,\eta] = 6[u_{11}\mathrm{D}_1\eta - u_{111}\eta] - 3[(u_1)^2\eta + u_2\eta],$$
$$W^2[u,\eta] = -3u_1\eta.$$

We showed that $\mathrm{U} = (x_2u_1 - x_1)\dfrac{\partial}{\partial u}$ yields a variational symmetry for L and determined $A^1 = x_2[3(u_{11})^2 - (u_1)^3 - 3u_1u_2]$, $A^2 = 3u$. Then for $\eta = x_2u_1 - x_1$, from Noether's theorem we obtain the conservation law

$$\mathrm{D}_1(W^1 - A^1) + \mathrm{D}_2(W^2 - A^2) = 0,$$

which here becomes

$$\begin{aligned}\mathrm{D}_1(x_2[3(u_{11})^2 - 2(u_1)^3 - 6u_1u_{111}] - 6u_{11} + 6x_1u_{111} + 3x_1[(u_1)^2 + u_2])\\ + \mathrm{D}_2(3[-x_2(u_1)^2 + x_1u_1 - u]) = 0.\end{aligned}$$

5.2.3 Equivalent Classes of Lie–Bäcklund Transformations; Lie–Bäcklund Symmetries

Noether's transformations of the form

$$\begin{aligned} x^* &= x + \epsilon\xi(x, u, \underset{1}{u}, \underset{2}{u}, \ldots, \underset{p}{u}) + O(\epsilon^2), \\ u^* &= u + \epsilon\eta(x, u, \underset{1}{u}, \underset{2}{u}, \ldots, \underset{p}{u}) + O(\epsilon^2), \end{aligned} \tag{5.24}$$

are clearly more general than those defining Lie groups of point transformations of earlier chapters since the infinitesimals ξ and η depend on derivatives $\underset{1}{u}, \underset{2}{u}, \ldots, \underset{p}{u}$. However, by requiring the contact conditions to be preserved as for point transformations in Chapter 2, one obtains the extended transformations of (5.24):

$$\begin{aligned} u_i^* &= u_i + \epsilon\eta_i^{(1)} + O(\epsilon^2), \\ &\vdots \\ u^*_{i_1i_2\cdots i_k} &= u_{i_1i_2\cdots i_k} + \epsilon\eta^{(k)}_{i_1i_2\cdots i_k} + O(\epsilon^2), \end{aligned}$$

where

$$\begin{aligned} \eta_i^{(1)} &= \mathrm{D}_i\eta - (\mathrm{D}_i\xi_j)u_j, \\ &\vdots \\ \eta^{(k)}_{i_1i_2\cdots i_k} &= \mathrm{D}_{i_k}\eta^{(k-1)}_{i_1i_2\cdots i_{k-1}} - (\mathrm{D}_{i_k}\xi_j)u_{i_1i_2\cdots i_{k-1}j}, \end{aligned}$$

$k = 1, 2, \ldots$. The kth extended infinitesimal generator is given by

$$\mathrm{U}^{(k)} = \xi_i\frac{\partial}{\partial x_i} + \eta^\gamma\frac{\partial}{\partial u^\gamma} + \eta_i^{(1)\gamma}\frac{\partial}{\partial u_i^\gamma} + \cdots + \eta^{(k)\gamma}_{i_1i_2\cdots i_k}\frac{\partial}{\partial u^\gamma_{i_1i_2\cdots i_k}}.$$

An important transformational property of (5.24) is revealed when we examine how a given function $u = f(x)$ is transformed under (5.24). Replacing u by $f(x)$ in (5.24), we have

$$x^* = x + \epsilon\xi(x, f(x), \underset{1}{f}(x), \underset{2}{f}(x), \ldots, \underset{p}{f}(x)) + O(\epsilon^2), \tag{5.25a}$$

$$u^* = f(x) + \epsilon\eta(x, f(x), \underset{1}{f}(x), \underset{2}{f}(x), \ldots, \underset{p}{f}(x)) + O(\epsilon^2). \tag{5.25b}$$

The dependence of u^* on x^* defines the image $f^*(x^*)$ of the function $f(x)$. One has to eliminate x from (5.25a,b) to obtain $f^*(x^*)$. Equation (5.25a) can be solved for x:

$$x = x^* - \epsilon\xi(x^*, f(x^*), \underset{1}{f}(x^*), \underset{2}{f}(x^*), \ldots, \underset{p}{f}(x^*)) + O(\epsilon^2). \tag{5.26}$$

Substituting (5.26) into (5.25b), we obtain

$$\begin{aligned} f^*(x^*) = f(x^*) + \epsilon[&\eta(x^*, f(x^*), \underset{1}{f}(x^*), \underset{2}{f}(x^*), \ldots, \underset{p}{f}(x^*)) \\ &- \frac{\partial f(x^*)}{\partial x_i^*}\xi_i(x^*, f(x^*), \underset{1}{f}(x^*), \underset{2}{f}(x^*), \ldots, \underset{p}{f}(x^*))] + O(\epsilon^2). \end{aligned} \tag{5.27}$$

Then if we replace x^* by x in (5.27), the image of $f(x)$ under transformation (5.24) is

$$\begin{aligned} f^*(x) = f(x) + \epsilon[&\eta(x, f(x), \underset{1}{f}(x), \underset{2}{f}(x), \ldots, \underset{p}{f}(x)) \\ &- \frac{\partial f(x)}{\partial x_i}\xi_i(x, f(x), \underset{1}{f}(x), \underset{2}{f}(x), \ldots, \underset{p}{f}(x))] + O(\epsilon^2). \end{aligned} \tag{5.28}$$

We now observe that the same image of $f(x)$ can also be obtained by a transformation leaving x invariant:

$$\begin{aligned} x^* &= x, \\ u^* &= u + \epsilon[\eta(x, u, \underset{1}{u}, \underset{2}{u}, \ldots, \underset{p}{u}) - u_i\xi_i(x, u, \underset{1}{u}, \underset{2}{u}, \ldots, \underset{p}{u})] + O(\epsilon^2). \end{aligned} \tag{5.29}$$

Thus two transformations (5.24) and (5.29) act on functions in the same way, resulting in:

Theorem 5.2.3-1. *Let $u = f(x)$ be a p-times differentiable function. The two transformations*

$$\begin{aligned} x^* &= x + \epsilon\xi(x, u, \underset{1}{u}, \underset{2}{u}, \ldots, \underset{p}{u}) + O(\epsilon^2), \\ u^* &= u + \epsilon\eta(x, u, \underset{1}{u}, \underset{2}{u}, \ldots, \underset{p}{u}) + O(\epsilon^2), \end{aligned} \tag{5.30}$$

and

$$\begin{aligned} x^* &= x, \\ u^* &= u + \epsilon[\eta(x, u, \underset{1}{u}, \underset{2}{u}, \ldots, \underset{p}{u}) - u_i\xi_i(x, u, \underset{1}{u}, \underset{2}{u}, \ldots, \underset{p}{u})] + O(\epsilon^2) \end{aligned} \tag{5.31}$$

are equivalent in the sense that both transform $f(x)$ to the same function $f^*(x)$ given by (5.28). *In this sense two infinitesimal generators*

$$\mathrm{U} = \xi_i(x, u, \underset{1}{u}, \underset{2}{u}, \ldots, \underset{p}{u})\frac{\partial}{\partial x^i} + \eta^\gamma(x, u, \underset{1}{u}, \underset{2}{u}, \ldots, \underset{p}{u})\frac{\partial}{\partial u^\gamma} \tag{5.32}$$

and

$$\hat{U} = [\eta^\gamma(x, u, \underset{1}{u}, \underset{2}{u}, \ldots, \underset{p}{u}) - u_i^\gamma \xi_i(x, u, \underset{1}{u}, \underset{2}{u}, \ldots, \underset{p}{u})]\frac{\partial}{\partial u^\gamma} \tag{5.33}$$

are equivalent.

Geometrically, transformation (5.31) maps a function in the "vertical" direction while transformation (5.30) includes "horizontal" movement as illustrated in Figure 5.2.3-1.

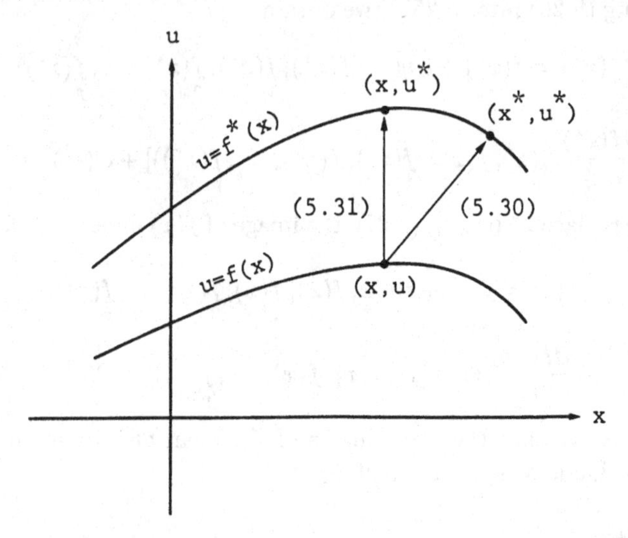

Figure 5.2.3-1. Comparison of the actions of (5.30) and (5.31).

An important implication of Theorem 5.2.3-1 is that one needs only to consider transformations of the form

$$x^* = x, \quad u^* = u + \epsilon\eta(x, u, \underset{1}{u}, \underset{2}{u}, \ldots, \underset{p}{u}) + O(\epsilon^2), \tag{5.34}$$

or infinitesimal generators of the form

$$\eta^\gamma(x, u, \underset{1}{u}, \underset{2}{u}, \ldots, \underset{p}{u})\frac{\partial}{\partial u^\gamma} \tag{5.35}$$

which leave x invariant. Clearly (5.34) includes (5.31) as a special case. This justifies our restriction to infinitesimal generators of the form (5.35) in the discussion of Noether's theorem. The extension formulae for (5.34), (5.35) are particularly simple, namely (5.20), (5.21).

It should be noted that going from (5.34) to (5.30), and hence from (5.35) to (5.32), is not unique. In fact one can show [see Exercise 2.3-16] that any decomposition

$$\eta(x, u, \underset{1}{u}, \underset{2}{u}, \ldots, \underset{p}{u}) = \hat{\eta}(x, u, \underset{1}{u}, \underset{2}{u}, \ldots, \underset{p}{u}) - u_i\hat{\xi}_i(x, u, \underset{1}{u}, \underset{2}{u}, \ldots, \underset{p}{u}) \tag{5.36}$$

of η of (5.34) into $\hat{\xi}$ and $\hat{\eta}$ yields a transformation $x^* = x + \epsilon\hat{\xi} + O(\epsilon^2)$, $u^* = u + \epsilon\hat{\eta} + O(\epsilon^2)$ equivalent to transformation (5.34). In other words, there are an infinite number of infinitesimal generators equivalent to (5.35). Each such transformation has a different set of "horizontal" components as shown in Figure 5.2.3-2. This means that an infinitesimal generator of the form (5.32) is never unique. For this reason and for ease of computations, the transformation form (5.35) is usually assumed. However if one is only interested in point transformations where ξ and η depend only on x and u, then the form (5.34) is too general unless η is assumed to depend linearly on the first derivatives of u and assumed to be independent of $\underset{2}{u}, \underset{3}{u}, \ldots$.

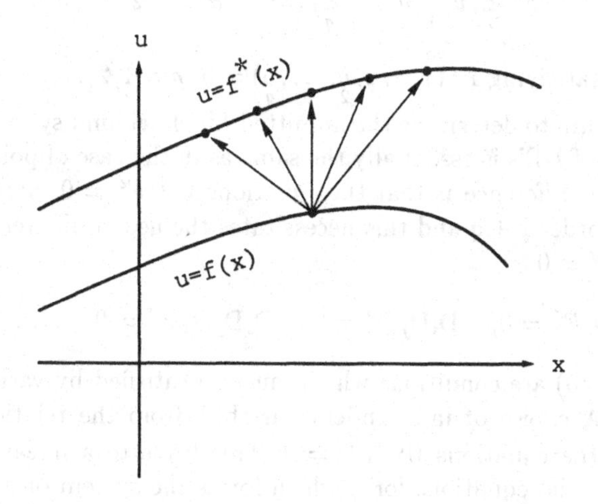

Figure 5.2.3-2. Action of the family of transformations equivalent to (5.34).

In view of (5.36), a scaling operator $\mathrm{U} = x\frac{\partial}{\partial x} - 2u\frac{\partial}{\partial u}$ is equivalent to $\hat{\mathrm{U}} = (-2u - xu_x)\frac{\partial}{\partial u}$ and in turn $\hat{\mathrm{U}}$ is equivalent to $\tilde{\mathrm{U}} = \frac{2u}{u_x}\frac{\partial}{\partial x} - xu_x\frac{\partial}{\partial u}$ provided u_x does not vanish. The total derivative operator $\mathrm{D}_k = \frac{\partial}{\partial x_k} + u_k\frac{\partial}{\partial u}$ is equivalent to the null operator $\hat{\mathrm{U}} = 0$ since $\eta - u_i\xi_i = u_k - u_k = 0$. Geometrically D_k just moves points along any given curve $u = f(x)$.

In earlier chapters a symmetry of a differential equation was defined to be a transformation which maps any solution of the differential equation to another solution of the same equation. From this point of view it is easy to extend the concept of symmetries of differential equations to Lie–Bäcklund transformations. The formulation becomes particularly simple for transformations that involve changes only in u since then the transformation itself represents the transformation of the solution: If the transformation

$$x^* = x, \quad u^* = u + \epsilon\eta(x, u, \underset{1}{u}, \underset{2}{u}, \ldots, \underset{p}{u}) + O(\epsilon^2), \tag{5.37}$$

generates a symmetry of a system of partial differential equations

$$F^\sigma(x, u, \underset{1}{u}, \underset{2}{u}, \ldots, \underset{q}{u}) = 0, \quad \sigma = 1, 2, \ldots, m, \tag{5.38}$$

then $u^*(x)$, defined by (5.37) for any solution $u(x)$ of (5.38), must also be a solution of (5.38). In terms of the infinitesimal η we have the following definition:

Definition 5.2.3-1. The transformation (5.37) with infinitesimal generator $\mathrm{U} = \eta^\gamma \frac{\partial}{\partial u^\gamma}$ defines a *Lie–Bäcklund symmetry* of (5.38) if and only if

$$\mathrm{U}^{(q)} F^\sigma(x, u, \underset{1}{u}, \underset{2}{u}, \ldots, \underset{q}{u}) = 0, \quad \sigma = 1, 2, \ldots, m, \tag{5.39}$$

for any $u(x)$ satisfying $F^\rho(x, u, \underset{1}{u}, \underset{2}{u}, \ldots, \underset{q}{u}) = 0$, $\rho = 1, 2, \ldots, m$.

The algorithm to determine the admitted Lie–Bäcklund symmetries of a given system of DE's is essentially the same as in the case of point symmetries. A minor difference is that the equations $\mathrm{U}^{(q)} F^\sigma = 0$ involve derivatives of u to order $p + q$ and this necessitates the use of differential consequences of $F^\sigma = 0$:

$$\mathrm{D}_i F^\sigma = 0, \quad \mathrm{D}_i \mathrm{D}_j F^\sigma = 0, \quad \mathrm{D}_i \mathrm{D}_j \mathrm{D}_k F^\sigma = 0, \quad \ldots. \tag{5.40}$$

Equations (5.40) are conditions which must be satisfied by variables x, u, u_i, $u_{ij}, \ldots$. A choice of independent variables from the relations (5.38), (5.40) splits the equations $\mathrm{U}^{(q)} F^\sigma = 0$. This leads to a linear system of partial differential equations for η which forms the system of *determining equations* for the Lie–Bäcklund symmetries of (5.38).

As an example for finding Lie–Bäcklund symmetries consider Burgers' equation

$$u_{xx} - u u_x - u_t = 0. \tag{5.41a}$$

It is convenient to use the notations u_1 for u_x, u_2 for u_{xx}, u_3 for $u_{xxx}, \ldots$, and u_{1t} for u_{xt}, u_{2t} for $u_{xxt}, \ldots$. Then (5.41a) becomes

$$u_2 - u u_1 - u_t = 0 \tag{5.41b}$$

with differential consequences

$$u_3 - u u_2 - (u_1)^2 - u_{1t} = 0,$$

$$u_4 - u u_3 - 3 u_1 u_2 - u_{2t} = 0, \quad \text{etc.}$$

We choose as independent variables $x, t, u, u_1, u_2, u_3, \ldots$. Then all derivatives of u with respect to t can be expressed in terms of these variables. We find all Lie–Bäcklund symmetries admitted by (5.41b) of the form

$$x^* = x, \quad u^* = u + \epsilon \eta(x, t, u, u_1, u_2, u_3) + O(\epsilon^2). \tag{5.42}$$

A symmetry of (5.41b) of the form (5.42) satisfies

$$\mathrm{U}^{(2)}(u_2 - uu_1 - u_t) = (\mathrm{D}_x)^2\eta - u\mathrm{D}_x\eta - u_1\eta - \mathrm{D}_t\eta = 0 \tag{5.43}$$

where u satisfies (5.41b) and its differential consequences. After eliminating dependent variables $u_t, u_{1t}, u_{2t}, u_{3t}$, (5.43) becomes

$$\begin{aligned} u_4 &\left[\frac{\partial^2\eta}{\partial u_3^2}u_4 + 2\frac{\partial^2\eta}{\partial u_2\partial u_3}u_3 + 2\frac{\partial^2\eta}{\partial u_1\partial u_3}u_2 + 2\frac{\partial^2\eta}{\partial u\partial u_3}u_1 + 2\frac{\partial^2\eta}{\partial x\partial u_3}\right] \\ &+ u_3\left[\frac{\partial^2\eta}{\partial u_2^2}u_3 + 2\frac{\partial^2\eta}{\partial u_1\partial u_2}u_2 + 2\frac{\partial^2\eta}{\partial u\partial u_2}u_1 + 2\frac{\partial^2\eta}{\partial x\partial u_2}\right] \\ &+ u_2\left[\frac{\partial^2\eta}{\partial u_1^2}u_2 + 2\frac{\partial^2\eta}{\partial u\partial u_1}u_1 + 2\frac{\partial^2\eta}{\partial x\partial u_1}\right] + u_1\left[\frac{\partial^2\eta}{\partial u^2}u_1 + 2\frac{\partial^2\eta}{\partial x\partial u}\right] \\ &+ \left[\frac{\partial^2\eta}{\partial x^2} - \frac{\partial\eta}{\partial x}u - \eta u_1 - \frac{\partial\eta}{\partial t} + \frac{\partial\eta}{\partial u_1}(u_1)^2 + 3\frac{\partial\eta}{\partial u_2}u_1u_2 \right. \\ &\left. + 4\frac{\partial\eta}{\partial u_3}u_1u_3 + 3\frac{\partial\eta}{\partial u_3}(u_2)^2\right] = 0. \end{aligned} \tag{5.44}$$

Equation (5.44) must hold for all values of $x, t, u, u_1, u_2, u_3, u_4$. Since η is independent of u_4 it follows that the coefficients of $(u_4)^2$, u_4 must vanish separately in (5.44) so that

$$\frac{\partial^2\eta}{\partial u_3^2} = 0, \quad \frac{\partial^2\eta}{\partial u_2\partial u_3}u_3 + \frac{\partial^2\eta}{\partial u_1\partial u_3}u_2 + \frac{\partial^2\eta}{\partial u\partial u_3}u_1 + \frac{\partial^2\eta}{\partial x\partial u_3} = 0.$$

Consequently

$$\eta = \alpha(t)u_3 + b(x, t, u, u_1, u_2), \tag{5.45}$$

where α and b are undetermined functions of their arguments. Substituting (5.45) into (5.44) and setting to zero the coefficients of $(u_3)^2$, u_3, we find

$$b = \left[-\frac{3}{2}\alpha(t)u + \frac{1}{2}\alpha'(t)x + \beta(t)\right]u_2 + d(x, t, u, u_1), \tag{5.46}$$

where β and d are arbitrary functions of their arguments. Substituting (5.45), (5.46) into (5.44) and setting to zero the coefficients of $(u_2)^2$, u_2, we get

$$\begin{aligned} d = -\frac{3}{2}\alpha(t)(u_1)^2 &+ \left[\frac{3}{4}\alpha(t)u^2 - \left(\beta(t) + \frac{1}{2}\alpha'(t)x\right)u \right. \\ &\left. + \frac{1}{8}\alpha''(t)x^2 + \frac{1}{2}\beta'(t)x + \gamma(t)\right]u_1 + f(x, t, u), \end{aligned} \tag{5.47}$$

where γ and f are arbitrary. Substituting (5.45)–(5.47) into (5.44) and setting to zero the coefficient of $(u_1)^2$, u_1, we obtain

$$f = -\frac{1}{4}\alpha'(t)u^2 + \left[\frac{1}{4}\alpha''(t)x + \frac{1}{2}\beta'(t)\right]u + \frac{3}{4}\alpha''(t)$$

$$-\frac{1}{8}\alpha'''(t)x^2 - \frac{1}{2}\beta''(t)x - \gamma'(t). \tag{5.48}$$

Finally substituting (5.45)–(5.48) into (5.44) and setting to zero the coefficients of u^2, u, and u^0, we are led to

$$\alpha(t) = p_3t^3 + p_2t^2 + p_1t + p_0,$$
$$\beta(t) = q_2t^2 + q_1t + q_0,$$
$$\gamma(t) = 3p_3t^2 + r_1t + r_0,$$

where $p_0, p_1, p_2, p_3, q_0, q_1, q_2, r_0, r_1$ are arbitrary constants. The infinitesimal generators corresponding to r_0, q_0, q_1, q_2, r_1 are:

$$\mathrm{U}_1 = u_1\frac{\partial}{\partial u}, \quad \mathrm{U}_2 = [u_2 - uu_1]\frac{\partial}{\partial u},$$

$$\mathrm{U}_3 = [2t(u_2 - uu_1) + xu_1 + u]\frac{\partial}{\partial u},$$

$$\mathrm{U}_4 = [t^2(u_2 - uu_1) + t(xu_1 + u) - x]\frac{\partial}{\partial u}, \quad \mathrm{U}_5 = [tu_1 - 1]\frac{\partial}{\partial u}. \tag{5.49}$$

Replacing u_2 by $u_t + uu_1$ in the coefficients of (5.49) and using the equivalency decomposition (5.36), we obtain the infinitesimal generators of the Lie group of *point* transformations admitted by Burgers' equation (5.41a):

$$\mathrm{X}_1 = \frac{\partial}{\partial x}, \quad \mathrm{X}_2 = \frac{\partial}{\partial t}, \quad \mathrm{X}_3 = x\frac{\partial}{\partial x} + 2t\frac{\partial}{\partial t} - u\frac{\partial}{\partial u},$$

$$\mathrm{X}_4 = xt\frac{\partial}{\partial x} + t^2\frac{\partial}{\partial t} + (x - tu)\frac{\partial}{\partial u}, \quad \mathrm{X}_5 = t\frac{\partial}{\partial x} + \frac{\partial}{\partial u}.$$

The infinitesimal generators corresponding to p_0, p_1, p_2, and p_3 correspond to higher order symmetries of Burgers' equation which are not equivalent to point symmetries since $\alpha(t) \not\equiv 0$:

$$\mathrm{U}_6 = [4u_3 - 6uu_2 - 6(u_1)^2 + 3u^2u_1]\frac{\partial}{\partial u},$$

$$\mathrm{U}_7 = [4tu_3 + (2x - 6tu)u_2 - 6t(u_1)^2 + (3tu^2 - 2xu)u_1 - u^2]\frac{\partial}{\partial u},$$

$$\mathrm{U}_8 = [4t^2u_3 + (4tx - 6t^2u)u_2 - 6t^2(u_1)^2 + (3t^2u^2 - 4txu + x^2)u_1 - 2tu^2 + 2xu + 6]\frac{\partial}{\partial u},$$

$$\mathrm{U}_9 = [4t^3u_3 + (6t^2x - 6t^3u)u_2 - 6t^3(u_1)^2 + (3t^3u^2 - 6t^2xu + 3tx^2 + 12t^2)u_1 - 3t^2u^2 + 6txu - 3x^2 - 6t]\frac{\partial}{\partial u}.$$

In the spirit of earlier chapters one can find *invariant solutions* arising from a Lie–Bäcklund symmetry. A solution $u = \Theta(x)$ of (5.38) is invariant under the action of Lie–Bäcklund symmetry (5.37) if and only if $u = \Theta(x)$ also satisfies the *invariant surface condition*

$$\eta(x, u, \underset{1}{u}, \underset{2}{u}, \ldots, \underset{p}{u}) = 0.$$

This leads us to the following:

Definition 5.2.3-2. $u = \Theta(x)$, with components $u^\gamma = \Theta^\gamma(x)$, $\gamma = 1, 2, \ldots, m$, is an *invariant solution* of (5.38) corresponding to the Lie–Bäcklund symmetry (5.37) admitted by the system of PDE's (5.38) if and only if

(i) $u^\nu = \Theta^\nu(x)$ is an invariant surface of (5.37) for each $\nu = 1, 2, \ldots, m$;

(ii) $u = \Theta(x)$ solves (5.38).

The situation simplifies if we seek invariant solutions arising from Lie–Bäcklund symmetries for a scalar evolution equation

$$u_t + G(x, t, u, u_1, \ldots, u_q) = 0$$

with two independent variables ($x_1 = x$, $x_2 = t$; $\dfrac{\partial u}{\partial t} = u_t$, $\dfrac{\partial^j u}{\partial x^j} = u_j$, $j = 1, 2, \ldots, q$). Here the invariant surface condition becomes

$$\eta(x, t, u, u_1, u_2, \ldots, u_p) = 0,$$

which is a pth order ODE with respect to independent variable x; t appears as a parameter in this ODE. The solution of the ODE is an invariant form

$$\phi(x, t, u, c_1(t), c_2(t), \ldots, c_p(t)) = 0$$

with arbitrary functions $c_1(t)$, $c_2(t), \ldots, c_p(t)$ as integration constants. These integration constants are determined by substituting this invariant form into the evolution equation.

5.2.4 Contact Transformations; Contact Symmetries

A *contact transformation* is a transformation of the form ($m = 1$)

$$x_j^\dagger = \phi_j(x, u, \underset{1}{u}), \tag{5.50a}$$

$$u^\dagger = \psi(x, u, \underset{1}{u}), \tag{5.50b}$$

$$u_j^\dagger = \psi_j(x, u, \underset{1}{u}), \tag{5.50c}$$

$j = 1, 2, \ldots, n$, which is one-to-one in some domain D in $(x, u, \underset{1}{u})$-space and preserves the contact condition $du = \underset{1}{u}\, dx$, i.e.,

$$du^\dagger = u_j^\dagger dx_j^\dagger.$$

It is assumed that $\{\phi_i, \psi\}$ depends essentially on $\underset{1}{u}$. (Otherwise a contact transformation is a point transformation.)

Let

$$\mathrm{D}_i = \frac{\partial}{\partial x_i} + u_i \frac{\partial}{\partial u} + u_{ij} \frac{\partial}{\partial u_j}$$

and let matrix

$$A = \begin{bmatrix} \mathrm{D}_1\phi_1 & \mathrm{D}_1\phi_2 & \cdots & \mathrm{D}_1\phi_n \\ \mathrm{D}_2\phi_1 & \mathrm{D}_2\phi_2 & \cdots & \mathrm{D}_2\phi_n \\ \vdots & \vdots & & \vdots \\ \mathrm{D}_n\phi_1 & \mathrm{D}_n\phi_2 & \cdots & \mathrm{D}_n\phi_n \end{bmatrix}$$

be such that A^{-1} exists. Then (5.50a–c) defines a contact transformation if

(i) $\{\psi_i(x, u, \underset{1}{u})\}$ satisfies

$$\begin{bmatrix} \psi_1 \\ \psi_2 \\ \vdots \\ \psi_n \end{bmatrix} = A^{-1} \begin{bmatrix} \mathrm{D}_1\psi \\ \mathrm{D}_2\psi \\ \vdots \\ \mathrm{D}_n\psi \end{bmatrix}$$

and

(ii) $\dfrac{\partial}{\partial u_{jk}}(A^{-1}\mathrm{D}_i\psi) = 0$, $i, j, k = 1, 2, \ldots, n$.

The following theorem results from (i) and (ii):

Theorem 5.2.4-1. [Lie (1890), Mayer (1875)]. *Equations* (5.50a–c) *define a contact transformation if and only if* $\{\phi_i, \psi, \psi_i\}$ *satisfies*

$$\frac{\partial \psi}{\partial u_i} - \psi_j \frac{\partial \phi_j}{\partial u_i} = 0, \tag{5.51a}$$

$$\frac{\partial \psi}{\partial x_i} + u_i \frac{\partial \psi}{\partial u} = \psi_j \left(\frac{\partial \phi_j}{\partial x_i} + u_i \frac{\partial \phi_j}{\partial u} \right), \tag{5.51b}$$

$i = 1, 2, \ldots, n$.

Proof. See Exercise 5.2-7. □

From Theorem 5.2.4-1 we see that either set of equations (5.51a) or (5.51b) can be used to determine $\{\psi_i\}$. Then the other set determines the conditions that $\{\phi_i, \psi\}$ must satisfy so that (5.50a,b) leads to a contact transformation (5.50a–c). The kth extended contact transformation acting on $(x, u, \underset{1}{u}, \ldots, \underset{k}{u})$-space is found in the same manner as discussed in Section 2.3.3 for point transformations.

An example of a contact transformation is the transformation given by

$$x_1^\dagger = \phi_1 = x_1 + u_2, \tag{5.52a}$$

$$x_2^\dagger = \phi_2 = x_2 + u_1, \tag{5.52b}$$

$$u^\dagger = \psi = u + u_1 u_2. \tag{5.52c}$$

Here

$$A = \begin{bmatrix} 1 + u_{12} & u_{11} \\ u_{22} & 1 + u_{12} \end{bmatrix},$$

$$A^{-1} = [(1 + u_{12})^2 - u_{11}u_{22}]^{-1} \begin{bmatrix} 1 + u_{12} & -u_{11} \\ -u_{22} & 1 + u_{12} \end{bmatrix}.$$

Then

$$\psi_1 = u_1, \quad \psi_2 = u_2.$$

Hence (5.52a–c) defines a contact transformation.

Definition 5.2.4-1. A one-parameter (ϵ) Lie group of contact transformations is defined by

$$x_j^* = x_j + \epsilon \xi_j(x, u, \underset{1}{u}) + O(\epsilon^2), \tag{5.53a}$$

$$u^* = u + \epsilon \eta(x, u, \underset{1}{u}) + O(\epsilon^2), \tag{5.53b}$$

$$u_j^* = u_j + \epsilon \eta_j^{(1)}(x, u, \underset{1}{u}) + O(\epsilon^2), \tag{5.53c}$$

$j = 1, 2, \ldots, n$, with infinitesimal generator

$$\xi_j(x, u, \underset{1}{u})\frac{\partial}{\partial x_j} + \eta(x, u, \underset{1}{u})\frac{\partial}{\partial u} + \eta_j^{(1)}(x, u, \underset{1}{u})\frac{\partial}{\partial u_j}, \tag{5.54}$$

provided the contact condition is preserved.

From Section 2.3.4, the preservation of the contact condition leads to

$$\begin{bmatrix} \eta_1^{(1)} \\ \eta_2^{(1)} \\ \vdots \\ \eta_n^{(1)} \end{bmatrix} = \begin{bmatrix} D_1\eta \\ D_2\eta \\ \vdots \\ D_n\eta \end{bmatrix} - B \begin{bmatrix} u_1 \\ u_2 \\ \vdots \\ u_n \end{bmatrix}$$

where

$$B = \begin{bmatrix} D_1\xi_1 & D_1\xi_2 & \cdots & D_1\xi_n \\ D_2\xi_1 & D_2\xi_2 & \cdots & D_2\xi_n \\ \vdots & \vdots & & \vdots \\ D_n\xi_1 & D_n\xi_2 & \cdots & D_n\xi_n \end{bmatrix},$$

i.e.,

$$\eta_j^{(1)} = \mathrm{D}_j\eta - (\mathrm{D}_j\xi_k)u_k, \quad j = 1, 2, \ldots, n.$$

Theorem 5.2.4-2. *Equations* (5.53a–c) *define a one-parameter Lie group of contact transformations if and only if* $\{\xi_j, \eta\}$ *satisfies*

$$\frac{\partial \eta}{\partial u_i} - \frac{\partial \xi_j}{\partial u_i} u_j = 0, \quad i = 1, 2, \ldots, n. \tag{5.55}$$

Proof.

$$\eta_j^{(1)} = \frac{\partial \eta}{\partial x_j} + \frac{\partial \eta}{\partial u} u_j + \frac{\partial \eta}{\partial u_i} u_{ij} - \left[\frac{\partial \xi_k}{\partial x_j} + \frac{\partial \xi_k}{\partial u} u_j . + \frac{\partial \xi_k}{\partial u_i} u_{ij}\right] u_k, \; j = 1, 2, \ldots, n.$$

Equations (5.53a–c) define a one-parameter Lie group of contact transformations if and only if $\frac{\partial \eta_j^{(1)}}{\partial u_{ik}} = 0$, $i, j, k = 1, 2, \ldots, n$. This leads to (5.55). □

Let the *characteristic function* W be defined by

$$W = \xi_j u_j - \eta. \tag{5.56}$$

Then the following theorem holds:

Theorem 5.2.4-3. *Let* (5.54) *be the infinitesimal generator of a one-parameter Lie group of contact transformations. In terms of the characteristic function* (5.56), *the infinitesimals are given by*

$$\xi_j = \frac{\partial W}{\partial u_j},$$

$$\eta = u_i \frac{\partial W}{\partial u_i} - W,$$

$$\eta_j^{(1)} = -\frac{\partial W}{\partial x_j} - u_j \frac{\partial W}{\partial u},$$

$j = 1, 2, \ldots, n.$

Proof. See Exercise 5.2-8. □

The following theorem relates Lie–Bäcklund transformations and contact transformation groups:

Theorem 5.2.4-4. Any *Lie–Bäcklund transformation with an infinitesimal generator of the form*

$$W(x, u, \underset{1}{u}) \frac{\partial}{\partial u}$$

is equivalent to a contact transformation group with infinitesimal generator

$$\xi_j(x, u, \underset{1}{u}) \frac{\partial}{\partial x_j} + \eta(x, u, \underset{1}{u}) \frac{\partial}{\partial u} + \eta_j^{(1)}(x, u, \underset{1}{u}) \frac{\partial}{\partial u_j}, \tag{5.57}$$

where

$$\xi_j = \frac{\partial W}{\partial u_j}, \tag{5.58a}$$

$$\eta = u_i \frac{\partial W}{\partial u_i} - W, \tag{5.58b}$$

$$\eta_j^{(1)} = -\frac{\partial W}{\partial x_j} - u_j \frac{\partial W}{\partial u}, \tag{5.58c}$$

$j = 1, 2, \ldots, n.$

Proof. Let η, ξ_j satisfy (5.58a,b). Then

$$\frac{\partial \eta}{\partial u_i} = \frac{\partial W}{\partial u_i} + u_j \frac{\partial^2 W}{\partial u_i \partial u_j} - \frac{\partial W}{\partial u_i} = u_j \frac{\partial^2 W}{\partial u_i \partial u_j} = u_j \frac{\partial \xi_j}{\partial u_i}, \quad i = 1, 2, \ldots, n.$$

Hence (5.55) is satisfied. Moreover

$$\eta - \xi_j u_j = u_i \frac{\partial W}{\partial u_i} - W - u_j \frac{\partial W}{\partial u_j} = -W;$$

$$\begin{aligned}
\eta_j^{(1)} &= \frac{\partial \eta}{\partial x_j} + \frac{\partial \eta}{\partial u} u_j - \left[\frac{\partial \xi_k}{\partial x_j} + \frac{\partial \xi_k}{\partial u} u_j\right] u_k \\
&= u_i \frac{\partial^2 W}{\partial x_j \partial u_i} - \frac{\partial W}{\partial x_j} + u_i u_j \frac{\partial^2 W}{\partial u \partial u_i} - u_j \frac{\partial W}{\partial u} \\
&\quad - \left[\frac{\partial^2 W}{\partial x_j \partial u_k} + \frac{\partial^2 W}{\partial u \partial u_k} u_j\right] u_k = -\frac{\partial W}{\partial x_j} - u_j \frac{\partial W}{\partial u}.
\end{aligned}$$

Hence (5.57), (5.58a–c) defines a contact transformation group equivalent to $W(x, u, \underset{1}{u}) \frac{\partial}{\partial u}$. □

Consequently any Lie–Bäcklund generator of the form

$$\eta(x, u, \underset{1}{u}) \frac{\partial}{\partial u} \tag{5.59}$$

is (uniquely) equivalent to an infinitesimal generator of a contact transformation group with $\eta(x, u, \underset{1}{u})$ playing the role of a characteristic function. The proof of the following theorem is left to Exercise 5.2-9:

Theorem 5.2.4-5. *A generator of the form* (5.59) *is equivalent to an infinitesimal generator of a Lie group of point transformations if and only if*

$$\frac{\partial^2 \eta}{\partial u_i \partial u_j} = 0, \quad i, j = 1, 2, \ldots, n.$$

Definition 5.2.4-2. A *proper Lie–Bäcklund symmetry* of a differential equation is a Lie–Bäcklund symmetry which is not equivalent to a point symmetry or contact symmetry.

5.2.5 Finding Variational Symmetries

Consider the problem of how to find the variational symmetries admitted by a given Lagrangian system. Some simple variational symmetries may be spotted by inspection. But at first sight it would appear to be a hopeless task to find variational symmetries from condition (5.22a). Fortunately condition (5.22a) forces a variational symmetry to be a Lie–Bäcklund symmetry of the corresponding Euler–Lagrange equations and thus variational symmetries can be determined by the algorithm in Section 5.2.3:

Theorem 5.2.5-1. *If the Lie–Bäcklund transformation*

$$x^* = x, \tag{5.60a}$$

$$u^* = u + \epsilon\eta(x, u, \underset{1}{u}, \underset{2}{u}, \ldots, \underset{p}{u}) + O(\epsilon^2), \tag{5.60b}$$

is a variational symmetry for the Lagrangian $L(x, u, \underset{1}{u}, \underset{2}{u}, \ldots, \underset{k}{u})$, *i.e. condition* (5.22a) *is satisfied, then* (5.60a,b) *is a Lie–Bäcklund symmetry admitted by the corresponding Euler–Lagrange equations* $\mathrm{E}_\gamma(L) = 0$, $\gamma = 1, 2, \ldots, m$.

Proof. Here we prove the theorem for the class of extremal solutions of the Euler–Lagrange equations. Let

$$\mathrm{D}_i = \frac{\partial}{\partial x_i} + u_i^\gamma \frac{\partial}{\partial u^\gamma} + u_{ij}^\gamma \frac{\partial}{\partial u_j^\gamma} + \cdots + u_{ii_1i_2\cdots i_\ell}^\gamma \frac{\partial}{\partial u_{i_1i_2\cdots i_\ell}^\gamma} + \cdots,$$

$i = 1, 2, \ldots, n$, and

$$\mathrm{U} = \eta^\gamma \frac{\partial}{\partial u^\gamma} + (\mathrm{D}_j\eta^\gamma)\frac{\partial}{\partial u_j^\gamma} + \cdots + (\mathrm{D}_{i_1}\mathrm{D}_{i_2}\cdots\mathrm{D}_{i_\ell}\eta^\gamma)\frac{\partial}{\partial u_{i_1i_2\cdots i_\ell}^\gamma} + \cdots,$$

denote total derivative operators and the extended infinitesimal generator, respectively. Consider the operator $e^{\epsilon\mathrm{U}}$ defined by

$$e^{\epsilon\mathrm{U}} = \mathrm{I} + \epsilon\mathrm{U} + \epsilon^2\frac{\mathrm{U}^2}{2} + \cdots + \frac{\epsilon^\ell\mathrm{U}^\ell}{\ell!} + \cdots = \sum_{\ell=0}^{\infty}\frac{\epsilon^\ell\mathrm{U}^\ell}{\ell!},$$

where I is the identity operator. Then

$$u^* = u + \epsilon\eta(x, u, \underset{1}{u}, \underset{2}{u}, \ldots, \underset{p}{u}) + O(\epsilon^2) = e^{\epsilon\mathrm{U}}u;$$

$$\underset{j}{u}^* = e^{\epsilon\mathrm{U}}\underset{j}{u}, \quad j = 1, 2, \ldots .$$

Hence

$$F(x^*, u^*, \underset{1}{u}^*, \underset{2}{u}^*, \ldots, \underset{k}{u}^*) = e^{\epsilon\mathrm{U}}F(x, u, \underset{1}{u}, \underset{2}{u}, \ldots, \underset{k}{u}).$$

Let

$$L^* = L(x, u^*, \underset{1}{u}^*, \underset{2}{u}^*, \ldots, \underset{k}{u}^*), \quad L = L(x, u, \underset{1}{u}, \underset{2}{u}, \ldots, \underset{k}{u}).$$

Then

$$J[u^*] - J[u] = \int_\Omega (L^* - L) dx = \int_\Omega (e^{\epsilon \mathrm{U}} - \mathrm{I}) L\, dx$$

$$= \int_\Omega \left(\epsilon \mathrm{U} + \frac{\epsilon^2 \mathrm{U}^2}{2} + \cdots + \frac{\epsilon^\ell \mathrm{U}^\ell}{\ell!} + \cdots \right) L\, dx.$$

It is left to Exercise 5.2-11 to show that operators U and D_i commute, i.e.

$$[\mathrm{U}, \mathrm{D}_i] = 0, \quad i = 1, 2, \ldots, n. \tag{5.61}$$

Since (5.60a,b) defines a variational symmetry it follows that for any $u(x)$

$$\mathrm{U}L = \mathrm{D}_i A^i$$

for some vector function $A(x, u, \underset{1}{u}, \underset{2}{u}, \ldots, \underset{r}{u})$. From (5.61) it follows that

$$\mathrm{U}^\ell L = \mathrm{D}_i(\mathrm{U}^{\ell-1} A^i), \quad \ell = 1, 2, \ldots.$$

Thus

$$L^* - L = \mathrm{D}_i C^i$$

where

$$C^i = \epsilon A^i + \frac{\epsilon^2}{2} \mathrm{U} A^i + \cdots + \frac{\epsilon^\ell \mathrm{U}^{\ell-1} A^i}{\ell!} + \cdots.$$

Since L^* and L differ by a divergence, from Theorem 5.2.1-4 we see that L and L^* have the same set of Euler–Lagrange equations, and hence a variational symmetry for L transforms an extremum of $J[u]$ into another extremum of $J[u]$. Consequently (5.60a,b) leaves the Euler–Lagrange equations for L invariant for any extremal solution of the Euler–Lagrange equations. [The theorem actually holds for all solutions of the Euler–Lagrange equations [cf. Olver (1986, p. 326)].] □

The converse of Theorem 5.2.5-1 is not true: Euler–Lagrange equations can admit Lie–Bäcklund symmetries which are not variational symmetries. An example when the converse is not true usually arises from a scaling symmetry admitted by the Euler–Lagrange equations which scales $J[u]$ such that $J[u^*] = cJ[u]$, $c \neq 1$. From experience, except for some scaling symmetries, a Lie–Bäcklund symmetry admitted by the Euler–Lagrange equations is invariably a variational symmetry, i.e. it yields a vector A for some conserved flux $A^i - W^i$.

In summary, if one finds all Lie–Bäcklund symmetries admitted by the Euler–Lagrange equations then in principle one can find all variational symmetries of the action integral $J[u]$ by checking which of these Lie–Bäcklund symmetries satisfy condition (5.22a).

5.2.6 NOETHER'S FORMULATION

As we have seen, to find conservation laws through a variational formulation it is unnecessary to consider symmetries which involve transformations of x. In particular the inclusion of such symmetry transformations yields no additional conservation laws.

Noether's original formulation involved symmetries which included transformations of x. For point symmetries her formulation often has the advantage of enabling one to find variational symmetries by inspection. Noether considered transformations of the form

$$x^* = x + \epsilon\xi(x, u, \underset{1}{u}, \underset{2}{u}, \ldots, \underset{p}{u}) + O(\epsilon^2), \tag{5.62a}$$

$$u^* = u + \epsilon\eta(x, u, \underset{1}{u}, \underset{2}{u}, \ldots, \underset{p}{u}) + O(\epsilon^2). \tag{5.62b}$$

She required a transformation of the form (5.62a,b) to leave the action integral $J[u]$ invariant for an arbitrary domain Ω:

$$\int_\Omega L(x, u(x), \underset{1}{u}(x), \underset{2}{u}(x), \ldots, \underset{k}{u}(x))dx$$

$$= \int_{\Omega^*} L(x^*, u^*(x^*), \underset{1}{u}^*(x^*), \underset{2}{u}^*(x^*), \ldots, \underset{k}{u}^*(x^*))dx^* \tag{5.63}$$

where $u^*(x^*)$ is the image of $u(x)$, obtained in the manner discussed in Section 5.2.3 through

$$x^* = x + \epsilon\xi(x, u(x), \underset{1}{u}(x), \underset{2}{u}(x), \ldots, \underset{p}{u}(x)) + O(\epsilon^2), \tag{5.64a}$$

$$u^* = u(x) + \epsilon\eta(x, u(x), \underset{1}{u}(x), \underset{2}{u}(x), \ldots, \underset{p}{u}(x)) + O(\epsilon^2), \tag{5.64b}$$

and domain Ω^* is the image of Ω obtained from (5.64a). For simplicity (5.63) is written as

$$\int_\Omega L[x, u(x)]dx = \int_{\Omega^*} L[x^*, u^*(x^*)]dx^*. \tag{5.65}$$

Changing the variables from x^* to x on the right-hand side of (5.65), one gets

$$\int_\Omega L[x, u(x)]dx = \int_\Omega L[x^*, u^*(x^*)]\ (dx^*/dx)\ dx \tag{5.66}$$

where (dx^*/dx) is the Jacobian for (5.64a):

$$(dx^*/dx) = \det\left|D_i x_j^*\right| = \begin{vmatrix} D_1x_1^* & D_1x_2^* & \cdots & D_1x_n^* \\ D_2x_1^* & D_2x_2^* & \cdots & D_2x_n^* \\ \vdots & \vdots & & \vdots \\ D_nx_1^* & D_nx_2^* & \cdots & D_nx_n^* \end{vmatrix}, \tag{5.67}$$

with D_i being the total derivative operator with respect to x_i. In (5.66) both x^* and $u^*(x^*)$ are functions of x defined by (5.64a,b). Since the arbitrary domain of integration is the same on both sides of (5.66), it follows that

$$L[x,u(x)] = L[x^*,u^*(x^*)]\,(dx^*/dx). \tag{5.68}$$

For a change of variables (5.64a), the Jacobian becomes $(dx^*/dx) = 1 + \epsilon \mathrm{D}_i\xi_i + O(\epsilon^2)$. Expanding (5.68) in ϵ and taking $O(\epsilon)$ terms, one gets

$$\mathrm{U}^{(k)}L + L\mathrm{D}_i\xi_i = 0 \tag{5.69}$$

where

$$\mathrm{U}^{(k)} = \xi_i\frac{\partial}{\partial x_i} + \eta^\gamma\frac{\partial}{\partial u^\gamma} + \eta_i^{(1)\gamma}\frac{\partial}{\partial u_i^\gamma} + \cdots + \eta_{i_1i_2\cdots i_k}^{(k)\gamma}\frac{\partial}{\partial u_{i_1i_2\cdots i_k}^\gamma}, \mathrm{D}_{i_\ell}\eta^{(\ell-1)\gamma}_{i_1i_2\cdots i_{\ell-1}},$$

with

$$\eta_i^{(1)\gamma} = \mathrm{D}_i\eta^\gamma - (\mathrm{D}_i\xi_j)u_j^\gamma,$$

$$\eta_{i_1i_2\cdots i_\ell}^{(\ell)\gamma} = \mathrm{D}_{i_\ell}\eta_{i_1i_2\cdots i_{\ell-1}}^{(\ell-1)\gamma} - (\mathrm{D}_{i_\ell}\xi_j)\,u_{i_1i_2\cdots i_{\ell-1}j}^\gamma,$$

$\ell = 1,2,\ldots,k$. Equation (5.69) is Noether's invariance condition for the action integral. One can show

$$\mathrm{U}^{(k)}L + L\mathrm{D}_i\xi_i = \hat{\mathrm{U}}^{(k)}L + \mathrm{D}_i(\xi_iL), \tag{5.70}$$

where $\hat{\mathrm{U}}^{(k)}$ is the kth extended "vertical" infinitesimal generator of (5.33) equivalent to $\mathrm{U}^{(k)}$. Since $\hat{\mathrm{U}}^{(k)}$ involves no transformation in x, we can use (5.22b) with $\hat{\eta} = \eta - u_j\xi_j$ replacing η:

$$\hat{\mathrm{U}}^{(k)}L = \mathrm{E}_\gamma(L)(\eta^\gamma - u_j^\gamma\xi_j) + \mathrm{D}_iW^i[u,\eta - u_j\xi_j]. \tag{5.71}$$

If $\mathrm{E}_\gamma(L) = 0$, $\gamma = 1,2,\ldots,m$, then (5.69) with (5.70), (5.71) leads to the conservation law:

$$\mathrm{D}_i\left(\xi_iL + W^i[u,\eta - u_j\xi_j]\right) = 0. \tag{5.72}$$

This is the original form of Noether's result and one can write down the conservation law once ξ and η are found which satisfy the invariance condition (5.69). Actually, as observed by Bessel-Hagen (1921), we see that for a conservation law to arise the right hand side of (5.69) need not vanish, but can be another divergence form, say D_iB^i. Then the following theorem holds:

Theorem 5.2.6-1. *Let* $\mathrm{U}^{(k)}$ *be the kth extension of the generator* $\mathrm{U} = \xi_i\dfrac{\partial}{\partial x_i} + \eta^\gamma\dfrac{\partial}{\partial u^\gamma}$ *of* (5.62a,b). *Suppose there exists a vector* $B(x,u,\underset{1}{u},\underset{2}{u},\ldots,\underset{q}{u})$ *satisfying the identity*

$$\mathrm{U}^{(k)}L + L\mathrm{D}_i\xi_i = \mathrm{D}_iB^i \tag{5.73}$$

for arbitrary $u(x)$. Then for any solution $u(x)$ of the Euler–Lagrange equations $\mathrm{E}_\gamma(L) = 0$, $\gamma = 1, 2, \ldots, m$, we have the conservation law

$$\mathrm{D}_i\left(\xi_i L + W^i[u, \eta - u_j\xi_j] - B^i\right) = 0. \tag{5.74}$$

To apply this theorem one needs to find transformations satisfying (5.65) or, more generally, (5.73). Though sometimes one can find such transformations by inspecting the action integral, in order to determine all transformations satisfying (5.73) one must first find all transformations (5.62a,b) admitted by the Euler–Lagrange equations for $p = 1, 2, \ldots$, and then check which infinitesimal generators satisfy criterion (5.73).

We are now able to show that Theorems 5.2.2-1 and 5.2.6-1 are equivalent, i.e. any conservation law obtained by Theorem 5.2.2-1 (Boyer's formulation) is also obtained by Theorem 5.2.6-1 (Noether's formulation) and vice-versa:

Theorem 5.2.6-2. *Let*

$$\hat{\eta}^\gamma = \eta^\gamma - u_j^\gamma \xi_j, \quad \gamma = 1, 2, \ldots, m.$$

Then

$$\hat{\mathrm{U}} = \hat{\eta}^\gamma \frac{\partial}{\partial u^\gamma}$$

satisfies condition (5.22a) *for some vector function A if and only if*

$$\mathrm{U} = \xi_j \frac{\partial}{\partial x_j} + \eta^\gamma \frac{\partial}{\partial u^\gamma}$$

satisfies condition (5.73) *for some vector function B.*

Proof. From (5.70) we have

$$\hat{\mathrm{U}}^{(k)} L = \mathrm{D}_i A^i$$

if and only if $B = A + \xi L$ satisfies

$$\mathrm{U}^{(k)} L + L\mathrm{D}_i\xi_i = \mathrm{D}_i B^i. \quad \square$$

As an example for comparing the two formulations for obtaining conservation laws consider the PDE ($x_1 = x$, $x_2 = t$)

$$\mathrm{E}(L) = (u_x)^2 u_{xx} - u_{tt} = 0 \tag{5.75}$$

which is the Euler–Lagrange equation for the Lagrangian

$$L = -\frac{1}{12}(u_x)^4 + \frac{1}{2}(u_t)^2. \tag{5.76}$$

Let

$$\hat{\mathrm{U}}^{(1)} = \hat{\eta}\frac{\partial}{\partial u} + \mathrm{D}_x\hat{\eta}\frac{\partial}{\partial u_x} + \mathrm{D}_t\hat{\eta}\frac{\partial}{\partial u_t}.$$

First we compute $W[u,\hat{\eta}]$ from $\hat{\mathrm{U}}^{(1)}$: Using integration by parts, we get

$$\begin{aligned}\hat{\mathrm{U}}^{(1)}L &= -\frac{1}{3}(u_x)^3(\mathrm{D}_x\hat{\eta}) + u_t(\mathrm{D}_t\hat{\eta})\\ &= \left[\mathrm{D}_x\left(-\frac{1}{3}(u_x)^3\hat{\eta}\right) + (u_x)^2u_{xx}\hat{\eta}\right] + [\mathrm{D}_t(u_t\hat{\eta}) - u_{tt}\hat{\eta}]\\ &= \mathrm{E}(L)\hat{\eta} + \mathrm{D}_xW^1 + \mathrm{D}_tW^2,\end{aligned}$$

so that

$$W^1[u,\hat{\eta}] = -\frac{1}{3}(u_x)^3\hat{\eta}, \quad W^2[u,\hat{\eta}] = u_t\hat{\eta}.$$

PDE (5.75) admits the following five once-extended infinitesimal generators of point symmetries:

$$\mathrm{X}_1^{(1)} = \frac{\partial}{\partial x}, \quad \mathrm{X}_2^{(1)} = \frac{\partial}{\partial t}, \quad \mathrm{X}_3^{(1)} = t\frac{\partial}{\partial u} + \frac{\partial}{\partial u_t},$$

$$\mathrm{X}_4^{(1)} = x\frac{\partial}{\partial x} + 2t\frac{\partial}{\partial t} - u_x\frac{\partial}{\partial u_x} - 2u_t\frac{\partial}{\partial u_t},$$

$$\mathrm{X}_5^{(1)} = -t\frac{\partial}{\partial t} + u\frac{\partial}{\partial u} + u_x\frac{\partial}{\partial u_x} + 2u_t\frac{\partial}{\partial u_t}.$$

Generators X_1 and X_2 correspond to invariance of (5.75) under translations in x and t, respectively; X_4 and X_5 correspond to invariance of (5.75) under scalings $x^* = \alpha x$, $t^* = \alpha^2 t$, $u^* = u$, and $x^* = x$, $t^* = \beta^{-1}t$, $u^* = \beta u$, respectively. Clearly the action integral $\int_\Omega L\,dx\,dt$ admits X_1 and X_2. If $x^* = \alpha x$, $t^* = \alpha^2\beta^{-1}t$, $u^* = \beta u$, then

$$\int_{\Omega^*} L^*\,dx^*dt^* = \alpha^{-1}\beta^3\int_\Omega L\,dx\,dt = \int_\Omega L\,dx\,dt$$

if and only if $\alpha = \beta^3$. Hence by inspection we see that the action integral corresponding to (5.76) admits $\mathrm{X}_5 + 3\mathrm{X}_4$ but not X_4 or X_5. Moreover it immediately follows that $B = 0$ in (5.73) for each of the generators X_1, X_2, and $\mathrm{X}_5 + 3\mathrm{X}_4$. For X_3, (5.73) becomes ($\xi_1 = \xi_2 = 0$)

$$\mathrm{X}_3^{(1)}L = u_t,$$

so that $B^1 = 0$, $B^2 = u$.

Using Noether's formulation, the resulting conservation laws $\mathrm{D}_xf^1 + \mathrm{D}_tf^2 = 0$ are:

(1) *From* X_1: $\hat{\eta} = \eta - u_j\xi_j = -u_x$;

$$\begin{aligned}f^1 &= \xi_1L + W^1[u,\hat{\eta}] = \frac{1}{4}(u_x)^4 + \frac{1}{2}(u_t)^2,\\ f^2 &= \xi_2L + W^2[u,\hat{\eta}] = -u_tu_x.\end{aligned}$$

(2) *From* X_2: $\hat{\eta} = -u_t$;

$$f^1 = \frac{1}{3}(u_x)^3 u_t, \quad f^2 = -\frac{1}{12}(u_x)^4 - \frac{1}{2}(u_t)^2.$$

(3) *From* X_3: $\hat{\eta} = t$;

$$f^1 = \xi_1 L + W^1[u,\hat{\eta}] - B^1 = -\frac{1}{3}t(u_x)^3,$$
$$f^2 = \xi_2 L + W^2[u,\hat{\eta}] - B^2 = tu_t - u.$$

(4) *From* $X_5 + 3X_4$: $\hat{\eta} = u - 3xu_x - 5tu_t$;

$$f^1 = \frac{3}{4}x(u_x)^4 + \frac{3}{2}x(u_t)^2 - \frac{1}{3}u(u_x)^3 + \frac{5}{3}t(u_x)^3 u_t,$$
$$f^2 = -\frac{5}{12}t(u_x)^4 - \frac{5}{2}t(u_t)^2 + uu_t - 3xu_x u_t.$$

Now consider Boyer's formulation: For X_1 and $X_5 + 3X_4$ the equivalent once-extended "vertical" infinitesimal generators are

$$\hat{U}_1^{(1)} = -u_x\frac{\partial}{\partial u} - u_{xx}\frac{\partial}{\partial u_x} - u_{xt}\frac{\partial}{\partial u_t},$$

$$\hat{V}^{(1)} = \hat{U}_5^{(1)} + 3\hat{U}_4^{(1)} = [u - 3xu_x - 5tu_t]\frac{\partial}{\partial u} - [2u_x + 3xu_{xx} + 5tu_{xt}]\frac{\partial}{\partial u_x}$$
$$- [4u_t + 3xu_{xt} + 5tu_{tt}]\frac{\partial}{\partial u_t}.$$

Then

$$\hat{U}_1^{(1)}L = \frac{1}{3}(u_x)^3 u_{xx} - u_t u_{xt} = D_x A^1 + D_t A^2$$

with

$$A^1 = \frac{1}{12}(u_x)^4 - \frac{1}{2}(u_t)^2, \quad A^2 = 0,$$

and

$$\hat{V}^{(1)}L = \frac{2}{3}(u_x)^4 + x(u_x)^3 u_{xx} + \frac{5}{3}t(u_x)^3 u_{xt}$$
$$- 4(u_t)^2 - 3xu_t u_{xt} - 5tu_t u_{tt} = D_x A^1 + D_t A^2$$

with

$$A^1 = x\left[\frac{1}{4}(u_x)^4 - \frac{3}{2}(u_t)^2\right], \quad A^2 = t\left[\frac{5}{12}(u_x)^4 - \frac{5}{2}(u_t)^2\right].$$

Using Boyer's formulation (Theorem 5.2.2-1) we obtain the same conservation laws as we get from Noether's formulation, i.e. $D_x f^1 + D_t f^2 = 0$ with

$$f^1 = W^1[u,\hat{\eta}] - A^1, \quad f^2 = W^2[u,\hat{\eta}] - A^2.$$

Note that for corresponding U_1, V we have $A^1 = -\xi_1 L$, $A^2 = -\xi_2 L$, as follows from the proof of Theorem 5.2.6-2 and the fact that $B^1 = B^2 = 0$.

In summary the difficulty in applying each formulation is that for a given symmetry admitted by the Euler–Lagrange equations there is no explicit formula for vector B (Noether's formulation) or vector A (Boyer's formulation). From the proof of Theorem 5.2.6-2 we see that for conservation laws corresponding to translational invariances in x_i (energy, momentum), rotational invariance (angular momentum), and scaling invariance, we have $B = 0$ for Noether's formulation but $A \neq 0$ for Boyer's formulation. Thus such simple conservation laws arise directly only through Noether's formulation.

5.2.7 Examples of Higher Order Conservation Laws

The study of nonlinear time evolution equations has become a very active subject during the last two decades. One can trace the origin of much of the current research activities to the discovery of an infinite sequence of higher order conservation laws for the Korteweg–de Vries (KdV) equation by Miura, Gardner, and Kruskal (1968). The conservation laws were constructed using the Miura transformation [Miura (1968)] which also led to the important discovery of the inverse scattering method for initial value problems of nonlinear evolution equations. The relationship between conservation laws and symmetries of such nonlinear evolution equations as the KdV equation and the sine-Gordon equation was studied by Steudel (1975a,b) and Kumei (1975, 1977). A very extensive study of conservation laws and symmetries is found in Olver (1986) which includes both Lagrangian and Hamiltonian formulations.. We illustrate here the use of Noether's theorem to obtain higher order conservation laws for the KdV equation

$$v_{xxx} + vv_x + v_t = 0. \tag{5.77a}$$

For convenience we use the notations v_1 for v_x, v_2 for v_{xx}, v_{1t} for v_{xt}, v_{2t} for v_{xxt}, etc., so that (5.77a) becomes

$$v_3 + vv_1 + v_t = 0. \tag{5.77b}$$

The KdV equation (5.77b) has no Lagrangian L. However if we make the substitution $v = u_1$ in (5.77b) then the transformed KdV equation

$$u_4 + u_1 u_2 + u_{1t} = 0 \tag{5.78}$$

is the Euler–Lagrange equation for

$$L = \frac{1}{2}(u_2)^2 - \frac{1}{6}(u_1)^3 - \frac{1}{2}u_1 u_t. \tag{5.79}$$

The KdV equation (5.77b) admits an infinite sequence of Lie–Bäcklund symmetries [Kumei (1977)]. The first few infinitesimal generators are:

$$\mathrm{V}_1 = [tv_1 - 1]\frac{\partial}{\partial v}, \quad \mathrm{V}_2 = [xv_1 + 3tv_t + 2v]\frac{\partial}{\partial v}, \quad \mathrm{V}_3 = v_1\frac{\partial}{\partial v},$$

$$\mathrm{V}_4 = v_t\frac{\partial}{\partial v}, \quad \mathrm{V}_5 = \left[\frac{3}{5}v_5 + vv_3 + 2v_1v_2 + \frac{1}{2}v^2v_1\right]\frac{\partial}{\partial v}. \tag{5.80}$$

Since $u = \int v dx$, the corresponding infinitesimal generators for the transformed KdV equation (5.78) are obtained by integrating with respect to x the coefficients of the infinitesimal generators (5.80) and then replacing v by u_1:

$$\mathrm{U}_1 = [tu_1 - x]\frac{\partial}{\partial u}, \quad \mathrm{U}_2 = [xu_1 + 3tu_t + u]\frac{\partial}{\partial u}, \quad \mathrm{U}_3 = u_1\frac{\partial}{\partial u},$$

$$\mathrm{U}_4 = u_t\frac{\partial}{\partial u}, \quad \mathrm{U}_5 = \left[\frac{3}{5}u_5 + u_1u_3 + \frac{1}{2}(u_2)^2 + \frac{1}{6}(u_1)^3\right]\frac{\partial}{\partial u}. \tag{5.81}$$

The first four infinitesimal generators of (5.81) correspond to invariance of (5.78) under Galilean transformations, scalings, and space and time translations. We use Boyer's formulation of Noether's theorem to obtain corresponding conservation laws. First of all for (5.79):

$$W^1[u,\eta] = u_2(\mathrm{D}_1\eta) - \left[u_3 + \frac{1}{2}(u_1)^2 + \frac{1}{2}u_t\right]\eta,$$

$$W^2[u,\eta] = -\frac{1}{2}u_1\eta.$$

The conservation law corresponding to U_1 was obtained in Section 5.2.2. One can easily show that U_2 is not a variational symmetry. The other conservation laws $\mathrm{D}_x f^1 + \mathrm{D}_t f^2 = 0$ where $f^1 = W^1 - A^1$, $f^2 = W^2 - A^2$, are obtained as follows:

(1) *From* U_3: $\mathrm{U}_3 L = \mathrm{D}_x L$. Hence $A^1 = L$, $A^2 = 0$;

$$f^1 = \frac{1}{2}(u_2)^2 - u_1u_3 - \frac{1}{3}(u_1)^3, \quad f^2 = -\frac{1}{2}(u_1)^2.$$

(2) *From* U_4: $\mathrm{U}_4 L = \mathrm{D}_t L$. Hence $A^1 = 0$, $A^2 = L$;

$$f^1 = u_2u_{1t} - u_3u_t - \frac{1}{2}(u_1)^2u_t - \frac{1}{2}(u_t)^2, \quad f^2 = -\frac{1}{2}(u_2)^2 + \frac{1}{6}(u_1)^3.$$

(3) *From* U_5:

$$\mathrm{U}_5 L = \left[\frac{3}{5}u_7 + u_1u_5 + 3u_2u_4 + 2(u_3)^2 + u_1(u_2)^2 + \frac{1}{2}(u_1)^2u_3\right]u_2$$

$$- \frac{1}{2}[(u_1)^2 + u_t] \left[\frac{3}{5}u_6 + u_1 u_4 + 2u_2 u_3 + \frac{1}{2}(u_1)^2 u_2\right]$$
$$- \frac{1}{2}u_1 \left[\frac{3}{5}u_{5t} + u_1 u_{3t} + u_{1t} u_3 + u_2 u_{2t} + \frac{1}{2}(u_1)^2 u_{1t}\right]. \qquad (5.82)$$

Grouping together terms involving the same powers in u and the same number of differentiations of u in (5.82), and integrating by parts, we find that $\mathrm{U}_5 L = \mathrm{D}_x A^1 + \mathrm{D}_t A^2$ where

$$\begin{aligned} A^1 = &\frac{1}{5}\Big[3u_2u_6 - 3u_3u_5 + \frac{3}{2}(u_4)^2 - \frac{5}{2}(u_1)^3 u_3 + \frac{5}{2}(u_1)^2(u_2)^2 \\ &- \frac{3}{2}(u_1)^2 u_5 + 8u_1u_2u_4 - 4u_1(u_3)^2 + 7(u_2)^2 u_3\Big] \\ &- \frac{1}{2}\Big[(u_1)^2 u_{2t} + u_1 u_t u_3 + \frac{1}{2}u_t(u_2)^2\Big] \\ &- \frac{3}{10}\left[u_t u_5 - u_{1t}u_4 + u_3 u_{2t} - u_2 u_{3t} + u_1 u_{4t}\right] \\ &- \frac{1}{12}(u_1)^3 u_t - \frac{1}{20}(u_1)^5, \\ A^2 = &\frac{1}{4}u_1(u_2)^2 - \frac{1}{24}(u_1)^4. \end{aligned}$$

[Note that the form of A is never unique since $\operatorname{div} A = \operatorname{div}(A + \operatorname{curl} E)$ for any vector function E.] Correspondingly

$$\begin{aligned} f^1 = &-\frac{3}{5}u_1u_2u_4 + \frac{1}{10}(u_2)^2u_3 - \frac{1}{4}(u_1)^2(u_2)^2 - \frac{1}{5}u_1(u_3)^2 - \frac{1}{6}(u_1)^3u_3 \\ &- \frac{1}{30}(u_1)^5 - \frac{3}{10}(u_4)^2 + \frac{1}{2}(u_1)^2u_{2t} + \frac{3}{10}[-u_{1t}u_4 + u_3u_{2t} \\ &- u_2u_{3t} + u_1u_{4t}], \end{aligned}$$

$$f^2 = -\left[\frac{3}{10}u_1u_5 + \frac{1}{2}(u_1)^2u_3 + \frac{1}{2}u_1(u_2)^2 + \frac{1}{24}(u_1)^4\right].$$

The substitution $v = u_1$ yields the corresponding conservation laws for the KdV equation (5.77b).

Exercises 5.2

1. Find Lagrangians for the following PDE's:

   (a) $u_{xt} + u_x u_{xx} - u_{yy} = 0$;

   (b) the system $u_{xx} + iu_t + kvu^2 = 0$, $v_{xx} - iv_t + kuv^2 = 0$ (nonlinear Schroedinger equations);

   (c) $u_{xx} - u_{tt} + (u^2)_{xx} + u_{xxxx} = 0$ (Boussinesq equation). [Let $u = v_x$. Integrate and write the equation as a fourth order PDE for v.]

2. Prove Theorem 5.2.1-2.

3. Use integration by parts to determine $W[u,\eta]$ of (5.22b) for the Lagrangians of Exercise 5.2-1.

4. (a) Find a Lagrangian which has the Euler–Lagrange equation

$$\frac{d^2u}{dt^2} + u + ku^3 = 0.$$

   (b) Obtain the conservation law associated with translational invariance in t by Noether's formulation (Theorem 5.2.6-1) and Boyer's formulation (Theorem 5.2.2-1).

5. Show that the indicated generator U generates a variational symmetry for the given Lagrangian L and obtain the corresponding conservation law:

   (a) $L = \frac{1}{2}u_x u_t - \cos u$, $\mathrm{U} = \left[u_{xxx} + \frac{1}{2}(u_x)^3\right]\frac{\partial}{\partial u}$;

   (b) $L = 2u_x v_x + i(uv_t - u_t v) - u^2v^2$,

$$\mathrm{U} = (u_{xxx} + 3uvu_x)\frac{\partial}{\partial u} + (v_{xxx} + 3vuv_x)\frac{\partial}{\partial v}.$$

6. The Kadomtsev–Petviashvili equation $u_{xt} + u_{yy} + (u^2)_{xx} + u_{xxxx} = 0$ admits an infinite-parameter Lie group of point transformations which involves three arbitrary functions [Tajiri, Nishitani, and Kawamoto (1982), Schwarz (1982)]. In particular it admits the infinitesimal generator

$$\mathrm{X} = Q'(t)y\frac{\partial}{\partial x} - 2Q(t)\frac{\partial}{\partial y} + \frac{1}{2}Q''(t)y\frac{\partial}{\partial u}$$

   where $Q(t)$ is an arbitrary function of t. Construct the corresponding conservation law for symmetry X. [Let $u = v_x$ and find the infinitesimal generator corresponding to X for the PDE with dependent variable v.]

7. Prove Theorem 5.2.4-1.

8. Prove Theorem 5.2.4-3.

9. Prove Theorem 5.2.4-5.

10. Find all infinitesimal generators of *contact* symmetries admitted by the Liouville equation $u_{xt} = e^u$ and construct the associated conservation laws.

11. Verify the commutation relations (5.61).

12. The Boussinesq equation of Exercise 5.2-1 can be written as a system

$$u_t = v_{xx}, \quad v_t = u_{xx} + u + u^2.$$

(a) Find a Lagrangian for this system.

(b) By inspection find three infinitesimal generators of point symmetries admitted by the system. Construct corresponding conservation laws.

(c) Show that

$$\mathrm{U} = \left[v_{xxxx} + \left(u - \frac{1}{2}\right) v_{xx} + u_x v_x\right] \frac{\partial}{\partial u}$$

$$+ \left[u_{xxxx} + \left(3u + \frac{1}{2}\right) u_{xx} + \frac{1}{2}(v_x)^2 + \frac{3}{2}(u_x)^2 + \frac{2}{3}u^3 - \frac{1}{2}u\right] \frac{\partial}{\partial v}$$

generates a variational symmetry and construct the associated conservation law.

5.3 Recursion Operators for Lie–Bäcklund Symmetries

In Section 5.2 we saw that the algorithm for finding Lie–Bäcklund symmetries admitted by differential equations is essentially the same as that for finding point symmetries. A difficulty in applying this algorithm is to determine a priori which derivatives could appear in the infinitesimal η:

$$\eta = \eta(x, u, \underset{1}{u}, \underset{2}{u}), \quad \eta = \eta(x, u, \underset{1}{u}, \underset{2}{u}, \underset{3}{u}), \ldots ?$$

More importantly, when a differential equation admits a Lie–Bäcklund symmetry, it usually admits an infinite sequence of such symmetries where successive elements of the sequence depend on higher derivatives of u. Olver (1977) introduced the concept of *recursion operators* to generate such infinite sequences of Lie–Bäcklund symmetries. We first examine the case of linear differential equations where recursion operators and corresponding infinite sequences of Lie–Bäcklund symmetries arise naturally for any admitted nontrivial point symmetry. We then consider the case of nonlinear differential equations.

5.3.1 Recursion Operators for Linear Differential Equations

In quantum mechanics a constant of the motion is associated with an infinitesimal generator admitted by Schroedinger's equation. In the standard formulation used in quantum mechanics the differential operator describing such an infinitesimal generator does not involve $\frac{\partial}{\partial u}$. We show that such an infinitesimal generator is equivalent to a Lie–Bäcklund symmetry $\mathrm{U} = \eta(x, u, \underset{1}{u}, \ldots, \underset{k}{u})\frac{\partial}{\partial u}$. In addition we show that for any linear differential equation an admitted Lie–Bäcklund symmetry directly leads to a recursion operator generating an infinite sequence of Lie–Bäcklund symmetries admitted by the equation.

Consider a system of linear differential equations

$$\mathrm{L}u = 0 \tag{5.83}$$

where L is a linear differential operator. We assume that (5.83) admits a nontrivial Lie–Bäcklund symmetry

$$\mathrm{U} = \eta^\gamma(x, u, \underset{1}{u}, \ldots, \underset{k}{u})\frac{\partial}{\partial u^\gamma}. \tag{5.84}$$

[Trivial Lie–Bäcklund symmetries admitted by linear differential equations correspond to $\mathrm{U} = u^\gamma \frac{\partial}{\partial u^\gamma}$ and $\mathrm{U} = f^\gamma(x)\frac{\partial}{\partial u^\gamma}$ where $u = f(x)$ solves (5.83).] Since (5.83) is linear it follows that (5.83) admits (5.84) if and only if η satisfies the determining equations

$$\mathrm{L}\eta = 0 \quad \text{when} \quad \mathrm{L}u = 0. \tag{5.85}$$

For linear equations, in most cases (but not always) η is linear in u, i.e.

$$\eta = \mathrm{R}u \tag{5.86}$$

for some linear differential operator R. [This must be the case for *point symmetries* admitted by a linear scalar PDE of second or higher order and by a linear scalar ODE of third or higher order [cf. Bluman (1989)].] For scalar u, (5.86) takes the form

$$\eta = r(x)u + r_i(x)u_i + r_{ij}(x)u_{ij} + \cdots + r_{i_1 i_2 \cdots i_k}(x)u_{i_1 i_2 \cdots i_k}, \tag{5.87a}$$

with

$$\mathrm{R} = r(x) + r_i(x)\mathrm{D}_i + r_{ij}(x)\mathrm{D}_i\mathrm{D}_j + \cdots + r_{i_1 i_2 \cdots i_k}(x)\mathrm{D}_{i_1}\mathrm{D}_{i_2}\cdots\mathrm{D}_{i_k}, \tag{5.87b}$$

for some functions $r(x)$, $r_i(x)$, $r_{ij}(x), \ldots, r_{i_1 i_2 \cdots i_k}(x)$. For a system of differential equations, (5.86) takes the form

$$\eta^\alpha = r^{\alpha\beta}(x)u^\beta + r_i^{\alpha\beta}(x)u_i^\beta + r_{ij}^{\alpha\beta}(x)u_{ij}^\beta + \cdots$$
$$+ r_{i_1 i_2 \cdots i_k}^{\alpha\beta}(x)u_{i_1 i_2 \cdots i_k}^\beta, \tag{5.88a}$$

where matrix differential operator R has matrix elements

$$\mathrm{R}^{\alpha\beta} = r^{\alpha\beta}(x) + r_i^{\alpha\beta}(x)\mathrm{D}_i + r_{ij}^{\alpha\beta}(x)\mathrm{D}_i\mathrm{D}_j + \cdots$$

$$+ r^{\alpha\beta}_{i_1 i_2 \cdots i_k}(x) \mathrm{D}_{i_1} \mathrm{D}_{i_2} \cdots \mathrm{D}_{i_k}, \tag{5.88b}$$

for some functions $r^{\alpha\beta}(x)$, $r_i^{\alpha\beta}(x)$, $r_{ij}^{\alpha\beta}(x), \ldots, r^{\alpha\beta}_{i_1 i_2 \cdots i_k}(x)$. Note that the operator R has no dependence on u or derivatives of u. When $\eta = \mathrm{R}u$ the determining equations (5.85) become

$$\mathrm{LR}u = 0 \quad \text{when} \quad \mathrm{L}u = 0, \tag{5.89}$$

i.e. if u solves $\mathrm{L}u = 0$ then $v = \mathrm{R}u$ solves $\mathrm{L}v = 0$. It immediately follows that if (5.83) admits (5.86) then it also admits

$$\eta = \mathrm{R}^s u, \quad s = 1, 2, \ldots . \tag{5.90}$$

More generally we have proved the following theorem:

Theorem 5.3.1-1. *If both* $\eta = \tilde{\eta}(x, u, \underset{1}{u}, \ldots, \underset{p}{u})$ *and* $\eta = \mathrm{R}u$, *where* R *is given by* (5.87b) *or* (5.88b), *solve the determining equations* (5.85), *then so do*

$$\eta = \mathrm{R}^s \tilde{\eta}, \quad s = 1, 2, \ldots . \tag{5.91}$$

Hence (5.83) *admits* (5.91).

We call R given by (5.87b) or (5.88b) a *recursion operator* for (5.83) corresponding to an admitted nontrivial Lie–Bäcklund symmetry of the form (5.86).

In both quantum mechanics and the study of group properties of special functions, symmetries of differential equations are considered exclusively in terms of their recursion operators R.

Since an operator $\mathrm{U} = \eta^\gamma \dfrac{\partial}{\partial u^\gamma}$ commutes with any total derivative operator D_i, one can prove the following theorem concerning commutation relations:

Theorem 5.3.1-2. *Let* $\eta_i = \mathrm{R}_i u$, $\mathrm{U}_i = \eta_i^\gamma \dfrac{\partial}{\partial u^\gamma}$ *where* R_i *is a linear differential operator of the form* (5.87b) *or* (5.88b), $i = 1, 2, \ldots, \ell$. *If* $[\mathrm{U}_i, \mathrm{U}_j] = \mathrm{U}_k$, *then* $[\mathrm{R}_i, \mathrm{R}_j] = -\mathrm{R}_k$.

Proof. See Exercise 5.3-1. □

As an example consider the Schroedinger equation for a harmonic ocillator ($x_1 = x$, $x_2 = t$):

$$\mathrm{L}u = (\mathrm{H} - i\mathrm{D}_t)u = \left(-\frac{1}{2}\mathrm{D}_x^2 + \frac{1}{2}x^2 - i\mathrm{D}_t\right) u = 0. \tag{5.92}$$

Equation (5.92) admits recursion operators $\mathrm{R}_1 = e^{it}(x + \mathrm{D}_x)$ and $\mathrm{R}_2 = e^{-it}(x - \mathrm{D}_x)$ and the trivial operator $\mathrm{R}_3 = 1$, with $[\mathrm{R}_1, \mathrm{R}_2] = 2\mathrm{R}_3$. The corresponding Lie–Bäcklund infinitesimal generators [cf. (5.86)]

$$\mathrm{U}_1 = e^{it}(xu + u_x)\frac{\partial}{\partial u}, \quad \mathrm{U}_2 = e^{-it}(xu - u_x)\frac{\partial}{\partial u}, \quad \mathrm{U}_3 = u\frac{\partial}{\partial u}, \tag{5.93}$$

satisfy the commutation relation $[\mathrm{U}_1, \mathrm{U}_2] = -2\mathrm{U}_3$. From Theorem 5.2.3-1 we see that the infinitesimal generators (5.93) are equivalent to the following infinitesimal generators of point symmetries:

$$\mathrm{X}_1 = e^{it}\left(-\frac{\partial}{\partial x} + xu\frac{\partial}{\partial u}\right), \quad \mathrm{X}_2 = e^{-it}\left(\frac{\partial}{\partial x} + xu\frac{\partial}{du}\right), \quad \mathrm{X}_3 = u\frac{\partial}{\partial u}.$$

From Theorem 5.3.1-1 we see that (5.91) trivially admits

$$[P(\mathrm{R}_1, \mathrm{R}_2)u]\frac{\partial}{\partial u}$$

for any polynomial function $P(a, b)$ in a and b.

As a second example consider the Schroedinger equation for the hydrogen atom:

$$\mathrm{L}u = \left(\frac{1}{2}\Delta + \frac{1}{r} + E\right)u = 0, \tag{5.94}$$

where

$$\Delta = \mathrm{D}_1^2 + \mathrm{D}_2^2 + \mathrm{D}_3^2,$$
$$r = \sqrt{(x_1)^2 + (x_2)^2 + (x_3)^2},$$
$$E = \text{const}.$$

It is well-known [Schiff (1968)] that (5.94) admits three infinitesimal generators corresponding to the Runge–Lenz vector

$$\vec{R} = \frac{1}{2}(\vec{p} \times \vec{L} - \vec{L} \times \vec{p}) - \frac{\vec{r}}{r}$$

where $\vec{p}$ and $\vec{L}$ are respectively linear momentum and angular momentum operators. The x_1-component of $\vec{R}$ is the recursion operator

$$\mathrm{R}_1 = -\left(x_1\mathrm{D}_3^2 - x_3\mathrm{D}_3\mathrm{D}_1 - x_2\mathrm{D}_1\mathrm{D}_2 + x_1\mathrm{D}_2^2 - \mathrm{D}_1 + \frac{x_1}{r}\right).$$

The Lie–Bäcklund symmetry corresponding to R_1 has infinitesimal generator

$$\mathrm{U}_1 = (\mathrm{R}_1 u)\frac{\partial}{\partial u} = -\left(x_1u_{33} - x_3u_{13} - x_2u_{12} + x_1u_{22} - u_1 + \frac{x_1}{r}u\right)\frac{\partial}{\partial u}. \tag{5.95}$$

Clearly (5.95) is a proper Lie–Bäcklund symmetry of (5.94) since it is not equivalent to any point symmetry or contact symmetry.

5.3.2 Recursion Operators for Nonlinear Differential Equations

Olver (1977) showed that if a nonlinear scalar PDE

$$F(x, u, \underset{1}{u}, \underset{2}{u}, \ldots, \underset{q}{u}) = 0 \tag{5.96}$$

of order $q \geq 2$ admits a *proper Lie–Bäcklund symmetry,* then it often admits an infinite sequence of proper Lie–Bäcklund symmetries which can be generated in terms of a recursion operator. In particular if (5.96) admits a proper Lie–Bäcklund symmetry

$$\mathrm{U}_1 = \eta_1 \frac{\partial}{\partial u}, \tag{5.97}$$

then it often admits proper Lie–Bäcklund symmetries

$$\mathrm{U}_i = \eta_i \frac{\partial}{\partial u}, \tag{5.98}$$

where

$$\eta_{i+1} = \mathrm{R}\eta_i, \quad i = 1, 2, \ldots, \tag{5.99}$$

in terms of some *recursion operator* R. We demonstrate that such an operator is realized as a recursion operator for a linear PDE associated with PDE (5.96) [Kumei (1981), Kapcov (1982)].

PDE (5.96) admits a Lie–Bäcklund symmetry

$$\mathrm{U} = \eta(x, u, \underset{1}{u}, \ldots, \underset{k}{u}) \frac{\partial}{\partial u} \tag{5.100}$$

if and only if η satisfies the determining equation

$$\mathrm{U}^{(q)} F = \mathrm{L}[u]\eta = 0 \tag{5.101}$$

for any u satisfying PDE (5.96), where, for a fixed u, $\mathrm{L}[u]$ is the linear operator

$$\mathrm{L}[u] = \frac{\partial F}{\partial u} + \frac{\partial F}{\partial u_i}\mathrm{D}_i + \cdots + \frac{\partial F}{\partial u_{i_1 i_2 \cdots i_q}}\mathrm{D}_{i_1}\mathrm{D}_{i_2}\cdots\mathrm{D}_{i_q}. \tag{5.102}$$

The *linearized equation* corresponding to (5.96) is by definition

$$\mathrm{L}[u]v = 0. \tag{5.103}$$

Equation (5.103) is a necessary condition for $u + \epsilon v$ to solve (5.96) when u solves (5.96). For any fixed solution u of (5.96), equation (5.103) is a linear PDE for v. An infinitesimal $\eta(x, u, \underset{1}{u}, \ldots, \underset{k}{u})$ of a Lie–Bäcklund symmetry admitted by (5.96) is obviously a special solution

$$v = \eta(x, u, \underset{1}{u}, \ldots, \underset{k}{u})$$

of (5.103).

Suppose the linear equation (5.103) admits a nontrivial symmetry with infinitesimal generator

$$(\mathrm{R}[u]v)\frac{\partial}{\partial v} \tag{5.104}$$

where for any fixed u solving (5.96), R[u] is of the form (5.87b), i.e.

$$\mathrm{R}[u] = r(x,u,\underset{1}{u},\ldots,\underset{\ell}{u}) + r_i(x,u,\underset{1}{u},\ldots,\underset{\ell}{u})\mathrm{D}_i + \cdots$$

$$+ r_{i_1 i_2 \cdots i_m}(x,u,\underset{1}{u},\ldots,\underset{\ell}{u})\mathrm{D}_{i_1}\mathrm{D}_{i_2}\cdots\mathrm{D}_{i_m}, \tag{5.105}$$

for some functions $r, r_i, \ldots, r_{i_1 i_2 \cdots i_m}$ of $(x,u,\underset{1}{u},\ldots,\underset{\ell}{u})$. From the above and Theorem 5.3.1-1 we have proved:

Theorem 5.3.2-1. *If PDE* (5.96) *admits* (5.100) *and its linearized PDE* (5.103) *admits* (5.104), *then*

$$v = (\mathrm{R}[u])^s \eta, \quad s = 1,2,\ldots,$$

solve PDE (5.103), *and hence* (5.96) *admits the infinite sequence of Lie–Bäcklund symmetries*

$$\mathrm{U}_s = ((\mathrm{R}[u])^s \eta)\frac{\partial}{\partial u}, \quad s = 1,2,\ldots. \tag{5.106}$$

A recursion operator R[u] satisfies the determining equation

$$\mathrm{L}[u]\mathrm{R}[u]v = 0 \tag{5.107}$$

for any (u,v) satisfying (5.96) and (5.103).

In applying Theorem 5.3.2-1 to a particular PDE (5.96) it is important to note that a priori we need not know which derivatives of u enter in the coefficients of R[u]. However m must be chosen. R[u] is found by applying Lie's algorithm. The situation is illustrated by the following example:

Consider the integrated Burgers' equation ($x_1 = x$, $x_2 = t$; $u_x = u_1$, $u_{xx} = u_2$, $v_x = v_1$, $v_{xx} = v_2$, $v_{xt} = v_{1t}$, etc.):

$$u_2 - \frac{1}{2}(u_1)^2 - u_t = 0. \tag{5.108}$$

The corresponding linearized PDE is

$$\mathrm{L}[u]v = 0 \tag{5.109a}$$

where

$$\mathrm{L}[u] = \mathrm{D}_x^2 - u_1\mathrm{D}_x - \mathrm{D}_t. \tag{5.109b}$$

Assume that (5.109a,b) admits a generator of the form (5.104) with

$$\mathrm{R}[u] = a + b\mathrm{D}_x + c\mathrm{D}_x^2, \tag{5.110}$$

where the coefficients a, b, c can depend on x, t, u, u_1, u_2, ... Then the determining equation is

$$\mathrm{L}[u]\mathrm{R}[u]v = v\mathrm{D}_x^2 a + 2v_1\mathrm{D}_x a + v_2 a + v_1\mathrm{D}_x^2 b + 2v_2\mathrm{D}_x b + v_3 b + v_2\mathrm{D}_x^2 c$$

$$+ 2v_3 D_x c + v_4 c - v u_1 D_x a - v_1 u_1 a - v_1 u_1 D_x b - v_2 u_1 b - v_2 u_1 D_x c$$

$$- v_3 u_1 c - v D_t a - v_t a - v_1 D_t b - v_{1t} b - v_2 D_t c - v_{2t} c = 0. \tag{5.111}$$

Equation (5.111) must hold for *any* solution (u, v) satisfying (5.108), (5.109a,b). Choosing v_3, v_2, v_1, and v as independent variables in (5.111), we find that (5.111) reduces to the following equations for determining a, b, and c:

$$D_x c = 0; \tag{5.112a}$$

$$2D_x b + 2u_2 c - D_t c = 0; \tag{5.112b}$$

$$2D_x a + D_x^2 b - u_1 D_x b + u_2 b - D_t b + u_3 c = 0; \tag{5.112c}$$

$$D_x^2 a - u_1 D_x a - D_t a = 0. \tag{5.112d}$$

Equations (5.112a–d) must hold for any solution u of (5.108). In solving (5.112a–d) we use PDE (5.108) and its differential consequences to eliminate derivatives of u with respect to t. From (5.112a) it follows that

$$c = c(t). \tag{5.113}$$

Then (5.112b) leads to

$$b = -c(t)u_1 + \frac{1}{2}xc'(t) + \alpha(t). \tag{5.114}$$

Substituting (5.113), (5.114) into (5.112c), we obtain

$$a = -\frac{1}{2}c(t)u_2 + \frac{1}{4}c(t)(u_1)^2 - \left[\frac{1}{4}c'(t)x + \frac{1}{2}\alpha(t)\right] u_1$$

$$+ \frac{1}{8}x^2 c''(t) + \frac{1}{2}x\alpha'(t) + \beta(t). \tag{5.115}$$

Then (5.112d) leads to

$$c'''(t) = 0, \quad \alpha''(t) = 0, \quad \beta'(t) = \frac{1}{4}c''(t). \tag{5.116}$$

Integration of (5.116) yields six integration constants. Hence (5.109a,b) admits six generators of the form (5.104) with $R[u] = R_i[u]$, $i = 1, 2, \ldots, 6$:

$$R_1[u] = 1, \quad R_2[u] = \frac{1}{2}u_1 - D_x,$$

$$R_3[u] = \frac{1}{2}(tu_1 - x) - tD_x,$$

$$R_4[u] = \frac{1}{4}[(u_1)^2 - 2u_2] - u_1 D_x + D_x^2,$$

$$R_5[u] = \frac{1}{4}[t(u_1)^2 - xu_1 - 2tu_2] - \left(tu_1 - \frac{1}{2}x\right) D_x + tD_x^2,$$

$$\mathrm{R}_6[u] = \frac{1}{4}[t^2(u_1)^2 - 2txu_1 - 2t^2u_2 + x^2 + 2t] - (t^2u_1 - xt)\mathrm{D}_x + t^2\mathrm{D}_x^2. \quad (5.117)$$

In (5.117) only $\mathrm{R}_2[u]$ and $\mathrm{R}_3[u]$ are independent generators leading to recursion operators since $\mathrm{R}_4 = (\mathrm{R}_2)^2$, $\mathrm{R}_5 = \mathrm{R}_2\mathrm{R}_3$, $\mathrm{R}_6 = (\mathrm{R}_3)^2$. Consequently in (5.110) we could have set $c = 0$, i.e. the recursion operators admitted by (5.109a,b) of the form (5.110) are embedded in recursion operators of the form $\mathrm{R}[u] = a + b\mathrm{D}_x$.

If PDE (5.108) admits a Lie–Bäcklund symmetry of the form (5.100) then (5.108) admits

$$[P(\mathrm{R}_2[u], \mathrm{R}_3[u])\eta)]\frac{\partial}{\partial u}$$

for any polynomial function $P(y, z)$ in y and z. Clearly (5.108) admits

$$\eta\frac{\partial}{\partial u} = u_1\frac{\partial}{\partial u}$$

corresponding to invariance under translations in x. Then (5.108) admits Lie–Bäcklund symmetries with infinitesimals given by

$$\mathrm{R}_2[u]u_1 = -u_2 + \frac{1}{2}(u_1)^2 = -u_t,$$

$$\mathrm{R}_3[u]u_1 = -tu_2 + \frac{1}{2}[t(u_1)^2 - xu_1] = -\left(tu_t + \frac{1}{2}xu_1\right),$$

corresponding to invariance of (5.108) under translations in t and scalings in x and t, respectively, and

$$(\mathrm{R}_2[u])^3u_1 = -u_{2t} + u_1u_{1t} - \frac{1}{4}[(u_1)^2 - 2u_2]u_t,$$

corresponding to a proper Lie–Bäcklund symmetry of (5.108).

5.3.3 Integro-Differential Recursion Operators

Equation (5.108) is an integrated form of Burgers' equation

$$\tilde{u}_2 - \tilde{u}\,\tilde{u}_1 - \tilde{u}_t = 0; \quad (5.118)$$

if u solves PDE (5.108) then

$$\tilde{u} = \mathrm{D}_x u \quad (5.119)$$

solves Burgers' equation (5.118). Consequently if

$$\eta\frac{\partial}{\partial u} \quad (5.120)$$

is a Lie–Bäcklund symmetry of (5.108) then

$$\tilde{\eta}\frac{\partial}{\partial \tilde{u}} = (D_x\eta)\frac{\partial}{\partial \tilde{u}} \tag{5.121}$$

is a Lie–Bäcklund symmetry of (5.118). From (5.119) and (5.121) we have

$$u = D_x^{-1}\tilde{u}, \tag{5.122a}$$

$$\eta = D_x^{-1}\tilde{\eta}, \tag{5.122b}$$

where the integral operator D_x^{-1} is the inverse operator to D_x satisfying

$$D_x D_x^{-1} f(x) = f(x), \quad D_x^{-1} D_x f(x) = f(x) \tag{5.123}$$

for any differentiable function $f(x)$ with compact support.

A symmetry

$$u^* = u + \epsilon R[u]\eta + O(\epsilon^2) \tag{5.124}$$

admitted by (5.108) clearly corresponds to a symmetry

$$\tilde{u}^* = \tilde{u} + \epsilon D_x R[D_x^{-1}\tilde{u}] D_x^{-1}\tilde{\eta} + O(\epsilon^2) \tag{5.125}$$

of (5.118). Hence if $R[u]$ is a recursion operator generating Lie–Bäcklund symmetries of the integrated Burgers' equation (5.108) then

$$\tilde{R}[\tilde{u}] = D_x R[D_x^{-1}\tilde{u}] D_x^{-1} \tag{5.126}$$

is a recursion operator generating Lie–Bäcklund symmetries of Burgers' equation (5.118). Corresponding to $R_2[u]$ and $R_3[u]$ of (5.117) we find

$$\tilde{R}_2[\tilde{u}] = -D_x + \frac{1}{2}\tilde{u} + \frac{1}{2}\tilde{u}_1 D_x^{-1},$$

$$\tilde{R}_3[\tilde{u}] = -tD_x + \frac{1}{2}(t\tilde{u} - x) + \frac{1}{2}(t\tilde{u}_1 - 1)D_x^{-1}. \tag{5.127}$$

The recursion operators (5.127) for Burgers' equation are not of the form (5.105) since they are integro-differential operators.

This leads us to consider *integro-differential recursion operators* of the form

$$R[u] = \sum_{j=0}^{m} r_j D_x^j + \sum_{j=1}^{n} r_{-j} D_x^{-j} \tag{5.128}$$

for scalar PDE's with independent variables x, t of evolution type

$$u_t + G(x, t, u, u_1, u_2, \ldots, u_q) = 0, \tag{5.129}$$

where $u_k = \frac{\partial^k u}{\partial x^k}$, $k = 1, 2, \ldots, q$, and D_x^{-j} denotes the j-fold product of D_x^{-1}, $j = 1, 2, \ldots$. The coefficients $\{r_j, r_{-j}\}$ can depend on x, t, u, u_1,

$u_2, \ldots$. Integro-differential recursion operators (5.128) are found by using Lie's algorithm as in the case for finding recursion operators of the form (5.105). Again the dependence of each coefficient of $\{r_j, r_{-j}\}$ on x, t, u, and derivatives of u is found in the course of solving the determining equations

$$\mathrm{L}[u]\mathrm{R}[u]v = 0 \tag{5.130}$$

where

$$\mathrm{L}[u] = \mathrm{D}_t + \frac{\partial G}{\partial u} + \frac{\partial G}{\partial u_1}\mathrm{D}_x + \frac{\partial G}{\partial u_2}\mathrm{D}_x^2 + \cdots + \frac{\partial G}{\partial u_q}\mathrm{D}_x^q \tag{5.131}$$

is the linear operator arising from the linearization of (5.129). Equation (5.130) must hold for any solution $(u(x), v(x))$ of (5.129) and its linearized equation

$$\mathrm{L}[u]v = 0.$$

As an example we find an integro-differential recursion operator for the Korteweg–de Vries (KdV) equation

$$u_t + uu_1 + u_3 = 0 \tag{5.132}$$

of the form

$$\mathrm{R}[u] = r\mathrm{D}_x^{-1} + a + b\mathrm{D}_x + c\mathrm{D}_x^2. \tag{5.133}$$

Here the linear operator is

$$\mathrm{L}[u] = \mathrm{D}_x^3 + u\mathrm{D}_x + u_1 + \mathrm{D}_t. \tag{5.134}$$

Using $\mathrm{L}[u]v = 0$ and choosing v_4, v_3, v_2, v_1, v, and $v_{-1} = \mathrm{D}_x^{-1}v$ as independent variables, we find that the determining equation $\mathrm{L}[u]\mathrm{R}[u]v = 0$ reduces to the following equations for finding a, b, c, and r:

$$\mathrm{D}_x c = 0; \tag{5.135a}$$

$$\mathrm{D}_x b + \mathrm{D}_x^2 c = 0; \tag{5.135b}$$

$$3\mathrm{D}_x a + 3\mathrm{D}_x^2 b + \mathrm{D}_x^3 c + u\mathrm{D}_x c - 2u_1 c + \mathrm{D}_t c = 0; \tag{5.135c}$$

$$3\mathrm{D}_x r + 3\mathrm{D}_x^2 a + \mathrm{D}_x^3 b - u_1 b + \mathrm{D}_t b - 3u_2 c = 0; \tag{5.135d}$$

$$3\mathrm{D}_x^2 r + \mathrm{D}_x^3 a + u\mathrm{D}_x a + \mathrm{D}_t a - u_2 b - u_3 c = 0; \tag{5.135e}$$

$$\mathrm{D}_x^3 r + u\mathrm{D}_x r + u_1 r + \mathrm{D}_t r = 0. \tag{5.135f}$$

From (5.135a,b) it follows that

$$c = c(t), \quad b = b(t).$$

Then (5.135c,d) successively lead to

$$a = \frac{2}{3}c(t)u - \frac{1}{3}c'(t)x + \alpha(t),$$

$$r = \frac{1}{3}b(t)u - \frac{1}{3}b'(t)x + \frac{1}{3}c(t)u_1 + \rho(t),$$

with arbitrary $\alpha(t)$, $\rho(t)$. Equations (5.135e,f) yield $b(t) = \rho(t) = 0$, $c(t) =$ const, $\alpha(t) =$ const. Finally we obtain the well-known integro-differential recursion operator

$$\mathrm{R}[u] = \mathrm{D}_x^2 + \frac{2}{3}u + \frac{1}{3}u_1\mathrm{D}_x^{-1} \tag{5.136}$$

for the KdV equation (5.132).

If we start with the point symmetry

$$\eta\frac{\partial}{\partial u} = -u_t\frac{\partial}{\partial u} = (uu_1 + u_3)\frac{\partial}{\partial u},$$

corresponding to invariance of the KdV equation (5.132) under translations in t, we obtain the infinite sequence (5.106) of proper Lie–Bäcklund symmetries admitted by (5.132). Note that the invariance of (5.132) under translations in t is related to its invariance under translations in x by

$$uu_1 + u_3 = \mathrm{R}[u]u_1.$$

On the other hand if we start with the point symmetry

$$\eta\frac{\partial}{\partial u} = (xu_1 + 3tu_t + 2u)\frac{\partial}{\partial u},$$

corresponding to scaling invariance of the KdV equation (5.132), then

$$\mathrm{R}[u]\eta = -xu_t + 4u_2 + t(3u_{2t} + 2uu_t - u_1u_2 - \frac{1}{2}u^2u_1)$$

$$+ \frac{4}{3}u^2 + \frac{1}{3}u_1\mathrm{D}_x^{-1}u,$$

and consequently the infinitesimal of the resulting infinitesimal generator depends on the integral of u. This is an example of a nonlocal symmetry generated through a recursion operator [cf. Kapcov (1982)]. Other types of nonlocal symmetries are considered in Chapter 7.

For the KdV equation we chose $n = -1$, $m = 2$ in (5.128). However in general a priori we need not specify n (n could be ∞) for a fixed choice of m in order to carry out Lie's algorithm. Then we successively solve a sequence of equations which result from the determining equation

$$\mathrm{L}[u]\mathrm{R}[u]v = 0$$

to determine $r_m, r_{m-1}, \ldots$. It turns out that $n = \infty$ for the integro-differential recursion operator for both the sine-Gordon equation [Exercise 5.3-8] and the modified KdV equation $u_t + u^2u_1 + u_3 = 0$ [Exercise 5.3-9].

5.3.4 Lie–Bäcklund Classification Problem

Consider again the nonlinear heat conduction equation ($u_1 = u_x$, $u_2 = u_{xx}, \ldots$)

$$u_t - \frac{\partial}{\partial x}(K(u)u_1) = 0. \tag{5.137}$$

We find all conductivities $K(u)$ for which PDE (5.137) admits proper Lie–Bäcklund symmetries [Bluman and Kumei (1980)]. We previously classified (5.137) with respect to point symmetries in Section 4.2.4 and with respect to point symmetries of an associated system in Section 4.3.4.

Suppose (5.137) admits a Lie–Bäcklund symmetry

$$\mathrm{U} = \eta(x, t, u, u_1, \ldots, u_p)\frac{\partial}{\partial u} \tag{5.138}$$

for some $p \geq 2$. For convenience let

$$\frac{\partial \eta}{\partial u} = \eta_0, \quad \frac{\partial \eta}{\partial u_i} = \eta_i, \quad \frac{\partial^2 \eta}{\partial u_i \partial u_j} = \eta_{ij}, \quad K' = \frac{dK}{du}, \quad K'' = \frac{d^2K}{du^2}.$$

PDE (5.137) admits (5.138) if and only if

$$\mathrm{U}^{(2)}\left(u_t - \frac{\partial}{\partial x}(K(u)u_1)\right) = \mathrm{D}_t\eta - K''(u_1)^2\eta$$

$$- 2K'u_1\mathrm{D}_x\eta - K'u_2\eta - K\mathrm{D}_x^2\eta = 0 \tag{5.139}$$

for any solution u of (5.137) where

$$\mathrm{D}_t\eta = \frac{\partial \eta}{\partial t} + \eta_0 u_t + \sum_{i=1}^{p} \eta_i u_{it};$$

$$\mathrm{D}_x\eta = \frac{\partial \eta}{\partial x} + \sum_{i=0}^{p} \eta_i u_{i+1};$$

$$\mathrm{D}_x^2\eta = \frac{\partial^2 \eta}{\partial x^2} + 2\sum_{i=0}^{p} \frac{\partial \eta_i}{\partial x} u_{i+1} + \sum_{i,j=0}^{p} \eta_{ij} u_{i+1} u_{j+1} + \sum_{i=0}^{p} \eta_i u_{i+2}.$$

After eliminating u_t and u_{it} in (5.139) by using (5.137), the determining equation (5.139) becomes a polynomial in u_{p+1}. The coefficients of each term in this polynomial must vanish. It turns out that η is independent of x and t. The vanishing of the coefficients of $(u_{p+1})^2$ and u_{p+1}, respectively, leads to equations

$$\eta_{pp} = 0, \quad pK'\eta_p u_1 = 2K\sum_{i=0}^{p-1} \eta_{ip} u_{i+1}. \tag{5.140}$$

Solving (5.140), we find that

$$\eta = \alpha K^{p/2} u_p + e(u, u_1, \ldots, u_{p-1}), \tag{5.141}$$

where e is an arbitrary function of its arguments, and $\alpha = \text{const}$. For some $p \geq 3$, $\alpha \neq 0$.

The substitution of (5.141) into (5.139) leads to a polynomial in u_p. The vanishing of its coefficients of $(u_p)^2$ and u_p, respectively, yields equations

$$\frac{\partial^2 e}{\partial u_{p-1}^2} = 0,$$

$$2K \sum_{i=0}^{p-2} u_{i+1} \frac{\partial^2 e}{\partial u_i \partial u_{p-1}} + (1-p)K' u_1 \frac{\partial e}{\partial u_{p-1}} - \frac{\alpha}{2} p(p+3) K' K^{p/2} u_2$$
$$+ \frac{\alpha}{4}[p^2 (K')^2 K^{p/2-1} - 2p(p+2) K'' K^{p/2}](u_1)^2 = 0. \tag{5.142}$$

From (5.142) we easily deduce that η is of the form

$$\eta = \alpha \left[K^{p/2} u_p + \frac{1}{4} p(p+3) K' K^{p/2-1} u_1 u_{p-1} \right] + f(u) u_{p-1}$$
$$+ g(u, u_1, \ldots, u_{p-2}), \tag{5.143}$$

where f and g are arbitrary functions of their respective arguments and, more importantly, the substitution of (5.143) into (5.142) yields a nontrivial solution for η only if the conductivity $K(u)$ satisfies the equation

$$2KK'' = 3(K')^2. \tag{5.144}$$

The solution of (5.144) is

$$K(u) = \frac{c}{(u+d)^2} \tag{5.145}$$

for arbitrary constants c and d. Without loss of generality we set $c = 1$, $d = 0$, and consider the PDE

$$u_t - \frac{\partial}{\partial x}(u^{-2} u_1) = 0. \tag{5.146}$$

Equation (5.146) *is the only PDE of the form* (5.137) *which could admit proper Lie–Bäcklund symmetries.*

It is left to Exercise 5.3-7 to show that PDE (5.146) has recursion operator

$$\mathrm{R}[u] = u^{-1} \mathrm{D}_x - 2u^{-2} u_1 + [2u^{-3}(u_1)^2 - u^{-2} u_2] \mathrm{D}_x^{-1} = \mathrm{D}_x^2 u^{-1} \mathrm{D}_x^{-1}. \tag{5.147}$$

If we start with the point symmetry

$$\eta\frac{\partial}{\partial u} = u_t\frac{\partial}{\partial u} = \mathrm{D}_x(u^{-2}u_1)\frac{\partial}{\partial u},$$

corresponding to invariance of (5.146) under translations in t, we obtain the following infinite sequence of proper Lie–Bäcklund symmetries admitted by (5.146):

$$\mathrm{U}_s = [(\mathrm{D}_x^2(u^{-1}\mathrm{D}_x)^{s-1})u^{-2}u_1]\frac{\partial}{\partial u}, \quad s = 1, 2, \ldots .$$

The linearization of (5.146) is discussed in Chapters 6 and 7.

Exercises 5.3

1. Prove Theorem 5.3.1-2.

2. Consider the linear PDE

$$x^2\frac{\partial^2 u}{\partial x^2} + x\frac{\partial u}{\partial x} + \left(x^2 - \frac{\partial^2}{\partial t^2}\right)u = 0. \tag{5.148}$$

   With $u = e^{\nu t}y(x,\nu)$, PDE (5.148) becomes Bessel's equation of order ν for $y(x,\nu)$:

$$\frac{d^2y}{dx^2} + \frac{1}{x}\frac{dy}{dx} + \left(1 - \frac{\nu^2}{x^2}\right)y = 0.$$

   (a) Determine recursion operators of the form $\mathrm{R} = a(x,t) + b(x,t)\mathrm{D}_x + c(x,t)\mathrm{D}_t$ which are admitted by (5.148), and compute their commutation relations.

   (b) Find appropriate linear combinations $\nu \pm 1$, say R_+ and R_-, of the recursion operators which can raise or lower the order ν by one:

$$\mathrm{R}_\pm e^{\nu t}y(x,\nu) = e^{(\nu\pm1)t}y(x,\nu\pm1)$$

   [Miller (1968)].

   (c) Obtain the infinitesimal generators $\eta\frac{\partial}{\partial u}$ corresponding to the recursion operators of (a). Find the equivalent infinitesimal generators of point symmetries.

3. Consider Legendre's equation

$$\left[(1-x^2)\frac{d^2}{dx^2} - 2x\frac{d}{dx} + n(n+1)\right]y(x,n) = 0. \tag{5.149}$$

   The previous problem suggests that in order to obtain recursion operators which shift integer n, we rewrite (5.149) in the form

$$\left[(1-x^2)\frac{\partial^2}{\partial x^2} - 2x\frac{\partial}{\partial x} + \frac{\partial}{\partial t}\left(\frac{\partial}{\partial t} + 1\right)\right]u = 0, \tag{5.150}$$

   by replacing n by $\frac{\partial}{\partial t}$ and $y(x,n)$ by $u = e^{nt}y(x,n)$.

(a) Find recursion operators of the form $R = a(x,t) + b(x,t)D_x + c(x,t)D_t$ which are admitted by PDE (5.150).

(b) The recursion operators $R_\pm$ with factors $e^{\pm t}$ will respectively raise and lower integer n, i.e. $R_\pm e^{nt} y(x,n) = e^{(n\pm 1)t} y(x, n \pm 1)$. Determine n and the ground state function $y(x,n) = P(x)$ such that $R_- e^{nt} P(x) = 0$. [If $n = 0$, $P(x) =$ const, the zeroth degree Legendre polynomial.]

(c) Show that $(R_+)^n P(x)$ generates the nth degree Legendre polynomial.

4. Consider Burgers' equation (5.118).

(a) Apply Lie's algorithm discussed in Section 5.3.3 to obtain its recursion operators (5.127). Assume the form $R[u] = rD_x^{-1} + a + bD_x$.

(b) Use (5.127) and $\overline{\eta} = u_1$ to generate its first few proper Lie–Bäcklund symmetries.

5. Consider the system of PDE's

$$u_2 + vu^2 + iu_t = 0, \tag{5.151a}$$

$$v^2 + uv^2 - iv_t = 0. \tag{5.151b}$$

Let $v = \overline{u}$ where a bar denotes the complex conjugate. Then (5.151a) becomes the cubic Schroedinger equation $u_2 + u^2\overline{u} + iu_t = 0$. Show that the matrix operator

$$R[u,v] = \begin{bmatrix} iD_x + iuD_x^{-1}v & iuD_x^{-1}u \\ -ivD_x^{-1}v & -i(D_x + vD_x^{-1}u) \end{bmatrix}$$

is a recursion operator for the system of PDE's (5.151a,b) [Ablowitz, Kaup, Newell, and Segur (1974)].

6. Show that a recursion operator $R[u]$ for a PDE $F(x, u, \underset{1}{u}, \underset{2}{u}, \ldots, \underset{q}{u}) = 0$ and its corresponding linear operator $L[u]$ commute: $[L[u], R[u]] = 0$ for any solution u of the PDE. In particular show that if $F = 0$ is an evolution equation, i.e. $\dfrac{\partial u}{\partial x_n} + G(x, u, \underset{1}{u}, \underset{2}{u}, \ldots, \underset{q}{u}) = 0$ where G involves no x_n-derivatives of u, then for the linear operator corresponding to G,

$$\tilde{L}[u] = \frac{\partial G}{\partial u} + \frac{\partial G}{\partial u_i} D_i + \cdots + \frac{\partial G}{\partial u_{i_1 i_2 \cdots i_q}} D_{i_1} D_{i_2} \cdots D_{i_q},$$

we have $[\tilde{L}[u], R[u]] + D_{x_n} R[u] = 0$ for any solution u of the evolution equation. $[D_{x_n} R[u]$ is an operator resulting from differentiating $R[u]$

with respect to x_n. For the KdV equation (5.132) [$x_n = t$],

$$\mathrm{D}_t\mathrm{R}[u] = \mathrm{D}_t\left(\mathrm{D}_x^2 + \frac{2}{3}u + \frac{1}{3}u_1\mathrm{D}_x^{-1}\right) = \frac{2}{3}u_t + \frac{1}{3}u_{1t}\mathrm{D}_x^{-1}$$

[Olver (1977)]].

7. If one is able to determine the first few Lie–Bäcklund symmetries admitted by a PDE, often one can make a reasonable guess on the form of a recursion operator R[u]. The validity of the form is then checked by the requirement $\mathrm{L}[u]\mathrm{R}[u]v = 0$ for any solution (u, v) of $F(x, u, \underset{1}{u}, \underset{2}{u}, \ldots, \underset{q}{u}) = 0$, $\mathrm{L}[u]v = 0$, or, equivalently, by showing that the commutation relation $[\mathrm{L}[u], \mathrm{R}[u]] = 0$ is satisfied for any solution u of $F(x, u, \underset{1}{u}, \underset{2}{u}, \ldots, \underset{q}{u}) = 0$. For each of the following PDE's find an operator R[u] satisfying $\eta_{i+1} = \mathrm{R}[u]\eta_i$, to within scalings of η_i, for the given η_i, $i = 1, 2$, and check its validity by both criteria:

   (a) Burgers' equation $u_2 - uu_1 - u_t = 0$ with

   $$\eta_1 = u_1, \quad \eta_2 = u_2 - uu_1,$$

   $$\eta_3 = u_3 - \frac{3}{2}uu_2 - \frac{3}{2}(u_1)^2 + \frac{3}{4}u^2u_1$$

   [Olver (1977)].

   (b) Nonlinear heat conduction equation $\frac{\partial}{\partial x}(u^{-2}u_1) - u_t = 0$ with

   $$\eta_1 = u^{-2}u_2 - 2u^{-3}(u_1)^2, \quad \eta_2 = u^{-3}u_3 - 9u^{-4}u_1u_2 + 12u^{-5}(u_1)^3,$$

   $$\eta_3 = u^{-4}u_4 - 14u^{-5}u_1u_3 - 10u^{-5}(u_2)^2 + 95u^{-6}(u_1)^2u_2 - 90u^{-7}(u_1)^4$$

   [Bluman and Kumei (1980)].

   (c) KdV equation $u_3 + uu_1 + u_t = 0$ with

   $$\eta_1 = u_1, \quad \eta_2 = u_3 + uu_1,$$

   $$\eta_3 = u_5 + \frac{5}{3}uu_3 + \frac{10}{3}u_1u_2 + \frac{5}{6}u^2u_1.$$

   (d) Modified KdV equation $u_3 + u^2u_1 + u_t = 0$ with

   $$\eta_1 = u_1, \quad \eta_2 = u_3 + u^2u_1,$$

   $$\eta_3 = u_5 + \frac{5}{3}u^2u_3 + \frac{20}{3}uu_1u_2 + \frac{5}{3}(u_1)^3 + \frac{5}{6}u^4u_1$$

   [Olver (1977)].

(e) Harry–Dym equation $u_t - \lambda u^3 u_3 = 0$ with

$$\eta_1 = u_1, \quad \eta_2 = u^3 u_3,$$

$$\eta_3 = u^5 u_5 + 5u^4 u_1 u_4 + 5u^4 u_2 u_3 + \frac{5}{2} u^3 (u_1)^2 u_3.$$

Assume that $u \to \text{const} \neq 0$, $u_i \to 0$ as $x \to \infty$, $i = 1, 2, \ldots$ [Leo, Leo, Soliani, Solombrino, and Martina (1983)].

8. Consider a recursion operator of the form

$$\mathrm{R}[u] = r_2 \mathrm{D}_x^2 + r_1 \mathrm{D}_x + r_0 + r_{-1} \mathrm{D}_x^{-1} + \cdots = \sum_{i=-\infty}^{2} r_i \mathrm{D}_x^i$$

for the sine-Gordon equation $u_{xt} = \sin u$.

(a) Show that the determining equation $\mathrm{L}[u]\mathrm{R}[u]v = 0$ is

$$\sum_{i=-\infty}^{2} [v_i \mathrm{D}_x \mathrm{D}_t r_{i-1} + \{\mathrm{D}_x^{i-1}(v \cos u)\} \mathrm{D}_x r_i + v_{i+1} \mathrm{D}_t r_i + \{\mathrm{D}_x^i (v \cos u)\} r_i - v_i (\cos u) r_i] = 0, \tag{5.152}$$

where $v_i = \mathrm{D}_x^i v$, $i = 2, 1, 0, -1, \ldots$.

(b) To solve (5.152), we need to first express the left hand side of (5.152) as a linear combination of $\{v_i\}$. This can be accomplished by expanding the terms $\mathrm{D}_x^{-1}(v \cos u)$, $\mathrm{D}_x^{-2}(v \cos u), \ldots$, appearing in (5.152), in terms of the Leibniz formula for integration

$$\mathrm{D}_x^{-j}(f(x)g(x)) = \sum_{i=0}^{\infty} (-1)^i \frac{(j+i-1)!}{i!(j-1)!} f_i g_{-j-i}, \tag{5.153}$$

$j = 1, 2, \ldots$, where $f_i = \mathrm{D}_x^i f$, $g_{-i} = \mathrm{D}_x^{-i} g$. Use the integration by parts formula

$$\mathrm{D}_x^{-1}(fg) = f g_{-1} - \mathrm{D}_x^{-1}(f_1 g_{-1}) = f g_{-1} - f_1 g_{-2} + \mathrm{D}_x^{-1}(f_2 g_{-2}) = \ldots$$

to verify (5.153).

(c) Set to zero the coefficients of $v_3, v_2, v_1, v_0, v_{-1}, v_{-2}$, respectively, in (5.152) and obtain the following equations:

$$\mathrm{D}_t r_2 = 0, \quad \mathrm{D}_t r_1 + \mathrm{D}_t \mathrm{D}_x r_2 = 0,$$

$$\mathrm{D}_t r_0 + \mathrm{D}_x \mathrm{D}_t r_1 + \cos u \mathrm{D}_x r_2 + 2(\mathrm{D}_x \cos u) r_2 = 0,$$

$$\mathrm{D}_t r_{-1} + \mathrm{D}_x \mathrm{D}_t r_0 + \mathrm{D}_x(r_1 \cos u) + \mathrm{D}_x(r_2 \mathrm{D}_x \cos u) = 0,$$
$$\mathrm{D}_t r_{-2} + \mathrm{D}_x \mathrm{D}_t r_{-1} + \cos u \mathrm{D}_x r_0 = 0,$$
$$\mathrm{D}_t r_{-3} + \mathrm{D}_x \mathrm{D}_t r_{-2} + \cos u \mathrm{D}_x r_{-1} - (\mathrm{D}_x \cos u) r_{-1} + \cos u \mathrm{D}_x r_0 = 0. \quad (5.154)$$

(d) Equations (5.154) can be solved successively for $r_2, r_1, r_0, \ldots, r_{-3}$. Assume that each r_i depends on t only through $u(x,t)$. Then the first two equations of (5.154) yield $r_2 = \alpha = \text{const}$, $r_1 = \beta = \text{const}$. Solve the remaining equations of (5.154) and show that for $\alpha \neq 0$ we are led to a recursion operator of the form

$$\mathrm{R}[u] = \mathrm{D}_x^2 + (u_1)^2 - u_1 u_2 \mathrm{D}_x^{-1} + u_1 u_3 \mathrm{D}_x^{-2} - u_1 u_4 \mathrm{D}_x^{-3} + \cdots. \quad (5.155)$$

(e) Expression (5.155) and the Leibniz formula (5.153) for $j = 1$ lead to

$$\mathrm{R}[u] = \mathrm{D}_x^2 + u_1 \mathrm{D}_x^{-1} u_1 \mathrm{D}_x. \quad (5.156)$$

Verify that (5.156) is a recursion operator for the sine-Gordon equation. [The recursion operator (5.156) was first found by Olver (1977).]

9. Apply the algorithm presented in the previous problem to obtain the recursion operator for the modified KdV equation $u_3 + u^2 u_1 + u_t = 0$.

10. The KdV equation $u_3 + u u_1 + u_t = 0$ admits a Galilean symmetry $\mathrm{U} = \eta \frac{\partial}{\partial u} = (t u_x - 1)\frac{\partial}{\partial u}$. Show that this symmetry, together with the recursion operator $\mathrm{R}[u]$ given by (5.136), generates nonlocal symmetries $\eta_i = (\mathrm{R}[u])^i \eta$, $i = 1, 2, \ldots$, since each η_i cannot be expressed in terms of x, t, u, and derivatives of u [Olver (1986)].

5.4 Discussion

In this chapter we generalized Lie groups of point transformations to Lie–Bäcklund transformations which are defined by infinitesimals depending on a finite number of derivatives of the dependent variables. We showed that a Lie–Bäcklund symmetry admitted by an action integral, i.e. a variational symmetry, leads to a conservation law (Noether's theorem).

If a nonlinear differential equation admits a Lie–Bäcklund symmetry, not equivalent to a point symmetry or contact symmetry, then usually it admits an infinite sequence of Lie–Bäcklund symmetries. We considered recursion operators which act as "ladder" operators generating such sequences starting from particular symmetries.

We can always compute Lie–Bäcklund symmetries of a given PDE by a simple generalization of Lie's algorithm. Since every Lie–Bäcklund symmetry is equivalent to one which leaves the independent variables invariant, we only need to compute infinitesimal generators of Lie–Bäcklund symmetries which do not operate on independent variables.

In general a global Lie–Bäcklund transformation acts on the infinite-dimensional space which includes all derivatives of dependent variables. This space becomes finite-dimensional if and only if the Lie–Bäcklund transformation is equivalent to a point transformation or contact transformation.

In Section 5.3 we showed that recursion operators are linear operators associated with symmetries of the linearized equations of given PDE's. This led us to a method, based on Lie's algorithm, for computing recursion operators.

Recursion operators for linear PDE's include, as a special case, ladder operators (raising and lowering operators) which are used in quantum mechanics [cf. Wybourne (1974)] and the theory of special functions [Talman (1968), Vilenkin (1968), Miller (1968)]. For linear PDE's the existence of recursion operators is closely related to the existence of separation of variables [Miller (1977)]. Winternitz, Smorodinsky, Uhlir, and Fris (1967) gave a method for finding recursion operators for time-independent Schroedinger equations. Anderson, Kumei, and Wulfman (1972) introduced the generalization of Lie's infinitesimal transformation method to include Lie–Bäcklund transformations and constructed recursion operators for linear PDE's.

For a system of PDE's which can be written in Hamiltonian form in two distinct ways, i.e. a bi-Hamiltonian system of PDE's, one can directly compute recursion operators [Magri (1978), Olver (1986)].

Nonlinear PDE's which admit recursion operators invariably seem to be related to linear PDE's. For certain nonlinear evolution equations one can transform initial value problems to inverse scattering problems involving linear systems of PDE's [Gardner, Greene, Kruskal, and Miura (1967), Lax (1968), Zakharov and Shabat (1971)]. For such nonlinear evolution equations there exist recursion operators which are related to linear operators arising in eigenvalue problems of the associated scattering problems [Ablowitz, Kaup, Newell, and Segur (1974)]. Konopelchenko (1987) reviews recent work on recursion operators and the integrability of nonlinear evolution equations. The classification problem of determining integrable nonlinear evolution equations is considered in Mikhailov, Shabat, and Yamilov (1987).

6

Construction of Mappings Relating Differential Equations

6.1 Introduction

In previous chapters we have considered the construction and use of infinitesimal transformations which leave a given differential equation invariant. These transformations map any solution of the given DE into another solution of the same DE. We have used infinitesimal transformations to reduce the order of ODE's and to construct invariant solutions and conservation laws for DE's.

In this chapter we use infinitesimal transformations to construct a transformation (*mapping*) which maps (transforms) a given DE into another DE (*target DE*) in the sense that any solution of the given DE is mapped into a solution of the target DE. In general such a mapping need not be a group transformation. If such a mapping exists it is necessary that any infinitesimal generator admitted by the given DE be mapped into an infinitesimal generator admitted by the target DE.

If the mapping from the given DE to the target DE is one-to-one (invertible) then the mapping must establish a one-to-one correspondence between infinitesimal generators of the given and target DE's. More precisely it is necessary that any Lie algebra of infinitesimal generators of the given DE be isomorphic to a Lie algebra of infinitesimal generators of the target DE. If the Lie algebra is of large dimension then the resulting severe restrictions on a mapping often lead to the discovery of the mapping.

If the mapping from the given DE to the target DE is allowed to be non-invertible then it is not necessary that there be a one-to-one correspondence between Lie algebras of infinitesimal generators of the given and target DE's. But such a non-invertible mapping must take any infinitesimal generator admitted by the given DE into an infinitesimal generator (which could be a null generator) admitted by the target equation. More precisely the mapping must establish a homomorphism between any Lie algebra of infinitesimal generators of the given DE and a Lie algebra of infinitesimal generators of the target DE.

The use of infinitesimal transformations to construct mappings which relate DE's is especially fruitful when a subalgebra of the Lie algebra of admitted infinitesimal generators uniquely determines the target DE. In this case if one constructs a mapping which transforms a subalgebra of the

Lie algebra of infinitesimal generators of the given DE into the subalgebra of infinitesimal generators uniquely determining the target DE then it automatically follows that the mapping transforms any solution of the given DE into a solution of the target DE.

Often for a given DE one does not want to determine whether it can be mapped into a specific target DE but whether it can be mapped into a member of a target class of DE's of interest. For example, can a given nonlinear system of DE's be mapped into a linear system of DE's? If the target class of DE's is uniquely determined by some Lie algebra of infinitesimal generators there should exist an algorithm to determine whether or not there exists a one-to-one mapping from the given DE to some member of the target class of DE's. Moreover this algorithm should construct the mapping if it exists. We establish specific algorithms to determine whether or not there exists an invertible mapping

(i) of a given nonlinear scalar PDE to a linear scalar PDE;

(ii) of a given nonlinear system of PDE's to a linear system of PDE's;

(iii) of a given linear scalar PDE with variable coefficients to a linear scalar PDE with constant coefficients.

These algorithms are constructive and lead to the mappings when they exist. We also consider the problem of finding the most general invertible mapping which can transform a given linear scalar PDE to a linear scalar PDE with constant coefficients.

6.2 Notations; Mappings of Infinitesimal Generators

For a *given system* of DE's, denoted by $R\{x,u\}$, with n independent variables $x=(x_1,x_2,\ldots,x_n)$ and m dependent variables $u=(u^1,u^2,\ldots,u^m)$, we use the following notations:

G_x: the group of all admitted continuous transformations;

L_x: the Lie algebra of G_x;

$\mathcal{G}_x$: a subgroup of continuous transformations, $\mathcal{G}_x \subset G_x$;

$\mathcal{L}_x$: the Lie algebra of $\mathcal{G}_x$, $\mathcal{L}_x \subset L_x$;

$T\mathcal{G}_x(\epsilon)$: a one-parameter subgroup, $T\mathcal{G}_x(\epsilon) \in \mathcal{G}_x$, given by

$$x_i^* = x_i + \epsilon\xi_i(x,u,\underset{1}{u},\ldots,\underset{k}{u}) + O(\epsilon^2), \tag{6.1a}$$

$$(u^\nu)^* = u^\nu + \epsilon\eta^\nu(x,u,\underset{1}{u},\ldots,\underset{k}{u}) + O(\epsilon^2); \tag{6.1b}$$

X: the infinitesimal generator corresponding to $Tg_x(\epsilon)$, $X \in \mathcal{L}_x$, given by

$$X = \xi_i \frac{\partial}{\partial x_i} + \eta^\nu \frac{\partial}{\partial u^\nu}. \tag{6.2}$$

For a *target system* of DE's, denoted by $S\{z, w\}$, with n independent variables $z = (z_1, z_2, \ldots, z_n)$ and m dependent variables $w = (w^1, w^2, \ldots, w^m)$, we use the following notations:

G_z: the group of all admitted continuous transformations;

L_z: the Lie algebra of G_z;

$\mathcal{G}_z$: a subgroup of continuous transformations, $\mathcal{G}_z \subset G_z$;

$\mathcal{L}_z$: the Lie algebra of $\mathcal{G}_z$, $\mathcal{L}_z \subset L_z$;

$Tg_z(\epsilon)$: a one-parameter subgroup, $Tg_z(\epsilon) \in \mathcal{G}_z$, given by

$$z_i^* = z_i + \epsilon\zeta_i(z, w, \underset{1}{w}, \ldots, \underset{k}{w}) + O(\epsilon^2), \tag{6.3a}$$

$$(w^\nu)^* = w^\nu + \epsilon\omega^\nu(z, w, \underset{1}{w}, \ldots, \underset{k}{w}) + O(\epsilon^2); \tag{6.3b}$$

Z: the infinitesimal generator corresponding to $Tg_z(\epsilon)$, $Z \in \mathcal{L}_z$, given by

$$Z = \zeta_i \frac{\partial}{\partial z_i} + \omega^\nu \frac{\partial}{\partial w^\nu}. \tag{6.4}$$

Note that continuous transformations of the form (6.1a,b), (6.3a,b) include Lie groups of point transformations (point symmetries), Lie groups of contact transformations (contact symmetries), and Lie–Bäcklund transformations (Lie–Bäcklund symmetries).

We let μ denote a mapping (assuming one exists) which transforms any solution $u = U(x)$ of $R\{x, u\}$ to a solution $w = W(z)$ of $S\{z, w\}$. In seeking μ, a priori we must restrict μ to a specific mapping form

$$z = \phi(x, u, \underset{1}{u}, \ldots, \underset{\ell}{u}), \tag{6.5a}$$

$$w = \psi(x, u, \underset{1}{u}, \ldots, \underset{\ell}{u}). \tag{6.5b}$$

We let $\mathcal{M}_\ell$ denote the class of mappings of the form (6.5a,b) which depend on at most the ℓth partial derivatives of u. For any $u = U(x)$, the mapping $\mu \in \mathcal{M}_\ell$, given by (6.5a,b), determines $w = W(z)$ and hence $\underset{1}{w}, \underset{2}{w}, \ldots$, in the same manner as in Section 2.3.5.

Any one-parameter subgroup $Tg_x(\epsilon) \in \mathcal{G}_x$ of the form (6.1a,b) induces, through the mapping $\mu \in \mathcal{M}_\ell$ of the form (6.5a,b), either a one-parameter subgroup $Tg_z(\epsilon) \in \mathcal{G}_z$ of the form (6.3a,b) or the identity transformation. The relationship between $Tg_x(\epsilon)$, μ, and $Tg_z(\epsilon)$ is illustrated in Figure 6.2-1.

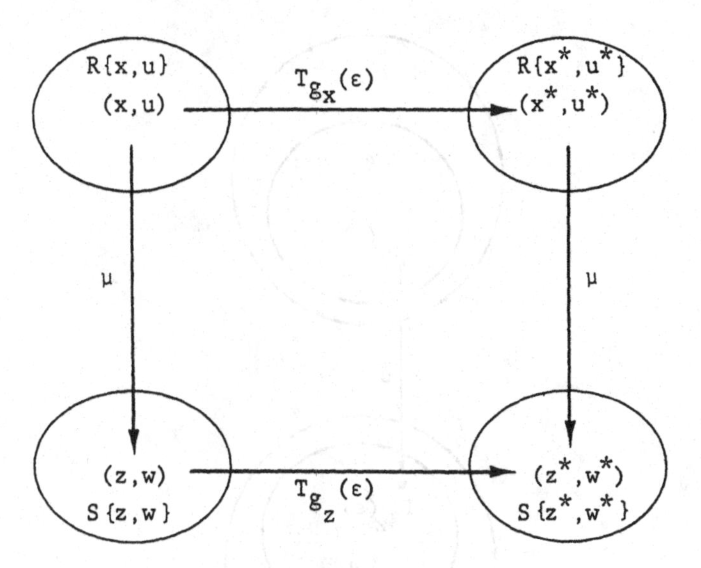

Figure 6.2-1

The relationship between G_x, $\mathcal{G}_x$, G_z, $\mathcal{G}_z$, and μ is illustrated in Figure 6.2-2.

The mapping μ must take any symmetry $Tg_x(\epsilon)$ of the form (6.1a,b) into a symmetry $Tg_z(\epsilon)$ of the form (6.3a,b) such that the composition transformations $\mu \circ Tg_x(\epsilon)$ and $Tg_z(\epsilon) \circ \mu$ yield the same action on (x,u)-space. Specifically

$$Tg_x(\epsilon)(x,u) = (x^*,u^*) = (x + \epsilon\xi + O(\epsilon^2), \quad u + \epsilon\eta + O(\epsilon^2)); \tag{6.6}$$

$$\mu(x,u) = (z,w) = (\phi(x,u,\underset{1}{u},\ldots,\underset{\ell}{u}), \ \psi(x,u,\underset{1}{u},\ldots,\underset{\ell}{u})), \tag{6.7}$$

and hence

$$\begin{aligned}\mu \circ Tg_x(\epsilon)(x,u) &= \mu(x^*,u^*) \\ &= (\phi(x^*,u^*,\underset{1}{u}^*,\ldots,\underset{\ell}{u}^*), \psi(x^*,u^*,\underset{1}{u}^*,\ldots,\underset{\ell}{u}^*))\end{aligned} \tag{6.8}$$

where

$$x^* = x + \epsilon\xi(x,u,\underset{1}{u},\ldots,\underset{k}{u}) + O(\epsilon^2),$$

$$u^* = u + \epsilon\eta(x,u,\underset{1}{u},\ldots,\underset{k}{u}) + O(\epsilon^2),$$

$$\underset{j}{u}^* = \underset{j}{u} + \epsilon\eta^{(j)}(x,u,\underset{1}{u},\ldots,\underset{k+j}{u}) + O(\epsilon^2), \quad j = 1,2,\ldots,\ell.$$

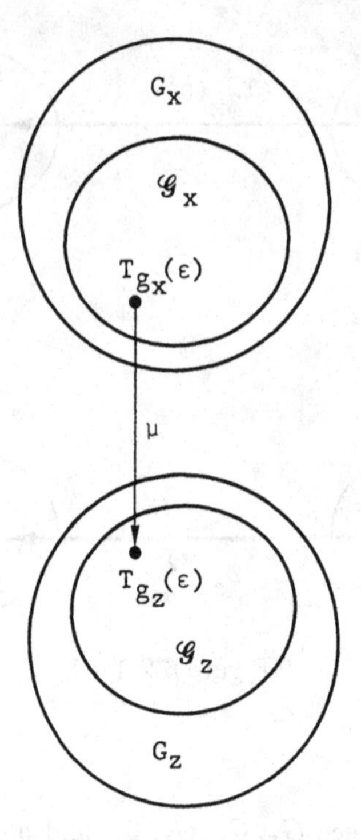

Figure 6.2-2

On the other hand,

$$\begin{aligned} Tg_z(\epsilon) \circ \mu(x,u) &= Tg_z(\epsilon)(z,w) = (z^*, w^*) \\ &= \ (z + \epsilon\zeta + O(\epsilon^2),\ w + \epsilon\omega + O(\epsilon^2)), \end{aligned} \tag{6.9}$$

where

$$\zeta = \zeta(z, w, \underset{1}{w}, \ldots, \underset{k}{w}) = \zeta(\phi, \psi, \underset{1}{\psi}, \ldots, \underset{k}{\psi})$$

and

$$\omega = \omega(z, w, \underset{1}{w}, \ldots, \underset{k}{w}) = \omega(\phi, \psi, \underset{1}{\psi}, \ldots, \underset{k}{\psi})$$

with

$$\phi = \phi(x, u, \underset{1}{u}, \ldots, \underset{\ell}{u}), \quad \psi = \psi(x, u, \underset{1}{u}, \ldots, \underset{\ell}{u}),$$

and

$$\underset{j}{\psi} = \underset{j}{\psi}(x, u, \underset{1}{u}, \ldots, \underset{\ell+j}{u}), \quad j = 1, 2, \ldots, k.$$

$\{\underset{j}{w} = \underset{j}{\psi}\}$, $j = 1, 2, \ldots, k$ is determined from the relations (6.5a,b) as mentioned earlier. Equating the $O(\epsilon)$ terms of (6.8) and (6.9) we obtain the following necessary conditions which the mapping μ, given by (6.5a,b), must satisfy:

$$\xi_i \frac{\partial \phi}{\partial x_i} + \eta^\nu \frac{\partial \phi}{\partial u^\nu} + \eta_i^{(1)\nu} \frac{\partial \phi}{\partial u_i^\nu} + \cdots + \eta_{i_1 i_2 \cdots i_\ell}^{(\ell)\nu} \frac{\partial \phi}{\partial u_{i_1 i_2 \cdots i_\ell}^\nu}$$

$$= \zeta(\phi, \psi, \underset{1}{\psi}, \ldots, \underset{k}{\psi}), \tag{6.10a}$$

$$\xi_i \frac{\partial \psi}{\partial x_i} + \eta^\nu \frac{\partial \psi}{\partial u^\nu} + \eta_i^{(1)\nu} \frac{\partial \psi}{\partial u_i^\nu} + \cdots + \eta_{i_1 i_2 \cdots i_\ell}^{(\ell)\nu} \frac{\partial \psi}{\partial u_{i_1 i_2 \cdots i_\ell}^\nu}$$

$$= \omega(\phi, \psi, \underset{1}{\psi}, \ldots, \underset{k}{\psi}). \tag{6.10b}$$

Equations (6.10a,b) can be expressed concisely in the form

$$\mathrm{X}^{(\ell)}\phi = \mathrm{Z}z, \tag{6.11a}$$

$$\mathrm{X}^{(\ell)}\psi = \mathrm{Z}w, \tag{6.11b}$$

where $\mathrm{X}^{(\ell)}$ is the ℓth extension of X. Note that if $\mathrm{X}^{(\ell)}\phi \equiv 0$, $\mathrm{X}^{(\ell)}\psi \equiv 0$, then $\mathrm{Z} \equiv 0$. In this case the mapping $\mu(x, u) = (\phi, \psi)$ is a differential invariant of $Tg_x(\epsilon)$ and hence induces the identity transformation $Tg_z(\epsilon) = I$ in (z, w)-space.

The relations (6.11a,b) play the essential role in all mapping algorithms we develop in this chapter. For there to be a mapping μ of the form (6.5a,b) from the given system of DE's $R\{x, u\}$ to the target system of DE's $S\{z, w\}$ it is necessary that each infinitesimal generator $\mathrm{X} \in \mathcal{L}_x$ corresponds to some infinitesimal generator $\mathrm{Z} \in \mathcal{L}_z$ and moreover through such a correspondence the components (ϕ, ψ) of μ must satisfy (6.11a,b). In effect the *mapping equations* (6.11a,b) are *necessary conditions* that μ must satisfy in terms of the symmetries of $R\{x, u\}$ and $S\{z, w\}$. In turn these conditions reduce considerably the nature of the dependence of the components (ϕ, ψ) of the mapping μ on $(x, u, \underset{1}{u}, \ldots, \underset{\ell}{u})$. Without such restrictions it is always tedious and often impossible to determine whether or not there exists a mapping $\mu \in \mathcal{M}_\ell$ from the given system of DE's $R\{x, u\}$ to the target system of DE's $S\{z, w\}$.

6.2.1 Theorems on Invertible Mappings

If one allows a non-invertible mapping from a given DE to a target DE then one must consider a particular class of mappings $\mathcal{M}_\ell$ of the form (6.5a,b) for some fixed $\ell = 1, 2, \ldots$. But if one seeks a one-to-one (invertible) mapping from a given DE to a target DE, then the following theorems on invertible mappings show that the class of possible mappings $\mathcal{M}_\ell$ is predetermined ($\mathcal{M}_\ell = \mathcal{M}_1$ if $m = 1$, $\mathcal{M}_\ell = \mathcal{M}_0$ if $m \geq 2$).

Theorem 6.2.1-1 [Case of one dependent variable u ($m = 1$)]. *If $m = 1$ then a mapping μ defines an invertible mapping from $(x, u, \underset{1}{u}, \underset{2}{u}, \ldots, \underset{p}{u})$-space to $(z, w, \underset{1}{w}, \underset{2}{w}, \ldots, \underset{p}{w})$-space for any fixed p if and only if μ is a* **one-to-one contact transformation** *of the form*

$$z = \phi(x, u, \underset{1}{u}), \tag{6.12a}$$

$$w = \psi(x, u, \underset{1}{u}), \tag{6.12b}$$

$$\underset{1}{w} = \underset{1}{\psi}(x, u, \underset{1}{u}) \tag{6.12c}$$

[cf. Section 5.2.4 for the conditions on (ϕ, ψ)].

This theorem is due to Bäcklund (1876). Note that if ϕ and ψ are independent of $\underset{1}{u}$ then (6.12a,b) defines a point transformation.

Theorem 6.2.1-2 [Case of more than one dependent variable u ($m \geq 2$)]. *If $m \geq 2$ then a mapping μ defines an invertible mapping from $(x, u, \underset{1}{u}, \underset{2}{u}, \ldots, \underset{p}{u})$-space to $(z, w, \underset{1}{w}, \underset{2}{w}, \ldots, \underset{p}{w})$-space for any fixed p if and only if μ is a* **one-to-one point transformation** *of the form*

$$z = \phi(x, u), \tag{6.13a}$$

$$w = \psi(x, u). \tag{6.13b}$$

This theorem and its proof are found in Müller and Matschat (1962).

6.3 Mapping of a Given DE to a Specific Target DE

Consider the problem of mapping a given DE to a specific target DE. Both are assumed to admit multi-parameter groups of continuous transformations defined by infinitesimal generators of one-parameter symmetry transformations of the forms (6.1a,b), (6.3a,b) respectively. We consider two types of such mapping problems through various examples:

(i) the mapping μ is non-invertible;

(ii) the mapping μ is an invertible point transformation.

6.3.1 Construction of Non-Invertible Mappings

(1) *The Mapping from the Heat Equation to Burgers' Equation (Derivation of the Hopf-Cole Transformation)*

Hopf (1950) and Cole (1951) independently showed that the mapping μ given by

$$z_1 = \phi_1 = x_1, \tag{6.14a}$$

$$z_2 = \phi_2 = x_2, \tag{6.14b}$$

$$w = \psi = -2u^{-1}\frac{\partial u}{\partial x_1}, \tag{6.14c}$$

transforms any solution $u = U(x_1, x_2)$ of the heat equation

$$\frac{\partial^2 u}{\partial x_1^2} = \frac{\partial u}{\partial x_2}, \tag{6.15}$$

to a solution $w = W(z_1, z_2)$ of Burgers' equation

$$\frac{\partial^2 w}{\partial z_1^2} = \frac{\partial w}{\partial z_2} + w\frac{\partial w}{\partial z_1}. \tag{6.16}$$

We now derive the Hopf–Cole transformation (6.14a–c) from a comparison of the infinitesimal generators of the Lie groups of point transformations admitted by the given PDE (6.15) and the target PDE (6.16) [Bluman (1974b)]. We use the notation $u_1 = \frac{\partial u}{\partial x_1}$, $u_2 = \frac{\partial u}{\partial x_2}$.

Point symmetries admitted by the given PDE (6.15) have infinitesimal generators

$$\begin{gathered}
\mathrm{X}_1 = \frac{\partial}{\partial x_1}, \quad \mathrm{X}_2 = \frac{\partial}{\partial x_2}, \quad \mathrm{X}_3 = x_1\frac{\partial}{\partial x_1} + 2x_2\frac{\partial}{\partial x_2}, \\
\mathrm{X}_4 = x_1 x_2\frac{\partial}{\partial x_1} + (x_2)^2\frac{\partial}{\partial x_2} - \left[\frac{1}{4}(x_1)^2 + \frac{1}{2}x_2\right]u\frac{\partial}{\partial u}, \\
\mathrm{X}_5 = x_2\frac{\partial}{\partial x_1} - \frac{1}{2}x_1 u\frac{\partial}{\partial u}, \quad \mathrm{X}_6 = u\frac{\partial}{\partial u}.
\end{gathered} \tag{6.17}$$

Point symmetries admitted by the target PDE (6.16) have infinitesimal generators

$$\begin{gathered}
\mathrm{Z}_1 = \frac{\partial}{\partial z_1}, \quad \mathrm{Z}_2 = \frac{\partial}{\partial z_2}, \quad \mathrm{Z}_3 = z_1\frac{\partial}{\partial z_1} + 2z_2\frac{\partial}{\partial z_2} - w\frac{\partial}{\partial w}, \\
\mathrm{Z}_4 = z_1 z_2\frac{\partial}{\partial z_1} + (z_2)^2\frac{\partial}{\partial z_2} + [z_1 - z_2 w]\frac{\partial}{\partial w}, \quad \mathrm{Z}_5 = z_2\frac{\partial}{\partial z_1} + \frac{\partial}{\partial w}.
\end{gathered} \tag{6.18}$$

Since the target PDE admits five infinitesimal generators of point symmetries and the given PDE admits six infinitesimal generators of point symmetries it immediately follows that there is no point transformation μ relating (6.15) and (6.16). Examining the coefficients of the infinitesimal generators (6.17) and (6.18), we see that the infinitesimal generators (6.17) admitted by (6.15) have the same action on (x_1, x_2)-space as the infinitesimal generators (6.18) admitted by (6.16) have on (z_1, z_2)-space. We also see that the infinitesimal generator X_6 must map into $Z_6 \equiv 0$. If we set $X_6 \equiv 0$, the commutator table for the Lie algebra $\mathcal{L}_x$ with basis set $\{X_1, X_2, \ldots, X_6\}$ becomes isomorphic to the commutator table for the Lie algebra $\mathcal{L}_z$ with basis set $\{Z_1, Z_2, \ldots, Z_5\}$, i.e. if we set $X_6 \equiv 0$, then

$$[Z_\alpha, Z_\beta] = C^\gamma_{\alpha\beta} Z_\gamma$$

and

$$[X_\alpha, X_\beta] = C^\gamma_{\alpha\beta} X_\gamma$$

with the *same* structure constants $\{C^\gamma_{\alpha\beta}\}$. These observations lead one to seek a transformation μ which could map (6.15) into (6.16). The common group action on independent variable space and the fact that the transformation must be non-invertible indicate that the simplest transformation μ which could map (6.15) into (6.16) is of the form

$$z_1 = x_1, \tag{6.19a}$$

$$z_2 = x_2, \tag{6.19b}$$

$$w = \psi(x_1, x_2, u, u_1, u_2). \tag{6.19c}$$

We now impose the necessary conditions (6.11b) on the mapping function ψ; the necessary conditions (6.11a) are already satisfied by a mapping μ of the form (6.19a–c). In order to do this we must first determine the once-extended infinitesimal generators $\{X_i^{(1)}\}$:

$$\begin{aligned}
X_1^{(1)} = X_1, \quad X_2^{(1)} = X_2, \quad X_3^{(1)} &= X_3 - u_1 \frac{\partial}{\partial u_1} - 2u_2 \frac{\partial}{\partial u_2}, \\
X_4^{(1)} = X_4 &- \left[\frac{1}{2} x_1 u + \frac{1}{4}(x_1)^2 u_1 + \frac{3}{2} x_2 u_1\right] \frac{\partial}{\partial u_1} \\
&- \left[\frac{1}{2} u + x_1 u_1 + \frac{1}{4}(x_1)^2 u_2 + \frac{5}{2} x_2 u_2\right] \frac{\partial}{\partial u_2}, \\
X_5^{(1)} = X_5 &- \frac{1}{2}[u + x_1 u_1] \frac{\partial}{\partial u_1} - \left[u_1 + \frac{1}{2} x_1 u_2\right] \frac{\partial}{\partial u_2}, \\
X_6^{(1)} = X_6 &+ u_1 \frac{\partial}{\partial u_1} + u_2 \frac{\partial}{\partial u_2}.
\end{aligned} \tag{6.20}$$

The necessary conditions (6.11b) become:

$$X_\alpha^{(1)} \psi = Z_\alpha w, \quad \alpha = 1, 2, \ldots, 6$$

with $Z_6 \equiv 0$. Then we obtain

$$\frac{\partial \psi}{\partial x_1} = 0, \tag{6.21a}$$

$$\frac{\partial \psi}{\partial x_2} = 0, \tag{6.21b}$$

$$x_1\frac{\partial \psi}{\partial x_1} + 2x_2\frac{\partial \psi}{\partial x_2} - u_1\frac{\partial \psi}{\partial u_1} - 2u_2\frac{\partial \psi}{\partial u_2} = -\psi, \tag{6.21c}$$

$$\begin{aligned} x_1x_2\frac{\partial \psi}{\partial x_1} + (x_2)^2\frac{\partial \psi}{\partial x_2} - \left[\frac{1}{4}(x_1)^2 + \frac{1}{2}x_2\right]u\frac{\partial \psi}{\partial u} \\ - \left[\frac{1}{2}x_1u + \frac{1}{4}(x_1)^2u_1 + \frac{3}{2}x_2u_1\right]\frac{\partial \psi}{\partial u_1} \\ - \left[\frac{1}{2}u + x_1u_1 + \frac{1}{4}(x_1)^2u_2 + \frac{5}{2}x_2u_2\right]\frac{\partial \psi}{\partial u_2} = x_1 - x_2\psi, \end{aligned} \tag{6.21d}$$

$$x_2\frac{\partial \psi}{\partial x_1} - \frac{1}{2}x_1u\frac{\partial \psi}{\partial u} - \frac{1}{2}[u + x_1u_1]\frac{\partial \psi}{\partial u_1} - \left[u_1 + \frac{1}{2}x_1u_2\right]\frac{\partial \psi}{\partial u_2} = 1. \tag{6.21e}$$

From (6.21a,b) it follows that $\psi = \psi(u, u_1, u_2)$. Then (6.21c–e) reduce to

$$u\frac{\partial \psi}{\partial u} + u_1\frac{\partial \psi}{\partial u_1} + u_2\frac{\partial \psi}{\partial u_2} = 0, \tag{6.22a}$$

$$1 + \frac{1}{2}u\frac{\partial \psi}{\partial u_1} + u_1\frac{\partial \psi}{\partial u_2} = 0, \tag{6.22b}$$

$$\psi - u_1\frac{\partial \psi}{\partial u_1} - 2u_2\frac{\partial \psi}{\partial u_2} = 0, \tag{6.22c}$$

$$u\frac{\partial \psi}{\partial u_2} = 0. \tag{6.22d}$$

The last equation leads to $\psi = \psi(u, u_1)$. Then it is easy to show that the unique solution of (6.22a–d) is the Hopf–Cole transformation

$$\psi = -\frac{2u_1}{u}.$$

(2) *The Mapping from the Modified KdV Equation to the KdV Equation (The Miura Transformation)*

Miura (1968) showed that the mapping μ given by

$$z_1 = x_1, \tag{6.23a}$$

$$z_2 = x_2, \tag{6.23b}$$

$$w = \psi = u^2 \pm i\sqrt{6}u_1, \tag{6.23c}$$

transforms any solution $u = U(x_1, x_2)$ of the modified KdV equation

$$\frac{\partial u}{\partial x_2} + u^2 \frac{\partial u}{\partial x_1} + \frac{\partial^3 u}{\partial x_1^3} = 0 \tag{6.24}$$

to a solution $w = W(z_1, z_2)$ of the KdV equation

$$\frac{\partial w}{\partial z_2} + w \frac{\partial w}{\partial z_1} + \frac{\partial^3 w}{\partial z_1^3} = 0. \tag{6.25}$$

Point symmetries admitted by the modified KdV equation (6.24) have infinitesimal generators

$$\mathrm{X}_1 = \frac{\partial}{\partial x_1}, \quad \mathrm{X}_2 = \frac{\partial}{\partial x_2}, \quad \mathrm{X}_3 = x_1 \frac{\partial}{\partial x_1} + 3x_2 \frac{\partial}{\partial x_2} - u \frac{\partial}{\partial u}. \tag{6.26}$$

Those admitted by the KdV equation (6.25) have infinitesimal generators

$$\mathrm{Z}_1 = \frac{\partial}{\partial z_1}, \quad \mathrm{Z}_2 = \frac{\partial}{\partial z_2}, \quad \mathrm{Z}_3 = z_1 \frac{\partial}{\partial z_1} + 3z_2 \frac{\partial}{\partial z_2} - 2w \frac{\partial}{\partial w},$$

$$\mathrm{Z}_4 = z_2 \frac{\partial}{\partial z_1} + \frac{\partial}{\partial w}. \tag{6.27}$$

Since the Lie algebras generated by (6.26) and (6.27) are not of the same dimension, it immediately follows that there exists no point transformation relating (6.24) and (6.25). Moreover as the group actions of X_1, X_2, X_3 and Z_1, Z_2, Z_3 are the same on their respective spaces of independent variables, we are led to consider the simplest form of mapping (with the modified KdV equation used as the given PDE):

$$z_1 = x_1, \tag{6.28a}$$

$$z_2 = x_2, \tag{6.28b}$$

$$w = \psi(x_1, x_2, u, u_1). \tag{6.28c}$$

The necessary conditions (6.11a) are automatically satisfied and conditions (6.11b) become ($\mathrm{X}_1^{(1)} = \mathrm{X}_1$, $\mathrm{X}_2^{(1)} = \mathrm{X}_2$, $\mathrm{X}_3^{(1)} = \mathrm{X}_3 - 2u_1 \frac{\partial}{\partial u_1} - 4u_2 \frac{\partial}{\partial u_2}$) :

$$\frac{\partial \psi}{\partial x_1} = 0, \tag{6.29a}$$

$$\frac{\partial \psi}{\partial x_2} = 0, \tag{6.29b}$$

$$u \frac{\partial \psi}{\partial u} + 2u_1 \frac{\partial \psi}{\partial u_1} = 2\psi. \tag{6.29c}$$

Consequently (6.28c) reduces to

$$\psi = u^2 F\left(\frac{u_1}{u^2}\right) \tag{6.30}$$

where $F(u_1/u^2)$ is an arbitrary function of u_1/u^2. The substitution of (6.30) into the KdV equation (6.25) leads to the Miura transformation (6.23a–c).

6.3.2 Construction of Invertible Point Mappings

We illustrate the construction of invertible point mappings by finding the point transformation which relates the cylindrical KdV equation and the KdV equation.

In 1979 Zal'mez [cf. Korobeinkov (1982)] showed that the transformation

$$z_1 = \frac{x_1}{\sqrt{x_2}}, \tag{6.31a}$$

$$z_2 = -\frac{2}{\sqrt{x_2}}, \tag{6.31b}$$

$$w = x_2 u - \frac{1}{2}x_1, \tag{6.31c}$$

transforms any solution $u = U(x_1, x_2)$ of the cylindrical KdV equation

$$\frac{\partial u}{\partial x_2} + \frac{1}{2}\frac{u}{x_2} + u\frac{\partial u}{\partial x_1} + \frac{\partial^3 u}{\partial x_1^3} = 0 \tag{6.32}$$

to a solution

$$w = W(z_1, z_2) = \frac{4}{(z_2)^2} U\left(-\frac{2z_1}{z_2}, \frac{4}{(z_2)^2}\right) + \frac{z_1}{z_2}$$

of the KdV equation

$$\frac{\partial w}{\partial z_2} + w\frac{\partial w}{\partial z_1} + \frac{\partial^3 w}{\partial z_1^3} = 0. \tag{6.33}$$

Point symmetries admitted by the cylindrical KdV equation (6.32) have infinitesimal generators

$$\mathrm{X}_1 = 2\sqrt{x_2}\frac{\partial}{\partial x_1} + \frac{1}{\sqrt{x_2}}\frac{\partial}{\partial u},$$

$$\mathrm{X}_2 = 2x_1\sqrt{x_2}\frac{\partial}{\partial x_1} + 4x_2\sqrt{x_2}\frac{\partial}{\partial x_2} + \left[\frac{x_1}{\sqrt{x_2}} - 4\sqrt{x_2}u\right]\frac{\partial}{\partial u}, \tag{6.34}$$

$$\mathrm{X}_3 = x_1\frac{\partial}{\partial x_1} + 3x_2\frac{\partial}{\partial x_2} - 2u\frac{\partial}{\partial u}, \quad \mathrm{X}_4 = \frac{\partial}{\partial x_1}.$$

The infinitesimal generators for point symmetries admitted by the KdV equation are given by (6.27). The Lie algebras arising from (6.27) and (6.34) are isomorphic. An isomorphism (it is not unique) leading to the same commutator table is given by the correspondence

$$\begin{aligned} \mathrm{X}_1 &\longleftrightarrow -\mathrm{Z}_1, \\ \mathrm{X}_2 &\longleftrightarrow \mathrm{Z}_2, \\ \mathrm{X}_3 &\longleftrightarrow -\tfrac{1}{2}\mathrm{Z}_3, \\ \mathrm{X}_4 &\longleftrightarrow \mathrm{Z}_4. \end{aligned} \tag{6.35}$$

This *suggests* the existence of a one-to-one point transformation μ mapping (6.32) to (6.33) of the form

$$z_1 = \phi_1(x_1, x_2, u), \tag{6.36a}$$

$$z_2 = \phi_2(x_1, x_2, u), \tag{6.36b}$$

$$w = \psi(x_1, x_2, u). \tag{6.36c}$$

The necessary conditions (6.11a,b) and the correspondence (6.35) yield the 12 equations

$$\mathrm{X}_1\phi_i = -\mathrm{Z}_1 z_i, \quad \mathrm{X}_2\phi_i = \mathrm{Z}_2 z_i, \quad \mathrm{X}_3\phi_i = -\frac{1}{2}\mathrm{Z}_3 z_i,$$

$$\mathrm{X}_4\phi_i = \mathrm{Z}_4 z_i, \quad i = 1, 2;$$

$$\mathrm{X}_1\psi = -\mathrm{Z}_1 w, \quad \mathrm{X}_2\psi = \mathrm{Z}_2 w, \quad \mathrm{X}_3\psi = -\frac{1}{2}\mathrm{Z}_3 w, \quad \mathrm{X}_4\psi = \mathrm{Z}_4 w.$$

The resulting explicit equations are:

$$2\sqrt{x_2}\frac{\partial\phi_1}{\partial x_1} + \frac{1}{\sqrt{x_2}}\frac{\partial\phi_1}{\partial u} = -1, \tag{6.37a}$$

$$2x_1\sqrt{x_2}\frac{\partial\phi_1}{\partial x_1} + 4x_2\sqrt{x_2}\frac{\partial\phi_1}{\partial x_2} + \left[\frac{x_1}{\sqrt{x_2}} - 4\sqrt{x_2}u\right]\frac{\partial\phi_1}{\partial u} = 0, \tag{6.37b}$$

$$x_1\frac{\partial\phi_1}{\partial x_1} + 3x_2\frac{\partial\phi_1}{\partial x_2} - 2u\frac{\partial\phi_1}{\partial u} = -\frac{1}{2}\phi_1, \tag{6.37c}$$

$$\frac{\partial\phi_1}{\partial x_1} = \phi_2, \tag{6.37d}$$

$$2\sqrt{x_2}\frac{\partial\phi_2}{\partial x_1} + \frac{1}{\sqrt{x_2}}\frac{\partial\phi_2}{\partial u} = 0, \tag{6.37e}$$

$$2x_1\sqrt{x_2}\frac{\partial\phi_2}{\partial x_1} + 4x_2\sqrt{x_2}\frac{\partial\phi_2}{\partial x_2} + \left[\frac{x_1}{\sqrt{x_2}} - 4\sqrt{x_2}u\right]\frac{\partial\phi_2}{\partial u} = 1, \tag{6.37f}$$

$$x_1\frac{\partial\phi_2}{\partial x_1} + 3x_2\frac{\partial\phi_2}{\partial x_2} - 2u\frac{\partial\phi_2}{\partial u} = -\frac{3}{2}\phi_2, \tag{6.37g}$$

$$\frac{\partial \phi_2}{\partial x_1} = 0, \tag{6.37h}$$

$$2\sqrt{x_2}\frac{\partial \psi}{\partial x_1} + \frac{1}{\sqrt{x_2}}\frac{\partial \psi}{\partial u} = 0, \tag{6.37i}$$

$$2x_1\sqrt{x_2}\frac{\partial \psi}{\partial x_1} + 4x_2\sqrt{x_2}\frac{\partial \psi}{\partial x_2} + \left[\frac{x_1}{\sqrt{x_2}} - 4\sqrt{x_2}u\right]\frac{\partial \psi}{\partial u} = 0, \tag{6.37j}$$

$$x_1\frac{\partial \psi}{\partial x_1} + 3x_2\frac{\partial \psi}{\partial x_2} - 2u\frac{\partial \psi}{\partial u} = \psi, \tag{6.37k}$$

$$\frac{\partial \psi}{\partial x_1} = 1. \tag{6.37l}$$

From (6.37e,h) we see that $\phi_2 = \phi_2(x_2)$. Then (6.37f,g) reduce to

$$4x_2\sqrt{x_2}\phi_2'(x_2) = 1, \quad 2x_2\phi_2'(x_2) = -\phi_2(x_2).$$

Hence

$$\phi_2 = -\frac{1}{2\sqrt{x_2}}. \tag{6.38}$$

Substituting (6.38) into (6.37d) and comparing the resulting equation with (6.37a) we see that

$$\frac{\partial \phi_1}{\partial u} = 0, \quad \frac{\partial \phi_1}{\partial x_1} = -\frac{1}{2\sqrt{x_2}}.$$

Then

$$\phi_1 = -\frac{x_1}{2\sqrt{x_2}} + A(x_2) \tag{6.39}$$

where $A(x_2)$ is arbitrary. The substitution of (6.39) successively into (6.37b) and (6.37c) leads to $A(x_2) \equiv 0$. Hence

$$\phi_1 = -\frac{x_1}{2\sqrt{x_2}}. \tag{6.40}$$

Equation (6.37l) leads to

$$\psi = x_1 + B(u, x_2) \tag{6.41}$$

where $B(u, x_2)$ is arbitrary. The substitution of (6.41) into (6.37i) leads to

$$\psi = x_1 - 2x_2u + C(x_2) \tag{6.42}$$

where $C(x_2)$ is arbitrary. The substitution of (6.42) successively into (6.37j) and (6.37k) leads to $C(x_2) \equiv 0$. Hence

$$\psi = x_1 - 2x_2u. \tag{6.43}$$

From (6.38), (6.40), (6.43) we see that the point transformation

$$z_1 = \phi_1 = -\frac{x_1}{2\sqrt{x_2}}, \tag{6.44a}$$

$$z_2 = \phi_2 = -\frac{1}{2\sqrt{x_2}}, \tag{6.44b}$$

$$w = \psi = x_1 - 2x_2 u, \tag{6.44c}$$

defines an invertible mapping ($x_2 > 0$) between the cylindrical KdV equation (6.32) and a target PDE whose admitted group of point transformations is the same as the group of point transformations admitted by the KdV equation. Now we have to determine the target PDE given by the mapping (6.44a–c). It is easy to show that if $u = U(x_1, x_2)$ satisfies the cylindrical KdV equation (6.32) then the mapping (6.44a–c) defines a solution

$$w = W(z_1, z_2) = \frac{z_1}{z_2} - \frac{1}{2(z_2)^2} U\left(\frac{z_1}{z_2}, \frac{1}{4(z_2)^2}\right)$$

of the target PDE

$$\frac{\partial w}{\partial z_2} + w\frac{\partial w}{\partial z_1} - \frac{1}{2}\frac{\partial^3 w}{\partial z_1^3} = 0. \tag{6.45}$$

Note that this target PDE is not the KdV equation (6.33). Why? The reason this could happen is that the KdV equation (6.33) is not the unique third order PDE which admits infinitesimal generators (6.27). There are other ways in addition to (6.35) of establishing the isomorphism between the respective Lie algebras formed by the generators (6.27) and (6.34). Trivially the scaling

$$\tilde{z}_1 = \lambda z_1, \tag{6.46a}$$

$$\tilde{z}_2 = -\frac{1}{2}\lambda^3 z_2, \tag{6.46b}$$

$$\tilde{w} = -2\lambda^{-2} w, \tag{6.46c}$$

maps (6.45) into the KdV equation

$$\frac{\partial \tilde{w}}{\partial \tilde{z}_2} + \tilde{w}\frac{\partial \tilde{w}}{\partial \tilde{z}_1} + \frac{\partial^3 \tilde{w}}{\partial \tilde{z}_1^3} = 0, \tag{6.47}$$

for *any* $\lambda \neq 0$. Under the scaling (6.46a–c), we have

$$
\begin{aligned}
Z_1 &\to \lambda\frac{\partial}{\partial \tilde{z}_1} = \lambda \tilde{Z}_1,\\
Z_2 &\to -\frac{1}{2}\lambda^3\frac{\partial}{\partial \tilde{z}_2} = -\frac{1}{2}\lambda^3\tilde{Z}_2,\\
Z_3 &\to \tilde{z}_1\frac{\partial}{\partial \tilde{z}_1} + 3\tilde{z}_2\frac{\partial}{\partial \tilde{z}_2} - 2\tilde{w}\frac{\partial}{\partial \tilde{w}} = \tilde{Z}_3,\\
Z_4 &\to -2\lambda^{-2}\left[\tilde{z}_2\frac{\partial}{\partial \tilde{z}_1} + \frac{\partial}{\partial \tilde{w}}\right] = -2\lambda^{-2}\tilde{Z}_4.
\end{aligned}
\tag{6.48}
$$

It is easy to check that the commutation relations $[Z_\alpha, Z_\beta] = C^\gamma_{\alpha\beta}Z_\gamma$, $[\tilde{Z}_\alpha, \tilde{Z}_\beta] = C^\gamma_{\alpha\beta}\tilde{Z}_\gamma$, $[X_\alpha, X_\beta] = C^\gamma_{\alpha\beta}X_\gamma$ all have the same structure constants $\{C^\gamma_{\alpha\beta}\}$. If we replace (6.35) by the correspondence

$$
\begin{aligned}
X_1 &\longleftrightarrow -\tilde{Z}_1 = -\lambda^{-1}Z_1,\\
X_2 &\longleftrightarrow \tilde{Z}_2 = -2\lambda^{-3}Z_2,\\
X_3 &\longleftrightarrow -\tfrac{1}{2}\tilde{Z}_3 = -\tfrac{1}{2}Z_3,\\
X_4 &\longleftrightarrow \tilde{Z}_4 = -\tfrac{1}{2}\lambda^2 Z_4,
\end{aligned}
\tag{6.49}
$$

for any constant $\lambda \neq 0$, then the resulting mapping

$$\tilde{z}_1 = -\frac{\lambda x_1}{2\sqrt{x_2}}, \tag{6.50a}$$

$$\tilde{z}_2 = \frac{\lambda^3}{4\sqrt{x_2}}, \tag{6.50b}$$

$$\tilde{w} = -2\lambda^{-2}[x_1 - 2x_2 u], \tag{6.50c}$$

transforms any solution $u = U(x_1, x_2)$ of the cylindrical KdV equation (6.32) to a solution

$$\tilde{w} = \tilde{W}(z_1, z_2) = \frac{\tilde{z}_1}{\tilde{z}_2} + \frac{\lambda^4}{4(\tilde{z}_2)^2}U\left(-\frac{\lambda^2\tilde{z}_1}{2\tilde{z}_2}, \frac{\lambda^6}{16(\tilde{z}_2)^2}\right)$$

of the KdV equation (6.47). The transformation (6.31a–c) corresponds to the special case $\lambda = -2$.

Exercises 6.3

1. (a) Find the most general second order PDE of the form
$$u_{xx} = K(x, t, u, u_x, u_t, u_{tt}, u_{xt})$$
which admits the point symmetries of Burgers' equation.
   (b) Find a transformation which maps any solution of the linear heat equation into a solution of $u_{xx} = \lambda(uu_x + u_t)$.

2. Verify (6.18).

3. From the commutation relations of the Lie algebra generated by (6.17):
   (a) Show that the necessary conditions used to obtain the Hopf–Cole transformation reduce to $\mathrm{X}_\alpha^{(1)}\psi = \mathrm{Z}_\alpha w$, $\alpha = 1, 2, 4$.
   (b) Accordingly reduce the sets of equations (6.21a–e), (6.22a–d).
   (c) Derive the Hopf–Cole transformation from these reduced sets of equations.

4. Show that there exists no mapping of the form
$$z_1 = x_1, \quad z_2 = x_2, \quad w = \psi(x_1, x_2, u, u_1, u_2),$$
which maps Burgers' equation $u_{11} = uu_1 + u_2$ to the heat equation $w_{11} = w_2$.

5. (a) Show that if a given PDE is invariant under translations in x_1 then a mapping μ from the given PDE to a target PDE of the form
$$z = \phi = x, \quad w = \psi(x, u, \underset{1}{u}, \ldots, \underset{k}{u}),$$
is such that $\dfrac{\partial \psi}{\partial x_1} \equiv 0$.
   (b) Find a corresponding result if the given PDE is invariant under scalings of u and the target PDE is invariant under scalings of w.

6. Find the most general third order PDE of the form
$$u_{xxx} = K(x, t, u, u_x, u_t, u_{xx}, u_{xt}, u_{tt}, u_{xxt}, u_{xtt}, u_{ttt})$$
which admits the point symmetries of the KdV equation.

7. Show that there exists no mapping of the form
$$z_1 = x_1, \quad z_2 = x_2, \quad w = \psi(x_1, x_2, u, u_1),$$
which maps the KdV equation $u_{111} + uu_1 + u_2 = 0$ to the modified KdV equation $w_{111} + w^2 w_1 + w_2 = 0$.

8. Verify (6.35).

6.4 Invertible Mappings of Nonlinear PDE's to Linear PDE's

Clearly an important mapping question is: When can a nonlinear system of DE's be related to a linear system of DE's? For example the well-known hodograph transformation maps any quasilinear system of first order PDE's with two independent and two dependent variables into a system of linear PDE's.

Kumei and Bluman (1982) give necessary and sufficient conditions under which a given nonlinear system of PDE's with $n \geq 2$ independent variables and $m \geq 1$ dependent variables can be transformed to a linear system of PDE's by some invertible mapping μ. This paper presents a simple algorithm to determine whether or not these conditions hold for a given nonlinear system of PDE's and, when these conditions hold, another simple algorithm to construct the mapping μ. The algorithm require no knowledge of a specific target linear system of PDE's: In the course of carrying out the calculations associated with the first algorithm a specific target linear system of PDE's emerges naturally as well as the necessary information to construct the mapping μ by the second algorithm.

From Theorems 6.2.1-1,2 it follows that for the case of a nonlinear scalar PDE ($m = 1$) the mapping μ must be an invertible *contact transformation* [cf. Section 5.2.4]

$$z = \phi(x, u, \underset{1}{u}), \tag{6.51a}$$

$$w = \psi(x, u, \underset{1}{u}), \tag{6.51b}$$

and for the case of a nonlinear system of PDE's ($m \geq 2$) the mapping μ must be an invertible *point transformation*

$$z = \phi(x, u), \tag{6.52a}$$

$$w = \psi(x, u). \tag{6.52b}$$

The key observation leading to our algorithms is that a linear system of PDE's $S\{z, w\}$, defined by a linear operator L,

$$\mathrm{L}w = g(z), \tag{6.53}$$

admits an infinite-parameter Lie group of point transformations

$$z^* = z, \tag{6.54a}$$

$$w^* = w + \epsilon\omega, \tag{6.54b}$$

where $\omega = f(z)$ is any function satisfying

$$\mathrm{L}f = 0. \tag{6.55}$$

Any contact transformation maps a contact transformation into another contact transformation; any point transformation maps a point transformation into another point transformation. Hence in the case of a scalar PDE ($m = 1$) the infinite-parameter Lie group of point transformations (6.54a,b) admitted by a target linear PDE (6.53) must correspond to an *infinite-parameter Lie group of contact transformations admitted by the given nonlinear PDE* $R\{x, u\}$ for an invertible mapping μ to exist; in the case of a system of PDE's ($m \geq 2$) the infinite-parameter Lie group of point transformations (6.54a,b) admitted by a target system of linear PDE's (6.53) must correspond to an *infinite-parameter Lie group of point transformations admitted by the given nonlinear system of PDE's* $R\{x, u\}$ for an invertible mapping μ to exist.

Consequently, in the case of a nonlinear *scalar* PDE ($m = 1$), *if the Lie algebra of infinitesimal generators of its admitted Lie group of contact transformations is at most finite-dimensional, then there exists no invertible mapping to any linear PDE;* in the case of a nonlinear *system* of PDE's ($m \geq 2$), *if the Lie algebra of infinitesimal generators of its admitted Lie group of point transformations is at most finite-dimensional, there exists no invertible mapping to any linear system of PDE's.*

6.4.1 Invertible Mappings of Nonlinear Systems of PDE's to Linear Systems of PDE's

Theorem 6.4.1-1 (Necessary conditions for the existence of an invertible mapping). *If there exists an invertible transformation μ which maps a given nonlinear system of PDE's $R\{x, u\}$ ($m \geq 2$) to a linear system of PDE's $S\{z, w\}$, then*

(i) *the mapping must be a point transformation of the form*

$$z_j = \phi_j(x, u), \quad j = 1, 2, \dots, n, \tag{6.56a}$$

$$w^\gamma = \psi^\gamma(x, u), \quad \gamma = 1, 2, \dots, m; \tag{6.56b}$$

(ii) *$R\{x, u\}$ must admit an infinite-parameter Lie group of point transformations having infinitesimal generator*

$$\mathrm{X} = \xi_i(x, u)\frac{\partial}{\partial x_i} + \eta^\nu(x, u)\frac{\partial}{\partial u^\nu} \tag{6.57}$$

with $\xi_i(x, u)$, $\eta^\nu(x, u)$ characterized by

$$\xi_i(x, u) = \sum_{\sigma=1}^{m} \alpha_i^\sigma(x, u) F^\sigma(x, u), \tag{6.58a}$$

$$\eta^\nu(x, u) = \sum_{\sigma=1}^{m} \beta_\nu^\sigma(x, u) F^\sigma(x, u), \tag{6.58b}$$

where $\alpha_i^\sigma(x,u)$, $\beta_\nu^\sigma(x,u)$, $i = 1,2,\ldots,n$, $\nu = 1,2,\ldots,m$, $\sigma = 1,2,\ldots,m$, *are some specific functions of* (x,u), *and* $F = (F^1, F^2, \ldots, F^m)$ *is an arbitrary solution of some linear system of PDE's*

$$\mathrm{L}[X]F = 0 \tag{6.59}$$

with $\mathrm{L}[X]$ *representing a linear differential operator depending on independent variables* $X = (X_1(x,u), X_2(x,u), \ldots, X_n(x,u))$.

Proof. Necessary condition (i) is Theorem 6.2.1-2 proved by Müller and Matschat (1962).

If a mapping μ of the form (6.56a,b) exists, then the resulting linear system of PDE's $S\{z,w\}$, represented by

$$\mathrm{L}[z]w = g(z), \tag{6.60}$$

for some linear differential operator $\mathrm{L}[z]$ depending on independent variables $z = (z_1, z_2, \ldots, z_n)$ with nonhomogeneous term $g(z) = (g^1(z), g^2(z), \ldots, g^m(z))$, admits

$$\mathrm{Z} = f^\gamma(z)\frac{\partial}{\partial w^\gamma} \tag{6.61}$$

where $f(z) = (f^1(z), f^2(z), \ldots, f^m(z))$ is an arbitrary solution of the homogeneous linear system

$$\mathrm{L}[z]f(z) = 0.$$

Corresponding to Z one must have an infinitesimal generator

$$\mathrm{X} = \xi_i(x,u)\frac{\partial}{\partial x_i} + \eta^\nu(x,u)\frac{\partial}{\partial u^\nu} \tag{6.62}$$

admitted by $R\{x,u\}$ such that the components ϕ_j, ψ^γ of the mapping μ and the coefficients $\xi_i(x,u)$, $\eta^\nu(x,u)$ of X satisfy relations (6.11a,b). In particular

$$\xi_i(x,u)\frac{\partial \phi_j}{\partial x_i} + \eta^\nu(x,u)\frac{\partial \phi_j}{\partial u^\nu} = 0, \quad j = 1,2,\ldots,n, \tag{6.63a}$$

$$\xi_i(x,u)\frac{\partial \psi^\gamma}{\partial x_i} + \eta^\nu(x,u)\frac{\partial \psi^\gamma}{\partial u^\nu} = f^\gamma(\phi), \quad \gamma = 1,2,\ldots,m. \tag{6.63b}$$

Since the Jacobian $\frac{\partial(\phi,\psi)}{\partial(x,u)} \neq 0$ for an invertible mapping μ, we can solve (6.63a,b) for $\xi_i(x,u)$, $\eta^\nu(x,u)$. They must be linear homogeneous in f:

$$\xi_i(x,u) = \sum_{\sigma=1}^{m} \alpha_i^\sigma(x,u)(x,u)f^\sigma(\phi(x,u)), \tag{6.64a}$$

$$\eta^\nu(x,u) = \sum_{\sigma=1}^{m} \beta_\nu^\sigma(x,u)f^\sigma(\phi(x,u)), \tag{6.64b}$$

where $\alpha_i^\sigma(x,u)$, $\beta_\nu^\sigma(x,u)$ are specific functions of (x,u). If we set $X(x,u) = \phi(x,u)$, then

$$\begin{aligned} F &= (F^1(x,u), F^2(x,u), \dots, F^m(x,u)) \\ &= (f^1(X(x,u)), f^2(X(x,u)), \dots, f^m(X(x,u))) \end{aligned}$$

satisfies

$$\mathrm{L}[X]F = 0. \quad \square$$

Theorem 6.4.1-2 (Sufficient conditions for the existence of an invertible mapping). *Let a given nonlinear system of PDE's $R\{x,u\}$ $(m \geq 2)$ admit an infinitesimal generator* (6.57) *whose coefficients are of the form* (6.58a,b) *with F being an arbitrary solution of a linear system* (6.59) *with specific independent variables*

$$X(x,u) = (X_1(x,u), X_2(x,u), \dots, X_n(x,u)).$$

If the linear homogeneous system of m first order PDE's for scalar Φ,

$$\alpha_i^\sigma(x,u)\frac{\partial \Phi}{\partial x_i} + \beta_\nu^\sigma(x,u)\frac{\partial \Phi}{\partial u^\nu} = 0, \quad \sigma = 1,2,\dots,m, \tag{6.65}$$

has an n functionally independent solutions

$$X_1(x,u), X_2(x,u), \dots, X_n(x,u),$$

and the linear system of m^2 first order PDE's

$$\alpha_i^\sigma(x,u)\frac{\partial \psi^\gamma}{\partial x_i} + \beta_\nu^\sigma(x,u)\frac{\partial \psi^\gamma}{\partial u^\nu} = \delta^{\gamma\sigma}, \tag{6.66}$$

where $\delta^{\gamma\sigma}$ is the Kronecker symbol, γ, $\sigma = 1,2,\dots,m$, has a solution

$$\psi = (\psi^1(x,u), \psi^2(x,u), \dots, \psi^m(x,u)),$$

then the invertible mapping μ given by

$$z_j = \phi_j(x,u) = X_j(x,u), \quad j = 1,2,\dots,n, \tag{6.67a}$$

$$w^\gamma = \psi^\gamma(x,u), \quad \gamma = 1,2,\dots,m, \tag{6.67b}$$

transforms $R\{x,u\}$ to a linear system of PDE's $S\{z,w\}$,

$$\mathrm{L}[z]w = g(z) \tag{6.68}$$

for some nonhomogeneous term $g(z)$.

Proof. By construction the mapping μ defined by (6.65), (6.66), (6.67a,b) is invertible. Let $f^\sigma(z) = F^\sigma(x,u)$, $\sigma = 1,2,\dots,m$. Then μ transforms X admitted by $R\{x,u\}$ to Z, admitted by a target system $S\{z,w\}$, of the form

$$\mathrm{Z} = f^\gamma(z)\frac{\partial}{\partial w^\gamma} \tag{6.69}$$

for any $f(z) = (f^1(z), f^2(z), \ldots, f^m(z))$ satisfying $\mathrm{L}[z]f(z) = 0$. Since the mapping μ is invertible it follows that any target system of PDE's admitting (6.69) must be of the form (6.68). □

We now consider two examples:

(1) *Linearization by a Hodograph Transformation.* The quasilinear system of first order PDE's ($m = 2$, $n = 2$) $R\{x, u\}$ given by

$$a(u^1,u^2)\frac{\partial u^1}{\partial x_1} + b(u^1,u^2)\frac{\partial u^1}{\partial x_2} + c(u^1,u^2)\frac{\partial u^2}{\partial x_1} + d(u^1,u^2)\frac{\partial u^2}{\partial x_2} = 0, \quad (6.70a)$$

$$p(u^1,u^2)\frac{\partial u^1}{\partial x_1} + q(u^1,u^2)\frac{\partial u^1}{\partial x_2} + r(u^1,u^2)\frac{\partial u^2}{\partial x_1} + s(u^1,u^2)\frac{\partial u^2}{\partial x_2} = 0, \quad (6.70b)$$

with Jacobian $\frac{\partial(u^1,u^2)}{\partial(x_1,x_2)} \neq 0$, admits an infinitesimal generator

$$\mathrm{X} = \xi_1\frac{\partial}{\partial x_1} + \xi_2\frac{\partial}{\partial x_2} \quad (6.71a)$$

where

$$\xi_1 = F^1(u^1,u^2), \quad \xi_2 = F^2(u^1,u^2), \quad (6.71b)$$

is an arbitrary solution of the linear system of PDE's

$$d(u^1,u^2)\frac{\partial F^1}{\partial u^1} - b(u^1,u^2)\frac{\partial F^1}{\partial u^2} - c(u^1,u^2)\frac{\partial F^2}{\partial u^1} + a(u^1,u^2)\frac{\partial F^2}{\partial u^2} = 0, \quad (6.72a)$$

$$s(u^1,u^2)\frac{\partial F^1}{\partial u^1} - q(u^1,u^2)\frac{\partial F^1}{\partial u^2} - r(u^1,u^2)\frac{\partial F^2}{\partial u^1} + p(u^1,u^2)\frac{\partial F^2}{\partial u^2} = 0. \quad (6.72b)$$

From equations (6.58a,b), (6.71b) we identify

$$\alpha_1^1 = \alpha_2^2 = 1, \quad \alpha_1^2 = \alpha_2^1 = 0, \quad \beta_1^1 = \beta_1^2 = \beta_2^1 = \beta_2^2 = 0.$$

From (6.72a,b) it follows that

$$X_1 = u^1, \quad X_2 = u^2.$$

Then equations (6.65) become

$$\frac{\partial \Phi}{\partial x_1} = 0, \quad \frac{\partial \Phi}{\partial x_2} = 0. \quad (6.73)$$

Clearly $\Phi = X_1 = u^1$, $\Phi = X_2 = u^2$ satisfy (6.73). Equations (6.66) become

$$\frac{\partial \psi^1}{\partial x_1} = 1, \quad \frac{\partial \psi^1}{\partial x_2} = 0, \quad \frac{\partial \psi^2}{\partial x_1} = 0, \quad \frac{\partial \psi^2}{\partial x_2} = 1. \quad (6.74)$$

Clearly a particular solution of (6.74) is

$$\psi^1 = x_1, \quad \psi^2 = x_2. \quad (6.75)$$

Then the invertible mapping μ given by

$$z_1 = u^1, \quad z_2 = u^2, \quad w^1 = x_1, \quad w^2 = x_2, \tag{6.76}$$

transforms the nonlinear system of PDE's (6.70a,b) to the linear system $S\{z,w\}$ given by

$$d(z_1,z_2)\frac{\partial w^1}{\partial z_1} - b(z_1,z_2)\frac{\partial w^1}{\partial z_2} - c(z_1,z_2)\frac{\partial w^2}{\partial z_1} + a(z_1,z_2)\frac{\partial w^2}{\partial z_2} = 0. \tag{6.77a}$$

$$s(z_1,z_2)\frac{\partial w^1}{\partial z_1} - q(z_1,z_2)\frac{\partial w^1}{\partial z_2} - r(z_1,z_2)\frac{\partial w^2}{\partial z_1} + p(z_1,z_2)\frac{\partial w^2}{\partial z_2} = 0. \tag{6.77b}$$

The mapping μ given by (6.76) is the well-known hodograph transformation which linearizes (6.70a,b). [If one had chosen a different solution than (6.75) for equations (6.74) then the resulting linear system $S\{z,w\}$ would contain a nonhomogeneous term $g(z)$ in (6.77a,b).]

(2) *Linearization of a Nonlinear Telegraph Equation.* The nonlinear telegraph equations $R\{x,u\}$ given by

$$\frac{\partial u^1}{\partial x_1} = \frac{\partial u^2}{\partial x_2}, \tag{6.78a}$$

$$\frac{\partial u^1}{\partial x_2} = (u^1)^2\frac{\partial u^2}{\partial x_1} + u^1(1-u^1), \tag{6.78b}$$

with Jacobian $\frac{\partial(u^1,u^2)}{\partial(x_1,x_2)} \neq 0$, admit an infinitesimal generator

$$\mathrm{X} = \xi_1\frac{\partial}{\partial x_1} + \xi_2\frac{\partial}{\partial x_2} + \eta^1\frac{\partial}{\partial u^1} + \eta^2\frac{\partial}{\partial u^2} \tag{6.79}$$

where

$$\xi_1 = F^1, \tag{6.80a}$$

$$\xi_2 = e^{-x_2}F^2, \tag{6.80b}$$

$$\eta^1 = e^{-x_2}u^1F^2, \tag{6.80c}$$

$$\eta^2 = F^1, \tag{6.80d}$$

and $F = (F^1, F^2)$ is an arbitrary solution of the linear system of PDE's

$$\frac{\partial F^2}{\partial X_2} - e^{-X_2}\frac{\partial F^1}{\partial X_1} = 0, \tag{6.81a}$$

$$\frac{\partial F^1}{\partial X_2} - e^{-X_2}\frac{\partial F^2}{\partial X_1} = 0, \tag{6.81b}$$

with

$$X_1 = x_1 - u^2, \quad X_2 = x_2 - \log u^1, \tag{6.82}$$

so that F depends only on (X_1, X_2).

Comparing (6.58a,b) and (6.80a–d), we identify

$$\alpha_1^1 = \beta_2^1 = 1, \ \alpha_2^2 = e^{-x_2}, \ \beta_1^2 = e^{-x_2}u^1, \ \alpha_1^2 = \alpha_2^1 = \beta_1^1 = \beta_2^2 = 0.$$

Then equations (6.65) become

$$\frac{\partial \Phi}{\partial x_1} + \frac{\partial \Phi}{\partial u^2} = 0, \quad \frac{\partial \Phi}{\partial x_2} + u^1\frac{\partial \Phi}{\partial u^1} = 0. \tag{6.83}$$

Clearly $\Phi = X_1 = x_1 - u^2$, $\Phi = x_2 - \log u^1$ satisfy (6.83). Equations (6.66) become

$$\begin{gathered}\frac{\partial \psi^1}{\partial x_1} + \frac{\partial \psi^1}{\partial u^2} = 1, \quad \frac{\partial \psi^1}{\partial x_2} + u^1\frac{\partial \psi^1}{\partial u^1} = 0, \\ \frac{\partial \psi^2}{\partial x_1} + \frac{\partial \psi^2}{\partial u^2} = 0, \quad \frac{\partial \psi^2}{\partial x_2} + u^1\frac{\partial \psi^2}{\partial u^1} = e^{x_2}.\end{gathered} \tag{6.84}$$

Clearly a particular solution of (6.84) is

$$\psi^1 = x_1, \quad \psi^2 = e^{x_2}. \tag{6.85}$$

Then the invertible mapping μ given by

$$z_1 = x_1 - u^2, \quad z_2 = x_2 - \log u^1, \quad w^1 = x_1, \quad w^2 = e^{x_2}, \tag{6.86}$$

transforms the nonlinear system of PDE's (6.78a,b) to the linear system $S\{z, w\}$ given by

$$\frac{\partial w^2}{\partial z_2} - e^{-z_2}\frac{\partial w^1}{\partial z_1} = 0, \tag{6.87a}$$

$$\frac{\partial w^1}{\partial z_2} - e^{-z_2}\frac{\partial w^2}{\partial z_1} = 0. \tag{6.87b}$$

Varley and Seymour (1985) found a hodograph-type transformation equivalent to (6.86) which linearized (6.78a,b).

6.4.2 Invertible Mappings of Nonlinear Scalar PDE's to Linear Scalar PDE's

We now consider the construction of invertible mappings which transform a nonlinear scalar PDE to a linear scalar PDE. Theorems 6.4.1-1,2 still hold for mapping a nonlinear scalar PDE to a linear scalar PDE if the nonlinear

PDE admits an infinite-parameter Lie group of point transformations which satisfies the criteria of these theorems.

As an example consider the nonlinear diffusion equation

$$\frac{\partial^2 u}{\partial x_1^2} = \left(\frac{\partial u}{\partial x_1}\right)^2 \frac{\partial u}{\partial x_2}. \tag{6.88}$$

One can show that (6.88) admits an infinite-parameter Lie group of point transformations with infinitesimal generator

$$\mathrm{X} = \xi_1 \frac{\partial}{\partial x_1} \tag{6.89}$$

where

$$\xi_1 = F \tag{6.90}$$

is an arbitrary solution of the linear heat equation

$$\frac{\partial^2 F}{\partial u^2} = \frac{\partial F}{\partial x_2} \tag{6.91}$$

such that F is independent of x_1. Hence Theorems 6.4.1-1,2 apply. From (6.58a,b) and (6.90) we identify

$$\alpha_1^1 = 1, \quad \alpha_2^1 = \beta_1^1 = 0.$$

From (6.91) it follows that

$$X_1 = u, \quad X_2 = x_2.$$

Then (6.65) becomes

$$\frac{\partial \Phi}{\partial x_1} = 0. \tag{6.92}$$

Clearly $\Phi = X_1 = u$, $\Phi = X_2 = x_2$, satisfy (6.92). Equation (6.66) for ψ becomes

$$\frac{\partial \psi}{\partial x_1} = 1. \tag{6.93}$$

A particular solution of (6.93) is

$$\psi = x_1. \tag{6.94}$$

Then the invertible mapping μ given by

$$z_1 = u, \quad z_2 = x_2, \quad w = x_1,$$

transforms (6.88) to

$$\frac{\partial^2 w}{\partial z_1^2} = \frac{\partial w}{\partial z_2}. \tag{6.95}$$

This result was previously found by using symmetry methods in a more ad-hoc manner [Bluman and Cole (1974, Section 2.15)].

Theorems 6.4.1-1,2 do not include the most general possibility of the existence of an invertible mapping of a nonlinear scalar PDE to a linear scalar PDE by means of a contact transformation:

Theorem 6.4.2-1 (Necessary Conditions for the Existence of a Mapping for a Scalar PDE). *If there exists an invertible transformation μ which maps a given nonlinear scalar PDE $R\{x,u\}$ $(m=1)$ to a linear scalar PDE $S\{z,w\}$, then*

(i) *the mapping μ must be a contact transformation of the form*

$$z_j = \phi_j(x,u,\underset{1}{u}), \tag{6.96a}$$

$$w = \psi(x,u,\underset{1}{u}), \tag{6.96b}$$

$$w_j = \psi_j(x,u,\underset{1}{u}), \tag{6.96c}$$

$j=1,2,\ldots,n;$

(ii) *$R\{x,u\}$ must admit an infinite-parameter Lie group of contact transformations with infinitesimal generator*

$$\mathrm{X} = \xi_i(x,u,\underset{1}{u})\frac{\partial}{\partial x_i} + \eta(x,u,\underset{1}{u})\frac{\partial}{\partial u} + \eta_i^{(1)}(x,u,\underset{1}{u})\frac{\partial}{\partial u_i} \tag{6.97}$$

with $\xi_i(x,u,\underset{1}{u})$, $\eta(x,u,\underset{1}{u})$, $\eta_i^{(1)}(x,u,\underset{1}{u})$ characterized by

$$\xi_i(x,u,\underset{1}{u}) = \alpha_i(x,u,\underset{1}{u})F(x,u,\underset{1}{u}) + \alpha_{ij}(x,u,\underset{1}{u})H_j(x,u,\underset{1}{u}), \tag{6.98a}$$

$$\eta(x,u,\underset{1}{u}) = \beta(x,u,\underset{1}{u})F(x,u,\underset{1}{u}) + \beta_j(x,u,\underset{1}{u})H_j(x,u,\underset{1}{u}), \tag{6.98b}$$

$$\eta_i^{(1)}(x,u,\underset{1}{u}) = \lambda_i(x,u,\underset{1}{u})F(x,u,\underset{1}{u}) + \lambda_{ij}(x,u,\underset{1}{u})H_j(x,u,\underset{1}{u}), \tag{6.98c}$$

where α_i, α_{ij}, β, β_j, λ_i, λ_{ij}, $i,j=1,2,\ldots,n$ are some specific functions of $(x,u,\underset{1}{u})$, $F(x,u,\underset{1}{u})$ is an arbitrary solution of some linear scalar PDE

$$\mathrm{L}[X]F = 0 \tag{6.99}$$

with $\mathrm{L}[X]$ representing a linear differential operator depending on independent variables

$$X = (X_1(x,u,\underset{1}{u}), X_2(x,u,\underset{1}{u}), \ldots, X_n(x,u,\underset{1}{u}))$$

of the same order as the order of PDE $R\{x,u\}$, and $H_j(x,u,\underset{1}{u})$ satisfies

$$H_j = \frac{\partial F}{\partial X_j}, \quad j=1,2,\ldots,n. \tag{6.100}$$

Proof. Necessary condition (i) is Theorem 6.2.1-1 due to Bäcklund (1876).

If a mapping μ of the form (6.96a–c) exists, then the resulting linear PDE $S\{z,w\}$, represented by

$$\mathrm{L}[z]w = g(z), \tag{6.101}$$

for some linear differential operator $\mathrm{L}[z]$ depending on independent variables $z=(z_1,z_2,\ldots,z_n)$ with nonhomogeneous term $g(z)$, admits an infinitesimal generator

$$\mathrm{Z} = f(z)\frac{\partial}{\partial w} \tag{6.102}$$

where $f(z)$ is an arbitrary solution of the linear homogeneous PDE

$$\mathrm{L}[z]f(z) = 0.$$

Corresponding to Z, PDE $R\{x,u\}$ must admit an infinitesimal generator of a contact symmetry

$$\mathrm{X} = \xi_i(x,u,\underset{1}{u})\frac{\partial}{\partial x_i} + \eta(x,u,\underset{1}{u})\frac{\partial}{\partial u} + \eta_i^{(1)}(x,u,\underset{1}{u})\frac{\partial}{\partial u_i} \tag{6.103}$$

whose coefficients ξ_i, η, $\eta_i^{(1)}$ and the components ϕ_j, ψ, ψ_j of the mapping μ satisfy relations (6.11a,b). In particular

$$\xi_i\frac{\partial\phi_j}{\partial x_i} + \eta\frac{\partial\phi_j}{\partial u} + \eta_i^{(1)}\frac{\partial\phi_j}{\partial u_i} = 0, \tag{6.104a}$$

$$\xi_i\frac{\partial\psi}{\partial x_i} + \eta\frac{\partial\psi}{\partial u} + \eta_i^{(1)}\frac{\partial\psi}{\partial u_i} = f(z), \tag{6.104b}$$

$$\xi_i\frac{\partial\psi_j}{\partial x_i} + \eta\frac{\partial\psi_j}{\partial u} + \eta_i^{(1)}\frac{\partial\psi_j}{\partial u_i} = \frac{\partial f(z)}{\partial z_j}, \tag{6.104c}$$

where the right-hand side of (6.104a–c) is evaluated at $z=\phi(x,u,\underset{1}{u})$ and $j=1,2,\ldots,n$. Since the Jacobian $\frac{\partial(\phi,\psi,\underset{1}{\psi})}{\partial(x,u,\underset{1}{u})} \neq 0$ for an invertible mapping μ, we can solve (6.104a–c) for ξ_i, η, $\eta_i^{(1)}$. They must be linear homogeneous in $f(z)$, $\frac{\partial f(z)}{\partial z_j}$:

$$\xi_i = \alpha_i(x,u,\underset{1}{u})f(z) + \alpha_{ij}(x,u,\underset{1}{u})\frac{\partial f(z)}{\partial z_j}, \tag{6.105a}$$

$$\eta = \beta(x,u,\underset{1}{u})f(z) + \beta_j(x,u,\underset{1}{u})\frac{\partial f(z)}{\partial z_j}, \tag{6.105b}$$

$$\eta_i^{(1)} = \lambda_i(x,u,\underset{1}{u})f(z) + \lambda_{ij}(x,u,\underset{1}{u})\frac{\partial f(z)}{\partial z_j}, \tag{6.105c}$$

where in the right-hand side of (6.105a–c) $z = \phi(x, u, \underset{1}{u})$ and α_i, α_{ij}, β, β_j, λ_i, λ_{ij} are specific functions of $(x, u, \underset{1}{u})$. If we set $X(x, u, \underset{1}{u}) = \phi(x, u, \underset{1}{u})$, then $F(x, u, \underset{1}{u}) = f(X(x, u, \underset{1}{u}))$ satisfies $\mathrm{L}[X]F = 0$. □

Theorem 6.4.2-2.. (Sufficient Conditions for the Existence of a Mapping for a Scalar PDE). *Let a given nonlinear scalar PDE* ($m = 1$) $R\{x, u\}$ *admit an infinitesimal generator* (6.97) *with coefficients of the form* (6.98a–c) *where F is an arbitrary solution of a linear PDE* (6.99) *of the same order as* $R\{x, u\}$ *and* $H_j = \frac{\partial F}{\partial X_j}$ *with specific independent variables*

$$X(x, u, \underset{1}{u}) = (X_1(x, u, \underset{1}{u}), X_2(x, u, \underset{1}{u}), \ldots, X_n(x, u, \underset{1}{u})). \tag{6.106}$$

Suppose the following four conditions holds:

(i) *the linear homogeneous system of $n+1$ first order PDE's satisfied by the scalar* $\Phi(x, u, \underset{1}{u})$, *namely,*

$$\alpha_i \frac{\partial \Phi}{\partial x_i} + \beta \frac{\partial \Phi}{\partial u} + \lambda_i \frac{\partial \Phi}{\partial u_i} = 0, \tag{6.107a}$$

$$\alpha_{ij} \frac{\partial \Phi}{\partial x_i} + \beta_j \frac{\partial \Phi}{\partial u} + \lambda_{ij} \frac{\partial \Phi}{\partial u_i} = 0, \quad j = 1, 2, \ldots, n, \tag{6.107b}$$

has

$$X_1(x, u, \underset{1}{u}), X_2(x, u, \underset{1}{u}), \ldots, X_n(x, u, \underset{1}{u}),$$

as n functionally independent solutions;

(ii) *the linear system of $n+1$ first order PDE's*

$$\alpha_i \frac{\partial \psi}{\partial x_i} + \beta \frac{\partial \psi}{\partial u} + \lambda_i \frac{\partial \psi}{\partial u_i} = 1, \tag{6.108a}$$

$$\alpha_{ij} \frac{\partial \psi}{\partial x_i} + \beta_j \frac{\partial \psi}{\partial u} + \lambda_{ij} \frac{\partial \psi}{\partial u_i} = 0, \quad j = 1, 2, \ldots, n, \tag{6.108b}$$

has a solution $\psi(x, u, \underset{1}{u})$;

(iii) *the linear system of $n(n+1)$ first order PDE's*

$$\alpha_i \frac{\partial \psi_j}{\partial x_i} + \beta \frac{\partial \phi_j}{\partial u} + \lambda_i \frac{\partial \psi_j}{\partial u_i} = 0, \tag{6.109a}$$

$$\alpha_{ik} \frac{\partial \psi_j}{\partial x_i} + \beta_k \frac{\partial \psi_j}{\partial u} + \lambda_{ik} \frac{\partial \psi_j}{\partial u_i} = \delta_{kj}, \tag{6.109b}$$

$j, k = 1, 2, \ldots, n$, *with Kronecker symbol* δ_{kj}, *has n functionally independent solutions*

$$\underset{1}{\psi}(x, u, \underset{1}{u}) = (\psi_1(x, u, \underset{1}{u}), \psi_2(x, u, \underset{1}{u}), \ldots, \psi_n(x, u, \underset{1}{u}));$$

(iv)

$$(z, w, \underset{1}{w}) = (X(x, u, \underset{1}{u}), \psi(x, u, \underset{1}{u}), \underset{1}{\psi}(x, u, \underset{1}{u}))$$

defines a contact transformation.

Then the invertible mapping μ given by

$$z_j = \phi_j(x, u, \underset{1}{u}) = X_j(x, u, \underset{1}{u}), \tag{6.110a}$$

$$w = \psi(x, u, \underset{1}{u}), \tag{6.110b}$$

$$w_j = \psi_j(x, u, \underset{1}{u}), \tag{6.110c}$$

$j = 1, 2, \ldots, n$, *transforms $R\{x, u\}$ to a linear PDE $S\{z, w\}$:*

$$\mathrm{L}[z]w = g(z), \tag{6.111}$$

for some nonhomogeneous term $g(z)$.

Proof. By construction the mapping μ defined by (6.107)–(6.109) is invertible. Let $f(z) = F(x, u, \underset{1}{u})$. Then μ transforms X admitted by $R\{x, u\}$ to Z, admitted by a target PDE $S\{z, w\}$, of the form

$$\mathrm{Z} = f(z)\frac{\partial}{\partial w} \tag{6.112}$$

for any $f(z)$ satisfying $\mathrm{L}[z]f(z) = 0$. Since the mapping μ is invertible it follows that any target PDE admitting (6.112) must be of the form (6.111). □

Using the properties of a contact transformation [Theorem 5.2.4-1] one can replace conditions (iii) and (iv) of Theorem 6.4.2-2 to determine $\underset{1}{\psi}$ by the following much simpler equations:

$$\frac{\partial \psi}{\partial u_i} - \psi_j \frac{\partial \phi_j}{\partial u_i} = 0, \tag{6.113}$$

$$\frac{\partial \psi}{\partial x_i} + u_i \frac{\partial \psi}{\partial u} = \psi_j \left(\frac{\partial \phi_j}{\partial x_i} + u_i \frac{\partial \phi_j}{\partial u} \right), \tag{6.114}$$

$i = 1, 2, \ldots, n$; $\phi_j = X_j$, $j = 1, 2, \ldots, n$, are determined from condition (i); and ψ is determined from condition (ii).

Now consider two examples of mapping nonlinear scalar PDE's to linear PDE's by contact transformations which are not point transformations:

(1) *Linearization by a Legendre Transformation.* The second order quasilinear PDE ($m = 1$, $n = 2$) $R\{x, u\}$ given by

$$a\left(\frac{\partial u}{\partial x_1}, \frac{\partial u}{\partial x_2}\right)\frac{\partial^2 u}{\partial x_1^2} + 2b\left(\frac{\partial u}{\partial x_1}, \frac{\partial u}{\partial x_2}\right)\frac{\partial^2 u}{\partial x_1 \partial x_2} + c\left(\frac{\partial u}{\partial x_1}, \frac{\partial u}{\partial x_2}\right)\frac{\partial^2 u}{\partial x_2^2} = 0,$$

or, equivalently,

$$a(u_1,u_2)u_{11}+2b(u_1,u_2)u_{12}+c(u_1,u_2)u_{22}=0, \tag{6.115}$$

admits a Lie–Bäcklund generator of the form

$$F(x,u,\underset{1}{u})\frac{\partial}{\partial u}, \tag{6.116}$$

where F satisfies the second order linear PDE

$$a(u_1,u_2)\frac{\partial^2 F}{\partial u_2^2}-2b(u_1,u_2)\frac{\partial^2 F}{\partial u_1 \partial u_2}+c(u_1,u_2)\frac{\partial^2 F}{\partial u_1^2}=0, \tag{6.117}$$

so that F is independent of (x_1,x_2,u).

To apply Theorems 6.4.2-1,2 to this example we must find the unique infinitesimal generator X of the contact symmetry corresponding to the Lie–Bäcklund symmetry (6.116) [cf. Theorem 5.2.4-4]. The characteristic function $W=-F$ leads to

$$\xi_1=-\frac{\partial F}{\partial u_1},\quad \xi_2=-\frac{\partial F}{\partial u_2},\quad \eta=F-u_1\frac{\partial F}{\partial u_1}-u_2\frac{\partial F}{\partial u_2},$$

$$\eta_1^{(1)}=\eta_2^{(1)}=0. \tag{6.118}$$

From equations (6.98a,b), (6.100), (6.117) it follows that

$$X_1=u_1,\quad X_2=u_2;$$

$$\alpha_1=\alpha_2=\alpha_{12}=\alpha_{21}=0,\quad \alpha_{11}=\alpha_{22}=-1,\quad \beta=1,\quad \beta_1=-u_1,$$

$$\beta_2=-u_2,\quad \lambda_1=\lambda_2=\lambda_{11}=\lambda_{12}=\lambda_{21}=\lambda_{22}=0.$$

Then equations (6.107a,b) become

$$\frac{\partial \Phi}{\partial u}=0,\quad \frac{\partial \Phi}{\partial x_1}+u_1\frac{\partial \Phi}{\partial u}=0,\quad \frac{\partial \Phi}{\partial x_2}+u_2\frac{\partial \Phi}{\partial u}=0. \tag{6.119}$$

Clearly $\Phi=X_1=u_1$, $\Phi=X_2=u_2$ satisfy (6.119). Equations (6.108a,b) become

$$\frac{\partial \psi}{\partial u}=1,\quad \frac{\partial \psi}{\partial x_1}+u_1\frac{\partial \psi}{\partial u}=0,\quad \frac{\partial \psi}{\partial x_2}+u_2\frac{\partial \psi}{\partial u}=0. \tag{6.120}$$

It is easy to determine that a particular solution of (6.120) is

$$\psi=u-x_1u_1-x_2u_2. \tag{6.121}$$

Then using (6.113) we obtain

$$\psi_1=-x_1,\quad \psi_2=-x_2. \tag{6.122}$$

Finally it is easy to check that equations (6.114) are satisfied. Thus the invertible mapping μ given by

$$z_1 = u_1, \quad z_2 = u_2, \quad w = u - x_1 u_1 - x_2 u_2, \quad w_1 = -x_1, \quad w_2 = -x_2, \tag{6.123}$$

transforms the nonlinear PDE (6.115a) to the linear PDE $S\{z, w\}$ given by

$$a(z_1, z_2)\frac{\partial^2 w}{\partial z_2^2} - 2b(z_1, z_2)\frac{\partial^2 w}{\partial z_1 \partial z_2} + c(z_1, z_2)\frac{\partial^2 w}{\partial z_1^2} = 0. \tag{6.124}$$

The mapping μ given by (6.123) is the well-known Legendre transformation which linearizes (6.115).

(2) *Linearization of an Equation Arising in a Fluid Flow Problem.* Sukharev (1967) showed that the system of first order PDE's

$$\frac{\partial v^2}{\partial x_2} + \frac{\partial v^1}{\partial x_1} = 0, \tag{6.125a}$$

$$v^1 \frac{\partial v^2}{\partial x_1} - (v^2)^p = 0, \tag{6.125b}$$

which describes a fluid flow through a long pipeline, admits an infinitesimal generator for an infinite-parameter Lie group of point transformations:

$$\mathrm{X} = g(v^2, x_2)\frac{\partial}{\partial x_1} + \left[(v^2)^p \frac{\partial}{\partial v^2} g(v^2, x_2)\right]\frac{\partial}{\partial v^1}, \tag{6.126}$$

where $g(v^2, x_2)$ is an arbitrary solution of the second order linear diffusion equation

$$\frac{\partial}{\partial v^2}\left((v^2)^p \frac{\partial g}{\partial v^2}\right) = \frac{\partial g}{\partial x_2}. \tag{6.127}$$

All other infinitesimal generators of point symmetries admitted by (6.125a,b) belong to a finite-dimensional Lie algebra. The use of (6.126) to linearize (6.125a,b) is left to Exercise 6.4-5.

Alternatively, the form of (6.126), (6.127) suggests that there exists a second order PDE, related to (6.125a,b), which can be linearized. To demonstrate this we introduce a potential function $u(x_1, x_2)$ such that

$$v^1 = -\frac{\partial u}{\partial x_2}, \quad v^2 = \frac{\partial u}{\partial x_1}, \tag{6.128}$$

so that (6.125a,b) reduces to the second order PDE

$$\frac{\partial u}{\partial x_2}\frac{\partial^2 u}{\partial x_1^2} + \left(\frac{\partial u}{\partial x_1}\right)^p = 0, \tag{6.129a}$$

or, equivalently,

$$u_2 u_{11} + (u_1)^p = 0. \tag{6.129b}$$

PDE (6.129b) admits a Lie–Bäcklund generator of the form

$$F(x, u, \underset{1}{u})\frac{\partial}{\partial u} \tag{6.130}$$

where F satisfies the second order linear PDE

$$(u_1)^p \frac{\partial^2 F}{\partial u_1^2} - \frac{\partial F}{\partial x_2} = 0 \tag{6.131}$$

and F is independent of (x_1, u, u_2). The infinitesimals of the unique contact transformation corresponding to (6.130), (6.131) with characteristic function $W = -F$, are given by

$$\xi_1 = -\frac{\partial F}{\partial u_1}, \quad \xi_2 = 0, \quad \eta = F - u_1 \frac{\partial F}{\partial u_1}, \quad \eta_1^{(1)} = 0, \quad \eta_2^{(1)} = \frac{\partial F}{\partial x_2}. \tag{6.132}$$

From equations (6.98a,b), (6.100), (6.131), it follows that

$$X_1 = u_1, \quad X_2 = x_2;$$

$$\alpha_1 = \alpha_2 = \alpha_{12} = \alpha_{21} = \alpha_{22} = 0, \quad \alpha_{11} = -1, \quad \beta = 1, \quad \beta_1 = -u_1,$$
$$\beta_2 = 0, \quad \lambda_1 = \lambda_2 = \lambda_{11} = \lambda_{12} = \lambda_{21} = 0, \quad \lambda_{22} = 1.$$

Then equations (6.107a,b) become

$$\frac{\partial \Phi}{\partial u} = 0, \quad \frac{\partial \Phi}{\partial x_1} + u_1 \frac{\partial \Phi}{\partial u} = 0, \quad \frac{\partial \Phi}{\partial u_2} = 0. \tag{6.133}$$

Clearly $\Phi = X_1 = u_1$, $\Phi = X_2 = x_2$ satisfy (6.133). Equations (6.108a,b) become

$$\frac{\partial \psi}{\partial u} = 1, \quad \frac{\partial \psi}{\partial x_1} + u_1 \frac{\partial \psi}{\partial u} = 0, \quad \frac{\partial \psi}{\partial u_2} = 0. \tag{6.134}$$

This leads to a particular solution

$$\psi = u - x_1 u_1. \tag{6.135}$$

Correspondingly (6.113), (6.114) reduce to

$$\psi_1 = -x_1, \quad \psi_2 = u_2. \tag{6.136}$$

Hence the invertible mapping μ given by

$$z_1 = u_1, \quad z_2 = x_2, \quad w = u - x_1 u_1, \quad w_1 = -x_1, \quad w_2 = u_2, \tag{6.137}$$

transforms the nonlinear PDE (6.129a) to the linear PDE $S\{z, w\}$ given by

$$(z_1)^p \frac{\partial^2 w}{\partial z_1^2} - \frac{\partial w}{\partial z_2} = 0. \tag{6.138}$$

The *non-invertible* transformation composed of μ^{-1} followed by non-invertible transformation (6.128) maps any solution of the second order linear PDE (6.138) to a solution of the original nonlinear system of PDE's (6.125a,b). This linearization first appeared in Kumei (1981).

Exercises 6.4

1. Show that the mapping μ defined by (6.65)–(6.67) is invertible.

2. (a) Find the infinitesimal generator of an infinite-parameter Lie group of point transformations admitted by

$$u_{xx} - \frac{1}{2}(u_x)^2 - u_t = 0. \tag{6.139}$$

(b) Find an invertible mapping relating (6.139) to a linear PDE [Kumei and Bluman (1982)].

3. Find the infinitesimal generator of an infinite-parameter Lie group of point transformations admitted by

$$u_{xx} - (u_x)^2 u_t = 0. \tag{6.140}$$

4. Consider the scalar PDE

$$\frac{\partial^2 u}{\partial x_1 \partial x_2} - \frac{\partial u}{\partial x_1}\frac{\partial u}{\partial x_2} - \frac{\partial u}{\partial x_1} - \frac{\partial u}{\partial x_2} = 0. \tag{6.141}$$

(a) Find the infinitesimal generator of an infinite-parameter Lie group of point transformations admitted by (6.141).

(b) Find an invertible mapping relating (6.141) to a linear PDE. This mapping was previously found by Thomas (1944).

5. Show that the infinitesimal generator (6.126) satisfies the criteria of Theorems 6.4.1-1,2, and hence find an invertible mapping which linearizes (6.125a,b).

6. (a) Show that the nonlinear heat conduction equation

$$\frac{\partial}{\partial x_1}\left[(u^1)^{-2}\frac{\partial u^1}{\partial x_1}\right] - \frac{\partial u^1}{\partial x_2} = 0 \tag{6.142}$$

does not admit an infinite-parameter Lie group of contact transformations.

(b) Consider the related system of PDE's [see Chapter 7]

$$\frac{\partial u^2}{\partial x_2} = (u^1)^{-2}\frac{\partial u^1}{\partial x_1}, \tag{6.143a}$$

$$\frac{\partial u^2}{\partial x_1} = u^1. \tag{6.143b}$$

Show that system (6.143a,b) admits the infinite-parameter Lie group of transformations with infinitesimal generator given by

$$\mathrm{X} = g(u^2, x_2)\frac{\partial}{\partial x_1} - \left[(u^1)^2 \frac{\partial}{\partial u^2} g(u^2, x_2)\right] \frac{\partial}{\partial u^1}, \tag{6.144}$$

where $g(u^2, x_2)$ satisfies the linear heat equation

$$\frac{\partial}{\partial u^2}\left(\frac{\partial g}{\partial u^2}\right) = \frac{\partial g}{\partial x_2}. \tag{6.145}$$

Linearize system (6.143a,b) by an invertible mapping.

(c) Transform (6.143)–(6.145) to (6.125)–(6.127) when $p = 0$.

(d) Show that

$$\mathrm{X} = g(u^2, x_2)\frac{\partial}{\partial x_1} \tag{6.146}$$

is admitted by

$$\left(\frac{\partial u^2}{\partial x_1}\right)^2 \frac{\partial u^2}{\partial x_2} = \frac{\partial^2 u^2}{\partial x_1^2} \tag{6.147}$$

when $g(u^2, x_2)$ satisfies the linear heat equation (6.145). [See Exercise 6.4-3.]

(e) Hence find a mapping μ which transforms any solution of the linear heat equation to a solution of the nonlinear heat conduction equation (6.142). Is this mapping invertible? [cf. Storm (1950), Rosen (1979), Bluman and Kumei (1980), Bluman, Kumei, and Reid (1988).]

7. Consider again Burgers' equation

$$\frac{\partial^2 u^1}{\partial x_1^2} - \frac{\partial u^1}{\partial x_2} - u^1\frac{\partial u^1}{\partial x_1} = 0. \tag{6.148}$$

(a) Show that (6.148) does not admit an infinite-parameter Lie group of contact transformations.

(b) Show that the related system of PDE's [see Chapter 7]

$$\frac{\partial u^2}{\partial x_1} = 2u^1, \tag{6.149a}$$

$$\frac{\partial u^2}{\partial x_2} = 2\frac{\partial u^1}{\partial x_1} - (u^1)^2, \tag{6.149b}$$

admits the infinite-parameter Lie group of transformations with infinitesimal generator

$$\mathrm{X} = e^{u^2/4}\left\{\left[2\frac{\partial g}{\partial x_1}(x_1, x_2) + g(x_1, x_2)u^1\right]\frac{\partial}{\partial u^1} + 4g(x_1, x_2)\frac{\partial}{\partial u^2}\right\} \tag{6.150}$$

where $g(x_1, x_2)$ satisfies the linear heat equation

$$\frac{\partial g}{\partial x_1} - \frac{\partial^2 g}{\partial x_2^2} = 0 \tag{6.151}$$

[Vinogradov and Krasil'shchik (1984), Kersten (1987)].

(c) Use Theorems 6.4.1-1,2 to linearize system (6.149a,b). [Let $F^1 = g(x_1, x_2)$, $F^2 = \frac{\partial g}{\partial x_1}$. Then $\alpha_1^1 = \alpha_1^2 = \alpha_2^1 = \alpha_2^2 = \beta_2^2 = 0$, $\beta_1^1 = u^1 e^{u^2/4}$, $\beta_1^2 = 2e^{u^2/4}$, $\beta_2^1 = 4e^{u^2/4}$. Show that $X_1 = x_1$, $X_2 = x_2$, $\psi^1 = -e^{-u^2/4}$, $\psi^2 = \frac{u^1}{2}e^{-u^2/4}$. Then the mapping $z_1 = x_1$, $z_2 = x_2$, $w^1 = -e^{-u^2/4}$, $w^2 = \frac{u^1}{2}e^{-u^2/4}$ transforms (6.150a,b) to

$$\frac{\partial w^1}{\partial z_2} = \frac{\partial w^2}{\partial z_1}, \quad \frac{\partial w^1}{\partial z_1} = w^2.$$

Then

$$\frac{\partial^2 w^1}{\partial z_1^2} - \frac{\partial w^1}{\partial z_2} = 0$$

and the Hopf–Cole transformation

$$u^1 = -2\frac{w^2}{w^1} = -2\frac{\partial w^1/\partial z_1}{w^1}$$

is derived as an application of Theorems 6.4.1-1,2.]

6.5 Invertible Mappings of Linear PDE's to Linear PDE's with Constant Coefficients

If a linear PDE has constant coefficients there is a whole arsenal of techniques to solve various boundary value problems posed for the equation. This gives rise to the natural questions:

(i) Can we transform a given linear PDE with variable coefficients to a linear PDE with constant coefficients by an invertible point transformation?

(ii) If this is possible, what is the most general form of such point transformations?

The answer to the second question is equivalent to finding the most general invertible point transformation which maps a given linear PDE with constant coefficients to another linear PDE with constant coefficients.

Both of these mapping questions can be fully formulated in terms of the infinitesimal generators of point symmetries admitted by a given linear PDE. In this section we establish necessary and sufficient conditions for mapping a variable coefficient linear PDE to a constant coefficient linear PDE. An algorithm is presented to determine whether such conditions hold for a given linear PDE and to determine the mapping when it exists. This work first appeared in Bluman (1983).

Consider a pth order linear PDE $R\{x,u\}$ with n independent variables $x=(x_1,x_2,\ldots,x_n)$ and dependent variable u:

$$a(x)u + a^i(x)u_i + \cdots + a^{i_1 i_2 \cdots i_p}(x)u_{i_1 i_2 \cdots i_p} = 0 \tag{6.152}$$

defined on domain $D \subset \mathbb{R}^n$. Our aim is to map (6.152) invertibly into a constant coefficient linear PDE $S\{z,w\}$, if possible, in terms of some new independent variables $z=(z_1,z_2,\ldots,z_n)$ and new dependent variable w:

$$bw + b^i w_i + \cdots + b^{i_1 i_2 \cdots i_p} w_{i_1 i_2 \cdots i_p} = 0. \tag{6.153}$$

The mapping μ must be of the form

$$z_i = \phi_i(x), \quad i = 1,2,\ldots,n, \tag{6.154a}$$

$$w = \psi(x,u) = G(x)u, \tag{6.154b}$$

in order to preserve linearity. We call $G(x)$ the *multiplier* of the mapping (6.154a,b). The mapping is invertible provided

$$\det\left|\frac{\partial \phi_i}{\partial x_j}\right| \neq 0 \ \text{ in } D. \tag{6.155}$$

A constant coefficient linear PDE $S\{z,w\}$ with n independent variables is invariant under the n-parameter Lie group of translations $\mathcal{G}_z$ of its independent variables. Hence it is necessary that $R\{x,u\}$ admit at least an n-parameter Lie group of point transformations in order to have an invertible mapping to a constant coefficient linear PDE. Moreover since $\mathcal{G}_z$ is Abelian and the mapping μ must preserve the commutation relations of $\mathcal{L}_z$, it is necessary that there be an n-parameter Abelian subgroup $\mathcal{G}_x$ of the Lie group of point transformations G_x admitted by $R\{x,u\}$. We give an algorithm where the mapping μ is found in the course of establishing the existence of an n-dimensional Abelian subalgebra $\mathcal{L}_x$ of L_x.

A constant coefficient linear PDE $S\{z, w\}$ admits n infinitesimal generators of translations given by

$$\mathrm{Z}_\alpha = \frac{\partial}{\partial z_\alpha}, \quad \alpha = 1, 2, \ldots, n. \tag{6.156}$$

In order for the mapping μ to exist the given linear PDE $R\{x, u\}$ must admit n infinitesimal generators of the form [cf. Theorem 4.2.3-7]

$$\mathrm{X}_\alpha = \xi_{\alpha j}(x)\frac{\partial}{\partial x_j} + f_\alpha(x)u\frac{\partial}{\partial u}, \quad \alpha = 1, 2, \ldots, n, \tag{6.157}$$

which satisfy the commutation relations

$$[\mathrm{X}_\alpha, \mathrm{X}_\beta] = 0, \quad \alpha, \beta = 1, 2, \ldots, n. \tag{6.158}$$

From (6.11a,b) the mapping μ given by (6.154a,b) satisfies the mapping equations

$$\mathrm{X}_\alpha \phi_i = \mathrm{Z}_\alpha z_i = \delta_{\alpha i}, \tag{6.159a}$$

$$\mathrm{X}_\alpha(G(x)u) = \mathrm{Z}_\alpha w = 0, \tag{6.159b}$$

$i, \alpha = 1, 2, \ldots, n$; $\delta_{\alpha i}$ is the Kronecker symbol.

From (6.159a,b) we obtain

$$\xi_{\alpha j}\frac{\partial \phi_i}{\partial x_j} = \delta_{\alpha i}, \quad i, \alpha = 1, 2, \ldots, n, \tag{6.160a}$$

$$\xi_{\alpha j}\frac{\partial G}{\partial x_j} + f_\alpha G = 0, \quad \alpha = 1, 2, \ldots, n. \tag{6.160b}$$

From (6.160a) it immediately follows that

$$\frac{\partial \phi_j}{\partial x_\alpha}\xi_{ji} = \delta_{\alpha i}, \quad i, \alpha = 1, 2, \ldots, n. \tag{6.160c}$$

Hence from (6.155), (6.160a), we see that μ is invertible if and only if

$$\det|\xi_{ij}(x)| \neq 0 \quad \text{in } D. \tag{6.161}$$

From the commutation relations (6.158) we obtain the *integrability conditions*

$$\xi_{\beta k}\frac{\partial \xi_{\alpha j}}{\partial x_k} = \xi_{\alpha k}\frac{\partial \xi_{\beta j}}{\partial x_k}, \quad \alpha, \beta, j = 1, 2, \ldots, n; \tag{6.162a}$$

$$\xi_{\beta k}\frac{\partial f_\alpha}{\partial x_k} = \xi_{\alpha k}\frac{\partial f_\beta}{\partial x_k}, \quad \alpha, \beta = 1, 2, \ldots, n. \tag{6.162b}$$

The following theorem shows that the necessary conditions (6.160a,b), (6.162a,b) are also sufficient conditions to determine the mapping μ.

Theorem 6.5.-1. *If $R\{x,u\}$ admits n infinitesimal generators* (6.157) *whose components $\{\xi_{ij}(x), f_i(x)\}$ satisfy equations* (6.162a,b) *and condition* (6.161) *then there exists a solution $\{\phi_i(x), G(x)\}$ of* (6.160a,b) *which defines an invertible mapping of the linear PDE $R\{x,u\}$ to a constant coefficient PDE $S\{z,w\}$.*

Proof. The proof is accomplished by showing that $\{\phi_i(x), G(x)\}$ solving (6.160a,b) whose coefficients are defined by (6.161), (6.162a,b), satisfies the integrability conditions

$$\frac{\partial^2\phi_i}{\partial x_j\partial x_k} = \frac{\partial^2\phi_i}{\partial x_k\partial x_j}, \quad i,j,k = 1,2,\ldots,n; \tag{6.163a}$$

$$\frac{\partial^2 G}{\partial x_j\partial x_k} = \frac{\partial^2 G}{\partial x_k\partial x_j}, \quad j,k = 1,2,\ldots,n. \tag{6.163b}$$

We show that (6.163a) holds and leave the verification of (6.163b) to Exercise 6.5-1:

Taking $\frac{\partial}{\partial x_k}$ of (6.160a), we obtain

$$\xi_{\alpha j}\frac{\partial^2\phi_i}{\partial x_k\partial x_j} = -\frac{\partial\xi_{\alpha j}}{\partial x_k}\frac{\partial\phi_i}{\partial x_j}.$$

Then

$$\frac{\partial\phi_\alpha}{\partial x_\ell}\xi_{\alpha j}\frac{\partial^2\phi_i}{\partial x_k\partial x_j} = -\frac{\partial\phi_\alpha}{\partial x_\ell}\frac{\partial\xi_{\alpha j}}{\partial x_k}\frac{\partial\phi_i}{\partial x_j},$$

and hence from (6.160c) we have

$$\frac{\partial^2\phi_i}{\partial x_k\partial x_\ell} = -\frac{\partial\phi_\alpha}{\partial x_\ell}\frac{\partial\xi_{\alpha j}}{\partial x_k}\frac{\partial\phi_i}{\partial x_j}. \tag{6.164}$$

From (6.160c) we get

$$\frac{\partial^2\phi_i}{\partial x_k\partial x_\ell} = \frac{\partial\phi_\alpha}{\partial x_k}\xi_{\alpha m}\frac{\partial^2\phi_i}{\partial x_m\partial x_\ell}. \tag{6.165}$$

Substituting (6.164) into the right-hand side of (6.165) and rearranging the order of the terms, we obtain

$$\frac{\partial^2\phi_i}{\partial x_k\partial x_\ell} = -\frac{\partial\phi_\alpha}{\partial x_k}\frac{\partial\phi_\beta}{\partial x_\ell}\xi_{\alpha m}\frac{\partial\xi_{\beta j}}{\partial x_m}\frac{\partial\phi_i}{\partial x_j}. \tag{6.166}$$

After substituting (6.162a) into the right-hand side of (6.166) and using (6.160c) we have

$$\frac{\partial^2\phi_i}{\partial x_k\partial_\ell} = -\frac{\partial\phi_\alpha}{\partial x_k}\frac{\partial\xi_{\alpha j}}{\partial x_\ell}\frac{\partial\phi_i}{\partial x_j}. \tag{6.167}$$

Comparing (6.164) and (6.167), we see that (6.163a) holds.

We now summarize the mapping algorithm which determines whether or not a linear PDE $R\{x,u\}$ can be mapped invertibly to a constant coefficient PDE $S\{z,w\}$ and finds the mapping when it exists:

(i) Find the determining equations for the infinitesimals of the Lie group of point transformations leaving $R\{x,u\}$ invariant. [It is unnecessary to solve the determining equations explicitly.]

(ii) Use the determining equations to check if the coefficients of $R\{x,u\}$ are such that the system of equations (6.162a,b) has a nontrivial solution where $\det|\xi_{ij}(x)| \neq 0$ in some domain D. If the system of equations (6.162a,b) only has trivial solutions where $\det|\xi_{ij}(x)| \equiv 0$ then no invertible mapping μ is possible.

(iii) Solve the system of equations (6.160a) to find $\phi(x) = (\phi_1(x), \phi_2(x), \ldots, \phi_n(x))$.

(iv) Find the multiplier $G(x)$ by solving the system of equations (6.160b) or, equivalently, by solving the system of equations

$$\frac{1}{G}\frac{\partial G}{\partial x_k} = -f_\alpha \frac{\partial \phi_\alpha}{\partial x_k}, \quad k = 1, 2, \ldots, n. \tag{6.168}$$

In the case of two independent variables, $n = 2$, we introduce the notation $\xi_1 = \xi_{11}$, $\xi_2 = \xi_{21}$, $\tau_1 = \xi_{12}$, $\tau_2 = \xi_{22}$. In this notation equations (6.162a,b) become

$$\xi_2 \frac{\partial \xi_1}{\partial x_1} + \tau_2 \frac{\partial \xi_1}{\partial x_2} = \xi_1 \frac{\partial \xi_2}{\partial x_1} + \tau_1 \frac{\partial \xi_2}{\partial x_2}, \tag{6.169a}$$

$$\tau_1 \frac{\partial \tau_2}{\partial x_2} + \xi_1 \frac{\partial \tau_2}{\partial x_1} = \tau_2 \frac{\partial \tau_1}{\partial x_2} + \xi_2 \frac{\partial \tau_1}{\partial x_1}, \tag{6.169b}$$

$$\xi_2 \frac{\partial f_1}{\partial x_1} + \tau_2 \frac{\partial f_1}{\partial x_2} = \xi_1 \frac{\partial f_2}{\partial x_1} + \tau_1 \frac{\partial f_2}{\partial x_2}; \tag{6.169c}$$

equations (6.160a) become the mapping equations

$$\xi_1 \frac{\partial \phi_1}{\partial x_1} + \tau_1 \frac{\partial \phi_1}{\partial x_2} = 1, \tag{6.170a}$$

$$\xi_2 \frac{\partial \phi_1}{\partial x_1} + \tau_2 \frac{\partial \phi_1}{\partial x_2} = 0, \tag{6.170b}$$

$$\xi_2 \frac{\partial \phi_2}{\partial x_1} + \tau_2 \frac{\partial \phi_2}{\partial x_2} = 1, \tag{6.170c}$$

$$\xi_1 \frac{\partial \phi_2}{\partial x_1} + \tau_1 \frac{\partial \phi_2}{\partial x_2} = 0; \tag{6.170d}$$

equations (6.168) become

$$\frac{1}{G}\frac{\partial G}{\partial x_1} = -\left[f_1 \frac{\partial \phi_1}{\partial x_1} + f_2 \frac{\partial \phi_2}{\partial x_1}\right], \tag{6.171a}$$

$$\frac{1}{G}\frac{\partial G}{\partial x_2} = -\left[f_1 \frac{\partial \phi_1}{\partial x_2} + f_2 \frac{\partial \phi_2}{\partial x_2}\right]; \tag{6.171b}$$

and the determinant condition (6.161) becomes

$$\xi_1 \tau_2 \neq \xi_2 \tau_1. \tag{6.172}$$

6.5.1 Examples of Mapping Variable Coefficient PDE's to Constant Coefficient PDE's

We now consider two examples where $R\{x,u\}$ has variable coefficients.

(1) *Parabolic Equation*

It is well-known that for any parabolic equation

$$\frac{\partial^2 v}{\partial x^2}+\alpha(x,y)\frac{\partial v}{\partial x}+\beta(x,y)\frac{\partial v}{\partial y}+\gamma(x,y)v=0, \tag{6.173}$$

there exists a point transformation of the form

$$x_1=x_1(x,y),$$
$$x_2=x_2(x,y),$$
$$u=H(x,y)v,$$

such that (6.173) becomes

$$\frac{\partial^2 u}{\partial x_1^2}+\frac{\partial u}{\partial x_2}+V(x_1,x_2)u=0, \tag{6.174}$$

for some function $V(x_1,x_2)$. We now determine $V(x_1,x_2)$ so that (6.174) can be mapped to a constant coefficient linear PDE $S\{z,w\}$ and find a mapping when this is possible.

If (6.174) admits

$$\xi_\alpha(x_1,x_2)\frac{\partial}{\partial x_1}+\tau_\alpha(x_1,x_2)\frac{\partial}{\partial x_2}+f_\alpha(x_1,x_2)u\frac{\partial}{\partial u},$$

then one can show that the determining equations for the infinitesimals $(\xi_\alpha,\ \tau_\alpha,\ f_\alpha)$ reduce to

$$\tau_\alpha(x_1,x_2)=\tau_\alpha(x_2), \tag{6.175a}$$

$$\xi_\alpha(x_1,x_2)=\frac{1}{2}x_1\tau_\alpha'(x_2)+A_\alpha(x_2), \tag{6.175b}$$

$$f_\alpha(x_1,x_2)=\frac{1}{8}(x_1)^2\tau_\alpha''(x_2)+\frac{1}{2}x_1A_\alpha'(x_2)+B_\alpha(x_2), \tag{6.175c}$$

with $\tau_\alpha(x_2)$, $A_\alpha(x_2)$, $B_\alpha(x_2)$, $V(x_1,x_2)$ satisfying

$$\frac{1}{8}(x_1)^2\tau_\alpha'''(x_2)+\frac{1}{4}\tau_\alpha''(x_2)+\frac{1}{2}x_1A_\alpha''(x_2)+B_\alpha'(x_2)$$
$$+\left[\frac{1}{2}x_1\tau_\alpha'(x_2)+A_\alpha(x_2)\right]\frac{\partial V}{\partial x_1}+\tau_\alpha(x_2)\frac{\partial V}{\partial x_2}+\tau_\alpha'(x_2)V=0, \tag{6.176}$$

$\alpha=1,2$. In order to satisfy (6.172) we must have

$$\tau_2(x_2)[x_1\tau_1'(x_2)+2A_1(x_2)]\neq\tau_1(x_2)[x_1\tau_2'(x_2)+2A_2(x_2)]. \tag{6.177}$$

Equations (6.169a,b) become

$$\tau_1(x_2)\tau_2'(x_2) = \tau_2(x_2)\tau_1'(x_2), \tag{6.178a}$$

$$A_2(x_2)\tau_1'(x_2) + 2A_1'(x_2)\tau_2(x_2) = A_1(x_2)\tau_2'(x_2) + 2A_2'(x_2)\tau_1(x_2), \tag{6.178b}$$

whose general solution is

$$\tau_1(x_2) = k\tau_2(x_2), \tag{6.179a}$$

$$A_1(x_2) = kA_2(x_2) + \ell[\tau_2(x_2)]^{1/2}, \tag{6.179b}$$

for arbitrary constants k, ℓ. In terms of (6.179a,b), the invertibility condition (6.177) becomes

$$\ell[\tau_2(x_2)]^{3/2} \neq 0 \tag{6.180}$$

and hence it is necessary that $\tau_2 \neq 0$, $\ell \neq 0$. The transformation

$$\tilde{z}_1 = \ell z_1,$$
$$\tilde{z}_2 = z_2 + kz_1,$$

maps any constant coefficient linear PDE $S\{z, w\}$ into another constant coefficient linear PDE. Let

$$\tilde{\mathrm{X}}_2 = \mathrm{X}_2,$$
$$\tilde{\mathrm{X}}_1 = \frac{1}{\ell}[\mathrm{X}_1 - k\mathrm{X}_2].$$

If $\tilde{\xi}_2 = \xi_2$, $\tilde{\tau}_2 = \tau_2$, $\tilde{\xi}_1 = \frac{1}{\ell}[\xi_1 - k\xi_2]$, $\tilde{\tau}_1 = \frac{1}{\ell}[\tau_1 - k\tau_2]$, $\tilde{\phi}_1 = \ell\phi_1$, $\tilde{\phi}_2 = \phi_2 + k\phi_1$, then ($\tilde{\xi}_1$, $\tilde{\tau}_1$, $\tilde{\xi}_2$, $\tilde{\tau}_2$, $\tilde{\phi}_1$, $\tilde{\phi}_2$) satisfy (6.170a–d) if and only if (ξ_1, τ_1, ξ_2, τ_2, ϕ_1, ϕ_2) do. Now relabel the quantities by unbarring all barred quantities. Without loss of generality, in (6.179a,b) we can set $k = 0$, $\ell = 1$, i.e.

$$\tau_1(x_2) = 0, \quad A_1(x_2) = [\tau_2(x_2)]^{1/2}. \tag{6.181}$$

Then for $\alpha = 1$ (6.176) becomes

$$x_1 A_1''(x_2) + 2\left[B_1'(x_2) + A_1(x_2)\frac{\partial V}{\partial x_1}\right] = 0 \tag{6.182}$$

and hence it is necessary that

$$\frac{\partial^3 V}{\partial x_1^3} = 0. \tag{6.183}$$

Now assume that $V(x_1, x_2)$ is of the form

$$V(x_1, x_2) = q_0(x_2) + q_1(x_2)x_1 + q_2(x_2)(x_1)^2. \tag{6.184}$$

Then from (6.182), (6.184) we get

$$A_1'' + 4q_2 A_1 = 0, \tag{6.185a}$$

$$B_1' = -q_1 A_1. \tag{6.185b}$$

The integrability condition (6.169c) and equations (6.175c), (6.181) lead to

$$A_1 A_2' - A_2 A_1' = 2B_1'\tau_2 = 2B_1'(A_1)^2. \tag{6.186}$$

Then

$$A_2 = 2B_1 A_1 \tag{6.187}$$

is a solution of (6.186). Finally (6.176) for $\alpha = 2$ leads to B_2 satisfying

$$B_2' = -\frac{1}{4}\tau_2'' - [q_0\tau_2]' - q_1 A_2. \tag{6.188}$$

If we know $\tau_2(x_2)$ then A_1, A_2, B_1, and B_2 can be determined from (6.181), (6.185a,b), (6.187), and (6.188). Hence *from Theorem* 6.5-1 *it follows that* (6.174) *can be mapped invertibly to a constant coefficient linear PDE if and only if* $V(x_1, x_2)$ *is of the form* (6.184).

Now we solve the mapping equations (6.170a–d), (6.171a,b) as follows: Equation (6.170d) leads to

$$\frac{\partial \phi_2}{\partial x_1} = 0,$$

and thus

$$\phi_2 = \phi_2(x_2). \tag{6.189}$$

Then (6.170c) yields

$$\phi_2' = \frac{1}{\tau_2(x_2)}. \tag{6.190}$$

From (6.181), (6.185a), and (6.190) we see that $T = \phi_2(x_2)$ is any solution of the ODE

$$2T'''T' - 3(T'')^2 - 16q_2(T')^2 = 0. \tag{6.191}$$

The solution of ODE (6.191) is left to Exercise 6.5-2. Equations (6.170a,b) lead to

$$\phi_1 = x_1(T')^{1/2} + D(x_2)$$

where

$$D'(x_2) = -A_2(T')^{3/2}.$$

Finally, equations (6.171a,b) lead to the multiplier $G(x)$:

$$\log G(x) = -\frac{1}{4}(x_1)^2\frac{A_1'}{A_1} + x_1\frac{B_1}{A_1} + W(x_2)$$

where

$$W' = B_2 T' - A_2 B_1 (T')^{3/2}.$$

Thus we have determined (ϕ_1, ϕ_2, G). One can then show that $S\{z, w\}$ is the constant coefficient linear PDE

$$\frac{\partial^2 w}{\partial z_1^2} + \frac{\partial w}{\partial z_2} + rw = 0$$

where the constant

$$r = \frac{1}{2}A_1'A_1 - (B_1)^2 + B_2 + q_0(A_1)^2.$$

(2) *Hyperbolic Equation*

Let $R\{x, u\}$ be the hyperbolic equation

$$\frac{\partial^2 u}{\partial x_1 \partial x_2} + \alpha(x_1, x_2)\frac{\partial u}{\partial x_1} + \beta(x_1, x_2)\frac{\partial u}{\partial x_2} + \gamma(x_1, x_2)u = 0. \tag{6.192}$$

PDE (6.192) admits

$$\xi_\lambda(x_1, x_2)\frac{\partial}{\partial x_1} + \tau_\lambda(x_1, x_2)\frac{\partial}{\partial x_2} + f_\lambda(x_1, x_2)u\frac{\partial}{\partial u}$$

if and only if $(\xi_\lambda, \tau_\lambda, f_\lambda)$ satisfy

$$\xi_\lambda = \xi_\lambda(x_1), \tag{6.193a}$$

$$\tau_\lambda = \tau_\lambda(x_2), \tag{6.193b}$$

$$\frac{\partial f_\lambda}{\partial x_1} = -\left[\beta\xi_\lambda'(x_1) + \frac{\partial \beta}{\partial x_1}\xi_\lambda(x_1) + \frac{\partial \beta}{\partial x_2}\tau_\lambda(x_2)\right], \tag{6.193c}$$

$$\frac{\partial f_\lambda}{\partial x_2} = -\left[\alpha\tau_\lambda'(x_2) + \frac{\partial \alpha}{\partial x_2}\tau_\lambda(x_2) + \frac{\partial \alpha}{\partial x_1}\xi_\lambda(x_1)\right], \tag{6.193d}$$

$$\frac{\partial^2 f_\lambda}{\partial x_1 \partial x_2} + \alpha\frac{\partial f_\lambda}{\partial x_1} + \beta\frac{\partial f_\lambda}{\partial x_2} + \gamma[\xi_\lambda'(x_1) + \tau_\lambda'(x_2)] + \frac{\partial \gamma}{\partial x_1}\xi_\lambda(x_1) + \frac{\partial \gamma}{\partial x_2}\tau_\lambda(x_2) = 0, \tag{6.193e}$$

$\lambda = 1, 2$. In order to satisfy (6.172) we must have

$$\xi_1(x_1)\tau_2(x_2) \neq \xi_2(x_1)\tau_1(x_2). \tag{6.194}$$

The integrability conditions (6.169a,b) lead to

$$\tau_2(x_2) = k\tau_1(x_2), \tag{6.195a}$$

$$\xi_1(x_1) = \ell\xi_2(x_1), \tag{6.195b}$$

for arbitrary constants k, ℓ. By letting

$$\tilde{z}_1 = z_1 + kz_2,$$

$$\tilde{z}_2 = z_2 + \ell z_1,$$

without loss of generality we can set $k = \ell = 0$, i.e.

$$\xi_1(x_1) = 0, \quad \tau_2(x_2) = 0. \tag{6.196}$$

Let $\tau_1(x_2)$ and $\xi_2(x_1)$ be arbitrary functions of their respective arguments. From (6.193c,d) and (6.196) we have

$$\frac{\partial f_1}{\partial x_1} = -\frac{\partial \beta}{\partial x_2}\tau_1(x_2), \tag{6.197a}$$

$$\frac{\partial f_1}{\partial x_2} = -\left[\alpha\tau_1'(x_2) + \frac{\partial \alpha}{\partial x_2}\tau_1(x_2)\right], \tag{6.197b}$$

$$\frac{\partial f_2}{\partial x_1} = -\left[\beta\xi_2'(x_1) + \frac{\partial \beta}{\partial x_1}\xi_2(x_1)\right], \tag{6.197c}$$

$$\frac{\partial f_2}{\partial x_2} = -\frac{\partial \alpha}{\partial x_1}\xi_2(x_1). \tag{6.197d}$$

The integrability condition (6.169c) and equations (6.197a,d) lead to the first necessary condition

$$\frac{\partial \beta}{\partial x_2} = \frac{\partial \alpha}{\partial x_1}. \tag{6.198}$$

Let

$$\delta = \frac{\partial \alpha}{\partial x_1} + \alpha\beta - \gamma. \tag{6.199}$$

Then after substituting (6.197a–d) into (6.193e) for $\lambda = 1, 2$, we obtain

$$\frac{\partial \delta}{\partial x_2}\tau_1(x_2) + \delta\tau_1'(x_2) = 0, \tag{6.200a}$$

$$\frac{\partial \delta}{\partial x_1}\xi_2(x_1) + \delta\xi_2'(x_1) = 0, \tag{6.200b}$$

which lead to the second necessary condition

$$\frac{\partial^2}{\partial x_1 \partial x_2}\log\delta = 0. \tag{6.201}$$

In particular $\delta(x_1, x_2)$ must be separable in the form

$$\delta(x_1, x_2) = mA(x_1)B(x_2) \tag{6.202}$$

for arbitrary $A(x_1)$, $B(x_2)$ where $m = 0$ if $\delta \equiv 0$, and $m = 1$ if $\delta \not\equiv 0$. If $m = 1$, then from (6.200a,b),

$$\tau_1(x_2) = \frac{1}{B(x_2)}, \tag{6.203a}$$

$$\xi_2(x_1) = \frac{1}{A(x_1)}. \tag{6.203b}$$

If $m = 0$, then $\tau_1(x_2)$ and $\xi_2(x_1)$ remain as arbitrary functions of their respective arguments. A solution of (6.197a–d) is

$$f_1 = -\alpha\tau_1(x_2), \tag{6.204a}$$

$$f_2 = -\beta\xi_2(x_1). \tag{6.204b}$$

Hence from Theorem 6.5-1 it follows that (6.192) can be mapped invertibly to a constant coefficient linear PDE if and only if its coefficients (α, β, γ) satisfy the equations

$$\frac{\partial \beta}{\partial x_2} = \frac{\partial \alpha}{\partial x_1},$$

$$\frac{\partial \alpha}{\partial x_1} + \alpha\beta - \gamma = mA(x_1)B(x_2)$$

for some functions $A(x_1)$, $B(x_2)$ and constant m.

For any (α, β, γ) satisfying (6.198), (6.199), and (6.202) we now construct a mapping μ to a constant coefficient linear PDE $S\{z, w\}$: The mapping equations (6.170a–d) lead to

$$\phi_1(x_1, x_2) = \phi_1(x_2) = \int \frac{1}{\tau_1(x_2)} dx_2,$$

$$\phi_2(x_1, x_2) = \phi_2(x_1) = \int \frac{1}{\xi_2(x_1)} dx_1.$$

Then (6.171a,b) leads to the multiplier $G(x)$ satisfying

$$\log G(x) = \int \alpha(x_1, x_2) dx_2.$$

Thus we have determined (ϕ_1, ϕ_2, G). The corresponding constant coefficient linear PDE $S\{z, w\}$ is given by

$$\frac{\partial^2 w}{\partial z_1 \partial z_2} - mw = 0. \tag{6.205}$$

In summary we have proved the following theorem:

Theorem 6.5.1-1. *The hyperbolic PDE* (6.192) *can be mapped invertibly by a point transformation to a constant coefficient linear PDE if and only if its coefficients* (α, β, γ) *satisfy*

$$\frac{\partial \beta}{\partial x_2} = \frac{\partial \alpha}{\partial x_1}, \quad \frac{\partial \alpha}{\partial x_1} + \alpha\beta - \gamma = mA(x_1)B(x_2)$$

for some functions $A(x_1)$, $B(x_2)$, *and constant* $m = 0$ *or* 1.

(i) *If* $m = 0$, *then the mapping* μ *given by*

$$z_1 = \phi_1(x_1, x_2) = x_1, \quad z_2 = \phi_2(x_1, x_2) = x_2,$$

$$w = \psi(x_1, x_2) = u \exp\left[\int \alpha(x_1, x_2) dx_2\right],$$

transforms PDE (6.192) *to the wave equation*

$$\frac{\partial^2 w}{\partial z_1 \partial z_2} = 0.$$

(ii) *If $m = 1$, then the mapping μ given by*

$$z_1 = \phi_1(x_1, x_2) = \int B(x_2)dx_2,$$

$$z_2 = \phi_2(x_1, x_2) = \int A(x_1)dx_1,$$

$$w = \psi(x_1, x_2) = u \exp\left[\int \alpha(x_1, x_2)dx_2\right],$$

transforms PDE (6.192) *to the Klein–Gordon equation*

$$\frac{\partial^2 w}{\partial z_1 \partial z_2} - w = 0.$$

6.5.2 Mapping Constant Coefficient PDE's to Constant Coefficient PDE's

Consider the problem of finding the most general invertible point transformation which maps a given constant coefficient linear PDE $R\{x, u\}$ to another constant coefficient linear PDE $S\{z, w\}$. To do this we modify the mapping algorithm developed in Section 6.5 by finding the general solution of equations (6.162a,b) and the general solution of the mapping equations (6.160a,b). This algorithm was applied to $R\{x, u\}$ given by the biharmonic equation

$$\left(\frac{\partial^2}{\partial x_1^2} + \frac{\partial^2}{\partial x_2^2}\right)^2 u = 0 \tag{6.206}$$

in Bluman and Gregory (1985). In this paper it was shown that the most general invertible point transformation which maps PDE (6.206) to another constant coefficient linear PDE is given by the conformal mapping

$$z_1 + iz_2 = \phi = \phi_1 + i\phi_2 = \frac{1}{A}\log\left[1 + A\left(\frac{aZ+b}{cZ+d}\right)\right], \tag{6.207a}$$

$$w = \psi = \left|\frac{d\phi}{dZ}\right| u, \tag{6.207b}$$

where $Z = x_1 + ix_2$, and (A, a, b, c, d) are arbitrary complex constants with $ad - bc \neq 0$. The corresponding real constant coefficient linear PDE is

$$\left(\frac{\partial^2}{\partial z_1^2} + \frac{\partial^2}{\partial z_2^2}\right)^2 w + [A^2 + (\overline{A})^2]\left(\frac{\partial^2}{\partial z_2^2} - \frac{\partial^2}{\partial z_1^2}\right) w$$

$$+ 2i[(\overline{A})^2 - A^2]\frac{\partial^2 w}{\partial z_1 \partial z_2} + |A|^4 w = 0, \tag{6.208}$$

where $\overline{A}$ is the complex conjugate of A.

In the limiting case $A = 0$, (6.207a,b) becomes

$$z_1 + iz_2 = \phi = \phi_1 + i\phi_2 = \frac{aZ+b}{cZ+d}, \tag{6.209a}$$

$$w = \psi = \left|\frac{d\phi}{dZ}\right| u, \tag{6.209b}$$

where (a, b, c, d) are arbitrary complex constants with $ad - bc \neq 0$, and PDE (6.208) is the biharmonic equation

$$\left(\frac{\partial^2}{\partial z_1^2} + \frac{\partial^2}{\partial z_2^2}\right)^2 w = 0. \tag{6.210}$$

Equations (6.209a,b) define the Lie group of point transformations admitted by the biharmonic equation (6.206). Equation (6.209a) is the group of bilinear transformations acting on (x_1, x_2)-space.

Exercises 6.5

1. Verify the integrability conditions (6.163b) of Theorem 6.5-1.
2. Reduce ODE (6.191) to a Riccati equation and two quadratures.
3. (a) Show that if $\gamma \equiv 0$ then PDE (6.192) can be mapped invertibly by a point transformation to the wave equation $\frac{\partial^2 w}{\partial z_1 \partial z_2} = 0$ if and only if α and β are of the form

   $$\alpha(x_1, x_2) = \frac{D'(x_2)}{C(x_1) + D(x_2)}, \quad \beta(x_1, x_2) = \frac{C'(x_1)}{C(x_1) + D(x_2)}$$

   where $C(x_1)$ and $D(x_2)$ are arbitrary differentiable functions of their respective arguments.

   (b) Find the mapping μ [Bluman (1983)].
4. Consider the class of hyperbolic equations given by

   $$c^2(x_1, x_2)\frac{\partial^2 u}{\partial x_1^2} - \frac{\partial^2 u}{\partial x_2^2} = 0.$$

   (a) Show that such a PDE can be mapped invertibly by a point transformation to the wave equation

   $$\frac{\partial^2 w}{\partial z_1^2} - \frac{\partial^2 w}{\partial z_2^2} = 0$$

   if and only if $c(x_1, x_2)$ is of the form

   $$c(x_1, x_2) = \frac{a_0 + 2a_1x_1 + a_2(x_1)^2}{b_0 + 2b_1x_2 + b_2(x_2)^2}$$

where the constants $(a_0, a_1, a_2, b_0, b_1, b_2)$ are related by

$$(a_1)^2 - a_0a_2 = (b_1)^2 - b_0b_2 = \Delta.$$

(b) Find the mapping μ [Hint: Distinguish between the cases $\Delta < 0$, $\Delta > 0$, $\Delta = 0$.] [Bluman (1983)].

5. Consider the hyperbolic equation

$$c^2(x_1)\frac{\partial^2 u}{\partial x_1^2} - \frac{\partial^2 u}{\partial x_2^2} = 0.$$

(a) Show that this PDE can be mapped invertibly by a point transformation to a constant coefficient linear PDE if and only if $c(x_1)$ satisfies the fourth order ODE

$$\left[\frac{c^2c'''}{2cc'' - (c')^2}\right]' = 0.$$

(b) Reduce this ODE to a first order ODE and three quadratures using the reduction algorithm of Chapter 3.

(c) Find the mapping μ [Bluman (1983)].

6. (a) Show that the elliptic equation

$$\frac{\partial^2 u}{\partial x_1^2} + \frac{\partial^2 u}{\partial x_2^2} + \alpha(x_1, x_2)\frac{\partial u}{\partial x_1} + \beta(x_1, x_2)\frac{\partial u}{\partial x_2} + \gamma(x_1, x_2)u = 0$$

can be mapped invertibly by a point transformation to a constant coefficient linear PDE if and only if its coefficients (α, β, γ) satisfy

$$\frac{\partial \beta}{\partial x_1} = \frac{\partial \alpha}{\partial x_2},$$

$$2\left(\frac{\partial \alpha}{\partial x_1} + \frac{\partial \beta}{\partial x_2}\right) + \alpha^2 + \beta^2 - 4\gamma = |K(Z)|^2$$

for some analytic function $K(Z)$ of the complex variable $Z = x_1 + ix_2$.

(b) Find the mapping μ and show that the constant coefficient PDE is equivalent to

(i) Laplace's equation

$$\frac{\partial^2 w}{\partial z_1^2} + \frac{\partial^2 w}{\partial z_2^2} = 0 \text{ if } K(Z) \equiv 0;$$

(ii) the Helmholtz equation

$$\frac{\partial^2 w}{\partial z_1^2} + \frac{\partial^2 w}{\partial z_2^2} - w = 0 \text{ if } K(Z) \not\equiv 0.$$

6.6 Discussion

In this chapter we showed how infinitesimal transformations can be used to determine whether or not a given differential equation can be mapped invertibly into a differential equation of a specific class of target differential equations provided that the specific class of target differential equations can be completely characterized in terms of symmetries. Algorithms were given to construct such mappings when they exist. In particular we fully considered the cases where

(i) the given differential equation is a nonlinear scalar PDE and the target class of differential equations is any linear scalar PDE;

(ii) the given differential equation is a nonlinear system of PDE's and the target class of differential equations is any linear system of PDE's;

(iii) the given differential equation is a linear scalar PDE with variable coefficients and the target class of differential equations is any linear scalar PDE with constant coefficients.

We also considered the problem of finding the most general invertible mapping which can transform a given linear scalar PDE with constant coefficients to another linear scalar PDE with constant coefficients.

Another type of mapping problem (*the equivalence problem*) involves finding the most general invertible point transformation (*the equivalence transformation*) which leaves a specific class of differential equations invariant. An example of such a problem is the following: Consider a wave equation of the form

$$\frac{\partial^2 u}{\partial t^2} - v(x)\frac{\partial^2 u}{\partial x^2} = 0 \quad (v(x) > 0).$$

Find the most general invertible point transformation, acting on the augmented (x, t, u, v)-space, which transforms any solution of this PDE into a solution of a wave equation of the same form, which becomes, after relabelling the transformed variables,

$$\frac{\partial^2 u}{\partial t^2} - V(x)\frac{\partial^2 u}{\partial x^2} = 0,$$

for some $V(x) > 0$. Here the equivalence problem involves finding all $V(x)$ equivalent to $v(x)$.

Clearly the set of all point transformations defined by the equivalence transformation forms a group of transformations. Consequently for a specific class of differential equations one can use Lie's algorithm to compute infinitesimal transformations for admitted equivalence transformations [cf. Ovsiannikov (1982)]. Here the Lie group acts on an augmented space of variables which includes the "dependent" variables associated with the equivalence problem as well as the independent and dependent variables of any member of a specific class of differential equations. After computing the infinitesimals of the equivalence transformation, one then determines the global equivalence transformation. It is important to note that one can use analytic continuation in the complex plane to extend the domain of validity of the parameters of the equivalence transformation obtained by Lie's algorithm.

In the next chapter we consider auxiliary systems of differential equations associated with a given system of one or more differential equations where at least one of the differential equations of the given system is written in a conserved form. This allows us to extend the applicability of Theorems 6.4.1-1,2 to linearize nonlinear PDE's by non-invertible mappings when the auxiliary system of PDE's admits an infinite-parameter Lie group of point transformations.

7

Potential Symmetries

7.1 Introduction

As defined previously, a symmetry group of a differential equation is a group which maps any solution of the differential equation to another solution of the differential equation. In previous chapters we considered symmetries defined by infinitesimal transformations whose infinitesimals depend on independent variables, dependent variables, and derivatives of dependent variables. Such symmetries are *local symmetries* since at any point x the infinitesimals are determined if $u(x)$ is sufficiently smooth in some neighborhood of x. In Chapters 5 and 6, by enlarging the classes of local symmetries admitted by given differential equations from point symmetries to contact symmetries, and, still more generally, to Lie–Bäcklund symmetries, we could find more conservation laws, construct mappings to related differential equations, and determine more invariant solutions.

In this chapter we further enlarge the classes of symmetries of differential equations by considering *nonlocal symmetries* whose infinitesimals, at any point x, depend on the global behavior of $u(x)$. In particular a symmetry is nonlocal if its infinitesimals depend on integrals of dependent variables. We systematically find nonlocal symmetries admitted by a given differential equation by realizing such symmetries as local symmetries which are admitted by an associated auxiliary system of differential equations.

For PDE's we establish a formulation which can be applied to a system of PDE's $R\{x,u\}$, with independent variables x and dependent variables u, when at least one PDE of the system can be written in a conserved form with respect to some choice of its variables. A conserved form naturally leads to auxiliary dependent variables v which are *potentials* and to an auxiliary system of PDE's $S\{x,u,v\}$. Most importantly $R\{x,u\}$ is embedded in $S\{x,u,v\}$: Any solution $(u(x),\ v(x))$ of $S\{x,u,v\}$ will define a solution $u(x)$ of $R\{x,u\}$ and to any solution $u(x)$ of $R\{x,u\}$ there corresponds a function $v(x)$ such that $(u(x),v(x))$ defines a solution of $S\{x,u,v\}$.

Suppose we find local symmetries defining a group G_S admitted by $S\{x,u,v\}$. Any symmetry in G_S maps any solution of $S\{x,u,v\}$ into another solution of $S\{x,u,v\}$ and hence maps any solution of $R\{x,u\}$ into another solution of $R\{x,u\}$. Consequently G_S induces symmetries admitted by $R\{x,u\}$. A *local* symmetry in G_S will induce a *nonlocal* symmetry admitted by $R\{x,u\}$ if the infinitesimals of variables (x,u) of $S\{x,u,v\}$ depend explicitly on the potential variables v. We call such a nonlocal symmetry a *potential symmetry* of $R\{x,u\}$. Potential symmetries admitted by $R\{x,u\}$ can be computed by Lie's algorithm since they are realized as local symmetries admitted by an auxiliary system $S\{x,u,v\}$. A local symmetry of $S\{x,u,v\}$ can be of point, contact, or Lie–Bäcklund type. The potential symmetries in our examples will arise from point symmetries.

A potential symmetry leads to the construction of solutions of a given system of PDE's $R\{x,u\}$ which cannot be obtained as invariant solutions of its local symmetries. Suppose a potential symmetry arises from a point symmetry of an auxiliary system $S\{x,u,v\}$. If $(u(x), v(x))$ is a corresponding invariant solution of $S\{x,u,v\}$, then the solution $u(x)$ of $R\{x,u\}$ is generally not an invariant solution of any point symmetry admitted by $R\{x,u\}$.

Moreover a potential symmetry can be used to construct the solution of a boundary value problem posed for a given system of PDE's $R\{x,u\}$. Assume that a BVP for $R\{x,u\}$ can be embedded in a BVP for an auxiliary system $S\{x,u,v\}$ in the following sense: If $(u(x), v(x))$ solves the BVP for $S\{x,u,v\}$, then $u(x)$ solves the BVP for $R\{x,u\}$. Then a point symmetry admitted by the BVP for $S\{x,u,v\}$ leads to the construction of the solution of the BVP for $R\{x,u\}$. Another important application of potential symmetries is their use in the construction of non-invertible mappings to relate nonlinear PDE's to linear PDE's.

We establish another formulation for a scalar ODE $R\{x,u\}$ which does not admit a point symmetry. We assume that $R\{x,u\}$ can be written in a conserved form through the use of a non-invertible transformation of u which introduces an auxiliary variable v. This leads to an auxiliary ODE $S\{x,v\}$. Any solution $v(x)$ of $S\{x,v\}$ will define a solution $u(x)$ of $R\{x,u\}$. A point symmetry admitted by $S\{x,v\}$ is called a *potential symmetry* of ODE $R\{x,u\}$. A potential symmetry of $R\{x,u\}$ essentially reduces the order of $R\{x,u\}$ since the corresponding point symmetry of $S\{x,v\}$ reduces the order of $S\{x,v\}$.

7.2 Potential Symmetries for Partial Differential Equations

Consider a scalar PDE $R\{x,u\}$ of order k which is written in a conserved form

$$\mathrm{D}_i f^i(x,u,\underset{1}{u},\ldots,\underset{k-1}{u}) = 0 \tag{7.1}$$

with independent variables $x = (x_1, x_2, \ldots, x_n)$ and a single dependent variable u;

$$\mathrm{D}_i = \frac{\partial}{\partial x_i} + u_i\frac{\partial}{\partial u} + u_{ij}\frac{\partial}{\partial u_j} + \cdots + u_{ii_1i_2\cdots i_{k-1}}\frac{\partial}{\partial u_{i_1i_2\cdots i_{k-1}}},$$

$i = 1, 2, \ldots, n$. Since PDE (7.1) is in a conserved form, there exists $\frac{1}{2}n(n-1)$ functions Ψ^{ij}, components of an antisymmetric tensor ($i < j$), such that (7.1) can be expressed in the form [Slebodzinski (1970)]

$$f^i(x, u, \underset{1}{u}, \ldots, \underset{k-1}{u}) = \sum_{i<j}(-1)^j \frac{\partial}{\partial x_j}\Psi^{ij} + \sum_{j<i}(-1)^{i-1}\frac{\partial}{\partial x_j}\Psi^{ji}, \tag{7.2}$$

$i, j = 1, 2, \ldots, n$.

Equations (7.2) define a system of n PDE's with $1 + \frac{1}{2}n(n-1)$ dependent variables (u, Ψ^{ij}). Consequently (7.2) is an underdetermined system of PDE's if $n \geq 3$. We can impose suitable constraints (effectively a change of gauge) on the functions Ψ^{ij} so that system (7.2) becomes a determined system of PDE's. This is accomplished by imposing the conditions

$$\Psi^{ij} = 0 \quad \text{for} \quad j \neq i + 1,$$

and introducing the *potentials* $v = (v^1, v^2, \ldots, v^{n-1})$ with

$$v^i = \Psi^{i,i+1}, \quad i = 1, 2, \ldots, n-1. \tag{7.3}$$

Then the system of PDE's (7.2), associated with $R\{x, u\}$ given by (7.1), becomes the following auxiliary system of PDE's $S\{x, u, v\}$:

$$f^1(x, u, \underset{1}{u}, \ldots, \underset{k-1}{u}) = \frac{\partial}{\partial x_2}v^1,$$

$$f^j(x, u, \underset{1}{u}, \ldots, \underset{k-1}{u}) = (-1)^{j-1}\left[\frac{\partial}{\partial x_{j+1}}v^j + \frac{\partial}{\partial x_{j-1}}v^{j-1}\right], \quad 1 < j < n,$$

$$f^n(x, u, \underset{1}{u}, \ldots, \underset{k-1}{u}) = (-1)^{n-1}\frac{\partial}{\partial x_{n-1}}v^{n-1}. \tag{7.4}$$

If $(u(x), v(x))$ is a solution of the system of PDE's $S\{x, u, v\}$ given by (7.4), then $u(x)$ solves PDE $R\{x, u\}$ given by (7.1).

If $n = 2$, let

$$f^1 = f(x, u, \underset{1}{u}, \ldots, \underset{k-1}{u}),$$

$$f^2 = -g(x, u, \underset{1}{u}, \ldots, \underset{k-1}{u}),$$

so that $R\{x, u\}$ becomes

$$D_1 f - D_2 g = 0. \tag{7.5}$$

Let the potential $\Psi^{12} = v^1 = v$. Consequently the auxiliary system $S\{x, u, v\}$ corresponding to conserved form (7.5) is given by

$$\frac{\partial v}{\partial x_2} = f(x, u, \underset{1}{u}, \ldots, \underset{k-1}{u}), \tag{7.6a}$$

$$\frac{\partial v}{\partial x_1} = g(x, u, \underset{1}{u}, \dots, \underset{k-1}{u}). \tag{7.6b}$$

Now assume that the auxiliary system $S\{x, u, v\}$ given by (7.4) admits a one-parameter (ϵ) Lie group of point transformations

$$x^* = X_S(x, u, v; \epsilon) = x + \epsilon\xi_S(x, u, v) + O(\epsilon^2), \tag{7.7a}$$

$$u^* = U_S(x, u, v; \epsilon) = u + \epsilon\eta_S(x, u, v) + O(\epsilon^2), \tag{7.7b}$$

$$v^* = V_S(x, u, v; \epsilon) = v + \epsilon\zeta_S(x, u, v) + O(\epsilon^2), \tag{7.7c}$$

with infinitesimals ξ_S, η_S, and ζ_S corresponding to x, u, and v, respectively. The infinitesimal generator corresponding to (7.7a–c) is denoted by

$$\mathrm{X}_S = \xi_{Si}(x, u, v)\frac{\partial}{\partial x_i} + \eta_S(x, u, v)\frac{\partial}{\partial u} + \zeta_S^\mu(x, u, v)\frac{\partial}{\partial v^\mu}; \tag{7.8}$$

$\xi_{Si}(x, u, v)$, $i = 1, 2, \dots, n$, denote the components of $\xi_S(x, u, v)$; $\zeta_S^\mu(x, u, v)$, $\mu = 1, 2, \dots, n-1$, denote the components of $\zeta_S(x, u, v)$. The group (7.7a–c) maps any solution of $S\{x, u, v\}$ into another solution of $S\{x, u, v\}$ and hence induces a mapping of any solution of $R\{x, u\}$ into another solution of $R\{x, u\}$. Thus the group (7.7a–c) is a symmetry group of PDE $R\{x, u\}$.

If the infinitesimals $(\xi_S(x, u, v), \eta_S(x, u, v))$ of (7.7a,b) do not depend explicitly on v, i.e.,

$$\frac{\partial \xi_{Si}}{\partial v^\mu} \equiv 0, \quad \frac{\partial \eta_S}{\partial v^\mu} \equiv 0, \quad i = 1, 2, \dots, n, \ \mu = 1, 2, \dots, n-1,$$

then (7.7a–c) only defines a point symmetry (7.7a,b) admitted by $R\{x, u\}$ with infinitesimal generator

$$\mathrm{X} = \xi_i(x, u)\frac{\partial}{\partial x_i} + \eta(x, u)\frac{\partial}{\partial u} \tag{7.9}$$

where

$$\xi_{Si} = \xi_i(x, u), \quad i = 1, 2, \dots, n;$$
$$\eta_S = \eta(x, u).$$

If the infinitesimals $(\xi_S(x, u, v), \eta_S(x, u, v))$ of (7.7a,b) do depend explicitly on v then (7.7a–c) defines a nonlocal symmetry X_S of $R\{x, u\}$. This symmetry is a nonlocal symmetry since the potentials v defined by the auxiliary system (7.4) appear only in derivative form in (7.4). This leads us to the following definition and to the proof of the subsequent theorem:

Definition 7.2-1. The point symmetry (7.7a–c) admitted by the auxiliary system of PDE's $S\{x, u, v\}$ [(7.4)] defines a *potential symmetry* admitted by $R\{x,u\}$ [(7.1)] if and only if the infinitesimals $(\xi_S(x,u,v), \eta_S(x,u,v))$ depend explicitly on v.

Theorem 7.2-1. *A potential symmetry of $R\{x,u\}$ is a nonlocal symmetry of $R\{x,u\}$.*

If $R\{x,u\}$ is a scalar evolution equation with two independent variables $x = (x_1, x_2)$ written in conserved form

$$D_2 u - D_1 f(x, u, u_1, \ldots, u_{k-1}) = 0 \tag{7.10}$$

where

$$u_p = \frac{\partial^p u}{\partial x_1^p}, \quad p = 1, 2, \ldots, k-1,$$

with associated auxiliary system $S\{x,u,v\}$ given by

$$\frac{\partial v}{\partial x_1} = u, \tag{7.11a}$$

$$\frac{\partial v}{\partial x_2} = f(x, u, u_1, \ldots, u_{k-1}), \tag{7.11b}$$

then a solution $(u(x), v(x))$ of $S\{x,u,v\}$ leads to a solution $v(x)$ of the evolution equation $T\{x,v\}$ given by

$$\frac{\partial v}{\partial x_2} = f(x, v_1, v_2, \ldots, v_k). \tag{7.12}$$

The following theorem establishes a one-to-one correspondence between point symmetries of $T\{x,v\}$ and point symmetries of $S\{x,u,v\}$:

Theorem 7.2-2. *A point symmetry of $S\{x,u,v\}$ [(7.11a,b)] induces a point symmetry of $T\{x,v\}$ [(7.12)] and, conversely, a point symmetry of $T\{x,v\}$ induces a point symmetry of $S\{x,u,v\}$.*

Proof. See Exercise 7.2-3. □

Note that equation (7.11a) defines a Bäcklund transformation [cf. Rogers and Shadwick (1982)] relating any solution of PDE (7.12) to a solution of PDE (7.10).

7.2.1 Examples of Potential Symmetries

(1) *Burgers' Equation*

Let $R\{x,u\}$ be Burgers' equation

$$\frac{\partial^2 u}{\partial x_1^2} - u\frac{\partial u}{\partial x_1} - \frac{\partial u}{\partial x_2} = 0. \tag{7.13}$$

Equation (7.13) can be written in the conserved form

$$\mathrm{D}_1\left(2\frac{\partial u}{\partial x_1}-u^2\right)-\mathrm{D}_2(2u)=0. \tag{7.14}$$

The auxiliary system $S\{x,u,v\}$ associated with (7.13) is given by

$$\frac{\partial v}{\partial x_1}=2u, \tag{7.15a}$$

$$\frac{\partial v}{\partial x_2}=2\frac{\partial u}{\partial x_1}-u^2. \tag{7.15b}$$

Vinogradov and Krasil'shchik (1984) [see also Kersten (1987)] showed that the system of PDE's (7.15a,b) admits the infinite-parameter Lie group of point transformations with infinitesimal generator

$$\mathrm{X}_S=e^{v/4}\left\{\left[\frac{2\partial\psi(x)}{\partial x_1}+\psi(x)u\right]\frac{\partial}{\partial u}+4\psi(x)\frac{\partial}{\partial v}\right\}, \tag{7.16a}$$

where $\psi(x)$ is any solution of the linear heat equation, i.e.,

$$\frac{\partial\psi}{\partial x_1}-\frac{\partial^2\psi}{\partial x_2^2}=0. \tag{7.16b}$$

The components of (7.16a) are

$$\xi_{S1}(x,u,v)=\xi_{S2}(x,u,v)\equiv 0,$$

$$\eta_S(x,u,v)=e^{v/4}\left[2\frac{\partial\psi(x)}{\partial x_1}+\psi(x)u\right],$$

$$\zeta_S^1(x,u,v)=4e^{v/4}\psi(x).$$

Consequently (7.16a,b) defines a potential symmetry of (7.13).

If $(u(x),v(x))$ satisfies (7.15a,b), then $u(x)$ solves Burgers' equation (7.13) and $v(x)$ solves $T\{x,v\}$ given by

$$\frac{\partial^2 v}{\partial x_1^2}-\frac{\partial v}{\partial x_2}-\frac{1}{4}\left(\frac{\partial v}{\partial x_1}\right)^2=0. \tag{7.17}$$

[PDE (7.17) is an integrated form of Burgers' equation: If $v(x)$ satisfies (7.17) then $u(x)=\frac{1}{2}\frac{\partial v}{\partial x_1}$ solves Burgers' equation.] It immediately follows from the form of (7.16a,b) that PDE (7.17) admits the infinite-parameter Lie group of point transformations defined by infinitesimal generator

$$\mathrm{Y}=e^{v/4}\psi(x)\frac{\partial}{\partial v}, \tag{7.18}$$

where $\psi(x)$ is any solution of (7.16b). Thus infinitesimal generator (7.16a,b) leads to a potential symmetry for Burgers' equation (7.13) and to a point symmetry of the integrated form of Burgers' equation (7.17).

(2) *Nonlinear Heat Conduction Equation*

Let $R\{x,u\}$ be the nonlinear heat conduction equation

$$\frac{\partial}{\partial x_1}\left(K(u)\frac{\partial u}{\partial x_1}\right) - \frac{\partial u}{\partial x_2} = 0. \tag{7.19}$$

As it is written, PDE (7.19) is already in a conserved form. Its associated auxiliary system $S\{x,u,v\}$ is given by

$$\frac{\partial v}{\partial x_1} = u, \tag{7.20a}$$

$$\frac{\partial v}{\partial x_2} = K(u)\frac{\partial u}{\partial x_1}. \tag{7.20b}$$

The infinitesimal generators for point symmetries admitted by (7.20a,b) were given in Section 4.3.4. From the forms of these infinitesimal generators we have [Bluman, Kumei, and Reid (1988)]:

Theorem 7.2.1-1. *The nonlinear heat conduction equation* (7.19) *admits a potential symmetry, corresponding to auxiliary system* (7.20a,b), *if and only if the conductivity* $K(u)$ *is of the form*

$$K(u) = \frac{1}{u^2+pu+q}\exp\left[r\int\frac{du}{u^2+pu+q}\right], \tag{7.21}$$

where p, q, *and* r *are arbitrary constants.*

The corresponding infinitesimal generators of potential symmetries admitted by (7.19) are listed below. Two cases arise:

(i) $\boxed{K(u) = \lambda(u+\kappa)^{-2},\ \lambda,\ \kappa \text{ arbitrary constants.}}$ Here (7.19) admits potential symmetries

$$\mathrm{X}_{S1} = -x_1(v+\kappa x_1)\frac{\partial}{\partial x_1} + (u+\kappa)(v+2\kappa x_1+x_1u)\frac{\partial}{\partial u}$$

$$+ [2\lambda x_2 + \kappa x_1(v+\kappa x_1)]\frac{\partial}{\partial v}; \tag{7.22a}$$

$$\mathrm{X}_{S2} = -x_1[(v+\kappa x_1)^2 + 2\lambda x_2]\frac{\partial}{\partial x_1} + 4\lambda(x_2)^2\frac{\partial}{\partial x_2}$$

$$+ (u+\kappa)[6\lambda x_2 + (v+\kappa x_1)^2 + 2x_1(u+\kappa)(v+\kappa x_1)]\frac{\partial}{\partial u}$$

$$+\left[\kappa x_1(v+\kappa x_1)^2+2\lambda x_2(2v+3\kappa x_1)\right]\frac{\partial}{\partial v}; \tag{7.22b}$$

$$X_{S3}=\psi(z,x_2)\frac{\partial}{\partial x_1}-(u+\kappa)^2\frac{\partial\psi(z,x_2)}{\partial z}\frac{\partial}{\partial u}-\kappa\psi(z,x_2)\frac{\partial}{\partial v}, \tag{7.22c}$$

where $z=v+\kappa x_1$, and $w=\psi(z,x_2)$ is an arbitrary solution of the linear heat equation

$$\lambda\frac{\partial^2 w}{\partial z^2}-\frac{\partial w}{\partial x_2}=0. \tag{7.22d}$$

(ii) $\boxed{K(u)=\dfrac{1}{u^2+pu+q}\exp\left[r\displaystyle\int\frac{du}{u^2+pu+q}\right]}$ (p, q, r arbitrary constants such that $p^2-4q-r^2\neq 0$).

Equation (7.19) admits the potential symmetry

$$X_S=v\frac{\partial}{\partial x_1}+(r-p)x_2\frac{\partial}{\partial x_2}-(u^2+pu+q)\frac{\partial}{\partial u}-(qx_1+pv)\frac{\partial}{\partial v}. \tag{7.23}$$

If $(u(x),v(x))$ satisfies (7.20a,b) then $u(x)$ solves the nonlinear heat conduction equation (7.19) and $v(x)$ solves $T\{x,v\}$ given by

$$K\left(\frac{\partial v}{\partial x_1}\right)\frac{\partial^2 v}{\partial x_1^2}-\frac{\partial v}{\partial x_2}=0. \tag{7.24}$$

[PDE (7.24) is an integrated form of the nonlinear heat conduction equation: If $v(x)$ satisfies (7.24) then $u(x)=\frac{\partial v}{\partial x_1}$ solves (7.19).] Consequently from the form of the infinitesimal generators admitted by (7.20a,b) [cf. Section 4.3.4] we obtain the following group classification of point symmetries admitted by (7.24) [$\frac{\partial v}{\partial x_1}=v_1$]:

(i) $\boxed{K(v_1)\text{ arbitrary}}$
Equation (7.24) admits

$$Y_1=\frac{\partial}{\partial v},\quad Y_2=\frac{\partial}{\partial x_1},\quad Y_3=\frac{\partial}{\partial x_2},$$
$$Y_4=x_1\frac{\partial}{\partial x_1}+2x_2\frac{\partial}{\partial x_2}+v\frac{\partial}{\partial v}. \tag{7.25}$$

(ii) $\boxed{K(v_1)=\lambda(v_1+\kappa)^\nu,\ \lambda,\ \kappa,\ \nu\ (\neq -2)\text{ arbitrary constants.}}$
In this case (7.24) admits infinitesimal generators (7.25) and

$$Y_5=x_1\frac{\partial}{\partial x_1}+\left[\left(1+\frac{2}{\nu}\right)v+\frac{2\kappa x_1}{\nu}\right]\frac{\partial}{\partial v}. \tag{7.26}$$

(iii) $\boxed{K(v_1)=\lambda(v_1+\kappa)^{-2},\ \lambda,\ \kappa\text{ arbitrary constants.}}$

Here (7.24) admits infinitesimal generators (7.25), (7.26) and

$$\mathrm{Y}_6 = -x_1(v+\kappa x_1)\frac{\partial}{\partial x_1} + [2\lambda x_2 + \kappa x_1(v+\kappa x_1)]\frac{\partial}{\partial v}; \tag{7.27a}$$

$$\mathrm{Y}_7 = -x_1[(v+\kappa x_1)^2 + \lambda x_2]\frac{\partial}{\partial x_1} + 4\lambda(x_2)^2\frac{\partial}{\partial x_2} + [\kappa x_1(v+\kappa x_1)^2 + 2\lambda x_2(2v+3\kappa x_1)]\frac{\partial}{\partial v}; \tag{7.27b}$$

$$\mathrm{Y}_\infty = \psi(z, x_2)\left[\frac{\partial}{\partial x_1} - \kappa\frac{\partial}{\partial v}\right], \tag{7.27c}$$

where $z = v + \kappa x_1$ and $w = \psi(z, x_2)$ is an arbitrary solution of (7.22d).

(iv) $\boxed{K(v_1) = \dfrac{1}{(v_1)^2 + pv_1 + q}\exp\left[r\displaystyle\int\frac{dv_1}{(v_1)^2+pv_1+q}\right]}$ (p, q, r arbitrary constants such that $p^2 - 4q - r^2 \neq 0$).

In this case (7.24) admits infinitesimal generators (7.25) and

$$\mathrm{Y}_5 = v\frac{\partial}{\partial x_1} + (r-p)x_2\frac{\partial}{\partial x_2} - (qx_1 + pv)\frac{\partial}{\partial v}. \tag{7.28}$$

The point group classification of (7.24) has also been derived by Akhatov, Gazizov, and Ibragimov (1987). If the conductivity $K(u)$ is given by (7.21) then the infinitesimal generators of point transformation groups admitted by (7.20a,b) induce potential symmetries for the heat conduction equation (7.19) and point symmetries of its integrated form (7.24).

(3) *Wave Equation for an Inhomogeneous Medium*

Suppose $R\{x,u\}$ is the wave equation for a variable speed $c(x_1)$:

$$\frac{\partial^2 u}{\partial x_2^2} - c^2(x_1)\frac{\partial^2 u}{\partial x_1^2} = 0. \tag{7.29}$$

PDE (7.29) can be written in the conserved form

$$\mathrm{D}_1\left(\frac{\partial u}{\partial x_1}\right) - \mathrm{D}_2\left(\frac{1}{c^2(x_1)}\frac{\partial u}{\partial x_2}\right) = 0. \tag{7.30}$$

The auxiliary system $S\{x,u,v\}$ associated with (7.30) is then

$$\frac{\partial v}{\partial x_1} = \frac{1}{c^2(x_1)}\frac{\partial u}{\partial x_2}, \tag{7.31a}$$

$$\frac{\partial v}{\partial x_2} = \frac{\partial u}{\partial x_1}. \tag{7.31b}$$

The infinitesimal generators for point symmetries admitted by (7.31a,b) were given in Section 4.3.4. By examining the forms of these infinitesimal generators we have the following theorem [Bluman, Kumei, and Reid (1988)]:

Theorem 7.2.1-2. *The wave equation* (7.29) *admits a potential symmetry, corresponding to auxiliary system* (7.31a,b), *if and only if its wave speed* $c(x_1)$ *satisfies the ODE*

$$cc'(c/c')'' = \text{const} = \mu. \tag{7.32}$$

We distinguish between the cases $\mu = 0$ and $\mu \neq 0$.

Case I. $\mu = 0$: To within arbitrary scalings and translations of x_1 in $c(x_1)$, for the following wave speeds $c(x_1)$ equation (7.29) admits the indicated potential symmetries:

(i) $c(x_1) = (x_1)^C$, C $(\neq 0, 1)$ an arbitrary constant:

$$\begin{aligned} \mathrm{X}_S = {} & 2x_1x_2\frac{\partial}{\partial x_1} + \left[(1-C)(x_2)^2 + \frac{(x_1)^{2-2C}}{1-C}\right]\frac{\partial}{\partial x_2} \\ & + [(2C-1)x_2u - x_1v]\frac{\partial}{\partial u} - [x_2v + (x_1)^{1-2C}u]\frac{\partial}{\partial v}. \end{aligned} \tag{7.33}$$

(ii) $c(x_1) = x_1$:

$$\begin{aligned} \mathrm{X}_S = {} & 2x_1x_2\frac{\partial}{\partial x_1} + 2\log x_1\frac{\partial}{\partial x_2} + [x_2u - x_1v]\frac{\partial}{\partial u} \\ & - [x_2v + (x_1)^{-1}u]\frac{\partial}{\partial v}. \end{aligned} \tag{7.34}$$

(iii) $c(x_1) = e^{x_1}$:

$$\begin{aligned} \mathrm{X}_S = {} & -4x_2\frac{\partial}{\partial x_1} + 2[(x_2)^2 + e^{-2x_1}]\frac{\partial}{\partial x_1} + 2[v - 2x_2u]\frac{\partial}{\partial u} \\ & + 2e^{-2x_1}u\frac{\partial}{\partial v}. \end{aligned} \tag{7.35}$$

Case II. $\mu \neq 0$: If $c(x_1)$ solves (7.32) with $\mu \neq 0$ then $c(x_1)$ reduces to one of the standard forms (4.156a–d). If $c(x_1)$ solves either

$$c' = \nu^{-1}\sin(\nu \log c) \tag{7.36a}$$

or

$$c' = \nu^{-1}\sinh(\nu \log c), \tag{7.36b}$$

then PDE (7.29) admits two potential symmetries

$$\mathrm{X}_{S\pm} = e^{\pm x_2}\left\{\frac{2c}{c'}\frac{\partial}{\partial x_1} \pm 2\left[\left(\frac{c}{c'}\right)' - 1\right]\frac{\partial}{\partial x_2}\right.$$

$$+\left(\left[2-\left(\frac{c}{c'}\right)'\right]u \mp \frac{c}{c'}v\right)\frac{\partial}{\partial u} - \left[\left(\frac{c}{c'}\right)' v \pm \frac{1}{cc'}u\right]\frac{\partial}{\partial v}\right\}. \tag{7.37}$$

If $(u(x), v(x))$ satisfies (7.31a,b) then $u(x)$ solves the wave equation (7.29) and $v(x)$ solves an associated wave equation

$$\frac{\partial}{\partial x_1}\left(c^2(x_1)\frac{\partial v}{\partial x_1}\right) - \frac{\partial^2 v}{\partial x_2^2} = 0. \tag{7.38}$$

Note that PDE (7.38), as it is written, is already in conserved form and leads to the same auxiliary system (7.31a,b) as PDE (7.29) with $u(x)$ now playing the role of the potential variable. Consequently by examining the forms of infinitesimal generators admitted by the auxiliary system $S\{x, u, v\}$ [(7.31a,b)] we see that if the wave speed $c(x_1)$ satisfies ODE (7.32) then both PDE's (7.29) and (7.38) admit potential symmetries.

7.2.2 Comparison of Point Symmetries of $R\{x, u\}$ and $S\{x, u, v\}$

Let $\mathcal{G}_R$ denote the Lie group of point symmetries admitted by $R\{x, u\}$, and let $\mathcal{G}_S$ denote the Lie group of point symmetries admitted by an auxiliary system $S\{x, u, v\}$. As was demonstrated for certain conductivities for the nonlinear heat equation and certain wave speeds for the wave equation, a point symmetry in $\mathcal{G}_S$ does not necessarily define a point symmetry in $\mathcal{G}_R$. Conversely a point symmetry in $\mathcal{G}_R$ may not correspond to a point symmetry in $\mathcal{G}_S$: It could happen that PDE $R\{x, u\}$ admits an infinitesimal generator

$$\mathrm{X} = \xi_i(x, u)\frac{\partial}{\partial x_i} + \eta(x, u)\frac{\partial}{\partial u}$$

but its auxiliary system $S\{x, u, v\}$ admits no infinitesimal generator of the form

$$\mathrm{X}_S = \tilde{\xi}_i(x, u)\frac{\partial}{\partial x_i} + \tilde{\eta}(x, u)\frac{\partial}{\partial u} + \zeta^\mu(x, u, v)\frac{\partial}{\partial v^\mu},$$

with $\tilde{\eta}(x, u) \equiv \eta(x, u)$, $\tilde{\xi}_i(x, u) \equiv \xi_i(x, u)$, $i = 1, 2, \ldots, n$. We show that this situation arises for both the nonlinear heat conduction equation and the wave equation.

(1) *Nonlinear Heat Conduction Equation*

If $K(u) = \lambda(u + \kappa)^{-4/3}$, then $S\{x, u, v\}$ given by (7.20a,b) admits no point symmetry corresponding to the infinitesimal generator

$$\mathrm{X} = (x_1)^2\frac{\partial}{\partial x_1} - 3x_1(u + \kappa)\frac{\partial}{\partial u}$$

admitted by $R\{x, u\}$ given by (7.19).

(2) *Wave Equation for an Inhomogeneous Medium*

In Section 4.2.4 we showed that the wave equation (7.29) admits a nontrivial four-parameter Lie group of point transformations $\mathcal{G}_R$ if and only if its wave speed $c(x_1)$ satisfies (4.78) or, equivalently, the fifth-order ODE

$$\left\{c^2\left[\frac{H'''}{2H'+H^2}+3\frac{[2(H')^3-2HH'H''-(H'')^2]}{[2H'+H^2]^2}\right]\right\}'=0, \tag{7.39}$$

where

$$H=c'/c.$$

In Section 4.3.4 we showed that the auxiliary system of PDE's (7.31a,b) associated with the wave equation (7.29) admits a nontrivial four-parameter Lie group of point transformations $\mathcal{G}_S$ if and only if its wave speed $c(x_1)$ satisfies (7.32) or, equivalently, the fourth-order ODE

$$[cc'(c/c')'']'=0. \tag{7.40}$$

One can show [Bluman and Kumei (1989)] that a wave speed $c(x_1)$ simultaneously satisfies ODE's (7.39) and (7.40) if and only if $c(x_1)$ satisfies either

$$(c/c')''=0 \tag{7.41}$$

or

$$c^2c'c'''+c(c')^2c''-c^2(c'')^2-\frac{1}{4}(c')^4=0. \tag{7.42}$$

The solution of (7.41) is

$$c(x_1)=(Ax_1+B)^C, \tag{7.43}$$

and the solution of (7.42) consists of two families of solutions given implicitly by

$$\sqrt{c(x_1)}-\arctan C\sqrt{c(x_1)}=Ax_1+B, \tag{7.44a}$$

and

$$2\sqrt{c(x_1)}+\log|(\sqrt{c(x_1)}-C)/(\sqrt{c(x_1)}+C)|=Ax_1+B, \tag{7.44b}$$

where A, B, and C are arbitrary constants.

A wave speed $c(x_1)$ simultaneously solves ODE's (7.41) and (7.42) if and only if either

$$c(x_1)=(Ax_1+B)^2 \tag{7.45a}$$

or

$$c(x_1)=(Ax_1+B)^{2/3}. \tag{7.45b}$$

If a wave speed $c(x_1)$ satisfies (7.39) and does not satisfy (7.41) or (7.42), then there exists an infinitesimal generator of a point symmetry admitted by $R\{x,u\}$ [(7.29)] which does not correspond to an infinitesimal generator of a point symmetry admitted by $S\{x,u,v\}$ [(7.31a,b)].

For example, the wave speed $c(x_1) = 1-(x_1)^2$ satisfies (7.39) but does not satisfy (7.41) or (7.42). In this case the wave equation $R\{x,u\}$ given by

$$\frac{\partial^2 u}{\partial x_2^2} - [1-(x_1)^2]^2\frac{\partial^2 u}{\partial x_1^2} = 0$$

admits a four-parameter Lie group of point transformations $\mathcal{G}_R$ with infinitesimal generators

$$\mathrm{X}_1 = u\frac{\partial}{\partial u},\quad \mathrm{X}_2 = \frac{\partial}{\partial x_2},\quad \mathrm{X}_3 = [1-(x_1)^2]\frac{\partial}{\partial x_1} - x_1 u\frac{\partial}{\partial u},$$

$$\mathrm{X}_4 = x_2(1-(x_1)^2)\frac{\partial}{\partial x_1} + \frac{1}{2}\log\left|\frac{x_1+1}{x_1-1}\right|\frac{\partial}{\partial x_2} - x_1 x_2 u\frac{\partial}{\partial u}.$$

But the auxiliary system $S\{x,u,v\}$ given by

$$\frac{\partial v}{\partial x_1} = \frac{1}{[1-(x_1)^2]^2}\frac{\partial u}{\partial x_2},$$
$$\frac{\partial v}{\partial x_2} = \frac{\partial u}{\partial x_1},$$

only admits the trivial two-parameter group $\mathcal{G}_S$ with infinitesimal generators

$$\mathrm{X}_{S1} = u\frac{\partial}{\partial u} + v\frac{\partial}{\partial v},\quad \mathrm{X}_{S2} = \frac{\partial}{\partial x_2}.$$

If the wave speed $c(x_1)$ is of the form (7.43) with $A=1$, $B=0$, i.e. $c(x_1) = x_1^C$ $(C \neq 0,1)$, then $R\{x,u\}$ admits $\mathcal{G}_R$ with infinitesimal generators

$$\mathrm{X}_1 = u\frac{\partial}{\partial u},\quad \mathrm{X}_2 = \frac{\partial}{\partial x_2},\quad \mathrm{X}_3 = x_1\frac{\partial}{\partial x_1} + (1-C)x_2\frac{\partial}{\partial x_2},$$
$$\mathrm{X}_4 = 2x_1x_2\frac{\partial}{\partial x_1} + \left[(1-C)(x_2)^2 + \frac{(x_1)^{2-2C}}{1-C}\right]\frac{\partial}{\partial x_2} + Cx_2u\frac{\partial}{\partial u}, \tag{7.46}$$

and the auxiliary system $S\{x,u,v\}$ admits $\mathcal{G}_S$ with infinitesimal generators

$$\mathrm{X}_{S1} = u\frac{\partial}{\partial u} + v\frac{\partial}{\partial v},\quad \mathrm{X}_{S2} = \frac{\partial}{\partial x_2},$$

$$\mathrm{X}_{S3} = x_1\frac{\partial}{\partial x_1} + (1-C)x_2\frac{\partial}{\partial x_2} - Cv\frac{\partial}{\partial v},$$

$$\mathrm{X}_{S4} = 2x_1x_2\frac{\partial}{\partial x_1} + \left[(1-C)(x_2)^2 + \frac{(x_1)^{2-2C}}{1-C}\right]\frac{\partial}{\partial x_2} \tag{7.47}$$

$$+\ [(2C-1)x_2u - x_1v]\frac{\partial}{\partial u} - [x_2v + (x_1)^{1-2C}u]\frac{\partial}{\partial v}.$$

Hence X_{Si} corresponds to X_i for $i = 1, 2, 3$ but X_{S4} does not correspond to X_4 since the coefficients of $\frac{\partial}{\partial u}$ in (7.46) and (7.47) are different:

$$\eta_{S4}(x, u, v) = (2C-1)x_2u - x_1v \neq \eta_4(x, u) = Cx_2u.$$

Recall that X_{S4} defines a potential symmetry (7.33) admitted by $R\{x, u\}$ when $C \neq 0, 1$. The limiting case $C = 1$ exhibits the same relationship between $\mathcal{G}_R$ and $\mathcal{G}_S$.

If the wave speed $c(x_1)$ is a solution of ODE (7.42) which is not of the forms (7.45a,b) then the four-parameter groups $\mathcal{G}_R$ and $\mathcal{G}_S$ have only two corresponding infinitesimal generators of point symmetries [Bluman and Kumei (1987)]:

$$X_1 = u\frac{\partial}{\partial u} \quad \text{and} \quad X_{S1} = u\frac{\partial}{\partial u} + v\frac{\partial}{\partial v},$$

$$X_2 = X_{S2} = \frac{\partial}{\partial x_2}.$$

For example, if $c(x_1)$ satisfies (7.44b) with $A = 1$, $B = 0$, then $\mathcal{G}_R$ has infinitesimal generators

$$X_1 = u\frac{\partial}{\partial u}, \quad X_2 = \frac{\partial}{\partial x_2}$$

$$X_\pm = e^{\pm x_2/2}(c(x_1)-1)^{-1/2}\left\{[c(x_1)]^{3/2}\frac{\partial}{\partial x_1} \mp \frac{\partial}{\partial x_2} + \frac{(c(x_1)-1)}{2}u\frac{\partial}{\partial u}\right\}, \tag{7.48}$$

whereas $\mathcal{G}_S$ has infinitesimal generators

$$X_{S1} = u\frac{\partial}{\partial u} + v\frac{\partial}{\partial v}, \quad X_{S2} = \frac{\partial}{\partial x_2},$$

$$\begin{aligned} X_{S\pm} = e^{\pm x_2}(c(x_1)-1)^{-1}\Big\{&4[c(x_1)]^{3/2}\frac{\partial}{\partial x_1} \mp 2(c(x_1)+1)\frac{\partial}{\partial x_2} \\ &+ [(3c(x_1)-1)u \mp 2[c(x_1)]^{3/2}v]\frac{\partial}{\partial u} \\ &+ [(3-c(x_1))v \mp 2[c(x_1)]^{-1/2}u]\frac{\partial}{\partial v}\Big\}. \end{aligned} \tag{7.49}$$

7.2.3 Types of Potential Symmetries of Linear PDE's

If $R\{x,u\}$ admits a potential symmetry X_S then a corresponding invariant solution $(u,v)=(\Theta_S(x),\phi_S(x))$ of the auxiliary system $S\{x,u,v\}$ yields a solution $u=\Theta_S(x)$ of $R\{x,u\}$. In general such a solution is not an invariant solution of its Lie group of point transformations $\mathcal{G}_R$.

Now let $R\{x,u\}$ be a linear PDE which admits point symmetries

$$X_1 = \xi_{1i}(x)\frac{\partial}{\partial x_i} + f_1(x)u\frac{\partial}{\partial u}, \tag{7.50}$$

$$X_2 = \xi_{2i}(x)\frac{\partial}{\partial x_i} + f_2(x)u\frac{\partial}{\partial u}, \tag{7.51}$$

where $\xi_1(x) \not\equiv 0$, $\xi_2(x) \not\equiv 0$, and a potential symmetry

$$X_S = \xi_{1i}(x)\frac{\partial}{\partial x_i} + (h(x)u + g_\nu(x)v^\nu)\frac{\partial}{\partial u}$$
$$+ (k_{\mu\nu}(x)v^\nu + \ell_\mu(x)u)\frac{\partial}{\partial v^\mu}. \tag{7.52}$$

Since invariant solutions $u=\Theta(x)$ of X_1 for $R\{x,u\}$ and an invariant solution $(u,v)=(\Theta_S(x),\phi_S(x))$ of X_S for $S\{x,u,v\}$ have the same similarity variables, we might expect that $\Theta_S(x)$ can be related to invariant solutions of X_1. We conjecture that $\Theta_S(x)$ can always be represented as a superposition of solutions of $R\{x,u\}$ obtained by applying the recursion operator

$$R_2 = \xi_{2i}(x)\frac{\partial}{\partial x_i} - f_2(x),$$

associated with some X_2, to invariant solutions $u=\Theta(x)$ of X_1 for $R\{x,u\}$ [cf. Section 5.3.1].

As an example let $R\{x,u\}$ be the wave equation

$$\frac{\partial^2 u}{\partial x_2^2} - (x_1)^4\frac{\partial^2 u}{\partial x_1^2} = 0. \tag{7.53}$$

PDE (7.53) admits point symmetries

$$X_1 = 2x_1x_2\frac{\partial}{\partial x_1} - \left[(x_1)^{-2} + (x_2)^2\right]\frac{\partial}{\partial x_2} + 2x_2u\frac{\partial}{\partial u}, \quad X_2 = \frac{\partial}{\partial x_2},$$

and the potential symmetry

$$X_S = 2x_1x_2\frac{\partial}{\partial x_1} - [(x_1)^{-2} + (x_2)^2]\frac{\partial}{\partial x_2} + (3x_2u - x_1v)\frac{\partial}{\partial u}$$
$$- [x_2v + (x_1)^{-3}u]\frac{\partial}{\partial v}$$

which is a point symmetry admitted by its auxiliary system $S\{x,u,v\}$ given by

$$\frac{\partial v}{\partial x_1} = (x_1)^{-4}\frac{\partial u}{\partial x_2}, \tag{7.54a}$$

$$\frac{\partial v}{\partial x_2} = \frac{\partial u}{\partial x_1}. \tag{7.54b}$$

The invariant form of invariant solutions $(u,v) = (\Theta_S(x), \phi_S(x))$ of X_S for $S\{x,u,v\}$ leads to the expression

$$u = x_1 F(z) + (x_1)^2 x_2 G(z) \tag{7.55}$$

for solutions of $R\{x,u\}$ with similarity variable

$$z = (x_1)^{-1} - x_1(x_2)^2,$$

where $F(z)$ and $G(z)$ are arbitrary [cf. Section 4.3.4]. The invariant form for invariant solutions $u = \Theta(x)$ of X_1 for $R\{x,u\}$ is given by

$$u = x_1 H(z) \tag{7.56}$$

for arbitrary $H(z)$. The recursion operator associated with X_2 is

$$R_2 = \frac{\partial}{\partial x_2}.$$

Then

$$R_2(x_1 H(z)) = -2(x_1)^2 x_2 H'(z). \tag{7.57}$$

A comparison of equations (7.55)–(7.57) suggests that the first term of (7.55) corresponds to an invariant solution, say $u = \Theta_1(x)$, of X_1 for $R\{x,u\}$. It also suggests that the second term of (7.55) corresponds to a solution of $R\{x,u\}$ obtained by applying the recursion operator $R_2 = \frac{\partial}{\partial x_2}$ to another invariant solution, say $u = \Theta_2(x)$, of X_1 for $R\{x,u\}$. In particular from (4.129a,b) it follows that (7.55) solves $R\{x,u\}$ if

$$u = \Theta_S(x) = \alpha x_1 + \beta (x_1)^2 x_2 z^{-2}$$

for arbitrary constants α, β. By substituting (7.56) into (7.53) we obtain invariant solutions $u = \Theta_1(x) = x_1$, $u = \Theta_2(x) = \frac{1}{2}x_1 z^{-1}$ of X_1 for $R\{x,u\}$. It is easy to see that

$$\Theta_S(x) = \alpha\Theta_1(x) + \beta R_2\Theta_2(x).$$

This example motivates the following definition:

Definition 7.2.3-1. Let $\xi_S(x)$ be the infinitesimal of x for a potential symmetry admitted by a linear PDE $R\{x,u\}$ with a linear auxiliary system $S\{x,u,v\}$. The potential symmetry with $\xi_S \not\equiv 0$ is a *potential symmetry of type I* if the infinitesimal $\xi(x)$ of x for any point symmetry of $\mathcal{G}_R$ is such that $\xi(x) \not\equiv \xi_S(x)$; otherwise the potential symmetry is a *potential symmetry of type II.*

It immediately follows that the similarity variables arising from a potential symmetry of type II of $R\{x,u\}$ are the similarity variables for some point symmetry of $\mathcal{G}_R$. Our previous example suggests that potential symmetries of type II are not useful for constructing essentially new solutions of $R\{x,u\}$.

For potential symmetries admitted by the wave equation (7.29) with auxiliary system (7.31a,b) we have the following classification theorem:

Theorem 7.2.3-1 [Bluman and Kumei (1987), Bluman, Kumei, and Reid (1988)].

(i) *If the wave speed* $c(x_1)$ *satisfies* $(c/c')'' = 0$, *then PDE* (7.29) *admits one potential symmetry of type II and admits no potential symmetries of type I.*

(ii) *If the wave speed* $c(x_1)$ *satisfies* $cc'(c/c')'' = \text{const} \neq 0$, *then PDE* (7.29) *admits two potential symmetries of type I and admits no potential symmetries of type II.*

7.2.4 Applications to Boundary Value Problems

In Section 4.4 we showed how point symmetries can be used to construct solutions of boundary value problems (initial value problems) posed for PDE's $R\{x,u\}$. If the Lie group of point transformations $\mathcal{G}_R$ admitted by $R\{x,u\}$ does not lead to the solution of the BVP, the solution might be obtained from symmetries by enlarging the group of $R\{x,u\}$ to include potential symmetries.

Suppose $S\{x,u,v\}$ is an auxiliary system leading to potential symmetries of $R\{x,u\}$. In order to use these potential symmetries to solve the given BVP for $R\{x,u\}$ it is necessary to embed the given BVP in a BVP posed for $S\{x,u,v\}$ so that if $(u(x), v(x))$ solves the BVP for $S\{x,u,v\}$, then $u(x)$ solves the BVP for $R\{x,u\}$.

As an example consider the following BVP (initial value problem) posed for the wave equation $R\{x,u\}$:

$$\frac{\partial^2 u}{\partial x_2^2} - c^2(x_1)\frac{\partial^2 u}{\partial x_1^2} = 0, \quad 0 < x_1 < \infty, \quad 0 < x_2 < \infty, \tag{7.58a}$$

$$u(x_1, 0) = U(x_1), \tag{7.58b}$$

$$\frac{\partial u}{\partial x_2}(x_1, 0) = W(x_1). \tag{7.58c}$$

BVP (7.58a–c) can be embedded in the following BVP for an auxiliary system $S\{x,u,v\}$:

$$\frac{\partial v}{\partial x_1} = \frac{1}{c^2(x_1)}\frac{\partial u}{\partial x_2}, \tag{7.59a}$$

$$\frac{\partial v}{\partial x_2} = \frac{\partial u}{\partial x_1}, \tag{7.59b}$$

$$u(x_1, 0) = U(x_1), \tag{7.59c}$$

$$v(x_1, 0) = V(x_1), \tag{7.59d}$$

$0 < x_1 < \infty$, $0 < x_2 < \infty$, for any $V(x_1)$ such that $V'(x_1) = \frac{W(x_1)}{c^2(x_1)}$.

For any bounded variable wave speed $c(x_1)$ the Lie group of point transformations $\mathcal{G}_R$ yields no analytical solutions of BVP (7.58a–c). On the other hand, in Section 4.4.3 we showed that for a bounded variable wave speed $c(x_1)$ which satisfies (4.259), the auxiliary system $S\{x, u, v\}$ admits point symmetries which yield an analytical solution of BVP (7.58a–c).

7.2.5 Non-Invertible Mappings of Nonlinear PDE's to Linear PDE's

In Section 6.4 we gave necessary and sufficient conditions under which nonlinear PDE's can be transformed to linear PDE's by invertible mappings [cf. Theorems 6.4.1-1,2, 6.4.2-1,2]. We showed that such an invertible mapping does not exist if a nonlinear scalar PDE does not admit an infinite-parameter Lie group of contact transformations and also does not exist if a nonlinear system of PDE's does not admit an infinite-parameter Lie group of point transformations. However if no such invertible mapping exists there remains the possibility of relating a given nonlinear PDE (either a scalar PDE or a system of PDE's) to a linear PDE by a non-invertible mapping. Potential symmetries can yield such non-invertible mappings.

Consider a nonlinear system of PDE's $R\{x, u\}$ which admits no infinite-parameter Lie group of transformations yielding the linearization of $R\{x, u\}$ by an invertible mapping. Assume that at least one PDE of $R\{x, u\}$ is in conserved form. Then $R\{x, u\}$ is embedded in the corresponding auxiliary system $S\{x, u, v\}$. Through this embedding the mapping between $R\{x, u\}$ and $S\{x, u, v\}$ is non-invertible since if $(u(x),\ v(x) + C)$ solves $S\{x, u, v\}$ then $u(x)$ solves $R\{x, u\}$ for any constant C. Suppose $S\{x, u, v\}$ admits an infinite-parameter Lie group of point transformations which leads to an invertible mapping of $S\{x, u, v\}$ to a linear system of PDE's. The composition of this invertible mapping and the non-invertible mapping between $S\{x, u, v\}$ and $R\{x, u\}$ yields a non-invertible mapping of this linear system of PDE's to $R\{x, u\}$. By construction this infinite-parameter Lie group of point transformations admitted by $S\{x, u, v\}$ is a potential symmetry of $R\{x, u\}$. Examples follow.

(1) *Burgers' Equation*

Suppose $R\{x, u\}$ is Burgers' equation (7.13). One can show that PDE (7.13) does not admit an infinite-parameter Lie group of contact transformations. Its associated auxiliary system $S\{x, u, v\}$ given by (7.15a,b) admits an infinite-parameter Lie group of point transformations (7.16a,b). This infinitesimal generator satisfies the criteria of Theorems 6.4.1-1,2 to linearize $S\{x,u,v\}$ by an invertible mapping. In turn this leads to the non-invertible Hopf–Cole transformation [cf. Exercise 6.4-7].

(2) *Nonlinear Heat Conduction Equation*

Let $R\{x,u\}$ be the nonlinear heat conduction equation

$$\frac{\partial}{\partial x_1}\left[u^{-2}\frac{\partial u}{\partial x_1}\right]-\frac{\partial u}{\partial x_2}=0. \tag{7.60}$$

PDE (7.60) does not admit an infinite-parameter Lie group of contact transformations but its associated auxiliary system $S\{x,u,v\}$ given by

$$\frac{\partial v}{\partial x_1}=u, \tag{7.61a}$$

$$\frac{\partial v}{\partial x_2}=u^{-2}\frac{\partial u}{\partial x_1}, \tag{7.61b}$$

admits an infinite-parameter Lie group of point transformations with infinitesimal generator

$$\mathrm{X}_S=\psi(v,x_2)\frac{\partial}{\partial x_1}-u^2\frac{\partial\psi(v,x_2)}{\partial v}\frac{\partial}{\partial u} \tag{7.62}$$

where $\psi(v,x_2)$ is an arbitrary function satisfying the linear heat equation

$$\frac{\partial^2\psi}{\partial v^2}-\frac{\partial\psi}{\partial x_2}=0. \tag{7.63}$$

Infinitesimal generator (7.62) satisfies the criteria of Theorems 6.4.1-1,2. One can easily obtain the invertible mapping

$$z_1=v,\quad z_2=x_2,\quad w^1=x_1,\quad w^2=\frac{1}{u},$$

which transforms any solution $(w^1(z_1,z_2),\, w^2(z_1,z_2))$ of the linear system of PDE's

$$\frac{\partial w^1}{\partial z_1}=w^2,$$

$$\frac{\partial w^1}{\partial z_2}=\frac{\partial w^2}{\partial z_1},$$

to a solution $(u(x_1,x_2),\, v(x_1,x_2))$ of the nonlinear system (7.61a,b) and hence to a solution $u(x_1,x_2)$ of (7.60).

(3) *Nonlinear Telegraph Equation*

Suppose $R\{x, u\}$ is the nonlinear telegraph equation [Varley and Seymour (1985)]

$$\frac{\partial^2 u}{\partial x_1^2} - \frac{\partial}{\partial x_2}\left[u^{-2}\frac{\partial u}{\partial x_2} + 1 - u^{-1}\right] = 0. \tag{7.64}$$

PDE (7.64) does not admit an infinite-parameter Lie group of contact transformations but its auxiliary system $S\{x, u, v\}$ given by

$$\frac{\partial v}{\partial x_1} = u^{-2}\frac{\partial u}{\partial x_2} + 1 - u^{-1}, \tag{7.65a}$$

$$\frac{\partial v}{\partial x_2} = \frac{\partial u}{\partial x_1}, \tag{7.65b}$$

admits an infinite-parameter Lie group of point transformations corresponding to infinitesimal generator (6.79)–(6.82) with $u^1 = u$, $u^2 = v$. This leads to the invertible mapping (6.86) which transforms $S\{x, u, v\}$ to the linear system (6.87a,b).

(4) *Thomas Equations*

Consider the nonlinear system of PDE's $R\{x, u\}$ given by

$$\frac{\partial u^1}{\partial x_1} - \frac{\partial u^2}{\partial x_2} = 0, \tag{7.66a}$$

$$\frac{\partial u^2}{\partial x_2} - u^1u^2 - u^1 - u^2 = 0, \tag{7.66b}$$

which describes a fluid flow through a reacting medium [Thomas (1944); see also Whitham (1974)] and also can be related to equations for two-wave interaction [Hasegawa (1974), Hashimoto (1974), Yoshikawa and Yamaguti (1974)]. The system (7.66a,b) does not admit an infinite-parameter Lie group of point transformations. Since PDE (7.66a) is in conserved form we can introduce a potential v such that

$$u^1 = \frac{\partial v}{\partial x_2}, \quad u^2 = \frac{\partial v}{\partial x_1}.$$

The associated auxiliary system $S\{x, u, v\}$ is given by

$$\frac{\partial v}{\partial x_1} = u^2, \tag{7.67a}$$

$$\frac{\partial v}{\partial x_2} = u^1, \tag{7.67b}$$

$$\frac{\partial u^2}{\partial x_2} - u^1u^2 - u^1 - u^2 = 0. \tag{7.67c}$$

One can show that (7.67a–c) admits an infinite-parameter Lie group of point transformations with infinitesimal generator

$$\mathrm{X}_S = e^v \left\{ \psi(x_1, x_2)\frac{\partial}{\partial v} + \left[\frac{\partial \psi(x_1,x_2)}{\partial x_2} + u^1\psi(x_1,x_2)\right]\frac{\partial}{\partial u^1} \right.$$

$$\left. + \left[\frac{\partial \psi(x_1,x_2)}{\partial x_1} + u^2\psi(x_1,x_2)\right]\frac{\partial}{\partial u^2} \right\} \tag{7.68}$$

where $\psi(x_1, x_2)$ is an arbitrary function satisfying the linear PDE

$$\frac{\partial^2 \psi}{\partial x_1 \partial x_2} - \frac{\partial \psi}{\partial x_1} - \frac{\partial \psi}{\partial x_2} = 0. \tag{7.69}$$

To apply Theorems 6.4.1-1,2, let $F^1 = \psi(x_1, x_2)$, $F^2 = \frac{\partial \psi}{\partial x_1}$, $F^3 = \frac{\partial \psi}{\partial x_2}$. Then from (6.58a,b) (with $v = u^3$) we have $\alpha_i^j = 0$, $i = 1, 2$ for $j = 1, 2, 3$ and $\beta_1^1 = u^1e^v$, $\beta_2^1 = u^2e^v$, $\beta_2^2 = \beta_1^3 = \beta_3^1 = e^v$, $\beta_1^2 = \beta_2^3 = \beta_3^2 = \beta_3^3 = 0$. From (6.65), (6.66) we obtain $X_1 = x_1$, $X_2 = x_2$, $\psi^1 = -e^{-v}$, $\psi^2 = e^{-v}u^2$, $\psi^3 = e^{-v}u^1$. Consequently the invertible mapping

$$z_1 = x_1, \quad z_2 = x_2, \quad w^1 = -e^{-v}, \quad w^2 = e^{-v}u^2, \quad w^3 = e^{-v}u^1,$$

transforms any solution $(w^1(z_1, z_2), w^2(z_1, z_2), w^3(z_1, z_2))$ of the linear system of PDE's

$$\frac{\partial w^1}{\partial z_1} = w^2, \tag{7.70a}$$

$$\frac{\partial w^1}{\partial z_2} = w^3, \tag{7.70b}$$

$$\frac{\partial w^2}{\partial z_2} = w^2 + w^3, \tag{7.70c}$$

to a solution $(u^1(x_1, x_2), u^2(x_1, x_2), v(x_1, x_2))$ of (7.67a–c) and hence to the solution

$$(u^1(x_1,x_2), u^2(x_1,x_2)) = -\left(\frac{w^3(x_1,x_2)}{w^1(x_1,x_2)}, \frac{w^2(x_1,x_2)}{w^1(x_1,x_2)}\right)$$

of (7.66a,b).

7.2.6 Conserved Forms

We have seen that potential symmetries significantly extend the applicability of infinitesimal transformation methods to PDE's. But in order to find a potential symmetry we must first have at least one PDE of a given system of PDE's $R\{x, u\}$ in conserved form. In principle any conserved form for $R\{x, u\}$ could lead to potential symmetries of $R\{x, u\}$. In our previous examples a conserved form was found easily by inspection. For Schroedinger's equation

$$-\frac{\partial^2 u}{\partial x_1} + V(x_1)u - i\frac{\partial u}{\partial x_2} = 0, \tag{7.71}$$

at first sight it is not obvious how to obtain a conserved form so that the associated auxiliary system of PDE's $S\{x, u, v\}$ is linear. However we can write (7.71) in the conserved form

$$\frac{\partial f}{\partial x_1} - \frac{\partial g}{\partial x_2} = 0$$

where

$$f = \omega(x_1)\frac{\partial u}{\partial x_1} - \omega'(x_1)u,$$
$$g = -i\omega(x_1)u,$$

with $\omega(x_1)$ being any function satisfying

$$\frac{\omega''(x_1)}{\omega(x_1)} = V(x_1).$$

The associated auxiliary system $S\{x, u, v\}$ is given by

$$\frac{\partial v}{\partial x_1} = -i\omega(x_1)u, \tag{7.72a}$$

$$\frac{\partial v}{\partial x_2} = \omega(x_1)\frac{\partial u}{\partial x_1} - \omega'(x_1)u. \tag{7.72b}$$

One can show that system (7.72a,b) yields potential symmetries for Schroedinger's equation (7.71) for a certain class of $V(x_1)$.

A systematic way of obtaining conserved forms is through Noether's theorem discussed in Chapter 5: If $R\{x, u\}$ can be derived from a variational formulation then essentially any point symmetry of $\mathcal{G}_R$ leads to a conserved form (conservation law) for $R\{x, u\}$. Each conservation law leads to a different auxiliary system $S\{x, u, v\}$.

We note that if $R\{x, u\}$ admits a potential symmetry through $S\{x, u, v\}$, and $S\{x, u, v\}$ can be derived from a variational principle, then the corresponding point symmetry admitted by $S\{x, u, v\}$ essentially leads to a conserved form for $S\{x, u, v\}$ through Noether's theorem, and in turn leads to a new auxiliary system $T\{x, u, v, w\}$ for $S\{x, u, v\}$ and hence for $R\{x, u\}$. Consequently point symmetries of $T\{x, u, v, w\}$ which are potential symmetries of $S\{x, u, v\}$ could lead to enlarging the class of nonlocal symmetries admitted by $R\{x, u\}$.

7.2.7 Inherited Symmetries

The concept of potential symmetries for a scalar PDE $R\{x, u\}$ can be extended to the reduced differential equations which arise when seeking invariant solutions of an auxiliary system $S\{x, u, v\}$. For motivation consider the nonlinear heat conduction equation

$$\frac{\partial}{\partial x_1}\left(K(u)\frac{\partial}{\partial x_1}\right) - \frac{\partial u}{\partial x_2} = 0. \tag{7.73}$$

For any $K(u)$ an important class of invariant solutions arises from the invariance of (7.73) under scalings $x_1^* = \alpha x_1$, $x_2^* = \alpha^2 x_2$, corresponding to the infinitesimal generator

$$\mathrm{X} = x_1\frac{\partial}{\partial x_1} + 2x_2\frac{\partial}{\partial x_2} \tag{7.74}$$

admitted by (7.73). These invariant solutions are of the form

$$u = U(z), \quad z = \frac{x_1}{\sqrt{x_2}},$$

where $U(z)$ satisfies the ODE

$$2\frac{d}{dz}\left(K(U)\frac{dU}{dz}\right) + z\frac{dU}{dz} = 0. \tag{7.75}$$

Now consider the problem of reducing the order of ODE (7.75). One can show that ODE (7.75) admits a one-parameter Lie group of point transformations if and only if

$$K(U) = \lambda(U + \kappa)^\nu, \tag{7.76}$$

for arbitrary constants λ, ν, κ. Hence using Lie's methods [cf. Chapter 3] we can reduce the order of (7.75) if and only if $K(u)$ is of the form (7.76).

This reduction of order arises directly from the structure of the Lie algebra of infinitesimal generators admitted by PDE (7.73). We consider this in the framework of the following theorem:

Theorem 7.2.7-1. *Let a scalar PDE $R\{x, u\}$ with two independent variables $x = (x_1, x_2)$ admit infinitesimal generators X and Y such that their commutator is*

$$[\mathrm{Y}, \mathrm{X}] = \mu \mathrm{X}, \tag{7.77}$$

for some constant μ. Let $R_{\mathrm{X}}\{z, U\}$, with independent variable (similarity variable) z and dependent variable $U = U(z)$, be the reduced ODE associated with invariant solutions corresponding to X for PDE $R\{x, u\}$. Then Y induces a one-parameter Lie group of point transformations admitted by ODE $R_{\mathrm{X}}\{z, U\}$.

Proof. The proof is obtained by following the procedure for reducing the order of an ODE which admits a two-parameter Lie group of transformations [cf. Sections 3.4.1,2]. It is left to Exercise 7.2-4. □

This theorem leads to the following definition:

Definition 7.2.7-1. The point symmetry of ODE $R_X\{z, U\}$, induced by the point symmetry Y of PDE $R\{x, u\}$, where Y satisfies (7.77), is an ***inherited point symmetry*** of ODE $R_X\{z, U\}$.

We remark that X and Y satisfying (7.77) generate a two-dimensional subalgebra $\mathcal{J}$ of the Lie algebra of infinitesimal generators admitted by $R\{x, u\}$; X generates a normal subalgebra of $\mathcal{J}$ [cf. Section 2.4.4].

Let $R\{x, u\}$ be PDE (7.73) and let $R_X\{z, U\}$ be ODE (7.75) with X given by (7.74). For arbitrary $K(u)$, PDE (7.73) admits X and $X_1 = \frac{\partial}{\partial x_1}$, $X_2 = \frac{\partial}{\partial x_2}$, but

$$[aX_1 + bX_2, X] = a\frac{\partial}{\partial x_1} + 2b\frac{\partial}{\partial x_2} \neq \mu X$$

for any constants a and b, and hence ODE (7.75) admits no inherited point symmetries. If $K(u) = \lambda(u + \kappa)^\nu$, PDE (7.73) admits X, X_1, X_2, and $Y = x_1\frac{\partial}{\partial x_1} + \frac{2}{\nu}(u + \kappa)\frac{\partial}{\partial u}$. Here $[Y, X] = 0$, and hence Y induces an inherited point symmetry of ODE (7.75). It is easy to see that the inherited point symmetry of ODE (7.75) is

$$Y = z\frac{\partial}{\partial z} + \frac{2}{\nu}(U + \kappa)\frac{\partial}{\partial U}. \tag{7.78}$$

Now consider the auxiliary system $S\{x, u, v\}$ of (7.73) given by

$$\frac{\partial v}{\partial x_1} = u, \tag{7.79a}$$

$$\frac{\partial v}{\partial x_2} = K(u)\frac{\partial u}{\partial x_1}, \tag{7.79b}$$

which admits

$$X_S = x_1\frac{\partial}{\partial x_1} + 2x_2\frac{\partial}{\partial x_2} + v\frac{\partial}{\partial v} \tag{7.80}$$

corresponding to $X = x_1\frac{\partial}{\partial x_1} + 2x_2\frac{\partial}{\partial x_2}$ admitted by (7.73). The point symmetry $X_T = X_S$ is also admitted by PDE $T\{x, v\}$ given by

$$\frac{\partial v}{\partial x_2} = K\left(\frac{\partial v}{\partial x_1}\right)\frac{\partial^2 v}{\partial x_1^2}. \tag{7.81}$$

The mapping

$$u = \frac{\partial v}{\partial x_1} \tag{7.82}$$

transforms any solution $v(x)$ of (7.81) to a solution $u = \frac{\partial v(x)}{\partial x_1}$ of (7.73). It is easy to see that X_S leads to invariant solutions

$$(u, v) = (U(z), \sqrt{x_2}V(z)), \quad z = \frac{x_1}{\sqrt{x_2}}, \tag{7.83}$$

of (7.79a,b) where $(U(z), V(z))$ satisfy the system of ODE's $S_{X_S}\{z, U, V\}$ given by

$$\frac{dV}{dz} = U, \tag{7.84a}$$

$$V - z\frac{dV}{dz} = 2K(U)\frac{dU}{dz}. \tag{7.84b}$$

Furthermore, from (7.83), (7.84a,b), we see that $u = U(z)$ leads to an invariant solution of $R\{x, u\}$ [(7.73)] corresponding to X and to an invariant solution of $T\{x, v\}$ [(7.81)] corresponding to X_T: If $(U(z), V(z))$ satisfies (7.84a,b) then $U(z)$ satisfies ODE $R_X\{z, U\}$ defined by (7.75), and $V(z)$ satisfies ODE $T_{X_T}\{z, V\}$ given by

$$2K\left(\frac{dV}{dz}\right)\frac{d^2V}{dz^2} + z\frac{dV}{dz} - V = 0. \tag{7.85}$$

Moreover the mapping (7.84a) transforms any solution of ODE (7.85) to a solution of ODE (7.75) and transforms the general solution of ODE (7.85) to the general solution of ODE (7.75). Hence ODE (7.75) is embedded in both the system of ODE's (7.84a,b) and ODE (7.85).

We have gained something from this embedding if (7.85) admits a one-parameter Lie group of point transformations for some $K(u) \neq \lambda(u+\kappa)^\nu$. Such a symmetry reduces the order of (7.85) and in this sense essentially reduces the order of ODE (7.75). To find such $K(u)$ we seek inherited point symmetries of ODE (7.85). In Section 7.2.1 we showed that if

$$K(u) = \frac{1}{u^2 + pu + q}\exp\left[r\int\frac{du}{u^2 + pu + q}\right], \tag{7.86}$$

then PDE (7.81) admits

$$Y_T = v\frac{\partial}{\partial x_1} + (r-p)x_2\frac{\partial}{\partial x_2} - (qx_1 + pv)\frac{\partial}{\partial v}. \tag{7.87}$$

The commutator of Y_T and X_T is

$$[Y_T, X_T] = 0,$$

and hence Y_T induces an inherited point symmetry of ODE (7.85). This inherited point symmetry is

$$\mathrm{Y}_T = \left[V + \frac{1}{2}(p-r)z\right]\frac{\partial}{\partial z} - \left[\frac{1}{2}(p+r)V + qz\right]\frac{\partial}{\partial V}. \tag{7.88}$$

Through (7.84a) the point symmetry (7.88) of ODE (7.85) induces the point symmetry

$$\mathrm{X}_S = \left[V + \frac{1}{2}(p-r)z\right]\frac{\partial}{\partial z} - \left[\frac{1}{2}(p+r)V + qz\right]\frac{\partial}{\partial V} - [q + pU + U^2]\frac{\partial}{\partial U} \tag{7.89}$$

of the system of ODE's (7.84a,b). In turn the point symmetry (7.89) of (7.84a,b) or, equivalently, the point symmetry (7.88) of (7.85), induces a nonlocal symmetry of ODE (7.75) since the infinitesimal of z depends explicitly on V which, as defined by (7.84a), cannot be expressed in terms of z, U, and derivatives of U to some finite order.

Exercises 7.2

1. Find potential symmetries of PDE (7.38) for wave speeds $c(x_1)$ satisfying
   (a) (7.32) when $\mu = 0$;
   (b) (7.36a,b).

2. The "usual" way of relating a scalar second order PDE $R\{x,u\}$ to a system of PDE's is to introduce additional dependent variables $v^i = \frac{\partial u}{\partial x_i}$, $i = 1, 2, \ldots, n$. This leads to a system of PDE's $\hat{S}\{x,u,v\}$. Show that a point symmetry admitted by $\hat{S}\{x,u,v\}$ induces a contact symmetry admitted by $R\{x,u\}$.

3. Let $R\{x,u\}$ be the scalar evolution equation

   $$\frac{\partial u}{\partial t} - \frac{\partial}{\partial x}G(x,t,u,u_1,\ldots,u_n) = 0$$

   where

   $$u_k = \frac{\partial^k u}{\partial x^k}, \quad k = 1, 2, \ldots, n$$

   with $n \geq 1$. The associated auxiliary system $S\{x,u,v\}$ is

   $$\frac{\partial v}{\partial x} = u, \quad \frac{\partial v}{\partial t} = G(x,t,u,u_1,\ldots,u_n).$$

It is easy to show that if $(u(x,t), v(x,t))$ solves $S\{x,u,v\}$ then $v(x,t)$ solves the evolution equation $T\{x,v\}$ given by

$$\frac{\partial v}{\partial t} = G(x,t,v_1,v_2,\ldots,v_{n+1}).$$

(a) Show that for a point symmetry

$$\mathrm{X}_T = \xi\frac{\partial}{\partial x} + \tau\frac{\partial}{\partial t} + \zeta\frac{\partial}{\partial v},$$

admitted by $T\{x,v\}$, we always have $\frac{\partial\tau}{\partial x} = \frac{\partial\tau}{\partial v} = 0$. Consequently show that a point symmetry admitted by $T\{x,v\}$ induces a point symmetry admitted by $S\{x,u,v\}$. As examples show that

(i) the point symmetry (7.18) admitted by (7.17) induces the point symmetry (7.16a,b) admitted by system (7.15a,b);

(ii) the point symmetries (7.27a–c) admitted by (7.24) for $K(v_1) = \lambda(v_1+\kappa)^{-2}$ induce the point symmetries (7.22a–c) admitted by system (7.20a,b).

(b) Show that for a point symmetry

$$\mathrm{X}_S = \xi\frac{\partial}{\partial x} + \tau\frac{\partial}{\partial t} + \eta\frac{\partial}{\partial u} + \zeta\frac{\partial}{\partial v},$$

admitted by $S\{x,u,v\}$, we always have $\frac{\partial\xi}{\partial u} = \frac{\partial\tau}{\partial u} = \frac{\partial\zeta}{\partial u} = 0$. Consequently show that a point symmetry admitted by $S\{x,u,v\}$ induces a point symmetry admitted by $T\{x,v\}$.

(c) Show that a point symmetry admitted by $R\{x,u\}$ does not necessarily induce a point symmetry admitted by $S\{x,u,v\}$.

4. Prove Theorem 7.2.7-1.

5. Verify (7.78).

6. Verify (7.88) and (7.89).

7. Consider PDE $R\{x,u\}$ given by

$$\frac{\partial u}{\partial x_2}\frac{\partial^2 u}{\partial x_1 \partial x_2} - \frac{\partial u}{\partial x_1}\frac{\partial^2 u}{\partial x_2^2} = \frac{\partial^3 u}{\partial x_2^3},$$

which arises from the Prandtl boundary layer equations [cf. Section 1.2]. The scaling symmetry

$$\mathrm{X} = 2x_1\frac{\partial}{\partial x_1} + x_2\frac{\partial}{\partial x_2} + u\frac{\partial}{\partial u}$$

reduces $R\{x,u\}$ to the Blasius equation $R_{\mathrm{X}}\{z,U\}$ given by

$$2\frac{d^3U}{dz^3} + U\frac{d^2U}{dz^2} = 0, \quad z = \frac{x_2}{\sqrt{x_1}}, \quad U = \frac{u}{\sqrt{x_1}}.$$

(a) Find all infinitesimal generators of point symmetries admitted by $R\{x,u\}$.

(b) Determine all infinitesimal generators of inherited point symmetries admitted by $R_X\{z,U\}$.

7.3 Potential Symmetries for Ordinary Differential Equations

Consider an nth order scalar ODE, $R\{x,u\}$, which does not admit a point symmetry. If $R\{x,u\}$ is obtained as a reduced ODE from a PDE in conserved form then an inherited point symmetry of an auxiliary ODE essentially reduces the order of $R\{x,u\}$ as discussed in Section 7.2.7. We present a procedure for essentially reducing the order of $R\{x,u\}$ without reference to any PDE. This procedure depends on being able to relate $R\{x,u\}$ to an nth order auxiliary ODE $S\{x,v\}$ such that

(i) a general solution of $S\{x,v\}$ yields a general solution of $R\{x,u\}$ through a mapping which connects $S\{x,v\}$ to $R\{x,u\}$;

(ii) $S\{x,v\}$ admits a point symmetry.

Let $R\{x,u\}$ be the nth order ODE ($n \geq 2$)

$$F(x,u,u_1,u_2,\ldots,u_n)=0, \tag{7.90}$$

where $u_k = \frac{d^k u}{dx^k}$, $k = 1,2,\ldots,n$. Assume there exists a transformation, defining an auxiliary variable v, of the form

$$u = f(x,v,v_1), \tag{7.91}$$

such that $R\{x,u\}$ can be expressed in the conserved form

$$\mathrm{D}G(x,u,u_1,\ldots,u_{n-1},v,v_1,\ldots,v_{n-1})=0 \tag{7.92}$$

for some function G where D is the total derivative operator

$$\mathrm{D}=\frac{\partial}{\partial x}+u_1\frac{\partial}{\partial u}+u_2\frac{\partial}{\partial u_1}+\cdots+u_n\frac{\partial}{\partial u_{n-1}}+v_1\frac{\partial}{\partial v}$$

$$+v_2\frac{\partial}{\partial v_1}+\cdots+v_n\frac{\partial}{\partial v_{n-1}}; \tag{7.93}$$

the function f must depend on v_1, i.e. $\frac{\partial f}{\partial v_1} \neq 0$. The nth order auxiliary ODE $S\{x,v\}$, related to (7.90), is given by

$$G(x,f,\mathrm{D}f,\ldots,\mathrm{D}^{n-1}f,v,v_1,\ldots,v_{n-1})=0. \tag{7.94}$$

Transformation (7.91) is a Bäcklund transformation which maps any solution $v(x)$ of $S\{x,v\}$ into a solution $u(x) = f(x, v(x), v'(x))$ of $R\{x,u\}$.

We now further assume that transformation (7.91) maps a general solution of $S\{x,v\}$ to a general solution of $R\{x,u\}$ and that $S\{x,v\}$ admits a one-parameter Lie group of point transformations corresponding to infinitesimal generator

$$X_S = \xi_S(x,v)\frac{\partial}{\partial x} + \zeta_S(x,v)\frac{\partial}{\partial v}. \tag{7.95}$$

This symmetry reduces the order of $S\{x,v\}$ by one and hence essentially reduces the order of $R\{x,u\}$ through the mapping (7.91). We call the point symmetry (7.95) of $S\{x,v\}$ a ***potential symmetry*** of $R\{x,u\}$.

7.3.1 An Example

As an example [Bluman and Reid (1988)] let $R\{x,u\}$ be the ODE considered in Section 7.2.7, namely

$$2\frac{d}{dx}\left(K(u)\frac{du}{dx}\right) + x\frac{du}{dx} = 0. \tag{7.96}$$

After a lengthy calculation one can show that ODE (7.96) admits a point symmetry if and only if

$$K(u) = \lambda(u+\kappa)^{\nu}, \tag{7.97}$$

for arbitrary constants λ, κ, ν. Hence if $K(u)$ is not of the form (7.97) then one cannot reduce the order of ODE (7.96) through a point symmetry.

We now seek potential symmetries of (7.96). Let

$$u = f(x, v, v_1) = v_1. \tag{7.98}$$

Then

$$x\frac{du}{dx} = \frac{d}{dx}[xv_1 - v].$$

Thus ODE (7.96) becomes the conserved form

$$\frac{d}{dx}[2K(u)u_1 + xv_1 - v] = 0, \tag{7.99}$$

and hence $G(x,u,u_1,v,v_1) = 2K(u)u_1 + xv_1 - v$. Thus our related auxiliary ODE $S\{x,v\}$ is given by

$$2K(v_1)v_2 + xv_1 - v = 0. \tag{7.100}$$

One can show that ODE (7.100) admits a point symmetry X_S if $K(u)$ is of the form (7.97) or

$$K(u) = \frac{1}{u^2+pu+q}\exp\left[\lambda\int\frac{du}{u^2+pu+q}\right], \tag{7.101}$$

for arbitrary constants p, q, λ. If $K(u)$ is of the form (7.101), then ODE (7.100) admits

$$X_S = \left[v + \frac{1}{2}(p-\lambda)x\right]\frac{\partial}{\partial x} - \left[\frac{1}{2}(p+\lambda)v + qx\right]\frac{\partial}{\partial v}. \tag{7.102}$$

The point symmetry (7.102) can be used to reduce the order of ODE (7.100).

For the rest of this subsection we restrict ourselves to the case $p = \lambda = 0$, $q = 1$, for which $K(u)$ is a bounded function

$$K(u) = \frac{1}{u^2+1}. \tag{7.103}$$

Then ODE (7.100) admits an infinitesimal generator of the rotation group, namely

$$X_S = v\frac{\partial}{\partial x} - x\frac{\partial}{\partial v}. \tag{7.104}$$

In terms of canonical coordinates

$$r = \sqrt{x^2+v^2},$$
$$\theta = \arctan\frac{v}{x},$$

ODE (7.100) becomes

$$2\frac{d^2\theta}{dr^2} + \left(\frac{4}{r} + r\right)\frac{d\theta}{dr} + (2r + r^3)\left(\frac{d\theta}{dr}\right)^3 = 0. \tag{7.105}$$

The substitution $P = \frac{d\theta}{dr}$ reduces ODE (7.105) to a Bernoulli equation with general solution

$$P(r, A) = \frac{A}{r\sqrt{r^2\exp(r^2/2) - A^2}}$$

where A is an arbitrary constant. Then the corresponding general solution of ODE (7.100) is given implicitly by

$$v = \sqrt{x^2+v^2}\sin\left[\int^{\sqrt{x^2+v^2}} P(\rho, A)d\rho + B\right]$$

where A and B are arbitrary constants. The corresponding general solution of ODE (7.96) is given implicitly by

$$u = u(x, A, B) = \tan\left[\int^r P(\rho, A)d\rho + B + \arctan[rP(r, A)]\right] \tag{7.106}$$

where

$$r\cos\left[\int^r P(\rho, A)d\rho + B\right] = x. \tag{7.107}$$

For a given problem the constants A and B are determined from boundary data. Fujita (1954) obtained solution (7.106), (7.107) after making a number of ingenious substitutions and transformations.

Exercises 7.3

1. For any point symmetry admitted by (7.100) find a corresponding Lie–Bäcklund symmetry admitted by (7.96). Show that the Lie–Bäcklund symmetry is equivalent to a contact symmetry admitted by (7.96). For $K(u) = \dfrac{1}{u^2+1}$, find the infinitesimal generator of the contact symmetry corresponding to (7.104).

7.4 Discussion

In this chapter we developed a theoretical framework to find potential symmetries of differential equations. Potential symmetries are nonlocal symmetries of partial differential equations. They are determined as local symmetries of associated auxiliary differential equations arising from conserved forms.

Finding potential symmetries of a given PDE (system of PDE's) involves two major steps:

(i) Determine a conserved form. This leads to auxiliary dependent variables (potentials) and an associated auxiliary system of PDE's.

(ii) Find infinitesimal generators of local symmetries admitted by the auxiliary system of PDE's. The form of an infinitesimal generator determines whether or not it defines a potential symmetry.

In principle different conserved forms could lead to different potential symmetries of a given PDE.

The introduction of potential symmetries extends the applicability of symmetry methods to obtain solutions of differential equations. Together with the algorithms developed in Chapter 6, the use of potential symmetries allows one to find non-invertible mappings which linearize a nonlinear scalar PDE or a nonlinear system of PDE's.

The material in this chapter on potential symmetries for partial differential equations is based upon the work of Bluman, Kumei, and Reid (1988).

Akhatov, Gazizov, and Ibragimov (1988) show that if one can find a Bäcklund transformation relating a given system of evolution equations to an auxiliary system of evolution equations with the same number of dependent variables, then local symmetries of the auxiliary system can lead to nonlocal symmetries of the given system.

In Krasil'shchik and Vinogradov (1984) [see also Vinogradov and Krasil' shchik (1984)] nonlocal symmetries are defined as local symmetries of an associated auxiliary system of differential equations whose integrability conditions lead to the given system of differential equations. A general form is assumed for the auxiliary system which involves unspecified functions. These unspecified functions are then determined in principle by demanding that the integrability conditions of the auxiliary system lead to the given system of differential equations. In order to apply their method it seems that one has to impose very strong assumptions on the form of the unspecified functions. Kersten (1987) considered this work in the context of exterior differential systems.

Special types of nonlocal symmetries for PDE's have been considered by Konopelchenko and Mokhnacev (1979, 1980), Kumei (1981), Kapcov (1982), and Pukhnachev (1987). In these works nonlocal symmetries are not realized as local symmetries of associated auxiliary PDE's.

The material in this chapter on potential symmetries for ordinary differential equations is based upon the work of Bluman and Reid (1988). Potential symmetries are sought for ODE's which admit no point symmetries. In order to find potential symmetries one introduces an auxiliary variable through a Bäcklund transformation so that a given ODE is expressed in conserved form. A local symmetry of an associated auxiliary ODE defines a potential symmetry of the given ODE. Since this local symmetry reduces the order of the auxiliary ODE, it essentially reduces the order of the given ODE.

References

Ablowitz, M.J., Kaup, D.J., Newell, A.C., and Segur, H. (1974). The inverse scattering transform-Fourier analysis for nonlinear problems. *Stud. Appl. Math.* **53**, 249–315.

Abramowitz, M. and Stegun, I.A. (Eds.) (1970). *Handbook of Mathematical Functions.* Dover, New York.

Aguirre, M. and Krause, J. (1985). Infinitesimal symmetry transformations. II. Some one-dimensional nonlinear systems. *J. Math. Phys.* **26**, 593–600.

Akhatov, I.S., Gazizov, R.K., and Ibragimov, N.H. (1987). Group classification of the equations of nonlinear filtration. *Sov. Math. Dokl.* **35**, 384–386.

Akhatov, I.S., Gazizov, R.K., and Ibragimov, N.H. (1988). Bäcklund transformations and nonlocal symmetries. *Sov. Math. Dokl.* **36**, 393–395.

Ames, W.F., Lohner, R.J., and Adams, E. (1981). Group properties of $u_{tt} = [f(u)u_x]_x$. *Int. J. Nonlin. Mech.* **16**, 439–447.

Anderson, R.L., Kumei, S., and Wulfman, C.E. (1972). Generalization of the concept of invariance of differential equations. *Phys. Rev. Lett.* **28**, 988–991.

Atkinson, F.V. and Peletier, L.A. (1974). Similarity solutions of the nonlinear diffusion equation. *Arch. Rat. Mech. Anal.* **54**, 373–392.

Bäcklund, A.V. (1876). Ueber Flächentransformationen. *Math. Ann.* **9**, 297–320.

Barenblatt, G.I. (1979). *Similarity, Self-Similarity, and Intermediate Asymptotics.* Consultants Bureau, New York.

Barenblatt, G.I. and Zel'dovich, Ya.B. (1972). Self-similar solutions as intermediate asymptotics. *Ann. Rev. Fluid Mech.* **4**, 285–312.

Baumann, G. and Nonnenmacher, T.F. (1987). Lie transformations, similarity reduction, and solutions for the nonlinear Madelung fluid equations with external potential. *J. Math. Phys.* **28**, 1250–1260.

Becker, H.A. (1976). *Dimensionless Parameters: Theory and Method.* Wiley, New York.

Bessel-Hagen, E. (1921). Über die Erhaltungssätze det Elektrodynamik. *Math. Ann.* **84**, 258–276.

Bianchi, L. (1918). *Lezioni sulla Teoria dei Gruppi Continui Finiti di Transformazioni.* Enrico Spoerri, Pisa.

Birkhoff, G. (1950). *Hydrodynamics — A Study in Logic, Fact and Similitude,* 1st ed. Princeton University Press, Princeton.

Bluman, G.W. (1967). Construction of solutions to partial differential equations by the use of transformation groups. Ph.D. Thesis, California Institute of Technology.

Bluman, G.W. (1971). Similarity solutions of the one-dimensional Fokker–Planck equation. *Int. J. Nonlin. Mech.* **6**, 143–153.

Bluman, G.W. (1974a). Applications of the general similarity solution of the heat equation to boundary value problems. *Quart. Appl. Math.* **31**, 403–415.

Bluman, G.W. (1974b). Use of group methods for relating linear and nonlinear partial differential equations. Proceedings of Symposium on Symmetry, Similarity and Group Theoretic Methods in Mechanics, Calgary, 203–218.

Bluman, G.W. (1983a). Dimensional analysis, modelling, and symmetry. *Int. J. Math. Educ. Sci. Technol.* **14**, 259–272.

Bluman, G.W. (1983b). On mapping partial differential equations to constant coefficient equations. *SIAM J. Appl. Math.* **43**, 1259–1273.

Bluman, G.W. (1989). Simplifying the form of Lie groups admitted by a given differential equation. *J. Math. Anal. Appl.*, to appear.

Bluman, G.W. and Cole, J.D. (1969). The general similarity solution of the heat equation. *J. Math. Mech.* **18**, 1025–1042.

Bluman, G.W. and Cole, J.D. (1974). *Similarity Methods for Differential Equations.* Appl. Math. Sci. No. 13, Springer-Verlag, New York.

Bluman, G.W. and Gregory, R.D. (1985). On transformations of the biharmonic equation. *Mathematika* **32**, 118–130.

Bluman, G.W. and Kumei, S. (1980). On the remarkable nonlinear diffusion equation $\frac{\partial}{\partial x}[a(u+b)^{-2}\frac{\partial u}{\partial x}] - \frac{\partial u}{\partial t} = 0$. *J. Math. Phys.* **21**, 1019–1023.

Bluman, G.W. and Kumei, S. (1987). On invariance properties of the wave equation. *J. Math. Phys.* **28**, 307–318.

Bluman, G.W. and Kumei, S. (1988). Exact solutions for wave equations of two-layered media with smooth transition. *J. Math. Phys.* **29**, 86–96.

Bluman, G.W. and Kumei, S. (1989). Use of group analysis in solving overdetermined systems of ordinary differential equations. *J. Math. Anal. Appl.*, **138**, 95–105.

Bluman, G.W., Kumei, S., and Reid, G.J. (1988). New classes of symmetries for partial differential equations. *J. Math. Phys.* **29**, 806–811; Erratum, *J. Math. Phys.* **29**, 2320.

Bluman, G.W. and Reid, G.J. (1988). New symmetries for ordinary differential equations. *IMA J. Appl. Math.* **40**, 87–94.

Boisvert, R.E., Ames, W.F., and Srivastava, U.N. (1983). Group properties and new solutions of Navier–Stokes equations. *J. Engrg. Math.* **17**, 203–221.

Boyer, T.H. (1967). Continuous symmetries and conserved currents. *Ann. Phys.* **42**, 445–466.

Bridgman, P.W. (1931). *Dimensional Analysis,* 2nd ed. Yale University Press, New Haven, Conn.

Buckingham, E. (1914). On physically similar systems; illustrations of the use of dimension equations. *Phys. Rev.* **4**, 345–376.

Buckingham, E. (1915a). The principle of similitude. *Nature* **96**, 396–397.

Buckingham, E. (1915b). Model experiments and the forms of empirical equations. *Trans. A.S.M.E.* **37**, 263–296.

Campbell, J.E. (1903). *Theory of Continuous Groups.* Oxford University Press, Oxford.

Cantwell, B.J. (1978). Similarity transformations for the two-dimensional, unsteady, stream-function equation. *J. Fluid Mech.* **85**, 257–271.

Cohen, A. (1911). *An Introduction to the Lie Theory of One-Parameter Groups, with Applications to the Solution of Differential Equations.* D.C. Heath, New York.

Cohn, P.M. (1965). *Lie Groups.* Cambridge Tracts in Math. and Math. Phys., No. 46, Cambridge University Press, Cambridge.

Cole, J.D. (1951). On a quasilinear parabolic equation occuring in aerodynamics. *Quart. Appl. Math.* **9**, 225–236.

Courant, R. and Hilbert, D. (1953). *Methods of Mathematical Physics,* Vol. I. Interscience, New York.

Curtis, W.D., Logan, J.D., and Parker, W.A. (1982). Dimensional analysis and the pi theorem. *Lin. Alg. Appl.* **47**, 117–126.

de Jong, F.J. (1967). *Dimensional Analysis for Economists.* North-Holland, Amsterdam.

Dickson, L.E. (1924). Differential equations from the group standpoint. *Annals of Math.* **25**, 287–378.

Dresner, L. (1983). *Similarity Solutions of Nonlinear Partial Differential Equations.* Research Notes in Math., No. 88, Pitman, Boston.

Eisenhart, L.P. (1933). *Continuous Groups of Transformations.* Princeton University Press, Princeton.

Friedman, A. and Kamin, S. (1980). The asymptotic behavior of gas in an n-dimensional porous medium. *Trans. Amer. Math. Soc.* **262**, 551–563.

Fujita, H. (1954). The exact pattern of a concentration-dependent diffusion in a semi-infinite medium, Part III. *Text. Res. J.* **24**, 234–240.

Galaktionov, V.A., Dorodnitsyn, V.A., Elenin, G.G., Kurdyumov, S.P., and Samarskii, A.A. (1988). A quasilinear heat equation with a source: peaking, localization, symmetry, exact solutions, asymptotics, structures. *J. Sov. Math.* **41**, 1222–1292.

Galaktionov, V.A. and Samarskii, A.A. (1984). Methods of constructing approximate self-similar solutions of nonlinear heat equations, IV. *Math. USSR, Sbornik* **49**, 125–150.

Gardner, C.S., Greene, J.M., Kruskal, M.D., and Miura, R.M. (1967). Method for solving the Korteweg–de Vries equation. *Phys. Rev. Lett.* **19**, 1095–1097.

Gilmore, R. (1974). *Lie Groups, Lie Algebras, and Some of Their Applications.* Wiley, New York.

Gonzalez–Gascon, F. and Gonzalez-Lopez, A. (1983). Symmetries of differential equations, IV. *J. Math. Phys.* **24**, 2006–2021.

Görtler, H. (1975). Zur Geschichte des π-Theorems. *ZAMM* **55**, 3–8.

Hansen, A.G. (1964). *Similarity Analyses of Boundary Value Problems in Engineering.* Prentice-Hall, Englewood Cliffs, N.J.

Hasegawa, A. (1974). Propagation of wave intensity shocks in nonlinear interaction of waves and particles. *Phys. Lett.* **47A**, 165–166.

Hashimoto, H. (1974). Exact solutions of a certain semi-linear system of partial differential equations related to a migrating predation problem. *Proc. Japan Acad.* **50**, 623–627.

Holmes, M.H. (1984). Comparison theorems and similarity solution approximations for a nonlinear diffusion equation arising in the study of soft tissue. *SIAM J. Appl. Math.* **44**, 545–556.

Hopf, E. (1950). The partial differential equation $u_t + uu_x = \mu u_{xx}$. *Comm. Pure Appl. Math.* **3**, 201–230.

Ibragimov, N.H. (1985). *Transformation Groups Applied to Mathematical Physics.* Reidel, Boston.

Kamenomostkaya (Kamin), S. (1973). The asymptotic behavior of the solution of the filtration equation. *Israel J. Math.* **14**, 76–87.

Kamin, S. (1975). Similarity solutions and the asymptotics of filtration equations. *Arch. Rat. Mech. Anal.* **60**, 171–183.

Kapcov, O.V. (1982). Extension of the symmetry of evolution equations. *Sov. Math. Dokl.* **25**, 173–176.

Kersten, P.H.M. (1987). *Infinitesimal Symmetries: a Computational Approach.* CWI Tract No. 34, Centrum voor Wiskunde en Informatica, Amsterdam.

Klamkin, M.S. (1962). On the transformation of a class of boundary value problems into initial value problems for ordinary differential equations. *SIAM Rev.* **4**, 43–47.

Konopelchenko, B.G. (1987). *Nonlinear Integrable Equations.* Lecture Notes in Physics No. 270, Springer-Verlag, Berlin.

Konopelchenko, B.G. and Mokhnachev, V.G. (1979). On the group-theoretical analysis of differential equations. *Sov. J. Nucl. Phys.* **30**, 288–292.

Konopelchenko, B.G. and Mokhnachev, V.G. (1980). On the group theoretical analysis of differential equations. *J. Phys.* **A13**, 3113–3124.

Korobeinkov, V.P. (1982). Certain types of solutions of Korteweg–de Vries–Burgers' equations for plane, cylindrical, and spherical waves. Proceedings IUTAM Symposium on Nonlinear Wave Deformations, Tallinn (in Russian).

Krasil'shchik, I.S. and Vinogradov, A.M. (1984). Nonlocal symmetries and the theory of coverings: an addendum to A.M. Vinogradov's 'Local symmetries and conservation laws'. *Acta Applic. Math.* **2**, 79–96.

Kumei, S. (1975). Invariance transformations, invariance group transformations, and invariance groups of the sine-Gordon equations. *J. Math. Phys.* **16**, 2461–2468.

Kumei, S. (1977). Group theoretical aspects of conservation laws on nonlinear dispersive waves: KdV-type equations and nonlinear Schrödinger equations. *J. Math. Phys.* **18**, 256–264.

Kumei, S. (1981). A Group Analysis of Nonlinear Differential Equations. Ph.D. Thesis, University of British Columbia.

Kumei, S. and Bluman, G.W. (1982). When nonlinear differential equations are equivalent to linear differential equations. *SIAM J. Appl. Math.* **42**, 1157–1173.

Kurth, K. (1972). *Dimensional Analysis and Group Theory in Astrophysics.* Pergamon Press, Oxford.

Lax, P.D. (1968). Integrals of nonlinear equations of evolution and solitary waves. *Comm. Pure Appl. Math.* **21**, 467–490.

Lefschetz, S. (1963). *Differential Equations: Geometric Theory,*. 2nd. ed. Interscience, New York.

Leo, M., Leo, R.A., Soliani, G., Solombrino, L., and Martina, L. (1983). Lie–Bäcklund symmetries for the Harry–Dym equations. *Phys. Rev.* **D27**, 1406–1408.

Lie, S. (1881). Über die Integration durch bestimmte Integrale von einer Klasse linearer partieller Differentialgleichungen. *Arch. for Math.* **6**, 328–368; also *Gesammelte Abhandlungen,* Vol. III, B.G. Teubner, Leipzig, 1922, 492–523.

Lie, S. (1890). *Theorie der Transformationsgruppen,* Vol. II. B.G. Teubner, Leipzig.

Lie, S. (1893). *Theorie der Transformationsgruppen,* Vol. III. B.G. Teubner, Leipzig.

Liu, Q. and Fang, F. (1986). Symmetry and invariant solution of the Schlögl model. *Physica* **139A**, 543–552.

Magri, F. (1978). A simple model of the integrable Hamiltonian equation. *J. Math. Phys.* **19**, 1156–1163.

Mayer, A. (1875). Directe Begründung der Theorie der Berührungstransformationen. *Math. Ann.* **8**, 304–312.

Mikhailov, A.V., Shabat, A.B., and Yamilov, R.I. (1987). The symmetry approach to the classification of non-linear equations. Complete lists of integrable systems. *Russian Math. Surveys* **42**, 1-63.

Milinazzo, F. (1974). Numerical Algorithms for the Solution of a Single Phase One-Dimensional Stefan Problem. Ph.D. Thesis, University of British Columbia.

Milinazzo, F. and Bluman, G.W. (1975). Numerical similarity solutions to Stefan problems. *ZAMM* **55**, 423–429.

Miller, W., Jr. (1968). *Lie Theory and Special Functions.* Academic Press, New York.

Miller, W., Jr. (1977). *Symmetry and Separation of Variables.* Addison-Wesley, Reading, Mass.

Miura, R.M. (1968). Korteweg–de Vries equation and generalizations. I. A remarkable explicit nonlinear transformation. *J. Math. Phys.* **9**, 1202–1204.

Miura, R.M., Gardner, C.S., and Kruskal, M.D. (1968). Korteweg–de Vries equation and generalizations. II. Existence of conservation laws and constants of motion. *J. Math. Phys.* **9**, 1204–1209.

Müller, E.A. and Matschat, K. (1962). Übër das Auffinden von Ähnlichkeitslösungen partieller Differentialgleichungssysteme unter Benutzung von Transformationsgruppen, mit Anwendungen auf Probleme der Strömungsphysik. *Miszellaneen der Angewandten Mechanik,* Berlin, 190–222.

Murota, K. (1985). Use of the concept of physical dimensions in the structural approach to systems analysis. *Japan J. Appl. Math.* **2**, 471–494.

Na, T.Y. (1967). Transforming boundary conditions to initial conditions for ordinary differential equations. *SIAM Rev.* **9**, 204–210.

Na, T.Y. (1979). *Computational Methods in Engineering Boundary Value Problems.* Academic Press, New York.

Neuringer, J.L. (1968). Green's function for an instantaneous line particle souce diffusing in a gravitational field and under the influence of a linear shear wind. *SIAM J. Appl. Math.* **16**, 834–842.

Newman, W.I. (1984). A Lyapunov functional for the evolution of solutions to the porous medium equation to self-similarity, I. *J. Math. Phys.* **25**, 3120–3123.

Noether, E. (1918). Invariante Variationsprobleme. *Nachr. König. Gesell. Wissen. Göttingen, Math.-Phys. Kl.,* 235–257.

Olver, P.J. (1977). Evolution equations possessing infinitely many symmetries. *J. Math. Phys.* **18**, 1212–1215.

Olver, P.J. (1986). *Applications of Lie Groups to Differential Equations.* GTM, No. 107, Springer-Verlag, New York.

Ovsiannikov, L.V. (1958). Groups and group-invariant solutions of differential equations. *Dokl. Akad. Nauk. USSR* **118**, 439–442 (in Russian).

Ovsiannikov, L.V. (1959). Group properties of the nonlinear heat conduction equation. *Dokl. Akad. Nauk. USSR* **125**, 492–495 (in Russian).

Ovsiannikov, L.V. (1962). *Group Properties of Differential Equations.* Novosibirsk (in Russian).

Ovsiannikov, L.V. (1982). *Group Analysis of Differential Equations.* Academic Press, New York.

Page, J.M. (1896). Note on singular solutions. *Am. J. of Math.* **XVIII**, 95–97.

Page, J.M. (1897). *Ordinary Differential Equations with an Introduction to Lie's Theory of the Group of One Parameter.* Macmillan, London.

Pukhnachev, V.V. (1987). Equivalence transformations and hidden symmetry of evolution equations. *Sov. Math. Dokl.* **35**, 555–558.

Rogers, C. and Shadwick, W.F. (1982). *Bäcklund Transformations and Their Applications.* Academic Press, New York.

Rosen, G. (1979). Nonlinear heat conduction in solid H_2. *Phys. Rev.* **B19**, 2398–2399.

Schiff, L.I. (1968), *Quantum Mechanics,* 3rd ed. McGraw-Hill, New York.

Schlichting, H. (1955). *Boundary Layer Theory.* McGraw-Hill, New York.

Schwarz, F. (1982). Symmetries of the two-dimensional Korteweg–de Vries equation. *J. Phys. Soc. Japan* **51**, 2387–2388.

Schwarz, F. (1985). Automatically determining symmetries of partial differential equations. *Computing* **34**, 91–106.

Schwarz, F. (1988). Symmetries of differential equations: from Sophus Lie to computer algebra. *SIAM Rev.* **30**, 450–481.

Sedov, L.I. (1959). *Similarity and Dimensional Methods,* 4th ed. Academic Press, New York.

Seshadri, R. and Na, T.Y. (1985). *Group Invariance in Engineering Boundary Value Problems.* Springer-Verlag, New York.

Slebodzinski, W. (1970). *Exterior Forms and Their Applications.* PWN, Warsaw.

Steudel, H. (1975a). Noether's theorem and higher conservation laws in ultrashort pulse progagation. *Ann. Physik* **32**, 205–216.

Steudel, H. (1975b). Noether's theorem and the conservation laws of the Korteweg–de Vries equation. *Ann. Physik* **32**, 445–455.

Storm, M.L. (1950). Heat conduction in simple metals. *J. Appl. Phys.* **22**, 940–951.

Sukharev, M.G. (1967). Invariant solutions of equations describing the motion of fluids and gases in long pipelines. *Dokl. Akad. Nauk. USSR* **175**, 781–784 (in Russian).

Tajiri, M. (1983). Similarity reductions of the one and two dimensional nonlinear Schrödinger equations. *J. Phys. Soc. Japan* **52**, 1908–1917.

Tajiri, M. and Hagiwara, M. (1983). Similarity solutions of the two-dimensional coupled nonlinear Schrödinger equation. *J. Phys. Soc. Japan* **52**, 3727–3734.

Tajiri, M., Nishitani, T., and Kawamoto, S. (1982). Similarity solutions of the Kadomtsev–Petviashvili equation. *J. Phys. Soc. Japan* **51**, 2350–2356.

Talman, J.D. (1968). *Special Functions; a Group Theoretic Approach.* W.A. Benjamin, New York.

Taylor, E.S. (1974). *Dimensional Analysis for Engineers.* Clarendon Press, Oxford.

Taylor, Sir G.I. (1950). The formation of a blast wave by a very intense explosion. II. The atomic explosion of 1945. *Proc. Roy. Soc.* **A201**, 175–186.

Thomas, H.C. (1944). Heterogeneous ion exchange in a flowing system. *J. Am. Chem. Soc.* **66**, 1664–1666.

Torrisi, M. and Valenti, A. (1985). Group properties and invariant solutions for infinitesimal transformations of a nonlinear wave equation. *Int. J. Nonlin. Mech.* **20**, 135–144.

Varley, E. and Seymour, B. (1985). Exact solutions for large amplitude waves in dispersive and dissipative systems. *Stud. Appl. Math.* **72**, 241–262.

Venikov, V.A. (1969). *Theory of Similarity and Simulation.* MacDonald Technical and Scientific, London.

Vilenkin, N.J. (1968). *Special Functions and the Theory of Group Representations.* Amer. Math. Soc., Providence, R.I.

Vinogradov, A.M. and Krasil'shchik, I.S. (1984). On the theory of nonlocal symmetries of nonlinear partial differential equations. *Sov. Math. Dokl.* **29**, 337–341.

Watson, G.N. (1922). *A Treatise on the Theory of Bessel Functions.* Cambridge University Press, Cambridge.

Whitham, G.B. (1974). *Linear and Nonlinear Waves.* Wiley, New York.

Winternitz, P., Smorodinsky, J.A., Uhlir, M., and Fris, I. (1967). Symmetry groups in classical and quantum mechanics. *Sov. J. Nucl. Phys.* **4**, 444–450.

Wulfman, C.E. (1979). Limit cycles as invariant functions of Lie groups. *J. Phys.* **A12**, L73–L75.

Wybourne, B.G. (1974). *Classical Groups for Physicists.* Wiley, New York.

Yoshikawa, A. and Yamaguti, M. (1974). On further properties of solutions to a certain semi-linear system of partial differential equations. *Publ. Res. Inst. Math. Sci., Kyoto Univ.* **9**, 577-595.

Zakharov, V.E. and Shabat, A.B. (1972). Exact theory of two-dimensional self-focusing and one-dimensional self-modulation of waves in nonlinear media. *JETP* **34**, 62–69.

Zel'dovich, Ya.B. (1956). The motion of a gas under the action of a short term pressure (shock). *Akust. Zh.* **2**, 28–38 (in Russian).

Author Index

Subject Index

Boldface indicates a key reference. References to exercises are bracketed.

M

T

Applied Mathematical Sciences

55. *Yosida:* Operational Calculus: A Theory of Hyperfunctions.
56. *Chang/Howes:* Nonlinear Singular Perturbation Phenomena: Theory and Applications.
57. *Reinhardt:* Analysis of Approximation Methods for Differential and Integral Equations.
58. *Dwoyer/Hussaini/Voigt (eds.):* Theoretical Approaches to Turbulence.
59. *Sanders/Verhulst:* Averaging Methods in Nonlinear Dynamical Systems.
60. *Ghil/Childress:* Topics in Geophysical Dynamics: Atmospheric Dynamics, Dynamo Theory and Climate Dynamics.
61. *Sattinger/Weaver:* Lie Groups and Algebras with Applications to Physics, Geometry, and Mechanics.
62. *LaSalle:* The Stability and Control of Discrete Processes.
63. *Grasman:* Asymptotic Methods of Relaxation Oscillations and Applications.
64. *Hsu:* Cell-to-Cell Mapping: A Method of Global Analysis for Nonlinear Systems.
65. *Rand/Armbruster:* Perturbation Methods, Bifurcation Theory and Computer Algebra.
66. *Hlaváček/Haslinger/Nečas/Lovíšek:* Solution of Variational Inequalities in Mechanics.
67. *Cercignani:* The Boltzmann Equation and Its Applications.
68. *Temam:* Infinite Dimensional Dynamical System in Mechanics and Physics.
69. *Golubitsky/Stewart/Schaeffer:* Singularities and Groups in Bifurcation Theory, Vol. II.
70. *Constantin/Foias/Nicolaenko/Temam:* Integral Manifolds and Inertial Manifolds for Dissipative Partial Differential Equations.
71. *Catlin:* Estimation, Control, and the Discrete Kalman Filter.
72. *Lochak/Meunier:* Multiphase Averaging for Classical Systems.
73. *Wiggins:* Global Bifurcations and Chaos.
74. *Mawhin/Willem:* Critical Point Theory and Hamiltonian Systems.
75. *Abraham/Marsden/Ratiu:* Manifolds, Tensor Analysis, and Applications, 2nd ed.
76. *Lagerstrom:* Matched Asymptotic Expansions: Ideas and Techniques.
77. *Aldous:* Probability Approximations via the Poisson Clumping Heuristic.
78. *Dacorogna:* Direct Methods in the Calculus of Variations.
79. *Hernández-Lerma:* Adaptive Markov Control Processes.
80. *Lawden:* Elliptic Functions and Applications.
81. *Bluman/Kumei:* Symmetries and Differential Equations.
82. *Kress:* Linear Integral Equations.
83. *釘Bebernes/Eberly:* Mathematical Problems from Combustion Theory.

Printed in the United States
by Baker & Taylor Publisher Services